LIQUID MEMBRANES

LIQUID MEMBRANES
PRINCIPLES AND APPLICATIONS IN CHEMICAL SEPARATIONS AND WASTEWATER TREATMENT

EDITED BY

VLADIMIR S. KISLIK

Casali Institute of Applied Chemistry
The Hebrew University of Jerusalem
Campus Givat Ram
Jerusalem 91904 Israel

ELSEVIER

Amsterdam • Boston • Heidelberg • London • New York • Oxford • Paris • San Diego • San Francisco • Singapore • Sydney • Tokyo

Elsevier
Radarweg 29, PO Box 211, 1000 AE Amsterdam, The Netherlands
Linacre House, Jordan Hill, Oxford OX2 8DP, UK

First edition 2010

British Library Cataloguing-in-Publication Data
A catalogue record for this book is available from the British Library.

Library of Congress Cataloging-in-Publication Data
A catalog record for this book is available from the Library of Congress

ISBN: 978-0-444-53218-3

Printed and bound in Great Britain
10 11 9 8 7 6 5 4 3 2 1

Contents

PREFACE

The chemical and engineering community is paying significant attention to the quest for technologies that would lead us to the goal of technological sustainability. A promising example with a lot of interest for process engineers is the strategy of process intensification. In this framework, an interesting and important case is the continuous growth of modern membrane engineering whose basic aspects satisfy the requirements of process intensification, which consists of innovative equipment, design, and process development methods that are expected to bring substantial improvements in chemical and any other manufacturing and processing, such as decreasing production costs, equipment size, energy consumption, waste generation, and improving remote control and process flexibility. Membrane operations, with their intrinsic characteristics of efficiency and operational simplicity, high selectivity and permeability for the transport of specific components, compatibility between different membrane operations in integrated systems, low energetic requirement, good stability under operating conditions and environment compatibility, easy control and scale-up, and large operational flexibility, represent an interesting answer for the rationalization of chemical productions.

Membrane separation is a relatively new and fast-growing field in supramolecular chemistry. It is not only an important process in biological systems, but becomes a large-scale industrial activity. For industrial applications, many synthetic membranes have been developed. Important conventional membrane technologies are microfiltration, ultrafiltration, electro- and hemodialysis, reverse osmosis, and gas separations. The main advantages are the high separation factors that can be achieved under mild conditions and the low energy requirements.

Liquids that are immiscible with the source (feed) and receiving (product) phases can also be used as membrane materials. They are defined as liquid membranes (LMs). This separation technology has grown very fast during the last decades. This book is dedicated to the science, engineering, and applications of the LM separation technologies in inorganic, organic, analytical chemistry, biochemical, biomedical engineering, and gas separations.

The book is written with two main objectives:

1. To provide comprehensive knowledge-based information on the principles and applications of a variety of LM separation processes. The book contains updated, useful, and systematized information. It contains a

critical analysis of new technologies published in the last 15 years. New directions of development in the field are presented.
2. To provide students and young researchers, new to separation science and technology, with a general overview of LM separations, critical analysis, classification, and grouping of many technologies, their theories and applications in different configurations of LM separations.

Several groups may benefit from this book. It can be used by scientists and engineers in the research and development of separation technologies who need more detailed and specialized information in this rapidly growing field. To students examining separation processes, LM separations, and membrane reactors, it will serve as a valuable textbook. The attempt to forge links between different methods and to unify general theoretical considerations of LM separations will bring some order in the understanding of the discipline.

Vladimir S. Kislik

List of Contributors

Dr. Alberto Figoli, PhD
Research Institute on Membrane Technology, ITM-CNR
c/o University of Calabria, Via P. Bucci 17/C
87030 Rende (CS) - Italy
Ph.: +39 0984 492027/2014
Fax: +39 0984 402103
E-mail: a.figoli@itm.cnr.it

Dr. Mousumi Chakraborty
Assistant Professor, Dept of Chemical Engineering
S.V. National Institute of Technology
Ichhanath, Surat -395007, India
Telephone No: +912612201642(0)
+912612253306(R), +919427473685 (M)
E-mail: mousumi_chakra@yahoo.com, mch@ched.svnit.ac.in

Dr. Pawel Dzygiel
Institute of Chemistry
University of Opole
Oleska 48, 46-052
Opole, Poland
Telephone: (48 77) 454 5841
Fax: (48 77) 441 0740
E-mail: Pawel.Dzygiel@uni.opole.pl, dzygielp@uni.opole.pl

Dr. Chiranjib Bhattacharjee
Professor, Dept of Chemical Engineering
Jadavpur University
Kolkata - 700032
India
Fax: +91 33 2414 6378
Phone: +91 33 2414 6666 (Ext 2306) (Off)
Mobile: +91 92305 62975, +91 98364 02118
E-mail: cbhattacharyya@chemical.jdvu.ac.in, chiranjib_b@yahoo.com

Dr. Siddhartha Datta
Professor, Dept of Chemical Engineering
Jadavpur University
Kolkata - 700032
IndiaPhone: +91 33 2431 1251 (R), +91 33 2414
E-mail: sdatta_che@rediffmail.com

Dr. Vladimir S. Kislik
Professor, Casali Institute of Applied Chemistry
The Hebrew University of Jerusalem
Campus Givat Ram
Jerusalem 91904 Israel
Telephone: 972 2 658 6559
Fax: 972 2 652 5280
Tel. & fax: 972 2 997 4918
E-mail: vkislik@vms.huji.ac.il; vkislik@bezeqint.net

Dr. Roman Tandlich
Institute for Water Research
Old Geology Building
Artillery Road
Rhodes University
P.O. Box 94
Grahamstown 6140
South Africa
Tel: 00-27-73-851-3210
Fax: 00-27-46-622-9427
E-mail: r.tandlich@ru.ac.zaroman@iwr.ru.ac.za

Dr. Piotr Wieczorek
Professor, Institute of Chemistry,
University of Opole, Oleska 48, 46-052
Opole, Poland
Telephone: (48 77) 454 5841 ext. 2550
fax: (48 77) 441 0740;
E-mail: Piotr.Wieczorek@uni.opole.pl

CHAPTER 1

Introduction, General Description, Definitions, and Classification. Overview

Vladimir S. Kislik

1. Introduction

A membrane is a semipermeable barrier between two phases. If one component of a mixture moves through the membrane faster than another mixture component, a separation can be accomplished. The basic properties of membrane operations make them ideal for industrial production: they are simple in concept and operation; they are modular and easy to scale-up; and they are low in energy consumption with a remarkable potential for an environmental impact, and energetic aspects.

Polymeric and inorganic membranes are used commercially for many applications including gas separations, water purification, particle filtration, and macromolecule separations [1-4].

If membranes are viewed as semipermeable phase separators, then the traditional concept of membranes as polymer films can be extended to include liquids. They are defined as liquid membranes (LMs). Liquid membrane system involves a liquid which is an immiscible with the source (feed) and receiving (product) solutions that serves as a semipermeable barrier between these two liquid and gas phases [5-7].

Liquid membrane systems are being studied extensively by researchers in such fields as analytical, inorganic, and organic chemistry; chemical engineering, biotechnology, and biomedical engineering; and wastewater treatment. Research and development activities within these disciplines involve diverse applications of liquid membrane technology, such as gas separations, recovery of valued or toxic metals, removal of organic compounds, development of sensing devices, and recovery of fermentation products and some other biological systems.

Institute of Applied Chemistry, the Hebrew University of Jerusalem, Campus Givat Ram, Jerusalem 91904, Israel

Liquid Membranes: Principles and Applications in Chemical Separations and Wastewater Treatment
DOI: 10.1016/B978-0-444-53218-3.00001-5

Highly integrated membrane processes, combining various membrane operations suitable for separation and conversion units, are an attractive opportunity because of the synergic effects that can be attained. Practically, there are a lot of opportunities for membrane separation processes in all areas of industry [8]. The most interesting developments for industrial membrane technologies are related to the possibility of integrating various membrane operations in the same industrial cycle, with overall important benefits in terms of product quality and plant compactness.

This chapter has the objective of introducing the reader to the basic definitions of the liquid membrane field, with classification and grouping of the technologies. An overview of the volume is also presented.

2. General Description of the LM Processes

The term liquid membrane transport includes processes incorporating liquid-liquid extraction (LLX) and membrane separation in one continuously operating device. It utilizes an extracting reagent solution, immiscible with water, stagnant or flowing between two aqueous solutions (or gases), the source or feed and receiving or strip phases. In most cases, the source and receiving phases are aqueous and the membrane organic, but the reverse configuration can also be used. A polymeric or inorganic microporous support (membrane) may be used as bearer (as in SLM) or barrier (as in many BLM technologies) or not used, as in ELM and layered BLM.

The commonly accepted mechanism for the transport of a solute in LM is solution-diffusion. The solute species dissolve in the liquid membrane and diffuse across the membrane due to an imposed concentration gradient. Different solutes have different solubilities and diffusion coefficients in a LM. The efficiency and selectivity of transport across the LM may be markedly enhanced by the presence of a mobile complexation agent (carrier) in the liquid membrane. Carrier in the membrane phase reacts rapidly and reversibly with the desired solute to form a complex. This process is known as facilitated or carrier-mediated liquid membrane separation. In many cases of LM transport, the facilitated transport is combined with coupling counter- or cotransport of different ions through LM. The coupling effect supplies the energy for uphill transport of the solute.

The general properties of liquid membrane systems have been a subject of extensive theoretical and experimental studies. Some general characteristics of LM processes are [5]:

(1) Liquid membrane separation is a rate process and the separation occurs due to a chemical potential gradient, not by equilibrium between phases.
(2) LM is defined based on its function, not the material used in fabrication.

Permeation is a general term for the concentration-driven membrane-based mass transport process. Differences in the permeability produce a separation between solutes at constant driving force. Because the diffusion coefficients in liquids are typically orders of magnitude higher than in polymers, a larger flux can be obtained with liquid membranes. Application of a pressure difference, an electric field, or temperature considerably intensifies the process, but these special methods are beyond the scope of this book.

3. Terminology and Classification

There are several different directions in LM separation classifications: according to module design configurations, according to transport mechanisms, according to applications, according to carrier type, and according to membrane support type. Below, these types of classifications are described and discussed briefly.

3.1. Classification according to module design configurations

According to configuration definition, three groups of liquid membranes are usually considered (see Fig. 1.1): bulk (BLM), supported or immobilized (SLM or ILM), and emulsion (ELM) liquid membrane transport. Some authors add to these definitions polymeric inclusion membranes, gel membranes, dual module hollow-fiber membranes, but, to my opinion, the first two types are the modifications of the SLM and the third is the modification of BLM. It will be discussed in detail in the respective chapters.

3.1.1. Bulk liquid membrane

Bulk liquid membrane (BLM) consists of a bulk aqueous feed and receiving phases separated by a bulk organic, water-immiscible liquid phase. The phases may be separating by microporous supports (see respective chapters) which separate the feed and receiving phases from the LM or module configuration may be without microporous supports (layered BLM). Many of the LM subject reviewers considered only layered BLM [7] and testified its transport and selectivity inefficiency to be a potential for the practical application. Many more technologies that were developed and tested in the last decade have to be included in the BLM group. These are similar BLM systems, such as hybrid liquid membrane (HLM) [9], hollow-fiber liquid membrane (HFCLM) [10], (HFLM) [11], pertraction [12-14], flowing liquid membranes (FLM) [15], membrane-based extraction and stripping [16-18], multimembrane hybrid system (MHS) [19], and membrane contactor systems [8, 20, 21]. All these systems are based on membrane-based nondispersive (as the means for blocking the organic reagent from mixing with the aqueous feed and strip solutions) selective

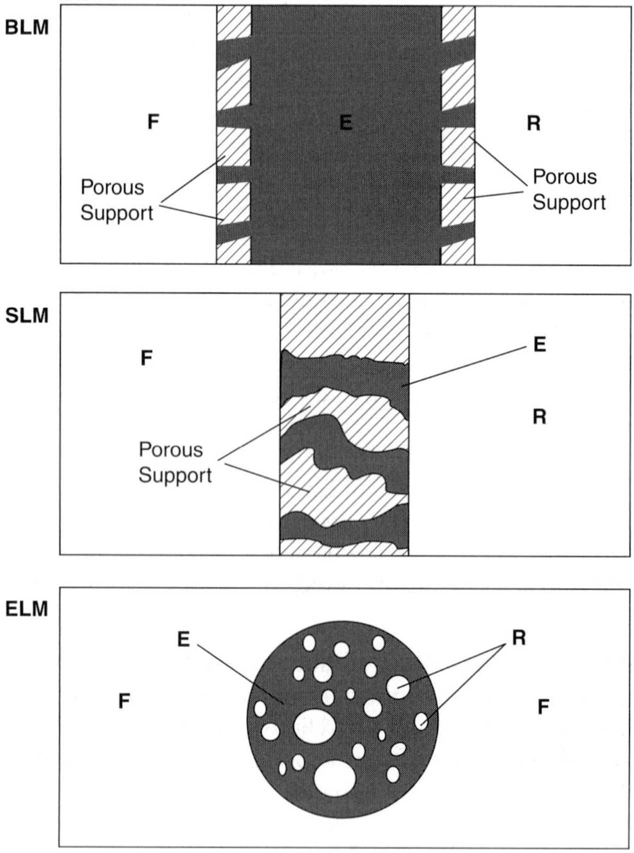

Figure 1.1 Three configurations of liquid membrane systems: bulk (BLM), supported (immobilized) (SLM or ILM), and emulsion (ELM). F is the source or feed phase, E is the liquid membrane, and R is the receiving phase.

extraction coupled to permselective diffusion of solute-extractant complexes and selective stripping of the solute in one continuous dynamic process. A great number of terms for similar bulk LM processes confuse the readers. The terms vary by membrane type used (hollow-fiber, flat neutral, ion-exchange sheets), or by module design.

Let us present some examples. The systems presented by the term membrane-based (or nondispersive) solvent extraction describe, as a rule, dynamic LM processes in which the equilibrium-based solvent extraction (forward and back) are only local processes taking place on the immiscible phases interfaces (on the surface of membrane support). The term pertraction or perstraction [2] spread over the supported and emulsion LMs, which is not accurate, because the SLM and ELM are steady state processes.

The term contactor systems present only membrane devices, mostly hollow-fiber, but not processes. The membrane in a contactor acts as a passive (not selective) barrier and as a means of bringing two immiscible fluid phases (such as gas and liquid, or an aqueous liquid and an organic liquid) in contact with each other without dispersion. The phase interface is immobilized at the membrane pore surface, with the pore volume occupied by one of the two fluid phases that are in contact. Contactor devices are used in many of the above-mentioned BLM systems (HLM, HFCLM, HFLM, FLM, pertraction, membrane-based extraction, MHS) as construction units. Sometimes, selective hydrophobic, hydrophilic, or ion-exchange membranes are used as barriers for additional selective separation in the devices similar to contactors.

Therefore, all above-mentioned bulk LM processes with water-immiscible liquid membrane solutions may be unified under the term bulk organic hybrid liquid membrane (BOHLM) systems. Bulk LM processes with water-soluble carriers [22] are defined as bulk aqueous hybrid liquid membrane (BAHLM) systems. These new technologies have the necessary transport and selectivity characteristics to have potential for commercial applications and are considered in detail in the respective chapters.

3.1.2. Supported or immobilized liquid membranes

Liquid impregnated (or immobilized) in the pores of a thin microporous solid support is defined as a supported liquid membrane (SLM or ILM). The SLM may be fabricated in different geometries. Flat sheet SLM is useful for research, but the surface area to volume ratio is too low for industrial applications. Spiral-wound and hollow-fiber SLMs have much higher surface areas of the LM modules (103 and 104 m^2/m^3, respectively [23]). The main problem of SLM technology is the stability: the chemical stability of the carrier, the mechanical stability of porous support, etc.

Related to the SLM systems are relatively new LM technologies, developed with the aim to improve stability parameters. These are gel LM [24, 25], ion-exchange membranes [26], swollen polymeric membranes [27], and polymeric inclusion membranes [28]. All these technologies are considered as modifications of the SLMs (details the reader can find in the respective chapter).

3.1.3. Emulsion liquid membranes

Emulsion liquid membrane (ELM) was invented by Li [29] in 1968. Receiving phase is emulsified in an immiscible liquid membrane. The emulsion is then dispersed in the feed solution and mass transfer from the feed to the internal receiving phase takes place. Liquid membranes may be either aqueous or organic solutions although the majority of publications describe water-in-oil emulsions.

The major problem with the ELMs is emulsion stability on the one hand and being easily broken to recover the internal phase, on the other. These two contradicting factors must be carefully balanced. Sometimes, the osmotic pressure gradient is problematic also [30]. The reader will find details in the corresponding chapters.

3.2. Classification according to transport mechanisms

According to the transport mechanisms, the LM techniques may be divided into six basic mechanisms of transport, schematically shown in Fig. 1.2.

3.2.1. Simple transport

In a simple transport (Fig. 1.2A), solute passes through due to its solubility in LM. Permeation stops when concentration equilibrium is reached. The solute does not react chemically with LM and is supposed to be in the same form in the feed (F), LM (E), and receiving, R, phases. As an example, some carboxylic and amino acids [19, 31], phenol [32] transport through xylene, decanol LM may be presented.

Uphill transport and selectivity can be achieved at reaction of the solute with components of the stripping solution (see Fig. 1.2B). Some authors relate this technique to the facilitated transport [33].

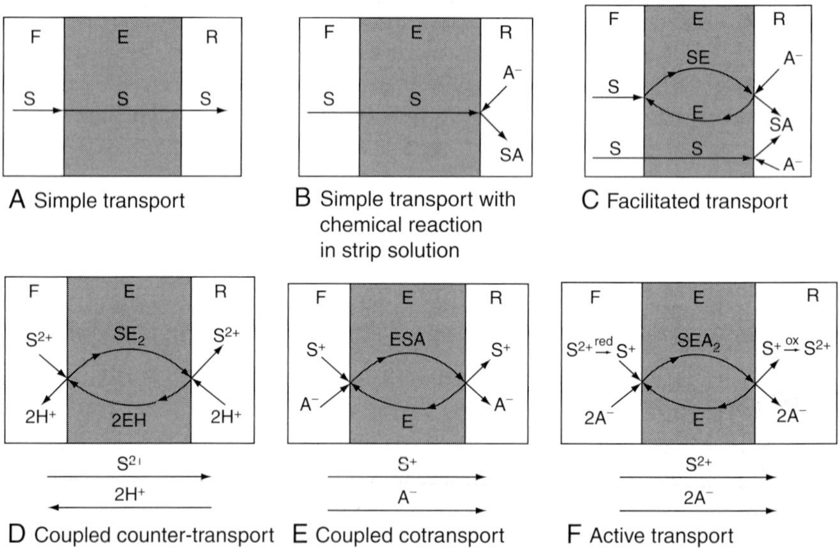

Figure 1.2 Schematic mechanisms of solute transport through the liquid membranes. S is solute to be separated; A are anions co-transported; E is liquid membrane, F is feed solution, and R is stripping solution; red is reduction; ox is oxidation. From Ref. [13] modified and reproduced with permission.

3.2.2. Facilitated or carrier-mediated transport

Carrier-assisted transport through liquid membranes is one of the important applications of supramolecular chemistry. The transport can be described by subsequent partitioning, complexation, and diffusion. Solute, partitioning (dissolving) in LM on a feed side-LM interface chemically reacts with a carrier, dissolved in the liquid membrane, to form complex. This complex reverse reacts on the LM-receiving side interface releasing the solute which partitioning to the receiving (strip) phase (see Fig. 1.2C).

Facilitated transport accelerates the transport. For example, trialkylphosphine sulfide increases the rate of phenol transport [32]. At the same time, the simple transport can take place also. Carriers for the selective transport of neutral molecules, anions, cations, or zwitterionic species have undergone intensive development in the last two decades.

3.2.3. Coupled counter- or cotransport

As examples of coupled countertransport (see Fig. 1.2D) and coupled cotransport (see Fig. 1.2E), the transport of titanium(IV) from low acidity (pH 1) and high acidity ($[H^+] = 7$ M) feed solutions, respectively, using the HLM system may be presented [9, 26]. The di-(2-ethylhexyl) phosphoric acid (DEHPA) carrier reacts with Ti(IV) ion to form complexes on the feed side at low acidity (pH region):

$$[\mathrm{Ti} \star 2\mathrm{H_2O}]^{4+}_{\mathrm{F}} + 4\overline{\mathrm{HL}}_{\mathrm{E}} \rightarrow [\overline{\mathrm{TiL_4} \star 2\mathrm{H_2O}}]_{\mathrm{E}} + 4\mathrm{H}^{+}_{\mathrm{F}} \quad (1)$$

at high (>7 mol/kg) acidity:

$$[\mathrm{Ti} \star 2\mathrm{HCl}]^{4+}_{\mathrm{F}} + 4\overline{\mathrm{HL}}_{\mathrm{E}} \rightarrow [\overline{\mathrm{TiL_4} \star 2\mathrm{HCl}}]_{\mathrm{E}} + 4\mathrm{H}^{+}_{\mathrm{F}} \quad (2)$$

and reversible reactions take place on the strip side at low acidity (pH region):

$$[\overline{\mathrm{TiL_4} \star 2\mathrm{H_2O}}]_{\mathrm{E}} + 2\mathrm{H}^{+}_{\mathrm{R}} \rightarrow [\mathrm{TiO}]^{2+}_{\mathrm{R}} + 4\overline{\mathrm{HL}}_{\mathrm{E}} + \mathrm{H_2O}_{\mathrm{R}} \quad (3)$$

at high acidity

$$[\overline{\mathrm{TiL_4} \star 2\mathrm{HCl}}]_{\mathrm{E}} + \mathrm{H_2O}_{\mathrm{R}} \rightarrow [\mathrm{TiO}]^{2+}_{\mathrm{R}} + 4\overline{\mathrm{HL}}_{\mathrm{E}} + 2\mathrm{Cl}^{-}_{\mathrm{R}} \quad (4)$$

Energy for the titanium uphill transport is gained from the coupled transport of protons in the opposite to titanium transport from the strip to the feed solutions. In the second case (high feed acidity) Cl anion cotransported with Ti(IV) cation in the same direction. In both cases, fluxes of titanium, protons, and chlorine anion are stoichiometrically coupled. As a rule, coupled transport used combining with the facilitated transport.

3.2.4. Active transport

Active transport (see Fig. 1.2F) is driven by oxidation-reduction, catalytic reactions, biochemical conversions on the membrane interfaces. As a rule, it is highly selective: no other species are transported at this type of transport.

In many cases, chemical reactions in LM are irreversible in active transport. As examples, copper transport by thioether [34] and picrate anions by ferrocene [35, 36] as carriers may be presented.

3.3. Classification according to applications

According to applications the LM techniques may be divided into:

(1) Metal separation-concentration
(2) Biotechnological products recovery-separation
(3) Pharmaceutical products recovery-separation
(4) Organic compounds separation, organic pollutants recovery from wastewaters
(5) Gas separations
(6) Fermentation or enzymatic conversion-recovery-separation (bioreactors)
(7) Analytical applications
(8) Wastewater treatment including biodegradation-separation techniques

3.4. Classification according to carrier type

(1) Water-immiscible, organic carriers
(2) Water-soluble polymers
(3) Electrostatic, ion-exchange carriers
(4) Neutral, but polarizable carriers

3.5. Classification according to membrane support type [37]

(1) Neutral hydrophobic, hydrophilic membranes
(2) Charged (ion-exchange) membranes
(3) Flat sheet, spiral module membranes
(4) Hollow-fiber membranes
(5) Capillary hollow-fiber membranes

Module design configurations are used as basic classification at developing of the chapters of this book. The application sections in every chapter are classified according to the application division. Some chapters, for example, gas separations and wastewater treatment, are added because these processes are very intensively researched during the last two decades and are developed in different module configurations.

4. Overview

The volume may be conditionally divided into four sections: general theory of the liquid membrane transport (Chapter 2); reviews of three basic LM configurations: SLM (Chapter 3), ELM (Chapter 4), and BLM

(Chapters 5 and 6) with theories and applications; specific LM applications: gases separations (Chapter 7) and wastewater treatment (Chapter 8); and perspectives in LM technologies development (Chapter 9).

The general theory chapter thoroughly describes the theory and analysis of various liquid membrane types and configurations. An attempt to unify theory of facilitated transport for different LM configurations is presented. The relationship of the chemical aspects of complexation reactions to the performance of facilitated transport is discussed. A procedures, which can be used to predict and optimize the facilitated transport, including measurements of the appropriate equilibrium, selectivity, driving forces of the transport with different configurations, are discussed. Transport is described in terms of partitioning, complexation, and diffusion. Most of the mechanistic studies were focused on diffusion-limited transport, in which diffusion of the solute-carrier complex through the membrane phase is the rate-limiting step for total transport. However, for some carriers, the rate-limiting step was found to be decomplexation at the membrane phase-receiving phase interface.

Factors which influence the effectiveness of membrane separation systems are summarized. These factors include the complexation/decomplexation kinetics, membrane thickness, complex diffusivity, anion type, solvent type, and the use of ionic additives.

Parameters, such as carrier and solvent properties, membrane support, temperature, etc., that influence transport kinetics are analyzed. Structural modifications and kinetic parameters of the carriers that improve the performance of LM are presented. Examples of carrier modifications are given.

In the next section, four chapters describe three main configurations of liquid membranes: supported, emulsion, and bulk LM. Each chapter is subdivided into theory and transport mechanisms, module design and experimental techniques, and applications in different fields of chemical, biochemical, environmental, and pharmaceutical separations.

In Chapter 3, P. Dzygiel and P. Wieczorek survey the applications of supported liquid membranes and their modifications (gel, polymer inclusion SLMs, integrated systems) in separations of metal ions, organics, gases, and contaminants in wastewater, in biochemical and biomedical processing. Choices of membrane support material, carriers and solvents which improve the transport kinetics and membrane stability in SLM system are discussed. The use of novel calix-his-crown ether carriers shows the potential for large-scale utilization in the future.

Applications in analytical, biotechnological, environmental, and stereo-isomer separations are reviewed. A few pilot-scale and industrial applications of the SLM processes are described.

M. Chakraborty, C. Bhattacharya, and S. Datta (Chapter 4) review recent advances in the theory and applications of ELM systems. Several mathematical models for the rheological curves are considered, and regions of applicability for the models are evaluated. The paper compares predictions of the reversible reaction model to the advancing front model for continuous flow ELM systems. The experimental data are discussed in terms of various parameters: feed phase acidity (pH), extracting, stripping agents and their concentrations, stirring rate, and temperature. Effects of surfactants, carriers, their concentrations, and external and internal phase compositions upon the properties of the extracting emulsions are discussed. Formulation of the emulsion membranes was optimized to provide emulsions with good stability during extraction, but which could be easily broken in an electrical coalescer under mild conditions. Preparation and splitting of emulsions for ELM systems is discussed. The possibility of employing microemulsions as liquid membranes to separate metals from contaminated water is explored.

Applications technologies in metal ion, inorganic species, hydrocarbons separations, biochemical and biomedical applications, and fine particles preparation using ELM are reviewed. Commercial applications include the removal of zinc, phenol, and cyanide from wastewaters. Potential applications in wastewater treatment, biochemical processing, rare earth metal extraction, radioactive material removal, and nickel recovery are described.

V. Kislik presents chapters devoted to BLM systems: bulk organic hybrid liquid membrane (BOHLM) which utilizes an organic solution of water-immiscible complexing agent (Chapter 5) and an aqueous solution of water-soluble complexing agent (BAHLM) (Chapter 6), flowing between ion-exchange and neutral microporous membranes. The membranes, which separate the carrier solution from feed and strip solutions, enable the transport of solutes (and water in the case of BAHLM), but block transfer of the carrier to the feed or to the strip phases.

Theoretical models (analytical and numerical), developed for simulation of the BOHLM and BAHLM transport kinetics, are based on independent experimental measurements of (a) individual mass-transfer coefficients of the solutes in boundary layers and (b) facilitating parameters of the liquid membrane (LMF potential) and IEM potential in the case of ion-exchange membrane (IEM) application. Satisfactory correlation between experimental and simulated data is achieved.

Selectivity parameters, needed for the BOHLM or BAHLM module design and their determination techniques, are analyzed. Selectivity can be controlled by adjusting the concentration, volume, and flow rate of the LM phase. Such control of the selectivity is one of the advantages of the bulk liquid membrane systems in comparison with other liquid membranes configurations and Donnan dialysis techniques. The idea of dynamic selectivity and determination techniques are presented and discussed.

Examples of the BOHLM modules—layered BLM, rotating disk, creeping film, HLM modules, MHS, FLM, HFLM, capillary liquid membrane modules, and membrane-based or nondispersive solvent extraction systems—are reviewed and compared. Carrier types and membrane supports used are analyzed.

Applications of the BOHLM systems (Chapter 5) in separation of (1) metals in the cationic and anionic forms in the weak and strong acidic aqueous solutions, (2) carboxylic and amino acid mixtures in aqueous solutions, (3) valuable drug species from the biochemical mixtures, and (4) potential application in catalysis and separation of valuable compounds (bioreactors) are reviewed.

Applications of the BAHLM technology (Chapter 6) in metal ions, salts separation, biotechnological, and isomers separations are reviewed. Commercially available membrane modules and equipment may be used in the BOHLM and BAHLM. It should be noted that Chapters 5 and 6 may also contain relevant information for other fields.

The specific applications section is classified and grouped according to the type of solutes separated: gas mixtures, conversion, degradation, separation, and purification of biochemical products (membrane bioreactors) at wastewater treatment, types of module tested and types of carrier used.

Gas separation (Chapter 7 by A. Figoli) covers a broad range of separation processes. This chapter deals with facilitated transport in liquid membranes. In particular, the envisaged goal is to provide an overview of the basic theory, the limitations, and advantages of the liquid membrane process in the gas field. The main efforts in this chapter are devoted to overcome the instability or lifetime of the liquid membranes which has limited their industrial application. The instability is mainly due to loss of carriers and/ or liquid phase from the membrane which influences the performance of the membrane itself. The different strategies employed in the years to improve the performance and stability of liquid membrane and the new directions to which address the future research are presented. New techniques such as the gelled SLM or by adding a thin top layer through interfacial polymerization reaction on the SLM are analyzed. Some examples of nonporous structures and microcapsule techniques, able to entrap more efficiently the carrier solvent are presented as studies in progress.

Applications of SLM, ELM, and BOHLM configurations for gas separations are reviewed. These are production of oxygen enriched air; carbon dioxide separation from various gas streams, including carbon dioxide from nitrogen, unsaturated hydrocarbons, and sugars from aqueous solutions; olefin, sulfur dioxide separation from various gas streams; hydrogen production and separation. Author presents state-of-the-art information for both the novice and practitioner.

Chapter 8, written by R. Tandlich, presents the field of wastewater treatment considering BLM, ELM, and SLM configurations of liquid membrane systems.

As a basic in BLMs for the wastewater treatment the author presents two-phase partitioning bioreactors. He presents the main criteria which must be considered in the selection of the LM solvent: biocompatibility (toxicity of the solvent to the microorganism), bioavailability (resistance of the solvent to biodegradation by the microorganism used), immiscibility in the aqueous phase, high solubility of pollutant in the solvent, favorable mass-transfer characteristics, etc. Biodegradation mechanisms and kinetics are discussed. Applications of bioreactors in wastewater treatment in laboratory, pilot, and industrial scale are reviewed. Potential applications are considered also.

The author discusses application of ELM, SLM, and polymer inclusion membrane techniques in separation of metal ions (precious metals, Cu, Ni, Zn, Pb, Cd, Cr(VI), Pu, Am, etc.) and organic pollutants (phenols and its derivatives, carboxylic acids, antibiotics, etc.) from wastewaters using laboratory, pilot, and industrial scale modules. Effects of experimental variables upon the solute flux for the various types of liquid membranes are analyzed. The author discusses potential and commercial aspects of liquid membrane technology in wastewater treatment.

Chapter 9, written by V. Kislik, is dedicated to potential directions in applications of different configurations of liquid membranes techniques in perspective research and development. While astonishing progress has been made over the past two decades in LM-based separation their potential for even more extensive industrial application remains unexploited in such fields as food/beverage processing, purification of chemical and biological products, wastewater reclamation, gaseous waste detoxification, hydrometallurgical processing, and production of gaseous and liquid fuels and petrochemicals.

Advanced directions in the fundamental LM science research, engineering, and applications are discussed in the aim to improve an industrial cycle, with overall important benefits of product quality and plant compactness.

A family of membranes in which structures are used not as intrinsic separation barriers, but as substrates for immobilization of catalysts (e.g., enzymes) or of specific complexing agents (e.g., affinity ligands) is under development. Novel polymeric materials of unique functionality or microstructure, inorganic (ceramic) semipermeable materials, novel ultrathin-barrier laminate structures comprised of both organic and refractory components, and interpenetrating multiphase structures with anomalous transport characteristics promises to yield LMs with superior chemical/thermal stability, fouling resistance, organic solvent resistance, and unusually high permeabilities and permselectivities. These developments should lead to new chemical synthesis processes and to novel and efficient strategies for industrial-scale purification of complex biological products.

Exciting opportunities also exist in design of production cycles by combining various LM operations suitable for separation and other separation/conversion units, thus realizing highly integrated membrane processes. Examples include BOHLM and BAHLM processes where membrane solvent extraction, integrated with affinity-complexation-ultrafiltration; selective LM transport/selective precipitation and extractive, membrane-moderated immobilized-cell biotransformation; LM technologies combined with electrochemistry. The production of ultrapure gases, the removal of trace concentrations of toxicants or high-value substances from liquid or gaseous streams, the development of novel chemical and biochemical sensors, and the synthesis of high-value chemical intermediates via membrane-immobilized catalysts in an electrochemical cell are among the many opportunity areas for ongoing membrane process research and development.

As potential directions for the BAHLM systems development, drug separations from biochemical mixtures, fermentation, catalysis and separation with enrichment of valuable compounds (BAHLM bioreactors), desalination of wastewater, and sea water and some integrated water-soluble complexing/filtration techniques are considered. It is suggested that the proposed BAHLM techniques may successfully and effectively replace the presenting separation systems with lower capital and operational costs.

Recent developments in LM module design, including rotational, vibrational membrane devices, pulsed-flow fluid management for polarization control, use of low-cost refractory monoliths as membrane supports, and use of electric potentials to minimize macrosolute polarization and fouling, may permit practical and economic application of membrane processes to liquid and gaseous streams which today are untreatable by such methods.

In summary, this chapter presents the entire breadth of LM technology with the intention of furthering research and industrial applications. The various types of liquid membrane configurations are surveyed and the advantages and disadvantages of each type are described. The tutorial section of this chapter also discusses typical experimental techniques and a survey of theoretical approaches.

REFERENCES

1. Mulder M. *Basic Principles of Membrane Technology*. Kluwer Academic: Norwell, MA, 1992.
2. Ho WSW, Sirka KK, Eds. *Membrane Handbook*. Chapman & Hall: New York, NY, 1992.
3. Osada Y, Makagawa T, Eds. *Membrane Science and Technology*. Marcel Dekker: New York, NY, 1992.
4. Noble RD, Stem SA, Eds. *Membrane Separation Technology*. Elsevier: New York, NY, 1995.
5. Noble RD, Way JD, Eds. *Liquid Membranes: Theory and Applications*. ACS Symposium Series 347. American Chemical Society: Washington, DC, 1987.
6. Araki T, Tsukube H, Eds. *Liquid Membranes: Chemical Applications*. CRC Press: Boca Raton, FL, 1990.
7. Bartsch RA, Way JD, Eds. *Chemical Separations with Liquid Membranes*. ACS Symposium Series 642. American Chemical Society: Washington, DC, 1996.

8. Drioli E, Romano M. Progress and new perspectives on integrated membrane operations for sustainable industrial growth. Ind. Eng. Chem. Res. 2001; 40: 1277.
9. Kislik V, Eyal A. Hybrid liquid membrane (HLM) system in separation technologies. J. Membr. Sci. 1996; 111: 259-272.
10. Majumdar S, Sirkar KK. Hollow-fiber contained liquid membrane. In: Ho WSW, Sirkar KK, Eds. *Membrane Handbook*. Van Nostrand Reinhold: New York, NY, 1992: 764-808.
11. Schlosser S, Sabolova E. Three-phase contactor with distributed U-shaped bundles of hollow-fibers for pertraction. J. Membr. Sci. 2002; 210(2): 331-347.
12. Zhivkova S, Dimitrov K, Kyuchoukov G, Boyadzhiev L. Separation of zinc and iron by pertraction in rotating film contactor with Kelex 100 as a carrier. Sep. Purif. Technol. 2004; 37: 9-16.
13. Schlosser S. Pertraction through liquid and polymeric membranes. In: Belafi-Bako K, Gubicza L, Mulder M, Eds. *Integration of Membrane Processes into Bioconversions*. Proceedings of the 16th European Membrane Society Annual Summer School, Veszprem, Hungary, August 1999. Kluwer Academic/Plenum Publishers: New York, NY, 2000: 73-100.
14. Wodzki R, Szczepanska G, Szczepanski P. Unsteady state pertraction and separation of cations in a liquid membrane system: Simple network and numerical model of competitive M^{2+}/H^{+} counter-transport. Sep. Purif. Technol. 2004; 36: 1-16.
15. Teramoto M, Takeuchi N, Maki T, Matsuyama H. Ethylene/ethane separation by facilitated transport membrane accompanied by permeation of aqueous silver nitrate solution. Sep. Purif. Technol. 2002; 28: 117-124.
16. Kubisova L, Sabolova E, Schlosser S, Martak J, Kertesz R. Mass-transfer in membrane based solvent extraction and stripping of 5-methyl-2-pyrazinecarboxylic acid and co-transport of sulphuric acid in HF contactors. Desalination 2004; 163: 27-38.
17. Kedem O, Bromberg L. Ion-exchange membranes in extraction processes. J. Membr. Sci. 1993; 78: 255-261.
18. Eyal A, Bressler EJ. Industrial separation of carboxylic and amino acids by liquid membranes: Applicability, process considerations, and potential advantages. Biotechnol. Bioeng. 1993; 41: 287-293.
19. Wodzki R, Nowaczyk J. Propionic and acetic acid pertraction through a multimembrane hybrid system containing TOPO or TBP. Sep. Purif. Technol. 2002; 26: 207-220.
20. Sengupta A. Degassing a liquid with a membrane contactor. US Patent 6,402,818, 2002.
21. Peterson PA, Runkle CJ, Sengupta A, Wiesler FE. Shell-less hollow fiber membrane fluid contactor. US Patent 6,149,817, 2000.
22. Eyal A, Kislik V. Aqueous hybrid liquid membrane A novel system for separation of solutes using water-soluble polymers as carriers. J. Membr. Sci. 1999; 161: 207-221.
23. Sirkar KK. Other new membrane processes. In: Ho WSW, Sirkar KK, Eds. *Membrane Handbook*. Van Nostrand Reinhold: New York, NY, 1992: 904-908.
24. Kesting RE. *Synthetic Polymeric Membranes: A Structural Perspective*. Wiley Interscience: New York, NY, 1985.
25. Tang M, Zhang R, Bowyer A, Eisenthal R, Hubble J. A reversible hydrogel membrane for controlling the delivery of macromolecules. Biotechnol. Bioeng. 2003; 82: 47-53.
26. Kislik V, Eyal A. Hybrid liquid membrane (HLM) and supported liquid membrane (SLM) based transport of titanium(IV). J. Membr. Sci. 1996; 111: 273-281.
27. Matson SL, Lee EK, Friesen DT, Kelly DJ. US Patent 4,737,166, 1988.
28. Sugiura M. J. Membr. Sci. 1992; 27: 269-276.
29. Li NN. Separating hydrocarbons with liquid membranes. US Patent 3,410,794, 1968.
30. Draxler J, Marr R. Chem. Eng. Process. 1986; 20: 319.
31. Schlosser S, Sabolova E. Transport of butyric acid through layered bulk liquid membranes. Chem. Pap. 1999; 53: 403-411.

32. Schlosser S, Rothova I, Frianova H. Hollow-fibre pertractor with bulk liquid membrane. J. Membr. Sci. 1993; 80: 99-106.
33. Chakraborty M, Bhattacharya C, Datta S. Study of the stability of (w/o)/w-type emulsion during the extraction of nickel (II) via emulsion liquid membrane. Sep. Sci. Technol. 2004; 39: 1-17.
34. Ohki A, Takagi M, Takeda T, Ueno K. Thioether-mediated copper transport through liquid membranes with the aid of redox reaction. J. Membr. Sci. 1983; 15: 231-244.
35. Shinbo T, Kurihara K, Kobatake Y, Kamo N. Active transport of picrate anion through organic liquid membrane. Nature (London) 1977; 270: 277-278.
36. Shinbo T, Sugiura M, Kamo N, Kobatake Y. Coupling between a redox reaction and ion transport in an artificial membrane system. J. Membr. Sci. 1981; 9: 1-11.
37. Michaels AS. Membranes, membrane processes, and their applications: Needs, unsolved problems, and challenges of the 1990s. Desalination 1990; 77: 5-34.

CHAPTER 2

Carrier-Facilitated Coupled Transport Through Liquid Membranes: General Theoretical Considerations and Influencing Parameters

Vladimir S. Kislik

1. Introduction

Solution-diffusion (with or without chemical reactions) is a commonly accepted mechanism for the transport of a solute in liquid membrane [1-7]. The rates of chemical changes and/or rates of diffusion may control all liquid membrane transport kinetics. Even at a simple LM transport, at partition of neutral molecules between two immiscible phases, there is a chemical change of the solute in its solvation environment. More drastic chemical changes of the solute species take place with the presence of carrier in LM when different chemical interactions (reversible or irreversible), formation of new coordination compound, dissociation or association, aggregation are possible. This is facilitated or carrier-mediated transport [2, 4-7]. The efficiency and selectivity of transport across the LM may be markedly enhanced. In many cases of LM transport, especially with cations or anions selective separations, facilitated transport is combined with stoichiometrically coupling countertransport of co-ions in the direction opposite to the solute, or cotransport of ions with the opposite ion charge to the solute in the same solute direction. The coupling effect supplies the energy for uphill transport of the solute.

At least one of the chemical or diffusion steps is slow enough to control the rate of the overall transport. So, analysis of mechanisms and kinetics of

Casali Institute of Applied Chemistry, The Hebrew University of Jerusalem, Campus Givat Ram, Jerusalem 91904, Israel

Liquid Membranes: Principles and Applications in Chemical Separations and Wastewater Treatment
DOI: 10.1016/B978-0-444-53218-3.00002-7

the chemical and diffusion steps of the overall LM transport system is needed to find the rate-controlling ones.

In this chapter, general considerations are presented in an attempt to advance the understanding of the LM science at facilitated, coupled transport which allows the optimization of solutes separations. Factors that influence the effectiveness and selectivity of separation are analyzed.

Active transport, driven by oxidation-reduction, catalytic, and bioconversion reactions on the liquid membrane interfaces will be considered in the respective chapters.

2. Mechanisms and Kinetics of Carrier-Facilitated Transport Through Liquid Membranes

The authors of hundreds of articles, published in this field, in trying to show the uniqueness of their works, have given new names and features to techniques and technologies that are similar to each other. This confuses and disorients readers, especially students and young researchers. The same is true for theories: hundreds of theories in this field need critical analysis and classification. In this chapter, recent aspects of carrier-facilitated, coupled transport through liquid membranes are reviewed with a classification and grouping of the theories.

2.1. Models of LM transport

The concept of LM transport is quite simple (see Figs 2.1 and 2.2): a solution E of an immiscible with aqueous solutions of feed (F) and receiving (R) solutions with or without component (carrier), chemically interacting with solutes transported, situates (1) as a thin layer of an emulsion globule (ELM), or (2) as a bulk layer (layered BLM), or (3) inside the pores of a thin microporous membrane support (SLM), or (4) as a stagnant layer between hollow fibers with flowing inside feed and receiving solutions (hollow-fiber contained liquid membrane, HFCLM), or (5) as a bulk solution, flowing between two membrane supports, which separate the LM from the feed and receiving phases (HLM, FLM, MHS, etc.; for details, see Chapter 5). A specific solute or solutes, driven by a chemical gradient, diffuse from the bulk F solution to the F/E interface, and are extracted from feed phase, due to their solubility in LM (E without carrier), and/or due to reversible chemical reaction with an extracting reagent (E with carrier component), or due to the irreversible reaction with catalytic reagent, with biochemical conversions components (using enzymes, whole cells, etc.) as a result of the thermodynamic conditions at the F/E interface. The solute or solute-LM

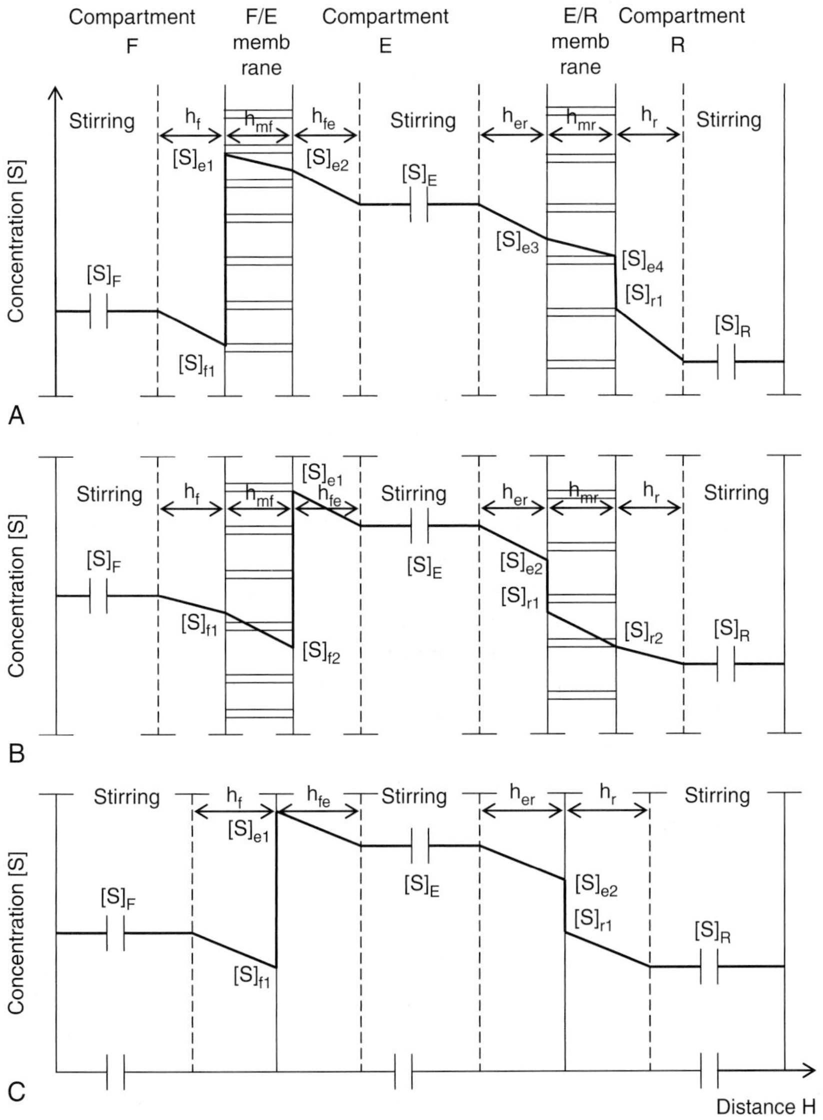

Figure 2.1 Concentration profiles for the transport of species S through (A) bulk liquid membrane (BLM) with hydrophobic membrane supports; (B) BLM with hydrophilic or ion-exchange membrane supports; (C) BLM without membrane support (layered BLM).

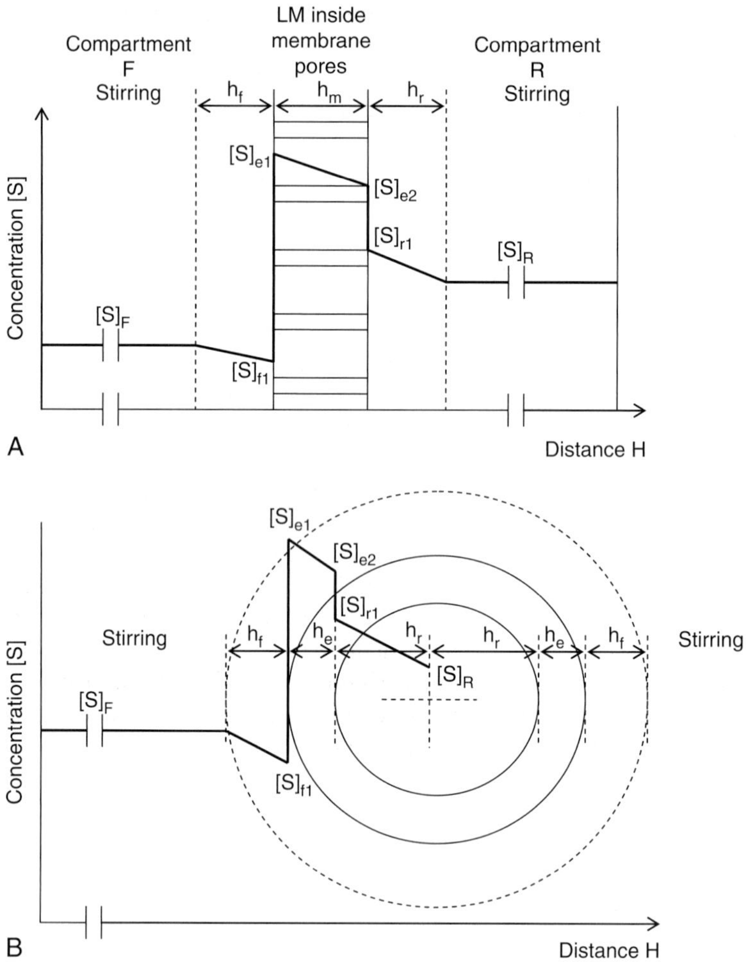

Figure 2.2 Concentration profiles for the transport of specie S through (A) supported liquid membrane (SLM) and (B) emulsion liquid membrane (ELM).

complex diffusing to the E/R interface is simultaneously decomplexed and stripped by the receiving phase due to the different thermodynamic conditions at the E/R interface and diffuses to the bulk R.

A universal model for all these types of transport does not exist, and the available knowledge concerning the specific interfacial processes should be taken into account in the description of real membrane process. There are two general approaches to modeling LM transport mechanisms: the differential and the integral approach. According to the differential approach [8–14], all phenomena taking place in the feed or in the strip phase, such as diffusion, chemical reactions, etc., are totally ignored. The measured transfer fluxes are dependent on phenomena occurring in the LM or at the surfaces of the membrane only. This approach is insufficient to explain the real transport in LM systems. The integral approach [1-7, 15-30] considers the three-liquid phase system to be a closed[1] multiphase system and, therefore, takes into consideration the processes and changes in all three liquids. Most models of the integral approach are very sophisticated because they assume many possible types of control, nonlinear equilibria, phase interactions, etc.

Kinetics of LM transport is a function of both the kinetics of the various chemical reactions occurring in the system and the diffusion rate of the various species that control the chemistry. To simplify the integral approach several models have been investigated. The irreversible thermodynamic method [31-33] is basically phenomenological and not particularly suitable for obtaining information at the molecular level. It is applicable to systems close to equilibrium which is not the case for most LM transport processes. The chemical kinetic approach is suitable for establishing the transport mechanism at the molecular level [34, 35]. The mechanisms of forward and backward extraction are the first and most important part of the whole LM transport process. This analysis can be realized in the steady-state approximation which is suitable for SLM [36, 37] and ELM [38]. In most cases of bulk LM transport, the nonsteady-state kinetic regimes have to be considered [28, 39, 40] and more general kinetic analysis is necessary. The combined chemical reactions' kinetics + diffusion method clearly shows the facilitated and coupling effects and other chemical events and diffusion constants.

Mechanistic studies of the processes are mainly focused on diffusion-limited transport. Recently, chemical reactions' kinetic aspects in membrane transport have been elucidated with new carriers for which the rate of decomplexation determines the rate of transport. Drastic chemical changes take place at facilitated transport: they can be described by subsequent partitioning, complexation and diffusion at the aqueous source solution/LM solution interface. At first two processes, the solvated water molecules can be removed from the solute ion; the carrier molecule can undergo an acid dissociation reaction; a new coordination compound, soluble in the organic phase, may be formed with chelating group of the carrier; carrier-solute complex can undergo changes in aggregation and so on. Inverse chemical processes can take place at decomplexation and partitioning at the LM-receiving aqueous-phase interface. At least one of the chemical steps of the overall reaction mechanisms may be slow enough, compared with the diffusion rate and overall transport kinetics would depend on the

[1]"A closed system is one with boundaries across it, through which no matter may pass, either in or out, but one in which other changes may occur, including expansion, contraction, internal diffusion, chemical reaction, heating, and cooling. An open system is one which undergoes all the changes allowed for a closed system and in addition it can lose and gain matter across its boundaries." [31]

rate of slow chemical reaction (or reactions). Very few tools have been developed up to date to investigate chemical changes occurring at liquid-liquid interfaces, and our knowledge is still limited and is based on indirect experimental evidence and speculations.

Most membrane separation systems involve stirring or continuous flow of the feed and receiving solutions to minimize the time for diffusion of dissolved species toward and away from the LM [1-7]. It follows that transport of species from the bulk of the phases to a region very close to interface can be considered instantaneous and the diffusion in the bulk of the phases can be neglected. But even the most vigorously stirred systems possess two thin films at the aqueous/organic interface that are essentially stagnant. These films, often referred to as diffusion films, Nernst films, diffusion layers, or boundary layers, vary from 50 to 500 μm thick [42] and can be crossed only by diffusion processes. The thickness of the diffusion films never go down to zero. The limiting value depends on the specific physicochemical properties of the liquids and specific hydrodynamic conditions. The two-film model is used here to describe the LM diffusional transport. For other models and theories of higher complexity, the reader can be referred to books [43, 44]. The time required for the diffusional crossing of the films by the solute may be longer or comparable with the time required for chemical reactions. So, diffusion across the boundary layers may control the overall kinetics of the LM transport.

Assuming that both the feed and receiving phases are rapidly mixed, virtually all membrane transport processes can be broken down into five parts: (1) diffusion through the thin aqueous feed film, (2) partition into LM and chemical complexation reactions at the feed-LM interface, (3) diffusion through the LM films on the feed and strip sides of LM flow, (4) a chemical decomplexation reaction at the LM-receiving phase interface and partition of the solute into receiving phase, and (5) diffusion through the aqueous receiving phase film. These five processes can be categorized as diffusion steps or chemical reaction steps. For many systems, the combined diffusion steps are rate determining and that the chemical reaction rate aspect is relatively insignificant. However, more and more investigations in LM separations show that the kinetics of interfacial chemical reactions governs the transport rate [39, 42]. We will discuss the influence of both diffusion and reaction rates at the membrane interfaces.

The basic idea of the chemical kinetics-diffusion method is that the solute presented in the different phases is considered as different chemical species obeying the laws of chemical kinetics. The general assumptions of the transport may be formulated as:

(1) Steady-state conditions of the solute transport through the phase interfaces: all fluxes are necessarily the same [36-38]. This assumption is related to SLM and ELM processes, in which the thickness of the

boundary (Nernst) films is very thin. For the BLM systems, the steady-state condition may be considered, when the BLM is hypothetically divided to two consecutive parts: feed-LM and LM-strip in one module. In this case, we can state the steady-state conditions for two parts of the module differently but at the same sampling time.

(2) The overall mass-transfer rate of solute can be controlled by any of the chemical reaction-diffusion resistances in the three-liquid phases.

(3) The overall mass-transfer resistance at steady state is the sum of individual mass-transfer resistances at diffusional regime through the boundary films and chemical reactions resistances at the phases' interface.

(4) Membrane supports have uniform pore size and wetting characteristics throughout membrane. Hydrophobic membrane pore is completely wetted by organic phase; hydrophilic or ion-exchange membrane pore is completely filled with aqueous phase.

(5) Thermodynamic local equilibrium at the uptake (feed-LM) and release (LM-strip) interfaces.

(6) The aqueous or organic stagnant boundary films diffusion resistances may be combined with the diffusion resistances of the same liquid films (aqueous or organic) inside the membrane pores (taking into account the membrane porosity and tortuosity) in one-dimensional series of diffusion resistances. This assumption is related to the BLMs with membrane supports only.

Let us consider the diffusion-chemical reactions stages of the LM transport from the most complex BLM systems to simpler ones. As can be seen in Fig. 2.1A transport of solutes or their complexes through the BLM with hydrophobic membrane supports consists of the following discrete steps:

(1a) Diffusion from the bulk feed through the feed-side stagnant boundary layer (h_f)

(2a) Partition into LM phase and interaction with carrier on the feed-side phases' F/E interface, as a result of thermodynamic conditions

(3a) Diffusion through the LM in the pores of the feed-side hydrophobic membrane support (h_{mf}), denoted by Helfferich [32-34] as "inter-diffusion"

(4a) Diffusion through the feed-side (h_{fe}) stagnant LM boundary layer

(5a) Diffusion through the strip-side (h_{er}) stagnant LM boundary layer

(6a) Interdiffusion through the strip-side membrane support (h_{mr})

(7a) Interaction with the stripping agent on the strip-side LM interface, as a result of different thermodynamic conditions, and partition into the strip phase

(8a) Diffusion through the strip-side stagnant boundary layer (h_r) to the bulk strip

According to assumption 6, the steps 3a-4a and 5a-6a may be combined to two consecutive individual mass-transfer steps.

Transport of solutes or their complexes through the BLM with hydrophilic or ion-exchange membrane supports consist of the following discrete steps (Fig. 2.1B):

(1b) Diffusion from the bulk feed through the feed-side stagnant boundary layer (h_f)
(2b) Interdiffusion through the feed phase filled pores of the feed-side hydrophilic or ion-exchange membrane (h_{mf})
(3b) Partition into LM phase and interaction with a carrier on the feed-side phases' F/E interface, as a result of thermodynamic conditions
(4b) Diffusion through the feed-side (h_{fe}) stagnant LM boundary layer
(5b) Diffusion through the strip-side (h_{er}) stagnant LM boundary layer
(6b) Interaction with the stripping agent on the strip-side LM interface, as a result of different thermodynamic conditions and partition into strip phase, filling membrane support pores
(7b) Interdiffusion through the strip-side membrane support (h_{mr})
(8b) Diffusion through the strip-side stagnant boundary layer (h_r) to the bulk strip

According to assumption 6, the steps 1b-2b and 7b-8b may be combined to two consecutive individual mass-transfer steps.

Now, let us simplify the system, excluding membrane supports, or steps 3a and 6a in Fig. 2.1A. We obtain the layered BLM system (see Fig. 2.1C).

Immobilizing of LM solution into the pores of thin microporous hydrophobic membrane support, separating the feed (source) and receiving (strip) phases, or excluding steps 4a, 5a and combining the steps 3a and 6a into one step of the solute-carrier complex interdiffusion through the membrane support, we obtain the SLM system (see Fig. 2.2A).

Adding the surfactant into LM solution, forming emulsion of receiving phase inside small droplets of LM and mixing them with the feed phase, or excluding of the steps 3a and 6a, and combining of the steps 4a and 5a into one, taking place in the very thin LM layer of the globule we obtain the ELM system (see Fig. 2.2B).

So, in all configurations we have (1) diffusion steps in aqueous feed and strip stagnant boundary layers, (2) diffusion of the complex solute-carrier in the LM phase and/or interdiffusion in the membrane support pores, (3) partitions between aqueous feed and organic LM phases at feed-LM interface (forward extraction) and between LM and aqueous strip phases at LM-strip interface (backward extraction), and (4) kinetics of chemical interactions with formation of solute-carrier complex (complexation) and destruction of complex (decomplexation).

2.2. Diffusion transport regime

For many systems, the diffusion steps are rate determining. The barriers to transport imposed by the need for diffusion across boundary layers can be minimized by decreasing film thickness, or by increasing the mobility of the diffusible species. Film thickness, up to a certain limiting value, is inversely related to mechanical energy supplied (e.g., by stirring) [45]. Viscosity and density of the liquids used as well as equipment geometry also affect film thickness [46] but interfacial films apparently cannot be completely eliminated.

2.2.1. Mathematical description of the diffusion transport

The diffusion flux J_s (M, g/cm^2/s) of the species is defined as the amount of matter passing perpendicularly through the unit area during the unit time. Different solutes have different solubilities and diffusion coefficients, D_s in a LM. In a steady-state permeation experiment, the flux of a species S through a membrane of thickness h is related to the concentration gradient through Fick's first law:

$$J_s = -D_s \frac{\partial S}{\partial h}. \tag{1}$$

High fluxes can be obtained when a large chemical potential (concentration gradient) is maintained over a thin membrane in which the diffusivity, D_s, of the species is high (as a rule, 10^{-5}–10^{-6} cm^2/s). This diffusion is expressed by Fick's second law:

$$\frac{\partial S}{\partial t} = D_s \frac{\partial^2 S}{\partial h^2}. \tag{2}$$

When steady state cannot be assumed, the concentration change with time must be considered.

For steady-state diffusion occurring across thin films, only one dimension can be considered and Eq. (1) is simplified:

$$J_s = -k_s([S_2] - [S_1]) = k_s([S_1] - [S_2]), \tag{3}$$

where

$$k_s = D_s/h \tag{4}$$

is individual mass-transfer coefficient dependant on the thickness of the diffusion film, which is constant (as the diffusion coefficient) at the process parameters used.

For the interdiffusion in the pores of membrane support, the individual mass-transfer coefficient is

$$k_{sm} = D_s \varepsilon_m / h_m \tau_m, \tag{5}$$

where h_m is the membrane support thickness, ϵ_m is the membrane porosity, and τ_m is the membrane tortuosity.

Following are descriptions of the diffusion steps of the LM systems:

(1) Diffusion through the aqueous boundary layer of the feed phase:

$$J_f = k_f([S_F] - [S_{f1}]). \tag{6}$$

This step is present in all configurations of LM: in BLM, SLM, and ELM (see Figs 2.1 and 2.2).

(2) Diffusion through the LM in the pores of membrane support on the feed side:

$$J_{fm} = k_{fm}([S_{e1}] - [S_{e2}]). \tag{7}$$

The step is present in the BLM with hydrophobic membrane support (step 3a in Fig. 2.1A) and in the SLM (Fig. 2.2A).

Or for the BLM with hydrophilic or ion-exchange membrane support (step 2b in Fig. 2.2B):

$$J_{fm} = k_{fm}([S_{f1}] - [S_{f2}]). \tag{8}$$

(3) Diffusion through the stagnant LM layer at the feed-side LM:

$$J_{fe} = k_{fe}([S_{e2}] - [S_E]). \tag{9}$$

This step is for BLMs with hydrophobic membrane supports (see Fig. 2.1A).

For the ELM (Fig. 2.2B):

$$J_{fe} = k_{fe}([S_{e1}] - [S_{e2}]). \tag{10}$$

Or for the BLMs with hydrophilic or ion-exchange membrane supports and BLM without membrane support (see Fig. 2.1B and C):

$$J_{fe} = k_{fe}([S_{e1}] - [S_E]). \tag{11}$$

(4) Diffusion through the stagnant LM layer at receiving side of the LM.

For the BLM with hydrophobic membrane support:

$$J_{er} = k_{er}([S_E] - [S_{e3}]). \tag{12}$$

For BLM with hydrophilic, ion-exchange membrane supports and without membrane support:

$$J_{er} = k_{er}([S_E] - [S_{e2}]). \tag{13}$$

(5) Diffusion through the pores of the strip-side membrane support.

For the BLM with hydrophobic support (Fig. 2.1A):

$$J_{\mathrm{mr}} = k_{\mathrm{mr}}([\mathrm{S}_{\mathrm{e3}}] - [\mathrm{S}_{\mathrm{e4}}]). \tag{14}$$

For the BLM with hydrophilic or ion-exchange membrane supports (Fig. 2.1B):

$$J_{\mathrm{mr}} = k_{\mathrm{mr}}([\mathrm{S}_{\mathrm{r1}}] - [\mathrm{S}_{\mathrm{r2}}]). \tag{15}$$

(6) Diffusion through the stagnant aqueous boundary layer of the receiving phase.

For the BLM with hydrophobic membrane support (Fig. 2.1A), the BLM without membrane support (Fig. 2.1C), the SLM (Fig. 2.2A), and the ELM (Fig. 2.2B):

$$J_{\mathrm{r}} = k_{\mathrm{r}}([\mathrm{S}_{\mathrm{r1}}] - [\mathrm{S}_{\mathrm{R}}]). \tag{16}$$

For the BLM with hydrophilic or ion-exchange membrane support (Fig. 2.1B):

$$J_{\mathrm{r}} = k_{\mathrm{r}}([\mathrm{S}_{\mathrm{r2}}] - [\mathrm{S}_{\mathrm{R}}]). \tag{17}$$

According to assumptions 1-4, the overall mass-transfer rate may be derived through the sums of individual step resistances and measured bulk phase concentrations $[\mathrm{S_F}]_i$, $[\mathrm{S_E}]_i$ (for BLM), and $[\mathrm{S_R}]_i$ (for details, see Section 2.4.3).

Decreased film and/or membrane support thicknesses are not the only way to increase diffusion rate. Structural features of phases can themselves alter diffusion. Thus, manipulation of carrier structural features offers minimal benefit for increasing transport rates.

There is another, more general description for time dependency of the solute fluxes. Using postulates of nonequilibrium thermodynamics [47], the general equation that relates the flux, J, of the solute to its concentration S and its derivative is

$$J_{\mathrm{S}} = US - D_{\mathrm{S}}\frac{\mathrm{d}S}{\mathrm{d}h}, \tag{18}$$

where U is the phase flow or stirring rate.

Referring to equation continuity, as h approaches zero, the steady-state layers are formed next to the phase's interface (but not for the bulk phase, where $h \gg 0$) and separation occurs by differential displacement permeation. According to Giddings' analysis [48] of such a system,

$$\ln\left[\left(S - \frac{J_0}{U}\right)/\left(S_0 - \frac{J_0}{U}\right)\right] = U\frac{h}{D}, \tag{19}$$

where S_0 is the initial solute concentration and J_0 is the initial flux.

Using Eq. (4) we obtain individual mass-transfer coefficient k_i for every layer at sampling time t_i:

$$k_i = U\left\{\ln\left[\left(S_i - \frac{J_{ss}}{U}\right)/\left(S_{i-1} - \frac{J_{ss}}{U}\right)\right]\right\}^{-1}, \tag{20}$$

where S_i and S_{i-1} are the concentrations of the solute in the bulk phase at time t_i and time of previous sampling, t_{i-1}, respectively, and J_{ss} is the flux at steady state.

Detailed application of this technique will be presented in Section 6 of this chapter and in Chapter 5.

2.2.2. Determination of diffusion coefficients

Several models have been devised to predict the role of various LM features on the solute flux [46, 49-57]. Different methods can be applied to determine diffusion coefficients independently from transport experiments, for example, determination of the lag-time [49], pulsed-field gradient NMR [50], and permeability measurements [42].

The diffusional process through a SLM is affected by the porosity and tortuosity of the polymeric support. Direct comparison of fluxes J and the corresponding diffusion coefficients D_m when using different supports is not possible and D_m has to be corrected for the membrane characteristics to obtain the bulk diffusion coefficient D_b [51]:

$$D_b = D_m \frac{\tau_m}{\varepsilon_m}. \tag{21}$$

The bulk diffusion coefficient D_b is derived by Stokes-Einstein relationship [52]. Simplest relationship in which the diffusion coefficient, D_b, is given:

$$D_b = kT/(6\pi\eta r) \tag{22}$$

and the Wilke-Chung relation [53]:

$$D_b = 7.4 \times 10^{-8}\left(\frac{[S]^{0.5}T}{\eta V^{0.6}}\right), \tag{23}$$

where k is the mass-transfer coefficient based on concentration, T is the temperature, η is the solvent viscosity, and r is the molecular radius.

This relationship is accurate for neutral molecules. For ionic species much more complex models are required, taking into account such factors as ionic charge, ionic strength, the presence of electric fields and others. In either case, the range of values for D_b is quite narrow; usually 10^{-5}-10^{-6} cm^2/s. Significant reduction of the diffusion process takes place when pores of the membrane are less than 10 times larger than the diffusing species.

Diffusion coefficient D_b can be obtained from lag-time experiments [51]. A lag-time is defined as the time required for the complex to diffuse

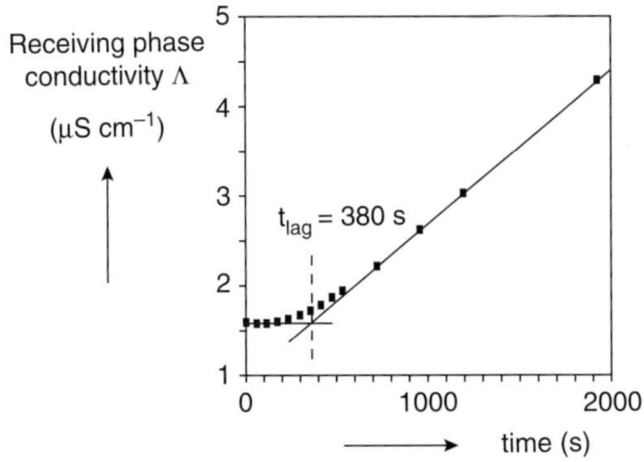

Figure 2.3 Lag-time experiment for the transport of $NaClO_4$ mediated by carrier. From Ref. [42] with permission.

across the membrane from the feed phase to the receiving phase, assuming dilute conditions:

$$t_{lag} = \frac{h_m^2}{6D_{lag}}. \tag{24}$$

Lag-times can be obtained from lag-time experiments (see Fig. 2.3). The resulting diffusion coefficient D_{1ag} has to be corrected for the tortuosity τ, to obtain the bulk diffusion coefficient ($D_b = D_{1ag}\tau$).

Determination of the diffusion coefficient by permeability experiments [42], when a liquid membrane is clammed between a feed and receiving phase, with a membrane solvent. At time $t = 0$, a carrier which is substituted with a chromophoric group is added to the feed phase ($[c_f]_0$). The carrier diffuses through the membrane and the increase of concentration in the receiving phase ($[c_r]_t$) is monitored by UV/Vis spectroscopy (A_m) as a function of time. The transport through the pores of the membrane is assumed to be rate limiting and Eq. (25) is derived:

$$\ln\frac{[c_f]_0 - [c_r]_t}{[c_f]_0} = -2\frac{A_m D_m}{V_r h_m}t, \tag{25}$$

The bulk diffusion coefficient is obtained by correction of D_m for the tortuosity and porosity of the support ($D_b = D_m\tau/\epsilon$).

Pulse field gradient (PFG) NMR spectroscopy can be applied to investigate self-diffusion of molecules in solution, through membranes, and through zeolites [52]. PFG NMR is a direct method to measure the mean square distance $\langle r^2(t)\rangle$ which is traveled by a tracer during a time period Δt. Under the conditions of free-isotropic diffusion in three dimensions (dilute solutions), the replacement is related to the self-diffusion coefficient D_{sd} by Eq. (26):

$$\langle r^2(t)\rangle = 6D_{sd}\Delta t. \tag{26}$$

With PFG NMR, the molecular displacements by self-diffusion can be measured to give the microscopic diffusion coefficient D_{sd} in the range of 10^{-6}–10^{-14} m^2/s. with an optimum between 10^{-9} and 10^{-13} m^2/s. The radial and axial diffusion coefficients in the pores were measured by changing the orientation of the magnetic field compared to the pores of the membrane. For Celgard the radial self-diffusion coefficient was about four times lower than the axial self-diffusion coefficient, while for Accurel no significant difference was found. These results indicate that not only the porosity and tortuosity of the support have a large influence on the diffusion process, but the morphology of the support as well.

2.3. Chemical reactions' kinetics regime transport

An increasing number of investigations report that chemical reaction kinetics, especially at the LM-receiving phase interface, play a sometimes critical role for overall transport kinetics [57-60]. When one or more of the chemical reactions are sufficiently slow in comparison with the rate of diffusion to and away from the interfaces, diffusion can be considered "instantaneous," and the solute transport kinetics occur in a kinetic regime. Kinetic studies of chemical reactions between solute and reagent (carrier) seek to elucidate the mechanisms of such reactions. Information on the mechanisms that control solvent exchange and complex formation is reported briefly below.

Two series of chemical reactions mechanisms and their kinetics have to be analyzed:

(1) Solute uptake at the aqueous feed phase-organic LM interface or partition and chemical interactions with solvent exchange and formation of solute-carrier complex (forward extraction or complexation) (step 2a or 3b)
(2) Solute release with chemical interactions between LM and aqueous strip phases at LM-strip interface with destruction of the complex (decomplexation or backward extraction) and partition of the solute between LM and aqueous strip phases (steps 7a and 6b)

(1) In the solvent exchange, the composition of the coordination sphere often changes, either because of the formation of complexes between the solutes and a complexing reagent, preferentially soluble in an organic phase, or because of the replacement of a ligand in the aqueous-phase solute complex with another more lipophilic one in the organic phase. Solvent exchange and complex formation are special cases of nucleophilic substitution reactions. The basic classification of nucleophilic substitutions is founded on the consideration that when a

new complex is formed through the breaking of a coordination bond with the first ligand (or water) and the formation of a new coordination bond with the organic ligand, the rupture and formation of the bonds can occur with different rates and through the formation of transition intermediates. The rate at which solvent molecules are exchanged between the primary solvation shell of an ion and the bulk solvent is of primary importance in the kinetics of complex formation from aquoions. In both water exchange and complex formation, a solvent molecule in the solvated ion is replaced with a new molecule (another water molecule or a ligand). Therefore, strong correlations exist between the kinetics and mechanisms of the two types of reactions. Observations [61] showed that the rates and the activation parameters for complex formation are similar to those for water exchange, with the complex formation rate constants usually about a factor of 10 lower than those for water exchange with little dependence on the identity of the ligand. This means that, at least as a first approximation, the complex formation mechanism can be described by the rapid equilibrium formation.

For ligand displacement reactions, very few generalizations can be made, since the reaction mechanisms tend to be specific to each chemical system. However, it has been experimentally observed, at least for aqueous-phase reactions, that the variation of the rates with the identity of the ligand correlates well with the variation in the thermodynamic stability of the complex. Therefore, whenever the complex is not extracted unchanged into the organic phase, thermodynamically very stable complexes can be expected to react slowly with the extractant. For example, rate of trivalent lanthanide or actinide cations extraction from aqueous solutions with weakly complexing ligands (Cl^- or NO_3^-) by diethylhexyl phosphoric acid (DEHPA) is very fast. On the other hand in the presence of polyaminocarboxylic acids, such as EDTA (powerful complexing agent), the extraction reaction proceeds only slowly. Ligand-substitution reactions with planar tetracoordinated complexes are often slow in comparison with the rate of diffusion through the interfacial diffusional films.

One more factor, the contact, interaction, and transfer of chemical species on the liquid-liquid interface of two immiscible phases have to be mentioned in the general consideration of chemical kinetics. Little direct information is available on physicochemical properties (interfacial tension, dielectric constant, viscosity, density, charge distribution, etc.) of the interface. The physical depth of the interfacial region can be estimated in the distance in which molecular and ionic forces have their influence. On the aqueous side (monolayers of charged or polar groups) this is several nanometers, on the organic side is the influence of Van der Waals forces. These interfacial zone interactions may slower exchange and complex formation reactions, but as a rule enough fast to be not rate controlling for the most chemical interactions on the interface. More detailed information about interfacial phenomena the reader may obtain in Refs [62, 63].

(2) All considerations, presented for the kinetics of forward extraction reactions, may be applied in the reverse mode to the chemical interactions at LM-strip interface with the complex destruction (decomplexation or backward extraction), the solute release and partition between LM and aqueous strip phases. It has to be taken into consideration two basic differences: quite different thermodynamic conditions on the LM-strip phase interface which may lead to different interaction mechanisms and kinetics and different, more slow kinetics of complex destruction (comparing to the complexation on the feed phase-LM interface), especially at destruction of aggregates, or oligomers which can be formed in the LM at high initial concentration of the solute [58, 64].

Until recently, the role of slow rates of cation release in LM transport was unclear. From ^{1}H NMR studies, it is known that the complexes can be kinetically stable [65, 66] and, as a consequence, decomplexation rates can be very slow. Influence of slow rates of alkali metal cations release was raised in transport through a BLM [58, 64, 67-69]. Recently, at cation transport experiments with different calix crown ether derivatives [42, 70], was proven that the rate of decomplexation can be rate controlling in the transport through SLMs.

2.3.1. Mathematical description of kinetic regime transport

The first goal of any kinetic study is to devise experiments that establish the algebraic form of the rate law and to evaluate the rate constants. Rate laws can be derived by measuring concentration variations as function of time or the initial rates as function of the initial concentrations. Unfortunately, there is no general method for finding the rate law and the reaction order from concentration measurements. Usually, a trial-and-error procedure is used, based upon intelligent guesses.

Experimental kinetic data derived by variables (concentration, temperature, nature of the solvent, presence of other solutes, structural variations of the reactants, etc.), refer to a reaction rate. Reaction mechanism is always only indirectly derived from primary data. Stoichiometry of the reaction, even when this is a simple one, cannot be directly related with its mechanism, and when the reaction occurs through a series of elementary steps, the possibility that the experimental rate law may be interpreted in terms of alternative mechanism increases. Therefore, to resolve ambiguities as much as possible, one must use all the physicochemical information available on the system. Particularly useful here is information on the structural relations between the reactants, the intermediate, and the reaction products.

Mathematical descriptions of simple rate laws, used, as a rule, by LM separations' investigators are presented below:

(1) Irreversible first-order reactions depend only on concentration of a solute S because the rate of the reverse reaction is always close to be equal to zero. Reaction and rate expressions:

$$\mathrm{S + E \xrightarrow{k} SE}, \tag{27}$$

$$-\frac{\mathrm{d[S]}}{\mathrm{d}t} = k\mathrm{[S]}, \tag{28}$$

$$\mathrm{[S]} = \mathrm{[S]_0 e}^{-kt}, \tag{29}$$

where [E] is the carrier concentration, $\mathrm{[S]_0}$ is the solute initial concentration, k is the reaction rate constant, and t is time.

(2) Reversible first-order reactions:

$$\mathrm{S + E} \underset{k_2}{\overset{k_1}{\longleftrightarrow}} \mathrm{SE}, \tag{30}$$

$$-\frac{\mathrm{d[S]}}{\mathrm{d}t} = k_1\mathrm{[S]} - k_2\mathrm{[SE]}, \tag{31}$$

At $t = 0$, $\mathrm{[S]} = \mathrm{[S]_0}$ and $\mathrm{[SE]} = 0$:

$$\mathrm{[S]} = \frac{\mathrm{[S]_0}}{k_1 + k_2}\left(k_2 + k_1 \mathrm{e}^{-(k_1+k_2)t}\right). \tag{32}$$

At $t = \infty$ (at equilibrium, eq), $k_1\mathrm{[S]_{eq}} = k_2\mathrm{[SE]_{eq}}$ and equilibrium constant K_{eq} is

$$K_{\mathrm{eq}} = \frac{k_1}{k_2} = \frac{\mathrm{[SE]_{eq}}}{\mathrm{[S]_{eq}}} \tag{33}$$

and

$$\ln\left(\frac{\mathrm{[S]} - \mathrm{[S]_{eq}}}{\mathrm{[S]_0} - \mathrm{[S]_{eq}}}\right) = -(k_1 + k_2)t. \tag{34}$$

The individual rate constants of the reaction can be evaluated from the slope of a plot, providing the equilibrium constant is available. Many distribution processes between immiscible liquid phases of noncharged species, as well as distribution of solute ions (e.g., metal ions) performed at very low solute concentrations, can be treated as first-order reversible reactions when the value of the equilibrium (partition) constant is not very high.

(3) Series first-order reactions may be referred to the cases when the mechanism goes through an intermediate S_{org}, for example, at an interfacially adsorption of the solute:

$$\mathrm{S_{aq}} \xrightarrow{k_1} \mathrm{S_{org}} + \mathrm{E} \xrightarrow{k_2} \mathrm{SE}, \tag{35}$$

$$\frac{d[S]_{org}}{dt} = k_1[S]_0 e^{-k_1 t} - k_2[S]_{org}. \quad (36)$$

At steady-state approximation $d[S]_{org}/dt = 0$:

$$[S]_{org} = [S]_0 \frac{k_1}{k_2} e^{-k_1 t} \quad (37)$$

and

$$[SE] = [S]_0 \left\{ 1 - \left(1 + \frac{k_1}{k_2} \right) e^{-k_1 t} \right\}, \quad (38)$$

$k_2 \gg k_1$ means that reactive intermediate (complex) is formed at low solute concentration which than interacted at increasing concentration to form the stable final complex aggregate $(SE)_n$ [64, 71]:

$$S_{aq} \xrightarrow{k_1} S_{org} + E \xrightarrow{k_2} SE \xrightarrow{k_3} (SE)_n. \quad (39)$$

Formation of the stable aggregates [72, 73] (e.g., by crosslinking) may cause slow rate of decomplexation at LM-strip phase interface and may be a rate-controlling regime of the LM transport.

2.3.2. Determination of kinetic parameters

2.3.2.1. Determination of α (dimensionless parameter which relates diffusion-limited transport to kinetically limited transport) In general for the layer:

$$\alpha = \frac{D}{kh} \quad (40)$$

and specifically for the LM in the membrane support:

$$\alpha_m = \frac{D_m}{kh_m \varepsilon_m}. \quad (41)$$

where h (or h_m) is the layer film (or membrane support) thickness, k is the rate of reaction, and ϵ_m is the porosity of the membrane support.

When the transport is purely limited by diffusion ($\alpha \sim 0$), parameters can be obtained from measurements of the flux as a function of concentration. In principle, all parameters can be derived by measuring the flux while varying concentrations and thickness of layers. In the case when the carrier at the feed phase interface of the membrane is fully loaded by solute the flux reaches its maximum value (J_{max}) [42]:

$$J_{max} = \left(\frac{D_m}{h_m} E_0 \right) \left(\frac{1}{1 + \alpha} \right), \quad (42)$$

where E_0 is the initial carrier concentration.

The first term in Eq. (42) describes the diffusion-limited flux, while the second term $l/(l + \alpha)$ is a correction factor for slow reaction kinetics. Because α is defined as the ratio $D_m/k_m h_m$, the value of h_m is known and of D_m can be determined independently, the value of k can be determined using Eq. (42). Direct determination of k and D_m from flux measurements as a function of the membrane thickness may be obtained varying the membrane thickness. As a result α is obtained also. If $\alpha < 1$, the transport is mainly limited by diffusion of the complex; while the transport is primarily controlled by the reaction rate in the case that $\alpha > 1$.

By combining Eq. (42) with relationship $D_m = D_{lag}\epsilon$, obtained by lag-time experiments (see Section 2.2.2), α may be obtained from D_{lag} and J_{max}:

$$\alpha = \frac{D_{lag}\varepsilon E_0}{h_m J_{max}} - 1. \quad (43)$$

Separation of the diffusional and kinetic terms is achieved by expressing the flux as the ratio of driving force and flux:

$$\frac{E_0}{J_{max}} = \frac{h_m}{D_m} + \frac{1}{\varepsilon\kappa}. \quad (44)$$

Consequently, a plot of E_0/J_{max} versus h_m gives a straight line with a slope $1/D_m$ and an intercept of $1/\epsilon\kappa$.

The value of α can be determined directly by calculation:

$$\alpha = \frac{1/\varepsilon\kappa}{h_m/D_m}. \quad (45)$$

2.3.2.2. Determination of activation energy Activation energy of transport gives information about the rate-limiting step in the transport process [9, 35, 39]. The activation energy for self-diffusion of a solvent often correlates well with the activation energy for diffusion of a solute species, since on a molecular level diffusion of a solute can be considered as a process in which either a solute or solvent (carrier) molecule jumps from solvent cavity to cavity (see Fig. 2.4). Since the activation energy for self-diffusion varies with the carrier and solvent used, it is important to determine the activation energy E_a at the transport process design.

The influence of the temperature on the transport rate is related by Arrhenius equation:

$$J = J_0 \exp(-E_a/RT). \quad (46)$$

The authors [36] varied the temperature to determine the activation energy E_a from an Eyring plot (slope of a curve $\ln J$ as a function of $1/T$, see Fig. 2.5). They indicated that E_a values below 20 kJ/mol are generally

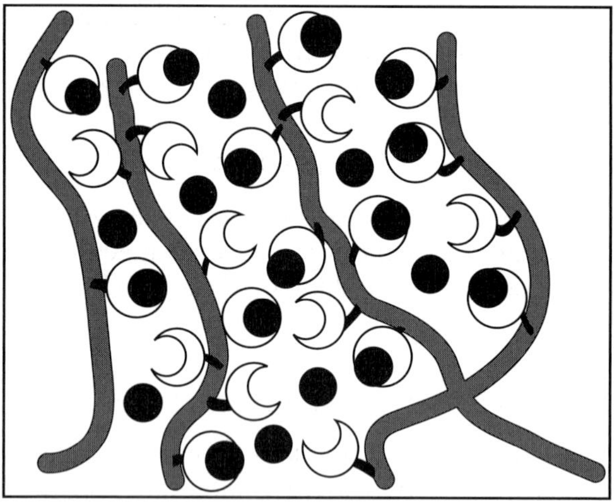

Figure 2.4 Facilitated transport through a complexing polymer matrix. From Ref. [41] with permission.

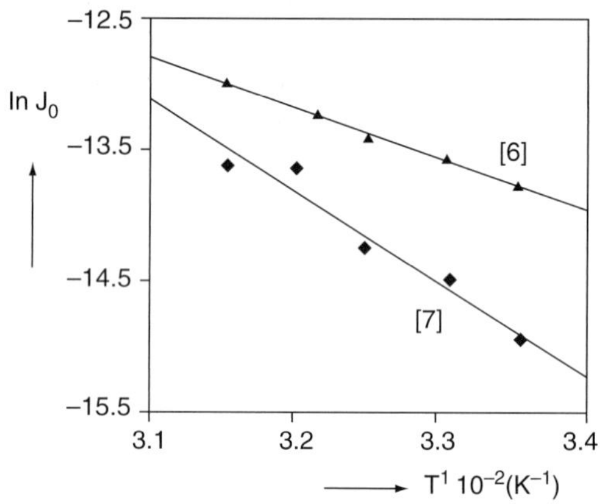

Figure 2.5 Eyring plots for $KClO_4$ transport by two {6} and {7} Calix-4-crown-5 carriers across a NPOE-Accurel membranes. From Ref. [42] with permission.

accepted as indicative of pure diffusion-limited transport. Generally, at activation energies above 40 kJ/mol, chemical reactions do play a role in the transport [74]. The authors [42] obtained activation energies for the transport of $KClO_4$ with two calix[4]crown-5 carriers, measuring the flux J_{max} as a function of operating temperature. Eyring plot of ln J_{max} as a function of $1/T$ showed values of the apparent activation energies: $E_a = 32 \pm 2$ kJ/mol and $E_a = 59 \pm 7$ kJ/mol. Consequently, they concluded that the transport of $KClO_4$ by first carrier {6} is diffusion limited and that of the second carrier {7} is determined by the slow kinetics of release.

The temperature dependency of the viscosity of organic solvents η has an Arrhenius-type behavior:

$$\eta(T) = \eta_0 \exp(E_a/RT). \tag{47}$$

Activation energy for self-diffusion in viscous flow may be calculated from Eq. (47) by measuring the kinematic viscosity as a function of temperature. The activation energy for the self-diffusion through NPOE membrane was 24 kJ/mol [42].

2.4. Mixed diffusional-kinetic transport regime

When both chemical reactions and film diffusion processes occur at rates that are comparable, the solvent extraction kinetics are said to take place in a mixed diffusional-kinetic regime, which, in engineering, is often referred to as "mass transfer with slow chemical reactions." This is the most complicated case, since the rate of extraction must be described in terms of both diffusional processes and chemical reactions, and a complete mathematical description can be obtained only by simultaneously solving the differential equations of diffusion and those of chemical kinetics. The unambiguous identification of the extraction rate regime (diffusional, kinetic, or mixed) is difficult from both the experimental and theoretical viewpoints [18, 19]. Experimental difficulties exist because a large set of different experimental information. Very broad range of several chemical and physical variables are needed. Unless simplifying assumptions can be used, frequently the differential equations have no analytical solutions, and boundary conditions have to be determined by specific experiments.

2.4.1. Identification of the rate-controlling transport regimes

The experimental identification of the regime that controls the transport kinetics is, in general, a problem that cannot be solved by reference to only one set of measurements. In some systems, a definite situation cannot be obtained even when the rate of transport is studied as a function of both hydrodynamic parameters (viscosity and density of the liquids, geometry of the module, stirring or flow rates) and concentrations of the chemical species involved in the transport. It is because the rates may sometimes show the same dependence on hydrodynamic and concentration parameters, even though the processes responsible for the rate are quite different. For a correct hypothesis on the type of regime that controls the transport kinetics, it is necessary to supplement kinetic investigations with other information concerning the biphasic system. This may be the interfacial tension, the solubility of the extractant in the aqueous phase, the composition of the solutes in solution, and so on. Below one can find criteria that are often used to distinguish between a diffusional regime and a kinetic regime:

(1) Comparison of the heat transfer and the mass-transfer coefficients.

If the same dependence of the heat transfer coefficient and the mass-transfer coefficient on the stirring or flowing rate of the phases is observed, the conclusion can be reached that the transport occurs in a diffusional regime.

(2) The reference substance method.

This method is based on the addition of another inert component with known diffusion rate. By following the simultaneous transfer of the species of interest and of the reference component as function of the hydrodynamic conditions, a diffusional regime will be indicated by a similar functional dependence, whereas a kinetic regime is indicated by a sharply different one.

Criteria 1 and 2 are complicated and may be used only occasionally to evaluate the transport regime.

(3) Evaluation of parameter α: relation between diffusion and reaction kinetics regimes (see Section 2.3.2.1).
(4) Evaluation of the activation energy of chemical reactions (see Section 2.3.2.2).

This criterion is not always very meaningful, since many chemical reactions occurring in separation processes exhibit activation energies of only a few kilocalories per mole, that is, have the same order of magnitude as those of diffusional processes.

(5) Dependency of the transport rate on the rate of stirring or flowing in all three phases.

This criterion is simplest, as proved by its widespread use and will be discussed here in more detail.

A typical curve of transport rate versus stirring or flowing rate is shown in Fig. 2.6. In general, a process occurring under the influence of diffusional contributions is characterized by an increase of the transport rate as long as the stirring or flowing rate of the phases is increased (Fig. 2.6, zone A). On the other hand, when the transport rate is close to be independent of the stirring rate it is sometimes possible to assume that the process occurs in a kinetic regime (Fig. 2.6, zone B). An increase in stirring or flowing rate produces a decrease in thickness of the diffusion films: the relationship is approximately linear. The rate of transport will increase with the rate of stirring or flowing, as long as a process is totally or partially diffusion controlled. When the thickness of the diffusion films is reduced to minimum, chemical reactions can be rate controlling, and the rate of transport becomes close to independent of the stirring rate.

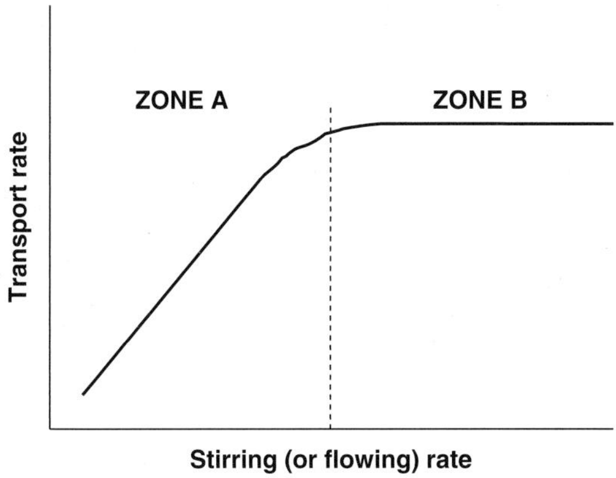

Figure 2.6 Typical curve of transport rate versus phases' stirring (flowing) rate at constant interfacial area.

Unfortunately, this kind of reasoning can lead to erroneous conclusions. Although zone A is certainly an indication that the process is controlled by diffusional processes, the opposite sometimes is not true for zone B: in spite of the increased stirring rate, it may happen that the thickness of the diffusion films never decreases below a sufficiently low value to make diffusion so fast that it can be completely neglected relative to the rate of the chemical reactions. This effect, sometimes called "slip effect," depends on the specific hydrodynamic conditions.

Sometimes an increase in the transport rate can take place with the increase in stirring rate when the system is in a kinetic regime. For example, in ELM systems the increase observed in zone A may indicate here an increase in the number of droplets of the dispersed phase (proportional to the overall interfacial area) and not a decrease in the thickness of the diffusion film. Moreover, the plateau region of zone B does not necessarily prove that the transport occurs in a kinetic regime: at high stirring rates the number of droplets of the dispersed phase eventually becomes constant, since the rate of drop formation equals the rate of drop coalescence. Lack of internal circulation and poor mixing can occur inside the droplets of the dispersed phase. This is particularly true in the ELM systems with the presence of strong surfactants and small droplets. Therefore, here also, a plateau region may simulate a diffusion-controlled regime. It is then apparent that criterion 5 can also lead to misleading conclusions.

Finally, it has to be emphasized that both the hydrodynamic parameters and the concentrations of the species involved simultaneously determine whether the transport regime is of kinetic, diffusional, or mixed diffusional-kinetic type. Therefore, it is not surprising that different investigators, who studied the same transport system in different hydrodynamic and concentration conditions, may have interpreted their results in terms of completely different transport regimes.

2.4.2. Basic parameters of transport regime

The basic parameters of a facilitated, coupled transport are related to properties of the solute, carrier, and its solvent and membrane supports. These are individual and overall mass-transfer coefficients (in diffusional and chemical reactions' kinetics regime), distribution constants, extraction, and coupling coefficients of forward extraction, K_{fp}, K_{fex}, and K_{fcp}, respectively, and backward extraction, K_{bp}, K_{bex}, and K_{bcp}, respectively; diffusion coefficients of complexes in all three phases, selectivity. Influence of these parameters on the solute transport in different configurations will be generally analyzed in the next sections and the respective chapters in detail.

2.4.3. Determination of transport parameters

2.4.3.1. For BLM configurations Let us consider the BLM module design as more complex system, and then discuss the specific things in the SLM and ELM systems.

Individual mass-transfer coefficients of solute species in the feed, carrier, and strip interfacial boundary layers are determined experimentally by feed, carrier, and strip flow rate variations, using Eq. (20).

For feed layer:

$$k_{(\mathrm{fe})_i} = U_{\mathrm{F}_j}\left\{-\ln\left[\frac{[\mathrm{S}_{\mathrm{F}_i}] - \dfrac{J_{\mathrm{F}_{ss}}}{U_{\mathrm{F}_i}}}{[\mathrm{S}]_{\mathrm{F}_{(i-1)}} - \dfrac{J_{\mathrm{F}_{ss}}}{U_{\mathrm{F}_i}}}\right]\right\}^{-1} \tag{48}$$

For carrier layers:

$$\frac{1}{2}k_{(\mathrm{ef})_i}(\text{or } k_{(\mathrm{er})_i}) = U_{\mathrm{E}_j}\left\{\pm\ln\left[\frac{[\mathrm{S}]_{\mathrm{E}_i} - \dfrac{J_{\mathrm{E}_{ss}}}{U_{\mathrm{E}j}}}{[\mathrm{S}]_{\mathrm{E}_{(i-1)}} - \dfrac{J_{\mathrm{E}_{ss}}}{U_{\mathrm{E}_i}}}\right]\right\}^{-1} \tag{49}$$

where "plus" (positive) is at increasing concentrations versus time, and "minus" (negative) is at decreasing ones.

For strip layer:

$$k_{(\mathrm{re})_i} = U_{\mathrm{R}_j}\left\{\ln\frac{[\mathrm{S}]_{R_i} - \dfrac{J_{\mathrm{R}_{ss}}}{U_{\mathrm{R}_j}}}{[\mathrm{S}]_{\mathrm{R}_{(i-1)}} - \dfrac{J_{\mathrm{R}_{ss}}}{U_{\mathrm{R}_j}}}\right\}^{-1} \tag{50}$$

where U_{F_j}, U_{E_j}, U_{R_j} are *j*th flow velocities, $J_{F_{ss}}, J_{E_{ss}}, J_{R_{ss}}$ are fluxes at steady state, $[S]_{F_i}$, $[S]_{E_i}$, $[S]_{R_i}$ are concentrations of solute species, sampled at time *i*, in the feed, carrier, and strip solutions, respectively.

Mass-transfer coefficients of solute species through the membrane may be calculated, using Eqs (4) and (5). D_S-"effective" diffusion coefficients of solute species in the feed, carrier, and strip solutions are evaluated by extrapolating the plots of $t_i = f(U)$ to $U \to 0$. The magnitude of D_S is far from the real diffusion coefficient of solute complexes in liquids, because of some assumptions, mentioned above. Equation (20) at $U \to 0$ becomes undefined, however, for calculation mass-transfer coefficients of solutes at visible flow rates, this parameter (coefficient) is quite applicable. The reduction in the area for diffusion by the impregnable sections of the microporous membrane is accounted for by ϵ_m whereas the increase in diffusion path length over membrane thickness in the tortuous membrane pores is compensated for by $\tau_{m.}$

Another way for determining individual mass-transfer coefficient of solute through the membrane, k_m, is experimenting with the same type of membranes but with different thicknesses, using Eq. (5).

Correlation factor between $[S]_i$ and J_{ss}/U_i in Eqs (48)–(50) has been checked in experiments with several metal ions transport [15, 58, 71] in the range of $U = 0.1$-1.5 cm^3/s. Results showed that $[S] > J_{ss}/U$ in 1.5-4.0 orders (for details see Chapter 5) It means that the J_{ss}/U ratio may be excluded from Eqs (48)–(50). Therefore, we obtain

$$k_{(fe)_i} = U_{F_i}\{-\ln([S]_{F_i})/([S]_{F_{(i-1)}})\}^{-1} \quad \text{or} \quad k_{(fe)_i} = U_{F_i}\{\ln([S]_{F_{(i-1)}})/([S]_{F_i})\}^{-1} \tag{51}$$

$$\frac{1}{2}k_{(ef)_i}(\text{or } k_{(er)_i}) = U_{E_j}\{\pm\ln[([S]_{E_i})/([S]_{E_{(i-1)}})]\}^{-1} \tag{52}$$

$$k_{(re)_i} = U_{R_i}\{\ln([S]_{R_i})/([S]_{R_{(i-1)}})\}^{-1} \tag{53}$$

Equations (51)–(53) are similar to those used by other researchers [3-8, 10-21, 25-29, 59-61]. Referenced authors obtained these equations by considering the basic Stokes-Einstein equation. We obtained the same equations as particular case from Eq. (16), based on kinetics of irreversible processes (nonequilibrium thermodynamics).

According to assumptions and concentration profiles, illustrated in Fig. 2.1A, solute being extracted through a hydrophobic flat membrane can be described by the solute flux from the bulk aqueous phase to the bulk organic phase in terms of individual mass-transfer coefficients at steady state and additivity of a one-dimensional series of diffusion resistances. Overall mass-transfer coefficient, $K_{F/E}$:

$$\frac{1}{K_{F/E}} = \frac{1}{k_f} + \frac{1}{K_{fex}k_{com}} + \frac{1}{K_{fex}k_{mf}} + \frac{1}{K_{fex}k_{fe}} \tag{54}$$

For the solute flux from the bulk organic phase to the bulk strip aqueous phase, the overall mass-transfer coefficient, $K_{E/R}$:

$$\frac{1}{K_{E/R}} = \frac{1}{k_{er}} + \frac{1}{k_{mr}} + \frac{K_{bex}}{k_{dcom}} + \frac{K_{bex}}{k_r} \tag{55}$$

where K_{fex} and K_{bex} are solute distribution coefficients at forward and backward extraction, respectively; k_f, k_{fe}, k_{er}, and k_r are solute individual mass-transfer coefficients in stagnant films; k_{mf} and k_{mr} are solute individual mass-transfer coefficients in membrane support pores (k_{mf} for the membrane on the feed side and k_{mr}—on the strip side); k_{com} and k_{dcom} are individual mass-transfer coefficients of chemical reactions (complexation) at the feed-LM interface and decomplexation at the LM-strip interface, respectively.

1, 3, and 4 components of the sum in Eq. (54) present the diffusional steps resistances and second component is the chemical kinetic regime at forward extraction. In Eq. (55) the diffusional steps are 1, 2, and 4, and third is kinetical step.

If the transport of the solute is from organic phase to aqueous phase, Eqs (54) and (55) remain unchanged. Only the signs in the flux expressions may change if the transport direction has changed. If the system is polar organic-nonpolar organic with the nonpolar organic present in the membrane pores, then the relations for aqueous-organic systems are valid with the polar organic phase used instead of the aqueous phase [2]. Similarly, for a biphasic aqueous system, relations for aqueous-organic systems can be used with the salt-containing aqueous phase instead of the aqueous phase.

When the microporous membrane is in the form of a hollow fiber (see Fig. 2.7), the interfacial areas on the two sides of the hollow fiber are different. The overall mass-transfer coefficient may be defined based on the surface area calculated using either the inside diameter (ID) or the outside diameter (OD) of the hollow fiber. For calculating an overall mass-transfer coefficient, the interfacial area should be based on the diameter where the aqueous-organic phase interface is located. Consider, for example, the aqueous feed and strip phases in hydrophobic fiber lumen (tube side) and organic LM phase on the shell side. The rate of solute extraction per unit fiber length with the aqueous-organic interface located on the fiber ID:

$$\frac{1}{K_{F/E}d_{ID}} = \frac{1}{k_f d_{ID}} + \frac{1}{k_{com}K_{fex}d_{ID}} + \frac{1}{k_{mf}K_{fex}(d_{OD}-d_{ID})} + \frac{1}{k_{fe}K_{fex}d_{OD}} \tag{56}$$

and

$$\frac{1}{K_{E/R}d_{OD}} = \frac{1}{k_{er}d_{OD}} + \frac{1}{k_{mr}(d_{OD}-d_{ID})} + \frac{K_{bex}}{k_{dcom}d_{ID}} + \frac{K_{bex}}{k_r d_{ID}} \tag{57}$$

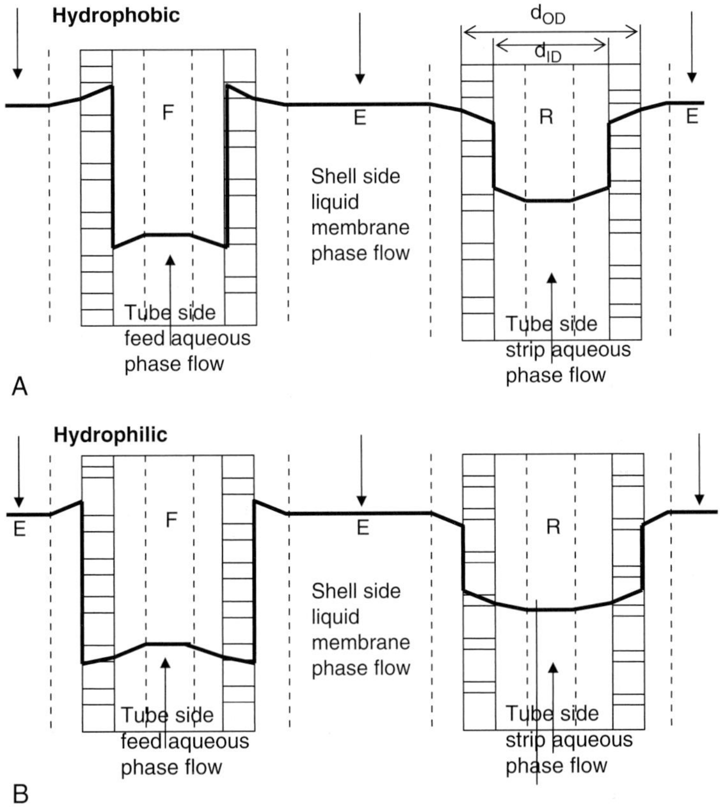

Figure 2.7 Solute concentration profiles for BLM transport using hollow fiber microporous membrane supports.

Here, aqueous-organic interface located on the fiber ID. Utility of such relations has been demonstrated in Ref. [78, 79]. Some authors provide relations without hollow fiber diameter corrections [80].

Hydrophilic or ion-exchange hollow fiber supports overall mass-transfer coefficients may be described by relations:

$$\frac{1}{K_{F/E}} = \frac{1}{k_f} + \frac{1}{k_{mf}} + \frac{1}{k_{com}K_{fex}} + \frac{1}{k_{fe}K_{fex}} \tag{58}$$

For the solute flux from the bulk organic phase to the bulk strip aqueous phase, the overall mass-transfer coefficient, $K_{E/R}$:

$$\frac{1}{K_{E/R}} = \frac{1}{k_{er}} + \frac{K_{bex}}{k_{dcom}} + \frac{K_{bex}}{k_{mr}} + \frac{K_{bex}}{k_r} \tag{59}$$

And for hollow fiber hydrophilic supports:

$$\frac{1}{K_{F/E}d_{OD}} = \frac{1}{k_f d_{ID}} + \frac{1}{k_{mf}(d_{OD} - d_{ID})} + \frac{1}{k_{com}K_{fex}d_{OD}} + \frac{1}{k_{fe}K_{fex}d_{OD}} \tag{60}$$

and

$$\frac{1}{K_{E/R}d_{OD}} = \frac{1}{k_{er}d_{OD}} + \frac{K_{bex}}{k_{dcom}d_{OD}} + \frac{K_{bex}}{k_{mr}(d_{OD} - d_{ID})} + \frac{K_{bex}}{k_r d_{ID}} \tag{61}$$

Here the aqueous-organic phase interface is located on the OD of the fiber.

2.4.3.2. For SLM configurations Now let us consider the SLM module design. The same as with the BLM systems, individual mass-transfer coefficients of solute species in the feed, and strip interfacial boundary layers are determined experimentally by feed and strip flow rate variations, using relations (51) and (53).

According to assumptions and concentration profiles, illustrated in Fig. 2.2A, solute being extracted through a hydrophobic flat membrane can be described by the solute flux from the bulk aqueous feed phase to the bulk aqueous strip phase in terms of individual mass-transfer coefficients at steady state and additivity of a one-dimensional series of diffusion resistances. Overall mass-transfer coefficient, K_{SLM}:

$$\frac{1}{K_{SLM}} = \frac{1}{k_f} + \frac{1}{K_{fex}k_{com}} + \frac{1}{K_{fex}k_{mf}} + \frac{K_{bex}}{K_{fex}k_{dcom}} + \frac{K_{bex}}{K_{fex}k_r} \tag{62}$$

For the SLM designed with hollow fiber module and feed phase in shell the overall mass-transfer equation:

$$\frac{1}{K_{SLM}d_{OD}} = \frac{1}{k_f d_{OD}} + \frac{1}{k_{com}K_{fex}d_{OD}} + \frac{1}{k_{mf}K_{fex}(d_{OD} - d_{ID})} + \frac{K_{bex}}{k_{dcom}K_{fex}d_{ID}} + \frac{K_{bex}}{k_r K_{fex}d_{ID}} \tag{63}$$

For the SLM with hollow fiber module and feed phase in tube:

$$\frac{1}{K_{SLM}d_{ID}} = \frac{1}{k_f d_{ID}} + \frac{1}{k_{com}K_{fex}d_{ID}} + \frac{1}{k_{mf}K_{fex}(d_{OD} - d_{ID})} + \frac{K_{bex}}{k_{dcom}K_{fex}d_{OD}} + \frac{K_{bex}}{k_r K_{fex}d_{OD}} \tag{64}$$

2.4.3.3. For ELM configurations Let us consider the ELM module design. The same as with the SLM systems, individual mass-transfer coefficients of solute species in the feed, and LM interfacial boundary layers are determined experimentally by feed stirring rate variations, using relations (51) and (53). According to assumptions and concentration profiles, illustrated in Fig. 2.2B, the solute flux from the bulk aqueous feed phase to the bulk aqueous strip phase can be described in terms of individual mass-transfer coefficients at steady state and additivity of a one-dimensional series of diffusion resistances. Overall mass-transfer coefficient, K_{ELM}:

$$\frac{1}{K_{\mathrm{ELM}}} = \frac{1}{k_{\mathrm{f}}} + \frac{1}{K_{\mathrm{fex}} k_{\mathrm{com}}} + \frac{1}{K_{\mathrm{fex}} k_{\mathrm{e}}} + \frac{K_{\mathrm{bex}}}{K_{\mathrm{fex}} k_{\mathrm{dcom}}} + \frac{K_{\mathrm{bex}}}{K_{\mathrm{fex}} k_{\mathrm{r}}} \tag{65}$$

3. Driving Forces in Facilitated, Coupled Liquid Membrane Transport

As it was described above the LM solute transport is characterized by diffusion of the solute to the feed-LM, F/E interface, due to concentration gradient of the system, by extraction from feed phase, due to its solubility in LM (LM without carrier), or due to reversible chemical reaction-complexation with an extracting reagent (LM with carrier component), or due to the irreversible chemical reaction with catalytic reagent, with biochemical conversions components (using enzymes, whole cells, etc.) as a result of the thermodynamic conditions at the F/E interface. The solute (at simple transport) or solute-LM complex diffusing to the LM-strip, E/R, interface due to the concentration gradient, is simultaneously decomplexed and stripped by the receiving phase due to the different thermodynamic conditions at the E/R interface and diffuses to the bulk strip, due to the concentration gradient.

In the simple transport flux of the solute that is not complexed with the carrier species is proportional to the concentration gradient of this free solute within the liquid membrane. The solutes to be transported are simply distributed over the phases by partition coefficient K_{fp}:

$$[\mathrm{S_{f1}}] \underset{k_{-1}}{\overset{k_1}{\longleftrightarrow}} [\mathrm{S_{e1}}] \tag{66}$$

$$K_{\mathrm{fp}} = [\mathrm{S_{e1}}]/[\mathrm{S_{f1}}] \tag{67}$$

This leads to the following relation for the flux J

$$J_{\mathrm{fe}} = k_{\mathrm{fe}} K_{\mathrm{fP}}([\mathrm{S_{f1}}] - [\mathrm{S_{e1}}]) \tag{68}$$

Here, the transport rates depend on the partition coefficient K_{fp} only. The solute concentration in the membrane can often be related to the gas phase partial pressure using Henry's law or a similar equilibrium relationship. At higher pressures, vapor-liquid equilibrium or gas-polymer absorption data are necessary to determine the concentration gradient in the membrane.

When the LM contains a carrier that is able to form a complex with the solute in the organic phase (see Eq. (39), steps 1, 2, and maybe 3):

$$K_{FE} = \frac{[SE]}{[S_{e1}][E]} \tag{69}$$

the forward extraction step becomes:

$$K_{fex} = K_{fP} K_{FE} = \frac{[SE]}{[S_{f1}][E]} \tag{70}$$

And equation for the flux:

$$J_{FE} = (K_P + F_{fex})([S_{f1}] - [S_{e1}]) \tag{71}$$

where facilitating factor of carrier at complex formation (forward extraction), F_{fex}:

$$F_{fex} = \frac{K_{fex}[E]_0}{\{(1 + K_{fex}[S_{fe}])(1 + K_{fex}[S_{e1}])} \tag{72}$$

The total flux is not directly proportional to the concentration gradient due to the existence of two transport mechanisms in the membrane; solution-diffusion and diffusion of the carrier-solute complex.

Now, let us consider the coupling effect, using the specific reactions of titanium ion transport (see reactions (1) and (2) in Chapter 1).

At low acidity (Eq. (1), pH region):

$$K_{F/E} = \frac{\left[\overline{TiL_4 \star 2H_2O}\right]_E \star [H]_F^4}{[Ti \star 2H_2O]_F \star \left[\overline{HL}\right]_E^4} = K_{fex} \star \left(\frac{[H]_F}{\left[\overline{HL}\right]_E}\right)^4 = \frac{K_{fex}}{K_{fcp}} \tag{73}$$

where K_{fcp} is the coupling coefficient (countertransport of proton) at forward extraction step.

At high acidity (Eq. (2), 7 M HCl):

$$K'_{F/E} = \frac{\left[\overline{TiL_4 \star 2HCl}\right]_E \star [H]_F^4}{[Ti \star 2HCl]_F \star \left[\overline{HL}\right]_E^4} = K'_{fex} \left(\frac{[H]_F}{\left[\overline{HL}\right]_E}\right)^4 = \frac{K'_{fex}}{K_{fcp}} \tag{73a}$$

Considering kinetics of solute release on the LM-receiving phase interface we can analyze the same chemical reactions (Eq. (39)) in an opposite direction (but at different thermodynamic conditions):

$$(SE)_n \xleftrightarrow{k_{-3}} SE \xleftrightarrow{k_{-2}} E + S_{org} \xleftrightarrow{k_{-1}} S_{aq} \tag{74}$$

Partition coefficient K_{bp} at backward release of the solute to aqueous receiving phase

$$K_{bP} = [S_{r1}] / \left[S_{e4(ore2)}\right] \tag{75}$$

Decomplexation constant

$$K_{ER} = \frac{[S_{e4(ore2)}][E]}{[SE]} \tag{76}$$

Backward extraction constant (in the direction of solute transport):

$$K_{bex} = K_{bP}K_{EF} = \frac{[S_{r1}][E]}{[SE]} \tag{77}$$

Facilitation factor of backward extraction at LM-receiving phase interface

$$F_{bex} = \frac{\{(1 + K_{bex}[S_{e4}])(1 + K_{bex}[S_{r1}])}{K_{bex}[E]_0} \tag{78}$$

And coupling constant at backward extraction step (see Eqs (3) and (4) in Chapter 1); at low acidity (Eq. (3)):

$$K_{E/R} = \frac{[\mathrm{TiO}]_R \star \overline{[\mathrm{HL}]}_E^4 \star [\mathrm{H_2O}]}{\overline{[\mathrm{TiL_4} \star 2\mathrm{H_2O}]}_E \star [\mathrm{H}]_R^2} = \frac{1}{K_{bex}} \star \frac{\overline{[\mathrm{HL}]}_E^4}{[\mathrm{H}]^2} = \frac{K_{bcp}}{K_{bex}} \tag{79}$$

where $[H_2O]$ is neglected.

And at high acidity (Eq. (4)):

$$K'_{E/R} = \frac{[\mathrm{TiO}]_R \star \overline{[\mathrm{HL}]}_E^4 \star [\mathrm{Cl}]_R^2}{\overline{[\mathrm{TiL_4} \star 2\mathrm{HCl}]}_E[\mathrm{H_2O}]} = \frac{1}{K'_{bex}} \star \overline{[\mathrm{HL}]}_E^4 \star [\mathrm{Cl}]_R^2 = \frac{K'_{bcp}}{K'_{bex}} \tag{79a}$$

Here we can see both coupling at countertransport of proton and cotransport of chlorine anion.

Considering total driving force coefficient of all LM transport system, K, we obtain:

$$K = \frac{K_{fex}}{K_{bex}} \star \frac{K_{bcp}}{K_{fcp}} = \frac{K_c}{K_d} \tag{80}$$

where

$$K_c = \frac{K_{fex}}{K_{bex}} \tag{81}$$

and

$$K_d = \frac{K_{fcp}}{K_{bcp}} \tag{82}$$

The solute transport is driven by solute concentration gradient, by Liquid membrane facilitation potential (LMF), K_c, and by Donnan equilibrium coupling, K_d. K_c is denoted as an internal LM carrier driving force coefficient, derived from extraction distribution constants for solute between feed-LM and LM-strip phases. K_d is denoted as an external driving force coefficient, derived from the coupling effect of the transport. At $K_c = 1$ ($K_{fex} = K_{bex}$) concentration of the solute in the carrier solution should be $[S]_R{}^{\star}K_{bex}$ (the system at equilibrium). Therefore, K_{bex} may be denoted as an irreversible coefficient for both closed and open LM systems (flowing feed, strip streams, buffered acidities, etc.). K_c is an "uphill pumping" border of the system.

Internal (carrier) driving force coefficients, K_{fex} and K_{bex}, or distribution coefficients, E_F and E_R, data [15] are determined by membrane-based extraction experiments. Membrane-based forward and backward extraction are carried out in two compartment modules, separated by the same membranes. The experiments lasted up to equilibrium conditions, when the concentration of solutes in every compartment is not changed with time.

The effect of carrier, contained in the terms F_{fex} and F_{bex} lowers the resistance of the liquid membrane [42, 65]. Basic features of carrier-mediated transport: (1) the flux is proportional to the carrier concentration; (2) the initial flux shows typical saturation behavior, at low solute concentrations, and close to independence of [S] at high solute concentrations; (3) the flux J shows a maximum with position depending on S_F and S_R. A very strong complexing carrier becomes fully loaded with solute even at low concentrations, and therefore does not produce a gradient over the LM. This leads to the conclusion that the best carrier is not necessarily the strongest complexing one. Apart from this thermodynamic reason, there may be also a kinetic reason: strong complexing agents frequently show a slow rate of decomplexation.

4. Selectivity

Let us consider the selectivity parameter on the example of metal ions separation. According to the transport model equations the selectivity of two solutes, for example, two metal species, S_{M_1/M_2}, is determined by relation:

$$S_{M_1/M_2} = \frac{K_{M_1}\left[S_F^0\right]_{M_1}}{K_{M_2}\left[S_F^0\right]_{M_2}} \tag{83}$$

where subscripts M_1 and M_2 refer to the two metal species; K_{M_1} and K_{M_2} are the total overall mass-transfer coefficients, $\left[S_F^0\right]_{M_1}$ and $\left[S_F^0\right]_{M_2}$ are the initial concentrations of two metal ions in the treated feed solution.

Introducing a separation factor, A, defined as a ratio of the total overall mass-transfer coefficients of the solutes (metal species):

$$A_{M_1/M_2} = K_{M_1}/K_{M_2} \tag{84}$$

an equation for the system selectivity is obtained

$$S_{M1/M2} = A_{M1/M2} \frac{[S_F^0]_{M1}}{[S_F^0]_{M2}} \tag{85}$$

Based on the principle of resistance additivity, the total overall mass-transfer coefficient, K_M, of every solute passing through the separation system, is related to the overall mass-transfer coefficients on the feed and strip sides as follows:

$$K_M = \frac{(K_{F/E}/K_{E/R})_M}{(K_{F/E} + K_{E/R})_M} \tag{86}$$

Thus, the separation factor of the two metal species is:

$$A_{M1/M2} = \frac{(K_{F/E} \star K_{E/R})_{M1} \star (K_{F/E} + K_{E/R})_{M2}}{(K_{F/E} \star K_{E/R})_{M2} \star (K_{F/E} + K_{E/R})_{M1}} \tag{87}$$

For evaluation of the selectivity of two metal species separation we can assume that in the same solution environment (water) the diffusion coefficients of these metal ions with the same charge have similar values and the diffusion coefficients of the metal-carrier complexes have similar values. Thus, we can represent separation factor as dependent only on the distribution coefficients at forward and backward extraction, determined experimentally through distribution coefficients at membrane-based equilibrium forward, $E_{F/E}$, and backward, $E_{E/R}$, extraction [15, 58]:

$$A_{M1/M2} \approx \frac{\left(\frac{E_{F/E}}{E_{E/R}}\right)_{M_1}}{\left(\frac{E_{F/E}}{E_{E/R}}\right)_{M_2}} \tag{88}$$

The distribution coefficient may be expressed as a function of the metal association (stability) constants in the LM solution, the association constants of metal ions with solvent environment in the feed and in the strip solutions and partition coefficients of the carrier and metal ion. In this case, the separation factor can be determined by stability constants of the metal complexes, formed with functional groups of carrier, if we assume that the metal ions are predominantly present (a) as free ions in the acid solution, so that complex concentrations can be disregarded and (b) as complexes in the LM solutions, so that free ion concentrations can be disregarded.

$$A_{M1/M2} \approx \frac{(\beta_{F/E}/\beta_{E/R})_{M1}}{(\beta_{F/E}/\beta_{E/R})_{M2}} \tag{89}$$

where $\beta_{F/E}$ and $\beta_{E/R}$ are stability constants of the metal-carrier complexes in equilibrium with the feed and strip solvent environment (as a rule acids for metal ions), respectively. Therefore, preliminary selectivity data of metal species separation may be evaluated without experimentation, if stability constants data are available in the literature.

From Eqs (62) and (63) it follows that high separation factors are favored when approaching conditions $(E_{F/E}/E_{E/R})_{M_1} \gg (E_{F/E}/E_{E/R})_{M_2}$ or $(\beta_{F/E}/\beta_{E/R})_{M_1} \gg (\beta_{F/E}/\beta_{E/R})_{M_2}$. On the other hand, the system loses its selectivity when distribution parameters of both metal species are either extremely high or extremely low [29, 37, 74]. Selectivity can be increased for BLM and partly for ELM systems by choosing a selective carrier with intermediate distribution data values, and adjusting its concentration, its volume, the acidity of the feed and strip solutions in such a way as to approach the above conditions [75, 76]. Another way to improve selectivity parameters is to choose some mixtures of the strong and relatively weak carriers [77]. In every case it has to be checked by experimentation.

5. Module Design Considerations for Separation Systems

Following, the reader can find general considerations at designing the HLM process for cadmium separation from wastewaters of the fertilizers industry [71], one of the BOHLM technologies, presented in Chapter 5 in detail.

Individual and overall mass-transfer coefficients are shown in Table 2.1, and concentration profiles of cadmium species in the feed, carrier, and strip solutions are shown in Fig. 2.8.

Comparison of the experimental and the simulated data shows that:

1. Variations of the feed and strip flow rates have little effect on the cadmium transport performance: the values of individual cadmium mass-transfer coefficients are similar at carrier or strip flow rates variations. Thus, diffusion of cadmium species through the feed and strip aqueous boundary layers does not control the transport rate. The rate-controlling steps could act as resistances to diffusion of the cadmium species in the carrier solution layers, especially in the membrane pores or the interfacial backward-extraction reaction kinetics.
2. Resistance to diffusion in the LM solution layers and membrane pores is not a rate-controlling step, since the overall mass-transfer coefficients on the LM-strip interface of the system are two orders less than that on the feed-LM side. Thus, we can conclude that the interfacial backward-extraction reaction rate is a rate-controlling step of cadmium transport in the system.

Table 2.1 Individual and overall mass-transfer coefficients, obtained using only extraction-backward-extraction driven transport equations

No.	Flow velocity (cm^3/s)	[Cd] flux from feed $J_F \times 10^{-10}$ ($mol/cm^2\,s$)	[Cd] flux to carrier $J_E \times 10^{-10}$ ($mol/cm^2\,s$)	[Cd] flux to strip $J_R \times 10^{-10}$ ($mol/cm^2\,s$)	$k_{c(fe)}$ (cm/s)	$k_{ef} = k_{c(er)} \times 10^{-2}$ (cm/s)	$k_{c(re)} \times 10^{-2}$ (cm/s)	$D_E \times 10^{-5}$ (cm^2/s)	$k_{c(mf)} = k_{c(mr)} \times 10^{-2}$ (cm/s)	$K_{cF} \times 10^{-3}$ (m/s)	$K_{cR} \times 10^{-5}$ (m/s)
1	2	3	4	5	6	7	8	9	10	11	12
1	0.083	2.12	1.87	0.17	4.4×10^{-1}	3.75	5.7			2.27	2.34
2	0.283	4.96	1.88	0.20	2.0	17.5	19	3.33	0.253	3.95	2.52
3	0.500	7.97	1.86	0.35	6.6	31	74			4.61	2.53
4	0.833	10.12	1.67	0.49	10.0	49.5	150			4.74	2.53

Notes: 1. Results, represented in columns 3 and 6, were obtained at various feed flow velocities (column 2) and fixed ($U = 0.5$ cm^3/s) carrier and strip solutions' flow velocities. Results in columns 4 and 7 were obtained at various carrier solution flow velocities and fixed ($U = 0.5$ cm^3/s) feed and strip solutions' flow velocities. Results in columns 5 and 8 were obtained at various strip solution flow velocities and fixed ($U = 0.5$ cm^3/s) feed and carrier solutions' flow velocities.
2. Coefficient D_E in the column 9 is defined as an "effective" diffusion coefficient (details and determination see Chapter 5).

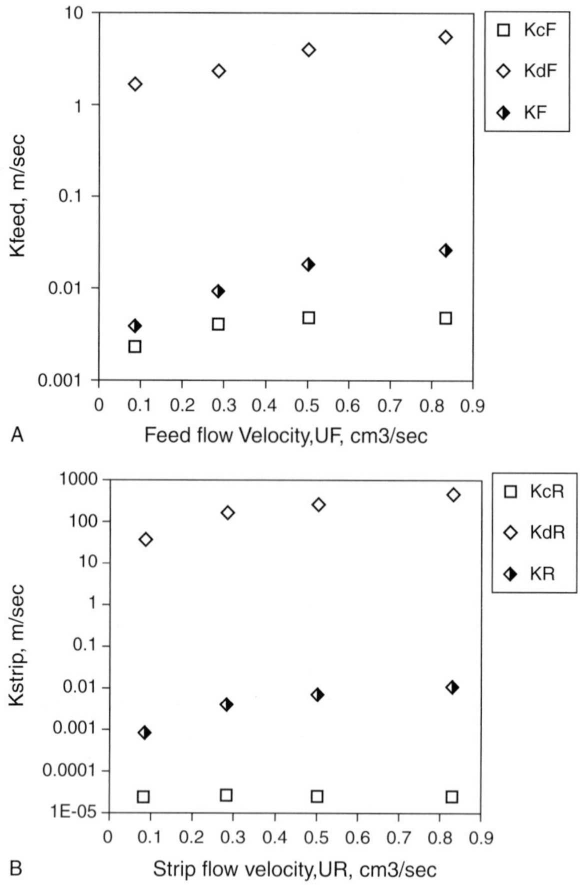

Figure 2.8 Effect of the feed (A) or strip (B) flow variations on the overall mass-transfer coefficients of cadmium species transport.

3. Discrepancies between experimentally obtained and theoretically calculated data for cadmium concentration in the strip phase are 10–150 times at feed or strip flow rate variations. These differences between the experimental and simulated data have the following explanation. According to the model, mass transfer of cadmium from the feed through the carrier to the strip solutions is dependent on the diffusion resistances: boundary layer resistances on the feed and strip sides, resistances of the free carrier and cadmium-carrier complex through the carrier solution boundary layers, including those in the pores of the membrane, and resistances due to interfacial reactions at the feed- and strip-side interfaces. In the model equations we took into consideration only mass-transfer relations, motivated by internal driving force (forward

extraction-backward-extraction distribution ratio, K_c). Mass-transfer relations, motivated by external (coupling) driving force (proton concentration gradients) between feed and strip phases, indicated by K_d coefficient, were not considered. Thus, resistances to the cadmium transport, due to diffusion kinetics of protons at the feed-membrane and strip-membrane interfaces, resistance to free carrier molecules diffusion through the boundary layers of the carrier solution and through the membrane pores opposite to cadmium direction were not taken into account.

There are two ways to evaluate individual mass-transfer coefficients of these processes:

1) by sampling and determining proton concentrations in the feed and strip phases, and free carrier concentration in the carrier solution during experiments with flow rate variations
2) by comparing the experimentally obtained cadmium concentration profiles with model predicted ones, which were calculated using only mass-transfer coefficients, K_c. We used the second, simpler way

Individual external mass-transfer coefficients, k_d (see Table 2.2) were evaluated using Eqs (51)–(53), where, $[Cd]_{F_i}$ $[Cd]_{E_i}$, and $[Cd]_{R_i}$ were taken from model calculated data and $[Cd]_{F_{i-1}}$, $[Cd]_{E_{i-1}}$, and $[Cd]_{R_{i-1}}$ were experimentally obtained data under the same conditions and sampling time, t_i. Correcting overall mass-transfer coefficients $K_{dF/E}$ on the feed side and $K_{dE/R}$ on the strip side were calculated by the following equations:

$$K_{dF/E} = \frac{k_{d(fe)} k_{d(mf+fe)}}{k_{d(fe)} + k_{d(mf+fe)}} \tag{90}$$

Table 2.2 Individual and overall mass-transfer coefficients, accounting for coupling effects of the cadmium transport (external driving force). Combined overall mass-transfer coefficients

No	Flow velocity (cm^3/s)	$k_{d(fe)}$ (m/s)	$k_{d(mf+ef)} = k_{d(mr+er)}$ (m/s)	$k_{d(re)}$ (m/s)	K_{dF} (m/s)	K_{dR} (m/s)	$K_F \times 10^{-2}$ (m/s)	$K_R \times 10^{-3}$ (m/s)
1	2	3	4	5	6	7	8	9
1	0.083	2.4	85.7	13.2	1.63	35.9	0.37	0.84
2	0.283	1.5	139.3	49.8	2.25	158.7	0.89	4.00
3	0.500	1.1	357.1	90	3.84	263.2	1.77	6.66
4	0.833	2.2	392.8	150	5.34	485.0	2.53	12.27

Note: For an explanation of the results, represented in columns 3-5, see note 1 of Table 2.1.

and

$$K_{dE/R} = \frac{k_{d(er)} k_{d(mr+er)}}{k_{d(er)} + k_{d(mr+er)}} \tag{91}$$

The overall mass-transfer coefficients of the LM system are calculated by Eqs (66) and (67), which are followed from Eq. (53):

$$K_{F/E} = \frac{K_{C(F/E)}}{K_{d(F/E)}} \tag{92}$$

and

$$K_{E/R} = \frac{K_{C(E/R)}}{K_{d(E/R)}} \tag{93}$$

Figure 2.8 shows the dependence of the cadmium transport internal driving force coefficients, $K_{c,}$ external driving force coefficients, K_d, and overall mass-transfer coefficients, K, on feed flow rate (Fig. 2.8A) or strip flow rate (Fig. 2.8B) variations. It is clearly seen that the resistivity to the diffusion of protons is much lower than that of cadmium species themselves.

Overall mass-transfer coefficients dependence on flow rates may be evaluated by the following equations:

$$K_{F/E} = 2.34e^{-4} + 3.70e^{-2U_F} - 7.92e^{-3U_F} \quad \text{at } R^2 = 0.991 \tag{94}$$

$$K_{E/R} = -1.41e^{-4} + 1.29e^{-2U_R} + 2.39e^{-3U_R} \quad \text{at } R^2 = 0.998 \tag{95}$$

Figure 2.9 shows examples of cadmium species concentration profiles in the feed, carrier, and strip solutions, obtained experimentally (dotted curves) and calculated, using model equations (continuous lines). There is a good correlation between experimental and simulated data. Comparing the simulated results with the experimental data, it appears that at higher flow rates, where boundary layers resistance becomes less important, the resistance to Cd-Cyanex 302 complexes diffusion inside membrane pores and/or membrane-strip interfacial reaction kinetics dominates as a rate-controlling steps for cadmium transport. In the first case, when diffusion of large organic complexes through the filled membrane pores is more limiting, we can make decisive improvements by changing the hydrophobic membranes to hydrophilic or cation exchange, due to make aqueous solution filled pores. The BOHLM (see Chapter 5) has this advantage in comparison with other liquid membrane technologies for many solutes (metals, carboxylic, amino acids, etc.).

Model analysis shows that the solute (cadmium) concentration enrichment cannot exceed the value of $[Cd]_R{}^\star K_{E/R}$ (cadmium concentration in the

Model calculated and experimental data

Time	Model calculated data			Experimental data		
	[Cd] in feed	[Cd] in carrier	[Cd] in strip	[Cd] in feed	[Cd] in carrier	[Cd] in strip
sec	mol/kg	mol/kg	mol/kg	mol/kg	mol/kg	mol/kg
0	4.89E–04	0	0	4.89E–04	0.00E+00	0.00E+00
3.60E+03	4.83E–04	3.17E–05	9.59E–09		2.42E–05	
7.20E+03	4.76E–04	6.24E–05	3.80E–08			
1.08E+04	4.70E–04	9.23E–05	8.45E–08		9.49E–05	
1.44E+04	4.65E–04	1.21E–04	1.49E–07		1.20E–04	
1.80E+04	4.59E–04	1.49E–04	2.30E–07		1.50E–04	
2.16E+04	4.53E–04	1.77E–04	3.27E–07		1.74E–04	
2.52E+04	4.48E–04	2.03E–04	4.41E–07	4.69E–04	1.90E–04	
2.88E+04	4.43E–04	2.29E–04	5.70E–07			
3.24E+04	4.37E–04	2.54E–04	7.14E–07			
3.60E+04	4.32E–04	2.78E–04	8.72E–07		2.70E–04	
3.96E+04	4.28E–04	3.02E–04	1.04E–06			
4.32E+04	4.23E–04	3.25E–04	1.23E–06			
4.68E+04	4.18E–04	3.47E–04	1.43E–06		3.30E–04	1.60E–06
5.04E+04	4.14E–04	3.69E–04	1.64E–06			
5.40E+04	4.09E–04	3.90E–04	1.86E–06	4.00E–04	3.80E–04	1.70E–06
5.76E+04	4.05E–04	4.10E–04	2.10E–06			
6.12E+04	4.01E–04	4.30E–04	2.35E–06			
6.48E+04	3.97E–04	4.49E–04	2.60E–06			
6.84E+04	3.93E–04	4.67E–04	2.87E–06			
7.20E+04	3.89E–04	4.85E–04	3.15E–06			
7.56E+04	3.85E–04	5.03E–04	3.44E–06			
7.92E+04	3.81E–04	5.20E–04	3.74E–06			
8.28E+04	3.78E–04	5.36E–04	4.04E–06			
8.64E+04	3.74E–04	5.52E–04	4.36E–06	3.70E–04	5.00E–04	4.40E–06
9.00E+04	3.71E–04	5.68E–04	4.68E–06			
9.36E+04	3.67E–04	5.83E–04	5.02E–06			
9.72E+04	3.64E–04	5.97E–04	5.36E–06			
1.01E+05	3.61E–04	6.12E–04	5.70E–06			
1.04E+05	3.58E–04	6.25E–04	6.06E–06			
1.08E+05	3.55E–04	6.39E–04	6.42E–06	3.48E–04	6.10E–04	6.20E–06
1.12E+05	3.52E–04	6.52E–04	6.79E–06			
1.15E+05	3.49E–04	6.64E–04	7.17E–06			
1.19E+05	3.46E–04	6.76E–04	7.55E–06			
1.22E+05	3.43E–04	6.88E–04	7.94E–06			
1.26E+05	3.41E–04	6.99E–04	8.33E–06	3.39E–02	6.70E–04	8.00E–06

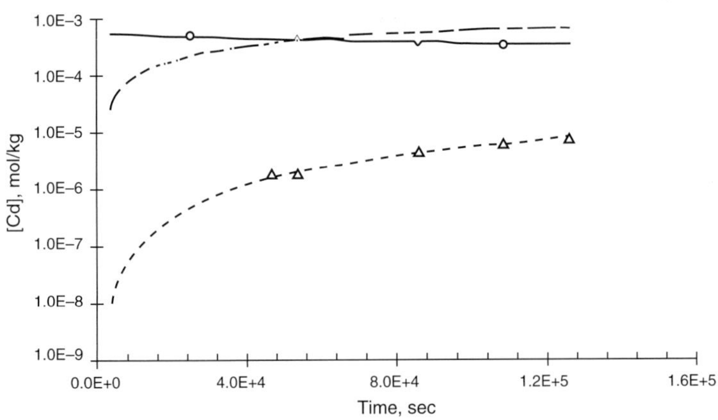

Figure 2.9 Cadmium transport concentration profiles. Comparison of the calculated (continuous lines) and experimentally obtained (dotted curves) data.

carrier phase ★ backward-extraction distribution coefficient), or $[Cd]_F{\star}K_{F/E}/K_{E/R}$ (for the SLM system). Thus, extraction distribution parameters control the enrichment ability of the LM process.

External, coupling driving force motivated cadmium transport perhaps is not a high factor in the system studied, because there is no a large proton concentration gradient between feed and strip aqueous phases.

Many researchers, proposing the LM processes for application, are based on the steady state of the system. Experimental and model simulation data show much higher mass-transfer rates through the HLM (for details see Chapter 5), with cadmium concentration in the carrier solution, reaching its maximum. At this stage, the both internal (extraction-backward-extraction distribution ratio) and external (coupling) driving forces motivate the cadmium transport in an optimal way.

At steady-state cadmium transport permeation is motivated mostly by an external driving force and the fluxes are about an order lower. A much more effective HLM module, with continuously flowing feed (open system), can be designed if the feed side membrane area S_F and the feed flow rate U_F enable us to obtain a fixed cadmium feed outlet concentration (e.g., 1 ppm) at a contact time, less than that at the maximum on the simulated concentration profile of the carrier solution.

More detailed analysis with the semiempirical model equations, developed for every configuration of LM systems, can be found in the respective Chapters 3–6. These considerations may be used in order to minimize experimental testing at the LM processes design.

Batch experiments are preferred at design of small systems requiring flexibility in their operation. Batch systems are more flexible, require less automation, give a longer residence time, and have a lower membrane area and lower capital cost compared with continuous operations. Stages-in-series design is preferred for a large plant with constant feed composition and throughput or where product residence time must be minimized. Tubular, hollow-fiber and spiral-wound modules most commonly used in various biochemical and food processing operations.

6. Factors, Affecting Carrier-Facilitated Coupling Transport

6.1. Carrier properties

Solute transport through the LM enhances by water-immiscible species, dissolved in the solvent (water-immiscible also). These species interact selectively with solutes and named carriers and transport is named as facilitated or membrane mediated or carrier enhanced, or depending on the authors' mood. The use of LM with carriers offers an alternative to solvent extraction for selective separation and concentration of solutes from dilute solutions. Most LM carriers are originally extractants developed for solvent extraction processes. The reader can find their descriptions in the Solvent Extraction Handbooks (e.g., [81]). There are many natural species, such as valinomycin or beauverin, which can be used as carriers. And very many synthetic carriers have been developed specially for LM. As an example, crown ether macrocycles, which selectively bind alkali and some other metal ions, have been synthesized and used as carriers in liquid membranes for the selective recognition of neutral, charged, or zwitter-ionic species [82, 83].

Performance of a LM is strongly related to the characteristics of a carrier. The main parameters of carriers in LM transport are:

1. High selectivity to species have to be separated
2. High capacity of the species have to be extracted
3. High ability of a carrier to complexate (to extract) a solute from an aqueous feed phase into LM phase at feed-LM interface (high extraction or distribution, or partition constant, $E_{F/E}$)
4. High ability of carrier-solute complex in a LM to be decomplexed and stripped from LM to an aqueous strip phase at the LM-strip interface (high decomplexation or stripping constant, $E_{E/R}$)
5. Rapid kinetics of formation (complexation) and destruction (decomplexation) of the complex on membrane interfaces
6. Rapid kinetics of diffusion of the carrier-solute complex through the LM (a measure of the diffusion rate, diffusion coefficient, D_{LM})
7. Stability of the carrier
8. No side reactions
9. No irreversible or degradation reactions
10. Low solubility of the carrier (and solvent) in the aqueous phases
11. No complexation (coextraction) of water
12. It should be easily regenerated
13. It should have suitable physical properties, such as density, viscosity, surface tension
14. Low toxicity for biological systems and low corrosivity
15. Reasonable price at industrial applications.

Different mechanisms of diffusion take place in LM: diffusion of the carrier-solute complex, diffusion of the uncomplexed carrier in the opposite direction, diffusion of the uncomplexed solutes. The last transport mechanism is not accessible to solutes that do not react with the carrier species. It is the complexation reaction that makes facilitated LM transport highly selective.

A great variety of carriers are used in the LM transport (see examples in the respective chapters). They may be divided to cation-, anion-exchangers, and neutral ligands. First group is the big number of organic acids and their derivatives and related proton donors. For example, some commercially available extractants: di(2-ethylhexyl)phosphoric acid (DEHPA), bis

(2,2,4-trimethyl-pentyl) phosphinic acid (Cyanex 272), some hydroxyoximes (LIX or Acorga series), oligoamide compounds, containing 8-hydroxyquinolyl groups (Kelex or LIX-26 series) [1-8, 15-24, 35, 38-40, 49-62, 65, 66]. Crown ethers and related macrocyclic multidentate ligands have a pronounced selectivity for cations (metal ions) [7, 41, 42, 66]. Their ability to selectively and reversibly bind metal ions may enable a LM to perform difficult separations. Wide variety of macrocyclic carriers exist and are developing in the last years [82, 83]. The complex stability, donor abilities of functional atoms (O, N, S, P, etc.), and/or groups of macrocyclic carriers are intensively studied in aim to develop some rules for predicting their carrier properties.

Second group is water-immiscible primary, secondary, tertiary amines, and their derivatives, quaternary amine salts and other proton acceptors (see respective chapters). Considerable effort has been devoted to the development of anion transporting agents, playing an important role in the biological and biochemical processes. Several amines, their derivatives, surfactants, lipophilic metal complexes, macrocycles with positively charged subunits are known and developing as anion carriers [22-24, 27-29, 44-46].

The third group is water-immiscible organic species with electron donor or acceptor properties, or solvating carriers. They include carbon-oxygen compounds (amides, ethers, ketones); phosphorus-oxygen compounds (tri-*n*-butylphosphate (TBP), dibutyl-phosphate (DBP) or -phosphonate (DBBP); phosphine oxides (tri-*n*-octylphosphine oxide (TOPO); phosphine sulfides (Cyanex 471); alkyl sulfides (dihexyl, diheptyl sulfides); nitrogen containing compounds (CLX 50), and so on [1-7, 81, 83]. All of them are known as selective extractants, but few of them are tested as carriers in LM processes.

Mixtures of carriers (ionic additives). Cotransport of opposite-charged ions is the most obvious way to maintain electroneutrality, but alternative means may be explored as additives to LM. In recent years, many studies have been conducted which examine the use of anionic membrane additives for maintenance of electroneutrality at cation transport [65, 84-89]. The anionic additives are typically lipophilic carboxylic, phosphoric, or sulfonic acids. Cation or neutral macrocyclic carriers coupled with anionic additives result in a synergistic transport of cations which exceeds that accomplished by each component individually. This synergism was demonstrated in Ref. [90]. The authors observed a 10- to 100-fold enhancement of copper extraction. Enhanced extraction is achieved by adding the anionic group to the cation coordination macrocycle.

Parthasarathy and Buffle [55] have systematically varied the chain length of a series of lipophilic carboxylic acids in a supported liquid membrane with 1,10-didecyldiaza-18-crown-6 as carrier. Chain lengths ranged from 10 to 18 carbons. Optimal Cu^{2+} transport was achieved with additives from 12 to 14 carbons in length, and lauric acid ($n = 12$) yielded the best results due to its decreased tendency to form precipitates with Cu^{2+}.

It appears that the ratio of anionic additive is the most critical measure [55, 88, 89]. While the optimal amount of additive varies from system to system, the potential benefit of the additives is well established.

To achieve high selectivity, a substrate-specific receptor must be present in the membrane phase, in which it can act as a carrier between source and receiving phase. Whereas in biological membranes this task is fulfilled by ionophores such as valinomycin (l), in artificial membranes we rely on the realm of synthetic macrocyclic receptors developed during the past two decade [83].

Some LM carriers tested meet serious problems: they show low rate of decomplexation which becomes a rate-limiting step of total LM transport. In aim to understand the phenomenon, mechanisms of Cd transport through the HLM [71] with di-(2,4,4-trimethyl-pentyl)monothiophosphinic acid, Cyanex 302, were investigated. Depending on Cd concentration in the organic phase, three different interaction stages were suggested. At low Cd concentration in the organic phase, cation-exchange extraction takes place with formation of tetrametric complexes as $CdL_2{\star}2HL$ (I stage). With increasing Cd concentration these nuclei grow in size by coordination bonding (extraction by solvation) with undissociated cadmium salt molecules, forming linear (or planar) aggregates [64] (II stage). Upon reaching a critical size, a structural reorganization occurs: polyhedral aggregates or "clusters" [72, 73] with polydentate, asymmetric bonds are forming (III stage). Some spectroscopic and chemical techniques showed results, which indirectly confirmed mechanisms, taking place at II and III stages. Kinetics of these aggregates decomposition and consequently back extraction of Cd to the strip phase is very slow. Of coarse, proposed aggregation mechanisms have to be proved, using direct techniques, for example, NMR, light scattering, zeta potential, and so on. The knowledge of the extraction mechanisms is important for designing the liquid membrane technology, processing feed solutions with significant metal ions concentrations. It is one of the directions for future investigations.

6.2. Solvent properties influencing transport

Nearly every liquid membrane system devised to date involves the solvation of carriers in an organic solvent. Because transport of any solute requires that it pass through this organic solvent, transport rates and selectivities depend heavily upon the properties of this solvent. Authors of the review [91] divided the free energy of LM transport into four components in a thermodynamic cycle: free energy of desolvation of the cation, free energy of desolvation of the anionic ligand, free energy for the gas phase interaction between cation and ligand, free energy of solvation of the solute/carrier complex. Under this method of describing membrane transport, three of the four components are intimately related to the nature of the organic solvent. Solvent characteristics influence the thickness of the Nernst films at the membrane interfaces, equilibrium constants for solute-carrier interaction in the membrane, partition coefficients, and the diffusivities of the species in the system [92]. Not only is the influence of the solvent type rather large, it is quite complex.

Many groups have investigated the suitability of various solvents for use in LM systems and have attempted to describe the relationship between solvent characteristics and transport properties [93-96]. Of all solvent properties, dielectric constant seems to be most predictable in its effect on transport [92]. For solvents, such as the halocarbons, transport usually decreases with increasing dielectric constants [93]. Figure 2.10 shows this trend for alkali metals binding by dicyclohexano-18-crown-6 in a number of alcohols. This trend holds true for many simple systems, but it breaks down under more complex conditions. Solvent donor number, molecule size, solvent viscosity, carrier solubility in the solvent, permanent and induced dipole moments, and heats of vaporization are important [94].

Solvent characteristics that influence the diffusion and extraction are found to be viscosity ($\acute{\eta}$) and polarity (€). For spherical solutes, the diffusion coefficient depends on the solvent according to the Stokes-Einstein relation (Eq. (22)). From this, it follows that the diffusion coefficient linearly increases with $T/\acute{\eta}$. Hence, the permeability increases linearly with the reciprocal viscosity of the membrane solvent [95]. Figure 2.11 shows relation of the diffusion coefficient D_m to the solvent viscosity.

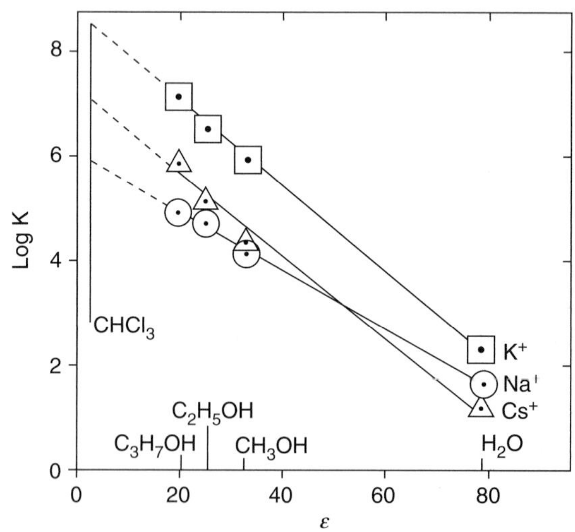

Figure 2.10 Relationship between carrier (macrocycle)-cation complex stability and solvents dielectric constant ε for several solvents. From Ref. [98] with permission.

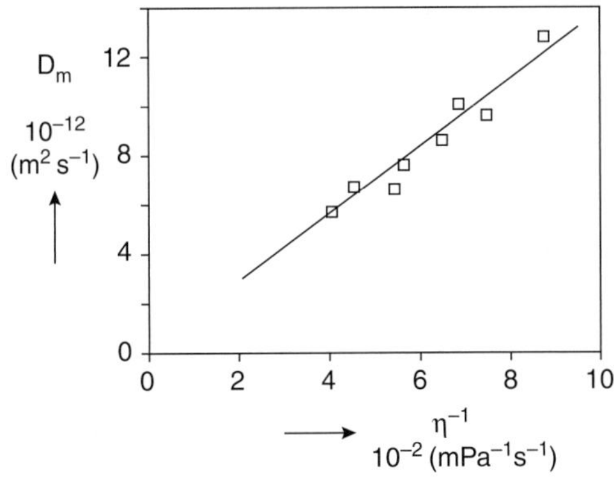

Figure 2.11 Relationship between the diffusion coefficient and solvent viscosity for the transport of $NaClO_4$. From Ref. [108] with permission.

The solvent effect on extraction constants is a combination of the influence on salt partition and association. Both processes are influenced by the polarity of the solvent [97]. The Kirkwood function describes the relation between polarity, €, and extraction constant: a more polar membrane solvent promotes extraction. In general, polar solvents favor salt partition, but the tendency toward complexation diminishes. Since the overall effect of solvent polarity on the extraction is positive, the polarity appears to affect the partition coefficient to a higher degree.

The use of mixed solvents introduces some significant advances in the LM transport [95, 96, 99]. Figure 2.12 shows the synergistic effect of binary mixtures of chloroform and nitrobenzene: maximum Na^+ transport occurs with an equimolar solution of the two solvents. Parthasarathy and Buffle [55] in a similar study report optimal transport of Cu^{2+} with an equimolar solution of phenylhexane and toluene.

Many equations have been developed for consideration of solvent properties in predicting transport rates [89], but their predictions still suffer from considerable inaccuracy. Moreover, the issue of solvent effect is a complex one and considerable room remains for further study of this subject. Certainly LM systems will continue to improve as our understanding of the complex influences of solvent properties increases.

6.3. Membrane support properties

Microporous membrane supports may be symmetric, asymmetric, or composite. They may have a uniform pore size or a distribution of pore sizes. They may be thick, thin, or ultrathin, with or without charges on external and internal surfaces.

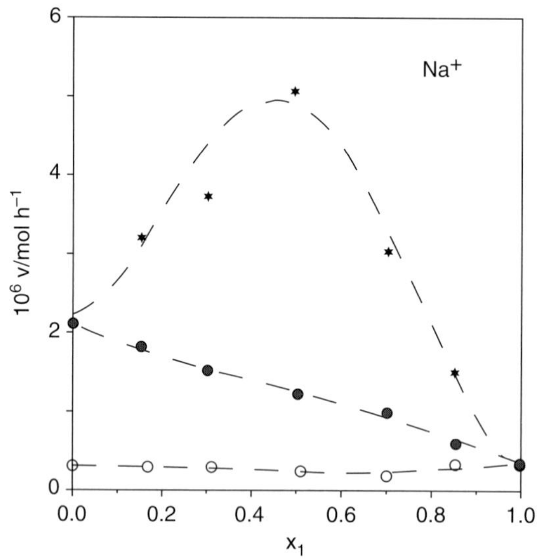

Figure 2.12 Synergistic effect of mixed solvents on Na^+ transport through a bulk liquid membrane. The source phase is 1 M NaCl. The receiving phase is distilled water and dibenzo-18-crown-6 is the carrier. Three mixed solvent systems were tested: (*) chloroform(l)-nitrobenzene(2); (•) dichloroethane(I)-nitrobenzene(2); and (○) chloroform (I)-dichloroethane(2). From Ref. [93] with permission.

Most membranes used as supports in liquid membrane technologies are polymeric in nature, although inorganic membranes have also become available. While many polymers have been examined for use as membrane materials, only a few are widely used. Detailed description of polymeric membranes can be found in reviews [100–106].

Membranes can be classified according to following characteristics:

1. Material of construction: polymer, ceramic (including glass and porcelain), metal
2. Structure: homogeneous, asymmetric, or composite
3. Method of manufacture: phase inversion, sintering, stretching, or track etching
4. Geometry: flat sheet, hollow fiber, or tubular
5. Hydrophobic or hydrophilic
6. Surface charge: neutral or charged (positive or negative)

Important membrane performance characteristics are (1) permeability, (2) selectivity, retention efficiency, (3) electrical resistance, (4) exchange capacity, (5) chemical resistance, (6) wetting behavior and swelling degree, (7) temperature limits, (8) mechanical strength, (9) cleanliness, and (10) adsorption properties.

Virtually the entire membrane manufacture today is based on laminate structures comprising a thin barrier layer deployed upon a much thicker, highly permeable support. Most are formed of compositionally homogeneous polysulfone, cellulose acetate, polyamides, and various fluoropolymers by phase inversion techniques in which ultrathin films of suitably permselective material are deposited on prefabricated porous support structures. Hydrophobic polymers as polyethylene, polypropylene, or polysulfone are often used as supports. A fairly comprehensive list of microporous and ultrafiltration commercial membranes and produced companies are presented in Refs [107-109]. A review on inorganic membranes has been given in Ref. [110].

For fast transport rates, the exchanging surface between the aqueous phases and the LM phase should be large. Therefore thin films with high porosity are used [111]. Commonly used commercial supports are Celgard and Accurel. The membrane thickness of Celgard 2500 is 25 μm with a porosity $\epsilon = 0.45$ and a tortuosity $\tau = 2.35$. Accurel has a thickness of 100 μm with a porosity $\epsilon = 0.64$ and a tortuosity $\tau = 2.1$.

Ion exchange membranes contain fixed anionic or cationic charged groups attached to the polymer backbone that are able to transport cations or anions. The specific properties of ion exchange membranes are all related to the presence of these charged groups. Amount, type, and distribution of ion exchange groups determine the most important membrane properties. Based on the type of fixed charge group, ion exchange membranes can be classified as strong acid and strong base or weak acid and weak base membranes. Strong acid cation-exchange membranes contain sulfone groups as charged ones. In weak acid membranes, carboxylic acid is the fixed charged group. Quaternary and tertiary amines are the fixed positive charged groups in strong and weak base anion exchange membranes, respectively.

The ion exchange capacity (IEC) is the number of fixed charges inside the ion exchange membrane per unit weight of dry polymer. The IEC is a crucial parameter which affects almost all other membrane properties. The IEC is expressed in milliequivalent of fixed groups per gram of dry membrane (meq/g membrane).

In cation-exchange membranes, the fixed negative charges are in electrical equilibrium with the mobile cations (counter-ions). The opposite relation exists in anion exchange membranes. The fixed charge density, expressed in milliequivalent of fixed groups per volume of water in the membrane (meq/l) strongly depends on the IEC and the swelling degree of the membrane: in the swollen state, the distance between the ion exchange groups increases thus reducing the fixed charge density. The transport of counter ions through the membrane is determined by the fixed charge density in the membrane and the difference between the concentration of the electrolyte solution in contact with that membrane. The concentration and type of the fixed ionic charges determine the permselectivity and the electrical resistance of the membrane.

When an ion exchange membrane is in contact with an electrolyte (salt solution), ions with the same charge (co-ions) as the fixed ions are excluded and cannot pass through the membrane, while the oppositely charged ions (counter-ions) can pass freely through the membrane. This effect is known as Donnan exclusion [112]. It reflects the ability of the membrane to discriminate between ions of opposite charge.

The electrical resistance of the membrane is an important property of ion-exchange membranes. The membrane resistance is determined by the IEC and the mobility of the ions within the membrane matrix. The electrical resistance is dependent on temperature and decreases with increasing temperature.

With respect to their structure and preparation procedure, the ion exchange membranes can be divided into two categories: homogeneous and heterogeneous. In homogenous ion-exchange membranes the fixed charge groups are evenly distributed over the entire membrane matrix. They are manufactured via polymerization and polycondensation of functional monomers. Heterogeneous membranes have distinct macroscopic uncharged polymer domains in the membrane matrix. The distinct difference between homogenous and heterogeneous ion exchange membranes also influences the properties of the specific membrane.

6.4. Coupling ions: Anion type

To maintain electroneutrality and solute uphill pumping many membrane carrier systems require a coupling (from the source or receiving phases) ion to be counter- or cotransported along with the solute ion. Because the coupling ion must also enter and cross the organic phase, it is bound to influence transport efficiency.

Proton or sometimes alkali metal cations are used for countertransport of cationic or cotransport of anionic solutes because of their good transport properties. It is not the case with the coupling anions. In fact, for K^+ transport by 18-crown-6 in a BLM, the anion effect differs by more than 100 [96]. Many studies of the anion effect on transport efficiency have been conducted [97–100]. The effects of anion hydration free energy, anion lipophilicity, and anion interactions with solvents have been mentioned, although anion hydration free energy seems to be the major determinant of transport efficiency. For example, transport of K^+ with dibenzo-18-crown-6 as a carrier, decreased in the order: picrate > PF_6^- > ClO_4^- > IO_4^- > BF_4^- > I^- > SCN^- > NO_3^- > Br^- > BrO_3^- > Cl^- > OH^- > F^- > acetate > SO_4^-. This order is almost identical to that for increasing anion hydration free energy. This example demonstrates the strong correlation between anion hydration and transport efficiency: larger anions are more easily dehydrated and thus more readily enter the membrane to facilitate transport.

While nearly all investigations of anion effects have focused on transport efficiency, a few recent works suggest a correlation between anion type and selectivity. At extraction of alkali metals into chloroform by dicyclohexano-18-crown-6 [101], selectivity for K^+ over both Rb^+ and Cs^+ decreases dramatically depending upon anion type in the order: $NO_3^- > SCN^- > ClO_4^-\ I^- > Br^-$. K^+/Cs^+ selectivity decreases from 16.0 for nitrate to 3.5 for bromide. The authors were unable to tie this trend to any particular parameter, although they discounted the possibility that it is correlated to anion radius, hydration enthalpy, or anion softness.

In research [89] the authors demonstrated that anions are capable not only of altering selectivities but indeed of reversing them. Typically, 18-crown-6 analogs show a strong preference for Hg^{2+} over Cd^{2+}. However, by using SCN^- as a counter-anion, completely reverse selectivity was obtained, resulting in highly selective transport of Cd^{2+} over Hg^{2+}. This result is due to the fact that the SCN ion forms coordination complexes with these cations, which lead to reverse selectivity. A similar result is present when Br^- is used as an anion for altering selectivity between Cd^{2+} and Zn^{2+}. These results suggest that careful consideration of coupling anion is crucial when designing a membrane system or when comparing results listed in the literature.

6.5. Influence of concentration polarization and fouling

Solute fluxes through the membrane and the membrane life time are primarily affected by the phenomena of concentration polarization and fouling. Polarization is an unavoidable consequence of the competition between convection and diffusion at a permselective barrier; while it cannot be eliminated, it can be mitigated by appropriate device design/fluid management strategies.

Fouling (e.g., microbial adhesion, gel layer formation, and solute adhesion) at the membrane surface is a more complex phenomenon involving polarization, irreversible adsorption of macrosolutes or colloid particulates to, and/or gradual buildup of an adherent and coherent layer of solid material on, the membrane surface. It is amenable to mitigation by appropriate selection or surface treatment of the membrane surface (to minimize adsorption) by suitable fluid management; or by employment of other forces to transport fouling solutes.

Although there are several types of fouling, the most problematic in many facilities is biofouling (also known as biofilm). Biofilms are complex structures that generally comprise a mixed community of microorganisms that are firmly attached to a surface of a membrane. Traditional methods for cleaning and controlling the development of biofilms rely on the use of chemicals, such as chlorine or chlorine-based compounds. Recently, nitric oxide (NO) treatment was proposed and technique developed for detection and removing of biofilms [113].

Large-scale membrane systems operate in a cyclic mode, where the normal run alternates with clean-in-place operation.

6.6. Influence of temperature

At higher temperatures, both the diffusion and decomplexation processes are accelerated, and transport rates increase. The diffusion coefficient increases, while the extraction coefficient K_{bex} decreases with increasing temperature.

7. Summary Remark

An attempt to unify the mass transport phenomena of liquid membrane separation underlying the basic LM configurations was presented in this chapter. The basic theory was developed in a simple physical-chemical-mathematical form and applied to the principal techniques in such a way to obtain comparable methods. Of course it is preliminary work: once we start forging links between different methods there will be spillover to further possibilities of integration.

REFERENCES

1. Mulder M, *Basic Principles of Membrane Technology*. Kluwer Academic Publishers, Norwell, MA, 1992.
2. Ho WSW, Sirkar KK, Eds., *Membrane Handbook*. Chapman & Hall, New York, NY, 1992.
3. Osada Y, Makagawa T, Eds., *Membrane Science and Technology*. Marcel Dekker, New York, NY, 1992.
4. Noble RD, Stem SA, Eds., *Membrane Separation Technology*. Elsevier, New York, NY, 1995.
5. Noble RD, Way JD, Eds., *Liquid Membranes: Theory and Applications*, ACS Symposium Series 347. American Chemical Society, Washington, DC, 1987.
6. Araki T, Tsukube H, Eds., *Liquid Membranes: Chemical Applications*. CRC Press, Boca Raton, FL, 1990.
7. Bartsch RA, Way JD, Eds., *Chemical Separations with Liquid Membranes*, ACS Symposium Series 642. American Chemical Society, Washington, DC, 1996.
8. Zhivkova S, Dimitrov K, Kyuchoukov G, Boyadzhiev L, Separation of zinc and iron by pertraction in rotating film contactor with Kelex 100 as a carrier. Sep. Pur. Technol. 2004; 37(1): 9-16.
9. Lazarova Z, Boyadzhiev L, Kinetic aspects of copper(II) transport across liquid membrane containing LIX-860 as a carrier. J. Membr. Sci. 1993; 78(3): 239-245.

10. Lazarova Z, Boyadzhiev L, Liquid film pertraction – A liquid membrane preconcentration technique. Talanta 1992; 39(8): 931-935.
11. Boyadzhiev L, Liquid pertraction or liquid membranes – State of the art. Sep. Sci. Technol. 1990; 25(3): 187-190.
12. Teramoto M, *et al.*, Development of spiral-type supported liquid membrane module for separation and concentration of metal Ions. Sep. Sci. Technol. 1987; 22(11): 2175-2201.
13. Matsuyama H, Teramoto M, *et al.*, Separation and concentration of heavy metal ions by spiral type flowing liquid membrane module. Water Treat. 1990; 5: 237-252.
14. Baker R, Blume I, Coupled transport membranes. In: Porter MC, Ed. *Handbook of Industrial Membrane Technology*. Noyes Publications, USA, 1990: 511-558.
15. Kislik V, Eyal A, Hybrid liquid membrane (HLM) system in separation technologies. J. Membr. Sci. 1996; 111: 259-272.
16. Majumdar S, Sirkar KK, Hollow-fiber contained liquid membrane. In: Ho WSW, Sirkar KK, Eds., *Membrane Handbook*. Van Nostrand Reinhold, New York, 1992: 764-808.
17. Schlosser S, Sabolova E, Three-phase contactor with distributed U-shaped bundles of hollow-fibers for pertraction. J Mem Sci 2002; 210(2): 331-347.
18. Schlosser S, Pertraction through liquid and polymeric membranes. In: Belafi-Bako K, Gubicza L, Mulder M, Eds., *Integration of Membrane Processes into Bioconversions*. Proceedings of the16th European Membrane Society Annual Summer School. Veszprem, Hungary, Aug. 1999. Kluwer Academic/Plenum Publishers, New York, NY, 2000: 73-100.
19. Wodzki R, Szczepanska G, Szczepanski P, Unsteady state pertraction and separation of cations in a liquid membrane system: Simple network and numerical model of competitive M^{2+}/H^{+} counter-transport. Sep. Pur. Technol. 2004; 36(1): 1-16.
20. Teramoto M, Takeuchi N, Maki T, Matsuyama H, Ethylene/ethane separation by facilitated transport membrane accompanied by permeation of aqueous silver nitrate solution. Sep. Pur. Technol. 2002; 28(2): 117-124.
21. Kubisova L, Sabolova E, Schlosser S, Martak J, Kertesz R, Mass-transfer in membrane based solvent extraction and stripping of 5-methyl-2-pyrazinecarboxylic acid and co-transport of sulphuric acid in HF contactors. Desalination 2004; 163: 27-38.
22. Schlosser S, Sabolova E, Transport of butyric acid through layered bulk liquid membranes. Chem. Pap. 1999; 53: 403-411.
23. Schlosser S, Rothova I, Frianova H, Hollow-fiber pertractor with bulk liquid membrane. J Mem Sci 1993; 80: 99-106.
24. Strzelbicki J, Schlosser S, Influence of surface-active substances on pertraction of cobalt (II) cations through bulk and emulsion liquid membranes. Hydrometallurgy 1989; 23: 67-75.
25. Majumdar S, Sirkar KK, Hollow-fiber contained liquid membrane. In: Ho WSW, Sirkar KK, Eds., *Membrane Handbook*. Van Nostrand Reinhold, New York, 1992: 764-808.
26. Sirkar KK, Other new membrane processes. In: Ho WSW, Sirkar KK, Eds., *Membrane Handbook*. Van Nostrand Reinhold, New York 1992: 904-908.
27. Vajda M, Sabolova E, Schlosser S, Mikulova E, Membrane-based extraction joined with membrane-based stripping in a circulating arrangement. II. Extraction of organic acids. Chem. Pap. 2003; 57(1): 3-10.
28. Wodzki R, Szczepanska G, Szczepanski P, Unsteady state pertraction and separation of cations in a liquid membrane system: Simple network and numerical model of competitive M^{2+}/H^{+} counter-transport. Sep. Pur. Technol. 2004; 36(1): 1-16.
29. Wodzki R, Nowaczyk J, Propionic and acetic acid pertraction through a multimembrane hybrid system containing TOPO or TBP. Sep. Pur. Technol. 2002; 26(2-3): 207-220.

30. Wodzki R, Sionkowski G, Pozniak G, Recovery and concentration of metal ions. IV. Up-hill transport of Zn(II) in a multimembrane hybrid system. Sep. Sci. Techol. 1999; 34: 627-649.
31. Haase R, *Thermodynamics of Irreversible Processes*. Dover Publications, New York, 1990: 108-111.
32. Helfferich F, *Ion Exchange*. McGraw-Hill, New York, 1962, Chapter 8.
33. Helfferich F, Ion-exchange kinetics. III. J Am. Chem. Soc. 1962; 66: 39-52.
34. Helfferich F, Ion-exchange kinetics. IV. J Am. Chem. Soc. 1963; 67: 1157-1164.
35. Fyles TM, mcMalik-Diemer VA, McGavin CA, Whitfield DM, Membrane transport systems. III. A mechanistic study of cation-proton coupled countertransport. Can. J. Chem. 1982; 60, 2259.
36. Szpakowska M, Nagy OB, Chemical kinetic approach to the mechanism of coupled transport of Cu(II) ions through bulk liquid membranes. J. Phys. Chem. A 1999; 103: 1553-1559.
37. Danesi PR, Horwitz EP, Vandergrift GF, Chiarizia R, Mass transfer rate through liquid membranes: Interfacial chemical reactions and diffusion as simultaneous permeability controlling factors. Sep. Sci. Technol. 1981; 16: 201-210.
38. Danesi P, Separation of metal species by supported liquid membranes. Sep. Sci. Technol. 1984-1985; 19(11 and 12): 857-864.
39. Teramoto M, Sakai T, Yanagawa K, Ohsuga M, Miyake Y, Modeling of the permeation of copper through liquid surfactant membranes. Sep. Sci. Technol. 1983; 18: 735-743.
40. Szpakowska M, Nagy OB, Non-steady state vs. steady state kinetic analysis of coupled ion transport through binary liquid membranes. J. Membr. Sci. 1993; 76: 27-38.
41. Michaels AS, Membranes, membrane processes, and their applications: Needs, unsolved problems, and challenges of the 1990s. Desalination, 1990; 77: 5-34.
42. Chrisstoffels LAJ, de Jong F, Reinhoudt DN, Mechanistic studies of carrier-mediated transport through supported liquid membranes. In Bartsch RA, Way JD, Eds., *Chemical Separations with Liquid Membranes*, ACS Symposium Series 642. American Chemical Society, Washington, DC, 1996; Chapter 3: 18-56.
43. Bird RB, Stewart WE, Lightfoot EN, *Transport Phenomena*, John Wiley & Sons, New York, 1960.
44. Cussler EL, *Diffusion, Mass Transfer in Fluid Systems*, Cambridge University Press, Cambridge, 1984; vol. 58: 44.
45. Ridberg J, Musikas C, Choppin GR, Eds., *Principles and Practices of Solvent Extraction*. Marcel Dekker, New York, 1992.
46. Izatt RM, Bruening RL, Bruening ML, lind HGC, Christensen JJ, Modeling diffusion-limited, neutral-macrocycle-mediated cation transport in supported liquid membranes. Anal. Chem. 1989; 61: 1140-1148.
47. Giddings JC, *Unified Separation Science*. Wiley, New York, 1991: 17-48.
48. Giddings JC, *Unified Separation Science*. Wiley, New York, 1991: 112-114.
49. Bromberg L, Levin G, Kedem O, Transport of metals through gelled supported liquid membranes containing carrier. J. Membr. Sci. 1992; 71, 41.
50. Zhu XX, MacDonald PM, Pulsed-gradient spin-echo NMR measurements of the diffusion coefficients of ketones in poly(methyl methacrylate). Macromolecules 1992; 25, 4345.
51. Kiani A, Bhave RR, Sirkar KK, Further studies on solvent extraction with immobilized interfaces in a microporous hydrophobic membrane. J. Membr. Sci. 1986; 26, 79.
52. Inoue Y, Mechanistic Principles of Liquid Membrane Transport. In: Araki T, Tsukube H, Eds., *Liquid Membranes: Chemical Applications*. CRC Press, Boca Raton, FL, 1990: 77-102.
53. Noble RD, Koval CA, Pellegrino LJ, Facilitated transport membrane systems. Chem. Eng. Prog. 1989: 58-70.

54. Basaran OA, Burban PM, Auvil SR, Facilitated transport with unequal carrier and complex diffusivities. Ind. Eng. Chem. Res. 1989; 28: 108-119.
55. Parthasarathy N, Buffle J, Capabilities of supported liquid membrane for metal speciation in natural waters: application to copper speciation. Anal. Chim. Acta 1994; 284: 649-659.
56. Nahir TM, Buck RP, Transport processes in membranes containing neutral ion carriers, positive ion complexes, negative mobile sites, and ion pairs. J. Phys. Chem. 1993; 97: 12363-12372.
57. Li Y, Gokel G, Hernandez L, Echegoyen L, Cation-exchange kinetics of sodium complexes with amide- and ester-substituted crown ethers in homogeneous solution. J. Am. Chem. Soc. 1994; 116: 3087-3096.
58. Kislik V, Eyal A, Hybrid liquid membrane (HLM) and supported liquid membrane (SLM) based transport of titanium (IV). J. Mem. Sci. 1996; 111: 273-281.
59. Juang R, Lo R, Kinetics of the coupled transport of vanadium(IV) from sulfate solutions through supported liquid membranes. Ind. Eng. Chem. Res. 1994; 33: 1011-1016.
60. Juang R, Permeation and separation of zinc and copper by supported liquid membranes using bis (2-ethylhexyl)phosphoric acid as a mobile carrier. Ind. Eng. Chem. Res. 1993; 32: 911-916.
61. Eigen M, Wilkins RW, *Mechanisms of Inorganic Reactions*, Vol. 49. Adv. Chem. Ser. American Chemical Society, Washington, DC, 1965.
62. Adamson AW, *Physical Chemistry of Surfaces*, Wiley, New York, 1982.
63. Davies T, Rideal EK, *Interfacial Phenomena*, Academic Press, New York, 1961.
64. Neuman RD, Nai-Fu-Zhou WuJ, Jons MA, Gaonkar AG, General model for aggregation of metal-extractant complexes in acidic organophosphorus solvent extraction systems. Sep. Sci. Technol. 1990; 25(13-15), 1655-1674.
65. Ghidini E, Ugozzoli F, Ungaro R, Harkema S, Abu El-Fadl A, Reinhoudt DN, Complexation of alkali metal cations by conformationally rigid, stereoisomeric calix[4] arene crown ethers: A quantitative evaluation of preorganization. J. Am. Chem. Soc. 1990; 112, 6979.
66. Reinhoudt DN, Dijkstra PJ, in't Veld PJA, Bugge KE, Harkema S, Ungaro R, Ghidini E, Kinetically stable complexes of alkali cations with rigidified calix[4]arenes. X-ray structure of a calixspherand sodium picrate complex. J. Am. Chem. Soc. 1987; 109, 4761.
67. J P Behr, M Kirch, J M Lehn, Carrier-mediated transport through bulk liquid membranes: Dependence of transport rates and selectivity on carrier properties in a diffusion-limited process. J. Am. Chem. Soc. 1985; 107, 241.
68. Fyles TM, On the rate-limiting steps in the membrane transport of cations across liquid membranes by dibenzo-18-crown-6 and lipophilic crown ether carboxylic acids. J. Membr. Sci. 1985; 24, 229.
69. Yoshida S, Hayano S, Kinetics of partition between aqueous solutions of salts and bulk liquid membranes containing neutral carriers. J. Membr. Sci. 1982; 11, 157.
70. Casnati A, Pochini A, Ungaro R, Bocchi C, Ugozzoli F, Egberink RJM, Struijk W, Lugtenberg RJM, de Jong F, Reinhoudt DN, 1,3-Alternate calix[4]arenecrown-5 conformers: New synthetic ionophores with better K+/Na+ selectivity than valinomycin. Chem. Eur. J. 1996; 2: 436-445.
71. Eyal AM, Kislik VS, Hybrid liquid membrane (HLM) process for metal recovery. In: Avicenne Joint Research Project Report: Membrane Recovery of Metal Pollutants from Waste Waters of the Fertilizers Industry (MERMEP), 1996: 25-29.
72. Johnson BFG, Rodgers A, Polyhedral rearrangements and fragmentation reactions in cluster complexes, In: Shriver DF *et al.* Eds., *The Chemistry of Metal Cluster Complexes*. VCH Publ., Weinheim, 1990: 303.
73. Echt O,Wan der Waals clusters: Metastable decay, coulomb explosion, and electron attachment. In: Benedek X. *et al.* Eds., *Elemental and Molecular Clusters*, Proceedings of the 13th International School, Erice, Italy, July 1987: 263.

74. O'Hara PA, Bohrer MP, Supported liquid membranes for copper transport. J. Membr. Sci. 1989; 44, 273.
75. Kedem O, Bromberg L, Ion-exchange membranes in extraction processes. J Mem Sci 1993; 78: 255-261.
76. Eyal A, Bressler EJ, Industrial separation of carboxylic and amino acids by liquid membranes: Applicability, process considerations, and potential advantages. Biotechnol. Bioeng. 1993; 41: 287-293.
77. Wodzki R, Nowaczyk J, Propionic and acetic acid pertraction through a multimembrane hybrid system containing TOPO or TBP. Sep. Pur. Technol. 2002; 26(2-3): 207-220.
78. Prasad R, Sirkar KK, Microporous membrane solvent extraction. Sep. Sci. Technol. 1987; 22: 619-640.
79. Prasad R, Sirkar KK, Dispersion-free solvent extraction with microporous hollow fiber modules. AIChE J. 1988; 34: 177-188.
80. D'Elia NA, Dahuron L, Cussler EL, Liquid-liquid extraction with microporous hollow fibers. J. Membr. Sci. 1986; 29: 309-319.
81. Lo TH, Baird MHI, Hanson C, Eds., *Handbook of Solvent Extraction*. Wiley, New York, 1983.
82. Nijenhuis WF, Buitnuis EG, *et al.* Calixcrowns as selective potassium cation carriers in Supported liquid membranes. J. Am. Chem. Soc. 1991; 113, 7963-7968.
83. Cooper SR, *Crown Compounds, Toward Future Applications*. VCH Publishers Inc, New York, 1992.
84. Ansari SA, Mohapatra PK, Prabhu DR, Manchanda VK, Evaluation of N,N,N′,N′-tetraoctyl-3-oxapentane-diamide (TODGA) as a mobile carrier in remediation of nuclear waste using supported liquid membrane. J Membr. Sci. 2007; 298, 1-2, 169-174.
85. Ma S-L, Wang X-Z, Fan C-H, Guo Q-L, Zhu W-X, Zhang J, The syntheses of phenol-containing diazamacrocycles and liquid membrane transports of alkali cations. J. Mol. Struct. 2007; 2, 46-52.
86. Ensor DD, McDonald GR, Pippin CG, Extraction of trivalent lanthanides by a mixture of didodecylnaphthalenesulfonic acid and a crown ether. Anal. Chem 1986; 58, 1814-1816.
87. Eugster R, Spichiger UE, Simon W, Membrane model for neutral-carrier-based membrane electrodes containing ionic sites. Anal. Chem. 1993; 65, 689-695.
88. Schaller U, Bakker E, Spichiger UE, Pretsch E, Ionic additives for ion-selective electrodes based on electrically charged carriers. Anal. Chem 1994; 66, 391-398.
89. Bakker E, Malinowska E, Schiller RD, Meyerhoff ME, Anion-selective membrane electrodes based on metalloporphyrins: The influence of lipophilic anionic and cationic sites on potentiometric selectivity. Talanta 1994; 41, 881-890.
90. Moyer BA, Case GN, Alexandratos SD, Kriger AA, Extraction of copper(II) from sulfuric acid by macrocycle-synergized cation exchange: Comparing a novel impregnated resin with its solvent-extraction analog. Anal. Chem. 1993; 65: 3389-3395.
91. Lamb JD, Izatt RM, Christensen JJ, Stability Constants of Cation-macrocycle Complexes and their Effect on Facilitated Membrane Transport. In: Izatt RM, Christensen JJ, Eds., *Progress in Macrocyclic Chemistry*, Vol. 2. Wiley, New York, 1981: 41-90.
92. Izatt RM, McBride DW, Brown PR, Lamb JD, Christensen JJ, The influence of halocarbon solvent on macrocycle-mediated cation transport through liquid membranes. J. Membr. Sci. 1986; 28: 69-76.
93. Izatt RM, Bruening RL, Bruening ML, Lamb JD, Macrocycle-mediated cation transport through liquid membranes. Isr. J. Chem. 1990; 30: 239-245.
94. Szpakowska M, Nagy OB, Application of the competitive preferential solvation theory to facilitated ion transport through binary liquid membranes. J. Phys. Chem. 1989; 93: 3851-3854.

95. Dernini S, Palmas S, Polcara AM, Maronglu B, Extraction and transport of sodium ion and potassium ion in a liquid membrane containing crown ethers: Effect of the mixed solvent. J. Chem. Eng. Data 1992; 37, 281-284.
96. Izatt RM, Bradshaw JS, Lamb JD, Bruening RL, Emulsion and Supported Liquid Membranes. In: Araki T, Tsukube H, Eds., *Liquid Membranes: Chemical Applications*. CRC Press, Boca Raton, FL, 1990: 123-140.
97. Christensen JJ, Lamb LD, Izatt SR, Starr SE, Weed GC, Astin MS, Stitt BD, Izatt RM, Effect of anion type on rate of facilitated transport of cations across liquid membranes via neutral macrocyclic carriers. J. Am. Chem. Soc. 1978; 100, 3219-3220.
98. Olsher U, Hankins MG, Kim YD, Bartsch RA, Anion effect on selectivity in crown ether extraction of alkali metal cations. J. Am. Chem. Soc. 1993; 115: 3370-3371.
99. Christensen JJ, Christensen SP, Biehl MP, Lowe SA, Lamb LD, Izatt RM, Effect of receiving phase anion on macrocycle-mediated cation transport rates and selectivities in water-toluene-water emulsion membranes. Sep. Sci. Technol. 1983; 18: 363-373.
100. Deblay P, Delepine S, Minier M, Renon H, Selection of organic phases for optimal stability and efficiency of flat-sheet supported liquid membranes. Sep. Sci. Technol. 1991; 26: 97-116.
101. Brown PR, Hallman JL, Whaley LW, Desai DH, Pugia MJ, Bartsch RA, Competitive, proton-coupled, alkali metal cation transport across polymer-supported liquid membranes containing sym-(decyl)-dibenzo-16-crown-5-oxyacetic acid: Variation of the alkyl 2-nitrophenyl ether membrane solvent. J. Membr. Sci. 1991; 56, 195-206.
102. Michaels AS, Membranes, membrane processes, and their applications: Needs, unsolved problems, and challenges of the 1990s. Desalination, 1990; 77, 5-34.
103. Lloyd DR, Meluch TB, Selection and evaluation of membrane materials for liquid separations. In: Lloyd DR, Ed., *Material Science of Synthetic Membranes*, ACS Symp. Ser. No 269, Washington DC, ACS, 1985.
104. Pusch W, Walch A, Synthetic membranes - Preparation, structure and application. Angew. Chem. Int. Ed. Engl. 1982; 21, 660-685.
105. Cabasso I, Hollow-fiber membranes. *Kirk-Othmer Encyclopedia of Chemical Technology*, 3rd ed., 1982; 12: 492-517.
106. Kesting RE, *Synthetic Polymeric Membranes*. New York, McGraw-Hill, 1971.
107. Goel V, Accomazzo MA, *et al.*, Deadend microfiltration: Application, design and cost. In: Ho WSW, Sirkar KK, Eds., *Membrane Handbook*. Chapman & Hall, New York, NY, 1992: 506-570.
108. Kulkarni SS, Funk EW, Li NN, Membranes. In: Ho WSW, Sirkar KK, Eds., *Membrane Handbook*. Chapman & Hall, New York, NY, 1992: 408-431.
109. Cheryan M, *Ultrafiltration Handbook*, Lancaster, PA, Technomic Publishing Co., 1986.
110. Hsieh HP, Inorganic membranes. AIChE Symp. Ser. 1988; 84, 1-18.
111. van Straaten-Nijenhuis WE, de Jong F, Reinhoudt DN, Effects of Crown Ethers and Cosolvent on the Activity and Enanthioselectivity of α-chymotrypsine in Organic Solvents. Reel. Trav. Chim. Pays-Bas 1993; 112, 317-326.
112. Donnan FG, Theory of membrane equilibria and membrane potencials in the presence of non-dialising electrolytes. A contribution to physical-chemical physiology. J. Mem. Sci. 1995; 100: 805-814.
113. Okeson K, Is nitric oxide the key to removing biofilms? Int. Desalination Water Reuse, 2008; 17: 30-32.

CHAPTER 3

Supported Liquid Membranes and Their Modifications: Definition, Classification, Theory, Stability, Application and Perspectives

Paweł Dżygiel *and* Piotr P. Wieczorek

Nomenclature

α_f	fraction of the transported substance that is extractable from the feed phase
α_s	fraction of the transported substance that is extractable from the strip phase
$\Delta C_{CS,m}$	concentration gradient of carrier-substance complex in the membrane
ε	porosity
γ_i	activity coefficient of the *i*th species
η	viscosity of the organic phase
μ_i	chemical potential of the *i*th species
θ	contact angle between the membrane pores and the membrane liquid
τ	tortuosity
Ab	antibody
Ab-Ag	antibody-antigen complex
Ag	antigen
C	concentration
C_f	total concentration in the feed phase
C_i	concentration of the *i*th species
C_s	total concentration in the strip phase
$C_{S,o}$	initial concentration of transported substance
D	diffusion coefficient
d	membrane thickness

Faculty of Chemistry, Opole University, Oleska 48, 45-052 Opole, Poland

Liquid Membranes: Principles and Applications in Chemical Separations and Wastewater Treatment
DOI: 10.1016/B978-0-444-53218-3.00003-9

D_{CS}	diffusion coefficient of carrier-substance complex
D_m	diffusion coefficient in membrane phase
D_o	diffusion coefficient in bulk solution
D_{SLM}	supported liquid membrane diffusion coefficient
E	extraction efficiency
E_e	concentration-enrichment factor
$E_{e(max)}$	maximum concentration-enrichment factor
J	flux
k	Boltzmann constant
K_a	dissociation constant
K_{aff}	affinity constant
K_{ext}	extraction constant
K_f	partition coefficient between organic phase and feed phase
k_m	mass-transfer coefficient
K_m	partition coefficient
K_s	partition coefficient between strip and membrane phase
L_m	ligand located in the membrane phase
n_f	total amount of extracted compound in the feed phase
n_s	total amount of extracted compound in the strip phase
n_w	total amount of extracted compound in the waste phase
P	permeability
P_c	transmembrane pressure
R	recovery
r	pore radius
r_s	molecular radius of the solute
SL_m	ligand-substance complex in membrane phase
S_m	substance located in the membrane phase
T	temperature
V_f	volume of the feed phase
V_s	volume of the strip phase

Abbreviations

AMPA	(aminomethyl)phosphonic acid
Armak	Armak, Chicago, MI, USA
Ashai Kasei	Ashai Kasei Corp., Tokyo, Japan
β-CD	β-cyclodextrin
BEHA	bis(2-ethylhexyl)-amine
BF_4	tetrafluoroborate
BMIM	1-butyl-3-methylimidazolium
Celanese	Celanese Plastic, Dallas, TX, USA
CTA	cellulose triacetate
DAE	diaminoethane
DBSA	dodecylbenzylsulfonic acid

DC18C6	dicyclohexano-18-crown-6
DEHPA	di-2-ethylhexyl phosphoric acid
DETA	diethylenetriamine
DEYA	diethylamine
DNB	dinitrobenzoyl
DNNS	dinonylnaphthalenesulfonic acid
DOP	dioctylphthalate
DOS	dioctylsebacate
DOTP	bis(2-ethylhexyl)terephthalate
DTPA	diethylenetriaminepentaacetic acid
Enka	Enka Produktgruppe Membrana, Wuppertal, Germany
Flow Lab.	Flow Laboratories, Rickmansworth, UK
FL-SLM	flat-sheet SLM
Gore	W. L. Gore & Associates, Inc., Newark, DE, USA
HF	hollow fiber
HF-SLM	hollow-fiber SLM
HLB	hydrophilic-lipophilic balance
Millipore	Millipore, Billerica, MA, USA
Mitzubishi Rayon	Mitzubishi Rayon Company, Otake, Japan
NPOE	*o*-nitrophenyloctylether
NTf_2	bis(trifluoromethanesulfonyl)imide
Nucelopore Corp.	Nucelopore Corp., Pleasanton, CA, USA
O/W	oil/water
OMIM	1-octyl-3-methylimidazolium
PE	polyethylene
PEHFSD	pseudoemulsion-based hollow-fiber strip dispersion
PF_6	hexafluorophosphate
PIM	polymer inclusion membranes
Polyplastics	Polyplastics Taiwan Co., Ltd, Taipei, Taiwan
PP	polypropylene
PPG	polypropylene glycol
PTFE	poly(tetrafluoroethylene)
PTSA	*p*-toluenesulfonic acid
PVC	poly(vinyl chloride)
PVDF	polyvinylidene fluoride
SDHLM	strip dispersion hybrid liquid membrane
SLM	supported liquid membrane
SPE	solid-phase extraction
Sumimoto	Sumimoto Chemical, Tokyo, Japan
TBEP	tri(butoxyethyl)phosphate
t-BuDC18C6	*t*-butyldicyclohexano-18-crown-6
TEHP	tris(2-ethylhexyl)phosphate
TOMA-Cl	trimethylammonium chloride
TOPO	trioctylphosphine oxide

1. Introduction

A supported liquid membrane (SLM) is one of the three-phase liquid membrane systems in which the membrane phase (liquid) is held by capillary forces in the pores of microporous polymeric or inorganic film. The immobilized liquid is a membrane phase and a microporous film serves as a support for the membrane. Usually SLMs are based on hydrophobic organic solvent immobilized in a polymeric membrane separating two aqueous solutions. In some cases, the arrangements are opposite and the pores in the support separating two nonaqueous phases are impregnated by water. The problem with this arrangement is that water has relatively high volatility and such membranes are not stable, but this problem can be solved using ionic liquids (ILs) as a membrane phase. An SLM can also be formed by immobilizing the membrane phase between two nonporous films which are permeable to transported substances and usually nonselective. The latter is much more stable, but is less suitable due to a higher mass-transfer resistance of the nonporous layer, because the diffusion coefficient (D) value in liquids is at least three or four orders of magnitude higher than in solid polymer membranes. Other types of liquid membranes, polymer inclusion membranes (PIMs) and gelled liquid membranes, have been investigated to improve SLM stability. These types of membranes are formed either by casting the solution of polymer, usually cellulose triacetate (CTA) or poly (vinyl chloride) (PVC), and plasticizer (solvent characterized by high viscosity) to form a thin, flexible and stable film (self-supporting PIM) or by liquid-phase gelation in the PVC pores of an SLM.

SLM was reported for the first time by Scholander [1] who used thin cellulose acetate filters impregnated with an aqueous hemoglobin solution for oxygen transport. A similar system was reported by Wittenberg [2] for studying the molecular mechanism of oxygen transport. In the 1960s and 1970s, the liquid membrane concept was mostly used in emulsion liquid membranes, when Li [3] patented their application for hydrocarbons separation. However in the beginning of the 1980s, there was an increase of research interest in the supported liquid membrane as SLMs are easier to implement into a continuous flow system. Since then SLMs have been used to solve an increasing number of separation problems, including metals and organic compounds separation and the resolution of stereoisomers. The unique flexibility and ease of preparation of SLMs in various configurations, despite some stability and lifetime problems, has resulted in their application in many, sometimes very different fields where selective and efficient separation methods are necessary: in hydrometallurgy, biotechnology, wastewater treatment, the capture of greenhouse gases, analytical and environmental chemistry, and in the pharmaceutical industry.

In this chapter, the principle, kinetic and transport mechanisms, stability, SLM design and configuration are presented. Moreover, selected applications of SLMs and future perspectives are discussed.

2. Supported Liquid Membrane Separation Technique—the Principle

Liquid membrane separation combines the solvent extraction and stripping processes (re-extraction) in a single step. The great potential for energy saving, low capital and operating cost, and the possibility to use expensive extractants, due to the small amounts of the membrane phase, make SLMs an area deserving special attention.

The principles and applications of SLM separation processes have been reviewed several times [4-7]. Briefly, in an SLM system an organic solvent is immobilized in the pores of a porous polymer or inorganic support material by capillary forces, separating two aqueous solutions: the feed (donor) and the strip (receiving, acceptor) phase (Fig. 3.1). The compounds are separated from the aqueous sample feed phase into an organic solvent immobilized in a support diffusing through the membrane phase, and then they are continuously back extracted to the other side of the membrane into the

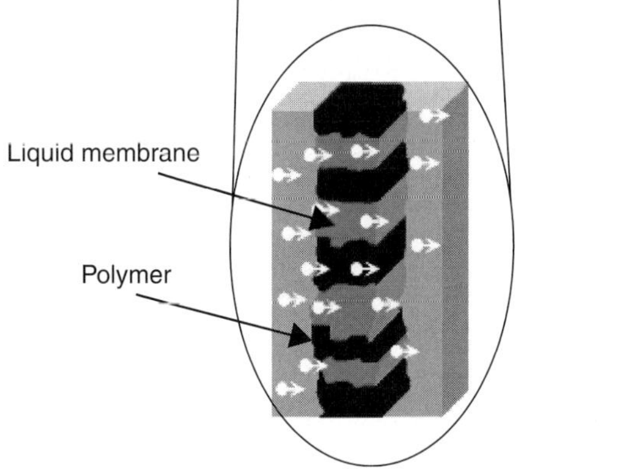

Figure 3.1 Scheme of a supported liquid membrane.

strip phase. The driving force is the difference in concentration of the compounds between the phases.

An SLM usually consists of an organic solvent immobilized in the pores of a hydrophobic microfiltration membrane. In many cases, the organic solvent contains a carrier which is able to selectively bind one of the components from the separated mixture (feed solution) which improves selectivity. The general term for mass transport through liquid membranes under the chemical potential gradient as a driving force is permeation. The permeants from liquid feed (donor solution) are transported through a nonporous, polymeric or liquid membrane phase and desorbed into another liquid phase. Schlosser [8] called this process pertraction, derived from the Latin "per-trahere," by analogy to the term extraction, which has been derived from the Latin "ex-trahere." The pertraction process can be seen as a combination of extraction and solvent stripping carried out simultaneously. While solvent extraction is an equilibrium process, pertraction is a dynamic, nonequilibrium diffusion process governed by the kinetics of the membrane transport. Therefore, the amount of transported compounds is not proportional to the amount of the organic, membrane phase as it is in extraction.

SLM separation systems can be classified into several different groups, according to their preparation methods, types of membrane support and membrane liquids, module types used and their application (for details, see Chapter 1).

Successful applications of SLMs are possible due to their advantages compared to other separation methods. The main advantages of SLMs are the small amounts of organic phase and extractant (carrier) used, one-step mass transfer, the possibility of achieving high separation factors, concentration of extracted compound(s) during separation, and low separation costs.

Nevertheless, there are some problems limiting the practical application of SLMs. The main problem is the stability of the liquid membrane, caused by leakage and/or losses of membrane phase components during transport process. However, by proper choice of the porous polymeric support, using organic solvents used as a membrane phase and membrane phase components, the instability can be significantly reduced.

3. Transport Mechanisms and Kinetics

The principal application of liquid membranes is to separate mixtures into their components and/or concentration (enrichment) of one or more of them. Such three-phase systems, when two miscible fluids are separated by a liquid which is immiscible with them, enable a mass transfer between these fluids. The efficiency of membrane transport for a particular compound depends on its partition coefficient between the different parts of a membrane system. Only compounds, which are easily extracted from the feed (donor) phase into the membrane phase and easily re-extracted from the membrane to the strip (acceptor, receiving) phase, are transported. Therefore, the separation of different compounds is based on the same principle as a liquid extraction followed by a back extraction. Only molecules with different physicochemical properties can be separated, even if they are of equal size. Separation and concentration are an entropy decreasing, that is, free energy increasing, process and do not happen spontaneously. To run a continuous separation or concentration a source of free energy is needed. For this reason, it is necessary to establish the difference in the chemical potentials of the mixture components on opposite sides of the membrane. In the case of mixtures two important properties have to be taken into account, namely solubility and diffusion. Note that the solubility coefficient of a single component becomes more complex due to the presence of other substances in the mixture. Additionally, the diffusion coefficient of any substance in a mixture may be a function of the concentration of all substances present in the membrane. In such a case, more than one substance will be transported through the membrane and fluxes of different substances may interact and affect the separation process. Therefore, many mathematical models, like solution-diffusion or simple network models, as well as the numerical model [9], have been used to express the mass transfer through liquid membrane systems.

3.1. Driving force and transport mechanisms

As mentioned above, membrane transport is a dynamic, nonequilibrium process. The transported compound has to dissolve in the organic, hydrophobic membrane phase and diffuse through it to enter the aqueous stripping phase. The mass transfer in this system takes place due to the difference in the chemical potential across the membrane as a driving force. The variation of chemical potential of component i can be expressed as

$$d\mu_i = RT \, d \ln c_i + RT \, d \ln \gamma_i, \tag{1}$$

where c_i is the concentration and γ_i is the activity coefficient.

The concentration profile in the SLM system is schematically shown in Fig. 3.2.

The transport of the substances from the feed solution to the strip side can be divided into the following steps: diffusion of substance S across the boundary aqueous layer in the feed (donor) phase, extraction (sorption) of substance on the donor/membrane phase interface, diffusion across the boundary layer on the feed (donor) side, convection transport in the liquid membrane zone, diffusion across the boundary layer on the strip (acceptor) phase of LM, re-extraction (desorption) on the membrane/strip phase

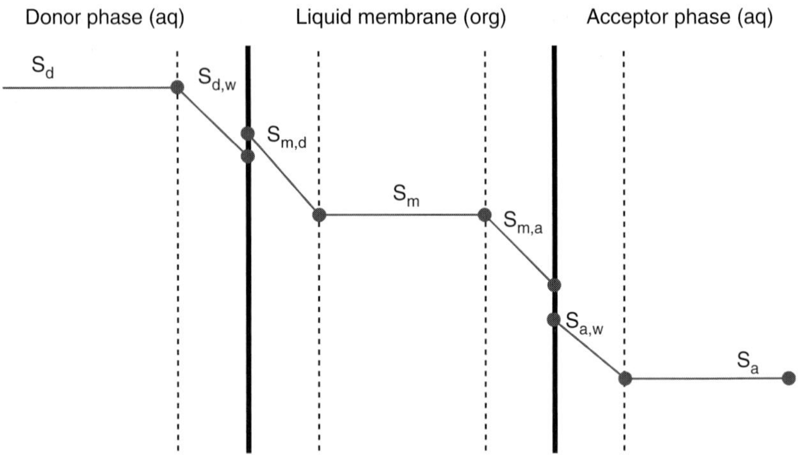

Figure 3.2 Concentration profile in SLM.

interface, and diffusion across the boundary layer in the strip (acceptor) phase.

The transport process itself is steady state, since the membrane surface area is small compared with the quantity of material to be transported and in the case where the activity coefficient is close to one the overall membrane flux (J) can be derived by applying Fick's first law:

$$J = k_{\mathrm{m}} \Delta C / d. \tag{2}$$

Here, k_{m} is the mass-transfer coefficient and is proportional to $K_{\mathrm{m}}D_{\mathrm{m}}$, where K_{m} is the partition coefficient and D_{m} is the diffusion coefficient. ΔC is the concentration difference between the strip and feed phases and d is the membrane thickness.

The membrane phase diffusion coefficient D_{m} for bulk liquid membrane is equal to the diffusion coefficient in bulk solution D_0. The diffusion coefficient is an important factor in the transport mechanism and is related to the viscosity by the empirical Stokes-Einstein equation [10]:

$$D_{\mathrm{m}} = D_{\mathrm{o}} = kT/(6\pi\eta r) \tag{3}$$

or

$$P = kTK_{\mathrm{f}}/(6\pi\eta r_{\mathrm{s}} d), \tag{4}$$

where D is the solute diffusion coefficient (cm^2/s), k is the Boltzmann constant, T is the absolute temperature in Kelvin, η is the viscosity of the organic phase (solvent), and r_{s} is the molecular radius of the solute (cm). P is the permeability of the solute (cm/s), d is the thickness of the membrane (cm), and K_{f} is the partition coefficient of the solute between the organic (membrane) phase and the water (feed) phase. The diffusion of solvents immobilized in membrane pores is much slower than in bulk solution [11].

For an SLM, the flux can be given by the equation:

$$J = D_{SLM} K_f \Delta C / d. \quad (5)$$

The D_{SLM} is the SLM diffusion coefficient and K_f is the distribution (partition) coefficient.

In an SLM, the effective diffusion coefficient has to be corrected according to the morphological characteristic of the porous polymeric membrane in which the liquid is immobilized. Therefore, the apparent diffusion coefficient D_{SLM} is related to the coefficient in bulk membrane phase D_m, through [12]

$$D_{SLM} = D_m \varepsilon / \tau, \quad (6)$$

where ε is the membrane porosity and τ is the tortuosity, related to the tortuosity factor β, defined as the average pore length/membrane thickness.

The separated substances can be transported across the membrane according to two main classes of transport mechanisms: simple permeation (passive diffusion) and carrier-mediated transport. Their occurrence significantly depends on the type and properties of transported compounds.

3.1.1. Simple permeation

In this transport mechanism, the transported substance dissolves in the organic membrane phase and diffuses to the receiving phase due to the concentration gradient between these two phases.

In simple pertraction (Fig. 3.3A), the permeant passes through the membrane due to its solubility in the organic phase. The transported compound is in the same form in both feed and strip phases, and the transport (permeation) stops when equilibrium concentration is reached, and it is impossible to concentrate the transported compound. As an example of this type of process, the separation of aliphatic from aromatic hydrocarbons can be given [8]. The transported molecules have to be uncharged in the feed phase for extraction to the hydrophobic membrane phase.

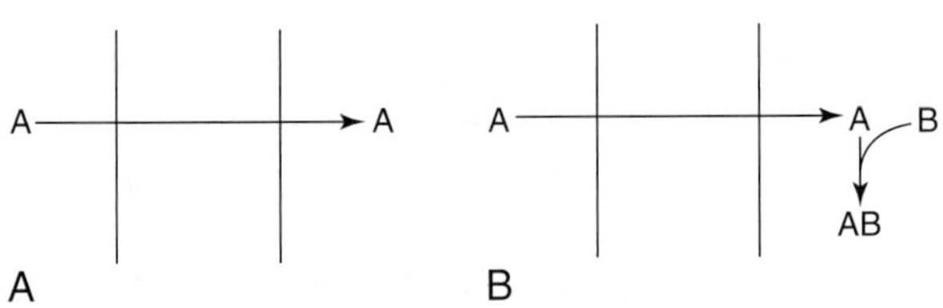

Figure 3.3 Transport mechanism: (A) simple pertraction and (B) pertraction with reaction in stripping phase.

The selectivity of separation and/or the pertraction capacity can be increased if the permeant undergoes a chemical reaction in the stripping phase. The separation effectiveness is significantly improved if the component to be removed is transformed almost quantitatively into an impermeable form in the receiving phase. Therefore, the product is insoluble in the membrane phase, which prevents it from diffusing back through the membrane (Fig. 3.3B). In this way, the concentration gradient can be maintained across the membrane, hence facilitating solute enrichment. Also, selectivity can be further enhanced by proper choice of the reaction window in the stripping solution, so that only the compounds of interest are ionized and trapped.

3.1.2. Carrier-mediated (facilitated) transport

Pertraction with SLMs utilizes as its principle of separation the high diffusion coefficient of solutes in liquids and the very high and different solubility of pertracted compounds in some liquids. When this solubility is too low, effective and selective transport is difficult to achieve, even if the trapping in the strip phase can be easily done. When there are various nontransportable, permanently charged compounds (with low solubility in the liquid membrane phase) such as metal ions, amino acids, or peptides, a chelating agent or carrier can be added to the membrane phase. The permeability and selectivity of SLM transport can then be increased by several orders of magnitude. The carrier should reversibly react with the pertracting compound and form a complex which can be transported through the organic membrane phase. In this situation, it is important that the carrier and its complex formed with the transported compound are soluble in the membrane but not in the feed and strip aqueous phases. Carrier-mediated transport can be limited by complexation reaction kinetics or by diffusion of the complex through the membrane phase. In many reports, the facilitated transport mechanism is analyzed using a complicated expression with membrane diffusion and a chemical reaction [13-16]. Facilitated transport through an SLM is usually described based on the idea that the reaction between compounds with the carrier takes place only at the membrane surface, a feed solution-liquid membrane interface. Such a simple mechanism in which the carrier stays in the membrane and two ion-exchange reactions take place on the water/membrane interfaces is called "Small Carrousel" [17]. Nevertheless, if the carrier is not very hydrophobic, it can leave the membrane interior and the chemical reaction takes place mainly in the aqueous phase. The mechanism in which the carrier moves from one aqueous phase through the membrane into another before returning and completing the cycle is called "Big Carrousel" [18]. When the ligand (L_m) is located in the membrane phase and is able to form a complex (SL_m) with the substrate (S_m), such reaction takes place in the membrane phase, characterized by the dissociation constant (K_a):

$$S_m + L_m \leftrightarrow SL_m, \tag{7}$$

$$K_a = [SL_m]/[S_m][L_m]. \tag{8}$$

If the association and dissociation processes are fast the transport is limited by diffusion and the flux decreases linearly with increasing the membrane thickness. Therefore, the flux can also be expressed using Fick's first law:

$$J = D_{CS}\Delta C_{CS,m}/d \tag{9}$$

and

$$\Delta C_{CS,m} = C_{S,0}K_{ext}, \tag{10}$$

where $\Delta C_{CS,m}$ is the concentration gradient of the carrier-substance complex, $C_{S,0}$ is the initial concentration of transported substance, and K_{ext} is the extraction constant and can be written as

$$K_{ext} = K_f K_a. \tag{11}$$

The possibility of using various compounds as a carrier resulted in different versions of the carrier-mediated transport mechanism such as simple carrier permeation in the strip phase, coupled cotransport, and coupled countertransport (Fig. 3.4).

In simple carrier permeation transport, the carrier reversibly reacts with extracted compound in the feed phase or at the feed-membrane interface, and forms an extractable complex. This complex is transported through the membrane phase in a similar way as in simple permeation without using a

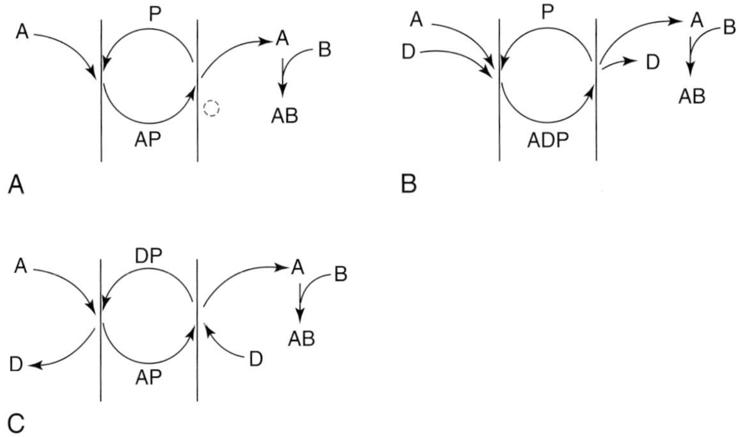

Figure 3.4 Carrier-mediated transport mechanism: (A) simple, (B) cotransport, and (C) countertransport.

carrier. At the membrane-strip phase interface, the compound can be trapped by another ligand used in the strip phase which forms a stronger and, in many cases, charged complex (Fig. 3.4A). Therefore, a permanent concentration difference is present through the membrane during the extraction time and a high enrichment factor can be achieved. For example, anionic surfactants can be transported by an ammonium compound [19] and the metal ions or organic acids by a variety of complexing agents [20].

In coupled transport, the complexation reaction between the carrier and extracted compounds takes place only at the water-membrane interface. The complex is formed at the feed-membrane interface and is transported through the membrane to the membrane-strip phase interface. An additional compound is transported together with this complex, in the same or opposite direction. At the membrane-strip phase interface, the decomposition of the complex takes place. In this type of transport, two substances are transported and at least two mass fluxes are realized. If both fluxes are in the same direction, coupled cotransport can be distinguished. If mass fluxes are in the opposite direction, a coupled countertransport occurs. Thus, the flux magnitude depends on the type and concentration of the carrier and transported substances.

When the carrier used in the membrane phase is in an uncharged form, the transported ionic substance can be extracted only as an ion pair, which is formed by adding to the feed phase a counterion or chelating agent. The carrier reacts with such a neutral ion pair giving a complex which is transported through the membrane (Fig. 3.4B). The concentration gradient of cotransported compound is the driving force in this type of transport. The uranyl sulfate anions transport by a tertiary amine carrier is an example of such coupled cotransport [21]. Transport of amino acids [22, 23] and peptide hydrochloride [24] by macrocyclic carriers is another example of cotransport involving ion-pair interaction with a neutral carrier. Thus, ionic substances and their counterions are transported in the same directions and their fluxes are stoichiometrically coupled.

Many permanently charged compounds, especially amino acids and their derivatives, can also be efficiently transported through liquid membranes by ionic carriers. In these cases, a gradient of counterions from the strip (receiving) phase to the feed (source) phase provides the driving force for the transport. Therefore, the amino acid carrier complex is transported through the membrane from the feed to the strip phase, and the counterions are transported in the opposite direction, that is, coupled countertransport (Fig. 3.4C). For transport of amino acids using Aliquat 336 as a carrier, a gradient of chloride ions from the strip to the feed phase provides a driving force for the mass transport [25]. Using DEHPA as a carrier for amino acid permeation, a pH gradient is created over the membrane [26]. The strip phase is kept strongly acidic ($pH \leq 1$) while $pH \approx 3$ for the feed phase.

3.2. Product recovery and enrichment

The quantitative and selective recovery of any substance is very important in a separation process. Parameters which are usually used for the description of membrane processes are pertraction efficiency and/or recovery. Additionally, especially in analytical applications, the enrichment factor is crucial.

As presented in the previous equations, the rate of mass transfer, and thus the pertraction efficiency, is proportional to the concentration difference ΔC over the membrane and can be expressed:

$$\Delta C = \alpha_f C_f - \alpha_s C_s K_s / K_f, \tag{12}$$

where α_f and α_s are the fractions of the transported substance, which is pertractable from the feed to the strip phases, respectively. C_f is the total concentration in the feed phase, while C_s is the total concentration in the strip phase. K_s is the partition coefficient for the substances between the strip and membrane phase, and K_f is the partition coefficient for the substances between the feed and membrane phase. While the feed and strip phases are mostly aqueous, both partition coefficients are similar. However, in some cases, the different ionic strengths of feed and strip phases can cause significant differences between K_s and K_f. Then, the concentration difference can be expressed as

$$\Delta C = \alpha_f C_f - \alpha_s C_s. \tag{13}$$

Usually, the membrane separation conditions are set to get α_f close to unity and α_s a very small value. The value C_s is zero from the beginning of the pertraction and increases successively to equilibrium, when concentration in the strip phase is the same as in the feed phase. The maximum concentration-enrichment factor (E_e) is obtained when ΔC has eventually reached zero:

$$E_{e(\max)} = (C_s/C_f)_{\max} = \alpha_f/\alpha_s. \tag{14}$$

The rate at which this condition is approached depends on many parameters, like diffusion coefficients in feed and membrane phases, partition coefficients, or membrane thickness. We can have two situations: membrane-controlled pertraction, when the diffusion of the transported compound through the liquid membrane is the limiting step, and feed-controlled pertraction, when the diffusion through the feed phase to the feed-membrane interface is the limiting step [27].

The pertraction efficiency, E, is defined as the fraction of substances transported from the feed (donor) phase to the strip (acceptor) phase. It is the measure of the rate of mass transfer through the membrane and is constant at specified pertraction time, phase composition, and ionic strength. The pertraction is expressed as

$$E = n_s/n_f = (C_s V_s)/(C_f V_f) \tag{15}$$

or

$$E_w = n_f - n_w / n_f, \tag{16}$$

where E_w is the pertraction efficiency; n_f, n_s, and n_w are the total amounts of pertracted compound in the feed, strip, and waste (outflux of the continuously pumped feed phase after contact with the membrane phase) phases, respectively. V_s and V_f are the volumes of the strip and feed phases. The value of E indicates how much of the compound is found in the strip phase, and E_w indicates how much of it is removed from the feed into the organic membrane phase. Thus, we can define the recovery R, as

$$R = E/E_w. \tag{17}$$

If E is equal to E_w, no substance is lost during the process and recovery is 100%, but if E is smaller than E_w some part of the pertracted compound is either adsorbed in the apparatus or left in the membrane phase. The latter effect can limit the SLM pertraction in some cases. In practice, the problem can be overcome by careful design of the experimental conditions.

4. Selectivity

The selectivity of the membrane process is the ability to transfer the compounds of interest but not the interfering compounds. The selectivity depends mostly on the membrane transport mechanisms and trapping method used. In simple permeation, the selectivity is not high and is governed by solubility differences between the sample components in the membrane phase. The pertraction efficiency and selectivity can be increased by adding specific carrier to the membrane phase.

The selectivity can be greatly enhanced by using specific antibody-antigen (Ab-Ag) interactions in the transport mechanism. Two main groups, mono- and polyclonal antibodies, can be used for selective liquid membrane pertraction of different compounds.

Another important problem is the enantiopurity of chiral pharmaceuticals and many other compounds with biological activity. Therefore, there is a great interest in developing methods that can help in stereoisomer separation and improve the stereoselectivity of separation methods. In this case, a specific chiral environment should be created to ensure enantioselectivity of separation.

4.1. Transport selectivity

The transport selectivity in the membrane pertraction process depends on the type of transport mechanism.

4.1.1. Selectivity of the simple permeation process

In simple permeation transport, the selectivity is not high and is governed by solubility differences between the sample components in the membrane phase. While the diffusion coefficient depends on the molecular radius of the solute, it could also affect the selectivity. The selectivity can be increased when the compounds are in different forms, active in the feed phase and inactive—by changing the pH or a specific reaction—in the strip phase. In this way, selective transport of acids and bases, for example, amines, can be achieved. The simplest way to improve selectivity and mass transfer is to adjust the pH of the aqueous phase, for example, the basic feed phase and acidic strip phase for selective amines extraction. If the pH of the feed (donor) phase is adjusted to a sufficiently high value, the transported amines are uncharged and are transported over the organic liquid that is used as a membrane phase. The strip (acceptor) phase on the other side of the membrane is an acidic solution or buffer with low pH. An amine molecule that diffuses through the membrane is immediately protonated at the membrane-strip interface, and is thus prevented from re-entering the membrane. The principle of SLM pertraction of basic compounds, for example, amines, is shown in Fig. 3.5.

It is clear that acidic compounds will be already charged (dissociated) in the alkaline feed solution and consequently not transported. Neutral components may be transported, but will distribute freely between all three phases. Macromolecules such as proteins will typically be charged, and therefore not transported. The pertraction rate of uncharged macromolecules will be very low, owing to their high radius and thus low diffusion coefficient. In summary, in the mentioned conditions, the SLM pertraction will be highly selective for small, basic compounds. The selectivity can be further tuned by careful selection of pH of the feed and strip phases. For example, it has been possible to discriminate aromatic amines, whose pK_a are around 5, and aliphatic amines with pH equal to or higher than 10.

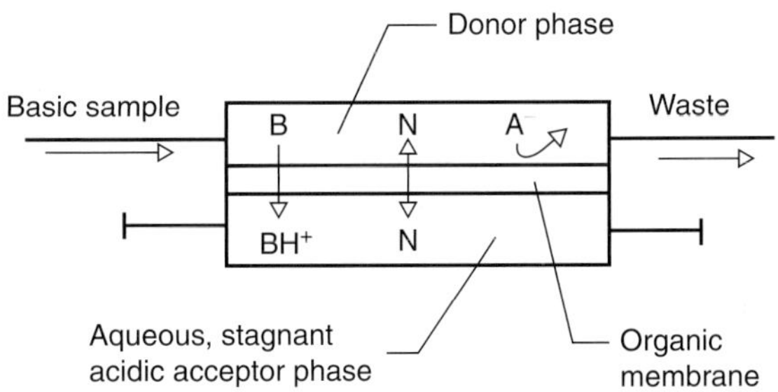

Figure 3.5 Schematic description of the SLM principle. From Ref. [27] with permission. © 2008 Elsevier.

Acidic compounds may be pertracted in a similar way to amines, but by reversing the pH conditions, acidic feed phase and basic strip phase.

Usually, nonpolar organic liquids are used as a membrane phase; therefore, polar compounds are less efficiently pertracted than hydrophobic ones. It is possible to increase the selectivity by making the liquids more polar or by adding ion-pairing or chelating reagents to the feed phase. One example of reagents added to the feed phase for metal ions pertraction is 8-hydroxyquinoline, which forms transportable complexes with many metals [20]. This complex is transferred through the membrane and the metal ions are trapped in the strip phase by another ligand, DTPA (diethylenetriaminepentaacetic acid), forming a stronger and nonpertractable charged complex.

4.1.2. Selectivity of carrier-mediated transport

The pertraction efficiency and selectivity can also be increased by adding specific carrier to the membrane phase. Various carrier molecules or ions can be incorporated in the membrane phase to enhance the selectivity and mass transfer. Most carriers used for this were originally developed for solvent extraction [28]. However, many new carriers were designed only for liquid membrane pertraction. Characteristic features of a good carrier for SLM pertraction are:

- Rapid kinetics of formation and decomposition of the complex on membrane interfaces
- No side reactions
- No irreversible or degradation reaction
- No copertraction of solvent
- Low solubility in the aqueous feed and strip phases
- Low toxicity to biomass, especially in the case of application in biological systems
- Acceptable price, especially in industrial applications

Taking into account the characteristics of a good carrier as listed above, various different carrier molecules or ions can be incorporated in the membrane phase to enhance selectivity. Many macrocyclic multidentate ligands such as crown ethers and kryptands are used (Fig. 3.6). Crown ethers and other macrocyclic compounds have a pronounced selectivity for metals [11, 29] and for amino acids and peptides [23].

TOPO (trioctylphosphine oxide), a compound that forms hydrogen bonds, is also used for carboxylic acids pertraction (Fig. 3.7). The pertraction efficiency of carboxylic acids of different polarities is strongly influenced by the content of the carrier in the membrane [30].

As the ionic carriers, mostly amines or carboxylic and phosphoric acids (Fig. 3.7) for metals, organic acids, and amines are typically used. A common carrier that has been used is Aliquat 336, a quaternary ammonium ion,

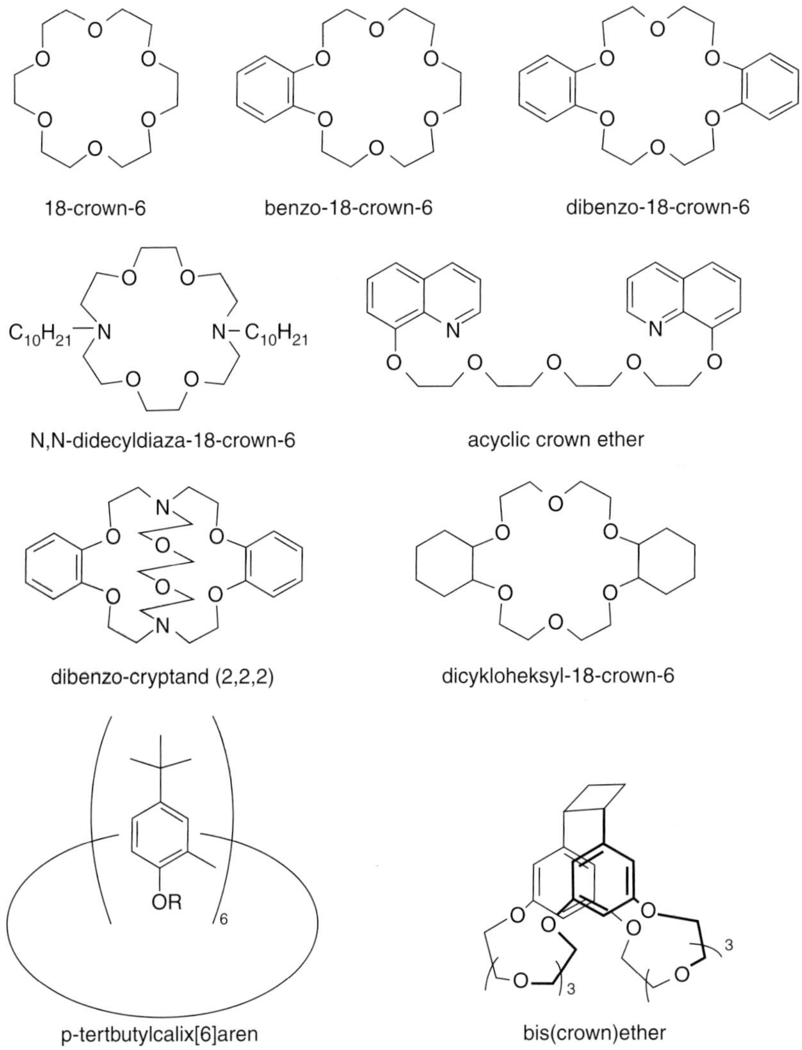

Figure 3.6 Macrocyclic carriers.

permanently positively charged in an ion pair with chloride. For metal extraction, the addition of thiocyanate ions to the donor is needed to form a negatively charged metal-thiocyanate complex, which can give an ion pair with the carrier [20]. Organic acids and amino acids are effectively transported with Aliquat 336 from basic solutions, where they are negatively charged [25]. Components that are not negatively charged in these conditions and cannot give an ion pair with cationic carrier are not extracted. The selectivity of organic and amino acids can be increased by carefully adjusting

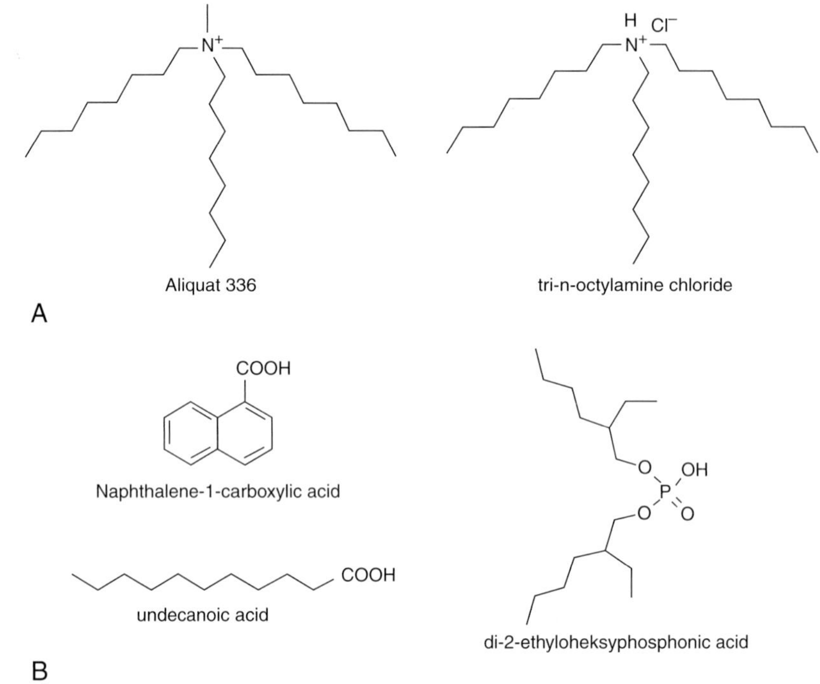

Figure 3.7 Ionic carriers: (A) cationic and (B) anionic.

the pH of the feed phase. The pK_a of transported compounds must differ by at least 2-3 orders of magnitude.

This ion-pair strategy was also used for transport of anionic surfactants with sulfuric acid anions, which cannot be protonated as carboxylic acids, and therefore not be transported by simple permeation [19]. SLM with Aliquat 336 as a carrier and chlorine ions gradient as a driving force was used for pertraction of glyphosate [*N*-(phosphonomethyl)glycine], a herbicide widely used all over the world, and its main metabolite, (aminomethyl) phosphonic acid (AMPA). The results show that it is possible to transport both glyphosate and AMPA in one step by adjusting the pH of the feed phase to higher than 10. When the pH of the feed phase is lower than 7, glyphosate could be transported selectively [31, 32].

There are a number of organic acids used as ionic carriers for metal ions and amines or amino acids pertraction. The mostly used anionic carrier for metal ions and amino acids is di-2-ethylhexyl phosphoric acid. The carrier is dissolved in the membrane phase as a dimmer, which reacts at the donor phase-membrane interface with amino acid cation, forming an ion pair and releasing a proton. For example, selective speciation of different chromium species (chromate and chromium ions) was achieved by the combination of two pertraction systems, one working with DEHPA for Cr^{3+} transport and the other with Aliquat 336 for chromate anions [33].

However, various carrier molecules or ions have been used in the membrane phase and many reagents in the strip phase for trapping, to enhance selectivity and mass transfer the pertraction, are often not selective enough. Such selective extraction could be obtained by utilizing soluble antibodies in the strip phase.

4.2. Immunological trapping

Selectivity can be greatly enhanced by using specific antibody-antigen (Ab-Ag) interactions in the pertraction mechanism. This mechanism was first used for the trace-level determination of pollutants in complex environmental matrices, using selective solid-phase extraction with immunosorbent as an extraction step. Antibodies can be immobilized on a solid support, for example, silica beads, to yield selective phases, called immunosorbents. These have been used for the extraction of many different compounds from various environmental matrices [34]. Exploring antibodies as specific reagents in liquid membrane pertractions promises a high degree of selectivity and enrichment. Utilizing membranes for matrix cleanup and antibodies for selective recognition is a powerful combination for selective pertraction. This extends the scope of SLM to include permanently neutral compounds that have not been feasible to enrich in standard SLM systems. A specific, for transported compounds, antibody (Ab) is introduced as trapping reagent into the SLM strip (acceptor) phase. If there are permanently neutral compounds in the feed (donor) phase, or the pH is adjusted so that the compounds (Ag) are uncharged, they are extractable into the organic membrane phase. After pertraction the uncharged compound diffuses through the membrane phase to the membrane-strip phase interface where it is re-extracted down the concentration gradient. The gradient is upheld by the binding of the transported compound (antigen, Ag) to the compound-specific antibody forming a nonpertractable antibody-antigen (Ab-Ag) complex in the strip phase. Thus antibody-antigen complex formation is the heart of the pertraction system.

Good trapping capacity is obtained because of the high affinity of the antigen toward its antibody. The schematic immuno-SLM pertraction system is shown in Fig. 3.8.

Immuno-SLM system was successfully applied as a sample preparation method for 4-nitrophenol and atrazine detection in water and juice samples. Polyclonal anti-4-nitrophenol antibodies were used for SLM immunopertraction from spiked reagent and wastewater samples [35]. The immuno-SLM was also used for selective pertraction of the popular herbicide atrazine as a model sample from tap and river water samples as well as from orange juice [36].

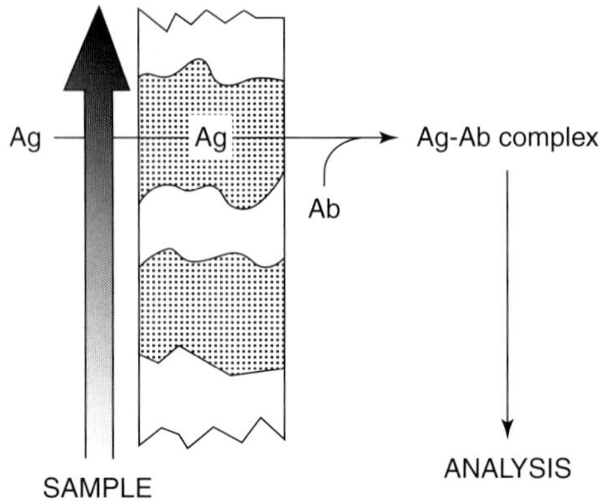

Figure 3.8 Immuno-SLM scheme.

4.3. Stereoselectivity

Almost all newly designed and biologically active substances, such as drugs or pesticides, are compounds with a strictly defined stereostructure. Therefore, there is a great interest in obtaining compounds with the required enantiopurity. This can be achieved by different approaches including stereoselective synthesis, biotransformation, or chiral separation. There is great pressure to develop methods that can help in the separation of stereoisomers. Different types of membrane processes, also involving SLM transport, have been applied for the separation of stereoisomers. Moreover, SLMs are sometimes additionally utilized as a convenient tool to examine the nature of stereoselective interaction with chiral carriers. To assure the transport stereoselectivity, and possible resolution of enantiomers, a chiral environment is required in the system. In SLMs, it can be achieved in two manners, namely by an application of a chiral organic liquid as a membrane phase and by introducing a chiral carrier into an achiral membrane [37].

The use of chiral membrane solvent to achieve transport stereoselectivity is not very widespread. Probably, the only report describing chiral liquid-phase-enhanced stereoselective transport is amino acid enantiomers separation by means of chiral alcohols, nopol and (2S)-(−)-methylbutan-1-ol [38] (see Fig. 3.9). It was shown that optical separation of six pairs of enantiomers of amino acids is possible in this way. However, the chiral discrimination (expressed as a flux ratio for both enantiomers and denoted as α) was moderate and the best result was 1.27 for serine. As a main conclusion, it was stated that the factor involved in chiral discrimination was an asymmetry of the amino acid molecule.

The most extensively examined method of stereoselective SLM separation is carrier-facilitated transport with chiral carriers. Different macrocyclic compounds, transition metal complexes, phosphates, lariat ethers, podands,

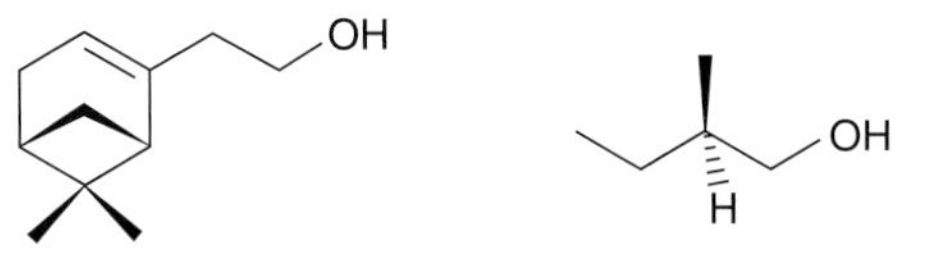

Figure 3.9 Chiral solvents.

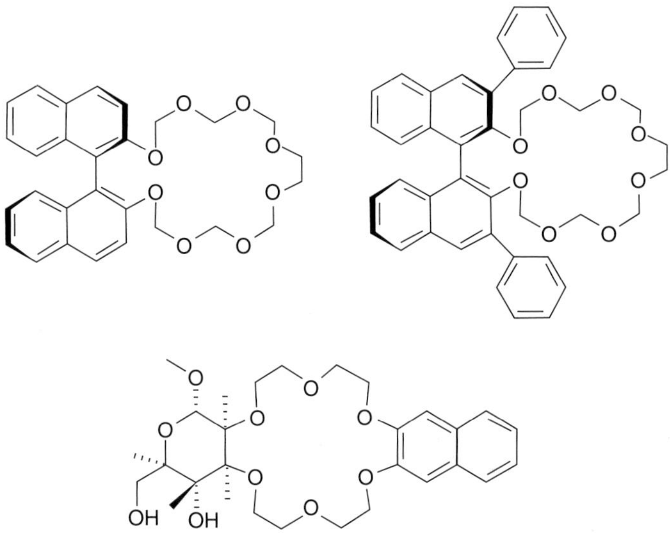

Figure 3.10 Selected macrocyclic chiral carriers.

amino acid derivatives, and macrocyclic pseudopeptides were examined as chiral carriers in liquid membranes (for selected examples, see Figs 3.10–3.12). Chiral crown ethers with naphthalene rings were applied as carriers for amino acid enantiomers transport with good selectivity, depending on the carrier and amino acid structure [39]. The other similar type carriers, optically active D-mannose derivatives of crown ethers, were examined in enantiomer separation of aromatic amino acids, in bulk and supported liquid membranes. By application of such carriers, a high separation factor of amino acid can be observed [40].

An interesting group of chiral carriers are those formed by species that utilize interactions between transported enantiomer and transition metal complexes. For instance, such a compound, acting as an additional chiral ligand for the copper central cation, is able to recognize an amino acid Cu(II) complex present in the feed phase. This double chiral carrier-amino acid-Cu (II) complex becomes diastereoisomeric and can be transported through a

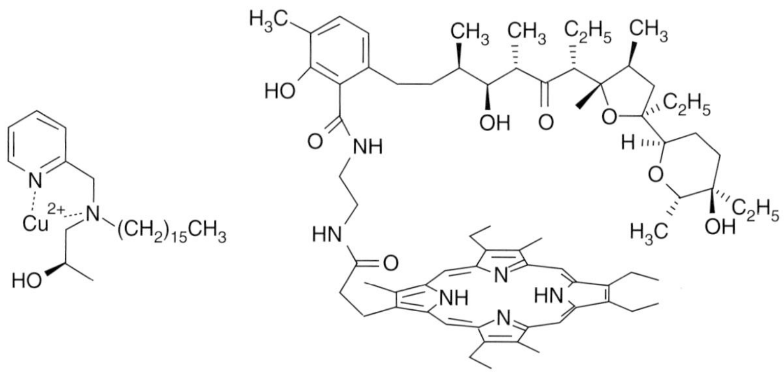

Figure 3.11 Hydrophobic Cu-chiral ligand complex and sapphyrin-lasalocid chiral conjugate carriers.

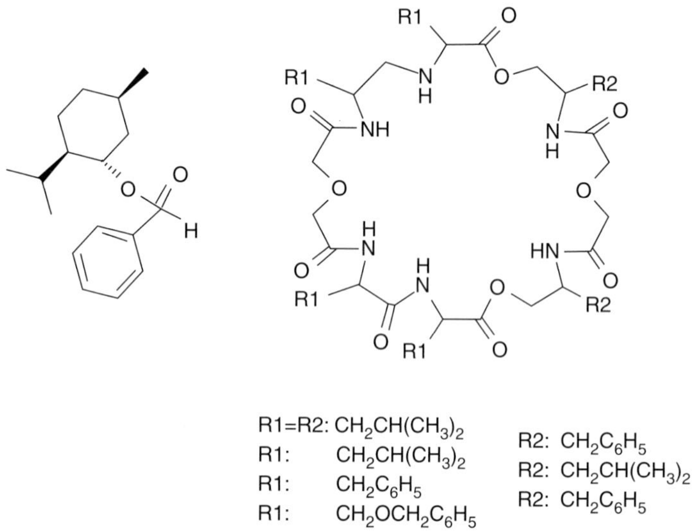

Figure 3.12 Chiral phosphoric acid derivative and macrocyclic pseudopeptide carriers.

liquid membrane in a cotransport and by this way both optical isomers are differentiated [41]. The other example of a structurally complicated chiral carrier is sapphyrin-lasalocid conjugate. It was applied in zwitterionic amino acid enantiomer separation in countercurrent transport, but only modest stereoselectivity, ranging between 1.5 and 2.0, was obtained [42].

Other classes of substances tested as potential chiral carriers in SLMs are dialkyl and monoalkyl phosphates, phosphonates, and phosphinates based on (−)-menthol and (−)-nopol [43]. The amino acids are transported through the membrane according to the countertransport mechanism and, in this case, the alcohol and aromatic carrier moieties are responsible for chiral interactions. The obtained fluxes were quite high, but the enantioselectivity was moderate. Chiral *N*-blocked amino acid derivatives and macrocyclic pseudopeptides are very interesting carriers for amino acid esters and salts separation [44, 45]. It was shown that they can act as effective transporters and chiral selectors for amino acid methyl esters. The selectivity obtained is moderate, however, and depends on the type of applied carrier and ester structure.

An interesting example taking the opposite approach was presented by Wieczorek *et al.* [24]. To separate a mixture of dipeptide diastereoisomers and their phosphonic analogues, achiral crown ethers were used as transport enhancers. In this case, the diastereomeric complex is formed between the chiral transported molecule and not the optically active carrier. The observed stereoselectivity depends on the peptide structure and was independent of the presence of carrier, but the application of carrier increased the transport rate of both diastereoisomers.

It can be concluded that stereoselective (and particularly enantioselective) separation can proceed by the simple application of a chiral organic phase or in most cases by incorporation of carriers in the membrane phase. Sometimes, their structure is very complex and these molecules can act as "real" receptors for the enantiomers. Different types of transport mechanisms are involved in the separation and the most popular one is cotransport. This is a result of the fact that most frequently used carriers are based on crown ethers' structure. The stereoselectivities by application of carrier-mediated SLM separation are very different and depend on the structure of the guest and host molecules. The magnitudes of the stereoselectivity are in most cases moderate but similar to other membrane-based separation techniques for stereoisomers.

5. Process and Membrane Units Design

5.1. Commonly used supports

In the SLM process, like in all membrane processes, the membrane plays a key role in the transport and separation efficiency. The permeation rate and separation efficiency depends strongly on the type of liquids and supports used for SLM construction. However, the transport properties depend on the type of liquids used as a membrane phase; the liquid membrane stability and mechanical stability depend, to a large extent, on the microstructure like pore shape, size, and tortuosity of the membrane used as a support. Therefore, many types of polymeric and inorganic microporous membrane supports are studied for the liquid membrane phase immobilization.

5.1.1. Polymeric support

Since the development of asymmetric synthetic polymer membranes, a number of membrane applications have been achieved. Most industrial polymers have been applied for nonporous and porous membrane formation. For immobilization of the liquid membrane phase, the microporous polymeric membranes are usually used and they are an integral part of the SLM manifold. In most cases, the choice of the polymeric support has an influence on the SLM stability, lifetime, and performance of the liquid membrane. It is thus very important to select a particular support for a given liquid membrane phase. The polymeric support should be characterized by high hydrophobicity, high porosity, small pore size, and proper tortuosity. The most important factors that influence the SLM system performance are physical stability of the support and the rate of the mass transfer through the membrane (flux of the solute). As a consequence, the support should be as thin as possible to maintain high fluxes, which depend on the diffusion pathway. However, thin supports are mechanically unstable. Often asymmetric porous membranes are used, for example, Fluoropore FG in which one microporous membrane (e.g., PTFE) is used for immobilization of organic phase and the second, thicker porous membrane is used to enhance the durability against physical stress. The character of the polymeric support also influences the liquid membrane stability; this issue will be presented below. As a polymeric support in flat sheet or hollow-fiber membrane configurations, polymers such as polypropylene (PP), polyethylene (PE), and poly(tetrafluoroethylene) (PTFE) are most frequently used. In Tables 3.1 and 3.2, characteristics of the commercially available microporous supports are presented together with the producer name.

5.1.2. Inorganic support

Although the design of new polymers for membrane formation is still in progress, the utilization of advanced inorganic membrane materials, such as ceramics, metals, porous metal oxides, and zeolites, is nowadays very important. The general advantages of inorganic membranes are mechanical and thermal stability, solvent and chemical resistance, sterilization ability, and biocompatibility. The progress in solid-state science allows preparation of new inorganic membrane materials. Sol-gel processing, plasma-enhanced chemical vapor deposition and hydrothermal synthesis are used for inorganic membrane formation. These membranes, as well as organic-inorganic composite membranes, are used in many processes, especially in nanofiltration, pervaporation, gas separation and catalytic membrane reactors [46-48]. Such membranes as well as hybrid organic-inorganic membranes have been successfully used for facilitated transport of solutes in liquid media. Hybrid membranes, prepared by incorporation of cellulose triacetate plasticized membranes with Aliquat 336 as a carrier, on inorganic silanes material prepared by sol-gel method, were used for platinum group ions transport. The results show that these membranes present higher selectivity toward

Table 3.1 Characteristic of some commercially available membranes used as a polymeric support in flat-sheet supported liquid membranes (FS-SLM)

Commercial name	Material	Producer	Thickness (μm)	Porosity (%)	Pore size (μm)
Celgard 2400	PP	Celanese	25	38	0.02
Celgard 2500	PP	Celanese	25	45	0.04
Accurel	PP	Enka	100	64	0.10
Accurel	PP	Enka	150	70	0.20 or 0.40
Accurel	PP	Enka	160	75	0.20
Accurel 1E-PP	PP	Enka	75	73	0.10-0.30
Accurel BS7C	PP	Armak	50	48	-
Duragard 2500	PP	Polyplastics	25	45	0.04
FP-DCH	PTFE	Flow Lab.	150	80	0.45
FHLP	PTFE	Millipore	60	85	0.50
FP-045	PTFE	Sumimoto	80	73	0.45
Millipore	PTFE	Millipore	125	68	10
Goretex	PTFE	Gore	60	78	0.20
Fluoropore FG	PTFE/PE	Millipore	60/115	70	0.20
Fluoropore FP-200	PTFE	Millipore	100	83	2.0
Fluoropore FP-045	PTFE	Millipore	80	75	0.45
Fluoropore FP-010	PTFE	Millipore	60	55	0.10
Nucelopore	Polycarbonate	Nucelopore Corp.	10	12	0.40

Table 3.2 Hollow-fiber polymeric membranes typically used as supports for SLM

Hollow fiber	Material	Producer	Inner diameter (mm)	Thickness (μm)	Porosity (%)	Pore size (μm)
Goretex TA001	PTFE	Gore	1.00	400	50	2
Trial manufacture	PE	Ashai Kasei	280	0.05	-	-
KPF-190M	PP	Mitzubishi Rayon	0.20	22	45	0.16
EHF-207T	PE	Mitzubishi Rayon	0.27	55	70	0.27

Pt(IV) than Pd(II) [49]. An interesting system, with asymmetric inorganic membranes, was used for selective metal ion separation. The membrane phase was a self-assembled monolayer of alkyl thiols as a hydrophobic phase for a trialkyl phosphate and phosphine oxide-based metal ion carrier. This organic mixture was attached on alumina porous supports with thin layers of gold. The thin membrane layer gave high fluxes and high selectivity, while metal ions transport was carrier limited [50].

Catalytically active supported ionic liquid membranes were used for propylene/propane vapor mixture separation. In this case, the ionic liquid was immobilized in the pores of an asymmetric ceramic support, displaying sufficient permeability, good selectivity, and long-term stability [51]. Porous inorganic membranes were also used as a support for chiral-selective liquid membranes. For this purpose, porous tubular ceramic membranes were impregnated with β-cyclodextrin polymer. Such SLMs were used for separation of enantiomers of racemic pharmaceutical chlorthalidone [52].

Although the inorganic supports show many advantages such as temperature stability, solvent, and mechanical resistance, there are not many papers dealing with the use of these membranes for SLM construction, probably due to the same problems as for polymeric supports: membrane stability and lifetime.

5.2. Organic solvents used in SLM

In making the choice for an organic liquid membrane solvent, several aspects should be taken into consideration. First of all, the organic liquid should be hydrophobic enough to ensure immiscibility with aqueous phases. Secondly, the solvent has to be characterized by low viscosity, which results in high mass transfer through the membrane. In this case, note that low viscosity decreases membrane stability. Surface tension between the organic phase and polymeric support is also essential due to the fact that the liquid membrane is kept in the support pores by capillary forces. The other important factor is solvent volatility (mostly depending on the solvent density), which should be low to keep the organic phase in the pores of the support. Also, interfacial tension and the surface contact angle between the aqueous and organic phase has to be considered. The low interfacial tension causes faster degradation of the liquid membrane, for example, by easier emulsion formation, but on the other hand it increases the mass transfer by facilitating contact between phases. It is important for the carrier-mediated transport that organic solvent easily solubilizes the carrier. The magnitude of the fluxes in this type of transport depends on the carrier concentration. The high solubility of the carrier causes the high mass transfer. For these reasons, the most commonly used organic solvents as the liquid membrane phase are hydrocarbons (aliphatic and aromatic), hydrophobic ethers and esters, long-chain alcohols, or mixtures of technical solvents, for example, kerosene (Table 3.3). Concluding it is not straightforward to select an organic solvent and the choice is a compromise among all the discussed properties and also the type of support.

5.3. Ionic liquids as membrane phase

A new approach in the use of organic solvents for preparation of SLMs is the application of ionic liquids. ILs are compounds composed of two parts: cationic organic moieties and an anionic, organic, or inorganic part. The cationic parts of most ionic liquids are organic based, including imidazolium,

Table 3.3 The selected properties of the commonly used organic solvents as a membrane phase

Organic solvent	Density $\times 10^3$ (kg/m^3)	Viscosity $\times 10^3$ (Pa s)	Surface tension $\times 10^3$ (N/m)	Solubility in water $\times 10^{-3}$ (kg/m^3)
Di-*n*-butyl phthalate	1.04	15.4	36.4	8.91
n-Amyl benzoate	0.95	3.42	32.7	8.06
Dodecane	0.75	1.50	24.9	0.07
Heptane	0.68	0.38	19.6	-
Toluene	0.87	0.54	27.9	-
Kerosene	0.79	1.24	25.3	-
o-Nitrophenyloctylether	1.04	12.5	33.9	0.16
Diphenyl methane	1.00	2.96	38.4	6.51
Dihexyl ether	0.79	1.87	-	-
1-Octanol	0.83	7.47	27.1	-

N-alkylpyridinium, tetraalkylammonium, and tetraalkylphosphonium ions. The common anionic parts of the ionic liquids are halides, nitrate, acetate, hexafluorophosphate ([PF_6]), tetrafluoroborate ([BF_4]), trifluoromethylsulfonate ([OTf]), and bis(trifluoromethanesulfonyl)imide ([NTf_2]). Ionic liquids possess exceptional physicochemical properties. Many of them remain in a liquid state at temperatures between 0 and 400 °C. Ionic liquids have low-to-negligible vapor pressures and in many cases show high thermal stability. They may be the most complex of all solvents because they are capable of virtually all possible types of interactions with solutes. Therefore, ionic liquids can solubilize a variety of organic and inorganic compounds. They can be designed to be immiscible or miscible with water and a number of organic solvents [53]. The physicochemical properties of ionic liquids are influenced by both their cationic and their anionic moieties. Han and Armstrong [54], in their review, exemplified several unique changes in the physiochemical properties when ionic liquids are used as solvents. For instance, both the densities and surface tensions of ILs change depending on the type of cations when the same anion is used as counterion and decrease if the length of alkyl chain increases. In contrast, the viscosities of the same group of ionic liquids with the same anion increase with an increase in alkyl chain length. The solubility of ILs depends on the cation and anion as well. For example, 1-butyl-3-methylimidazolium chloride ([BMIM][Cl]) and [BMIM][BF_4] are soluble in water, while [BMIM][PF^6] and [BMIM][NTf_2] are immiscible with water. Increasing the length of the alkyl chain on the cation lowers the solubility of ILs with [BF4] anions. 1-octyl-3-methylimidazolium tetrafluoroborate ([OMIM][BF_4]) is immiscible with water [55]. Combinations of different possible cations and anions result in a large number of ionic liquids with different properties. Therefore, ionic liquids often are referred to as "tailor-made or tunable materials" [56]. There are several excellent reviews on this topic covering various aspects of the use of ionic liquids including their applications in separation science [53, 57-59].

Due to these special properties, ionic liquids can be used as a hydrophobic or hydrophilic membrane phase. Consequently, they can be immobilized both on hydrophobic and on hydrophilic supports. For example, for immobilization of the methylimidazolium-based ionic liquid used for the separation of water-soluble ions such as sodium chloride or thymol blue, the hydrophilic polyvinylidene fluoride, polyethersulfone, and nylon were used [60]. Recently, several interesting reports were published concerning applications of SLM separation in which ionic liquids were used as a membrane phase. This type of SLM was used for separation of gases [61, 62], hydrocarbons, including aromatic and aliphatic [63, 64], organic sulfur and nitrogen compounds [65], and other organic compounds like alcohols, amines, or ketones [66, 67]. Despite the fact that ionic liquids are widely used in SLM due to their unusual selectivities, high extraction efficiencies, durability, and resistance to thermal degradation, more research is necessary on their long-term stability, recyclability, toxicity, and reduction of their water solubility.

5.4. Membrane units (module design)

The main purpose for using the module is to hold the porous polymer soaked with liquid phase in a manner which ensures its mechanical stability, large effective surface area, and free flow of the feed and receiving phases. There are several types of configurations of SLMs which can fulfill this requirement and they are determined by the shape of applied module (see Fig. 3.13). The most popular ones are hollow-fiber SLM (HF-SLM, Fig. 3.13A) and flat-sheet SLM (FL-SLM, Fig. 3.13B). Sometimes other shapes of SLM configurations are also utilized, for example, spiral-wound SLMs (Fig. 3.13C).

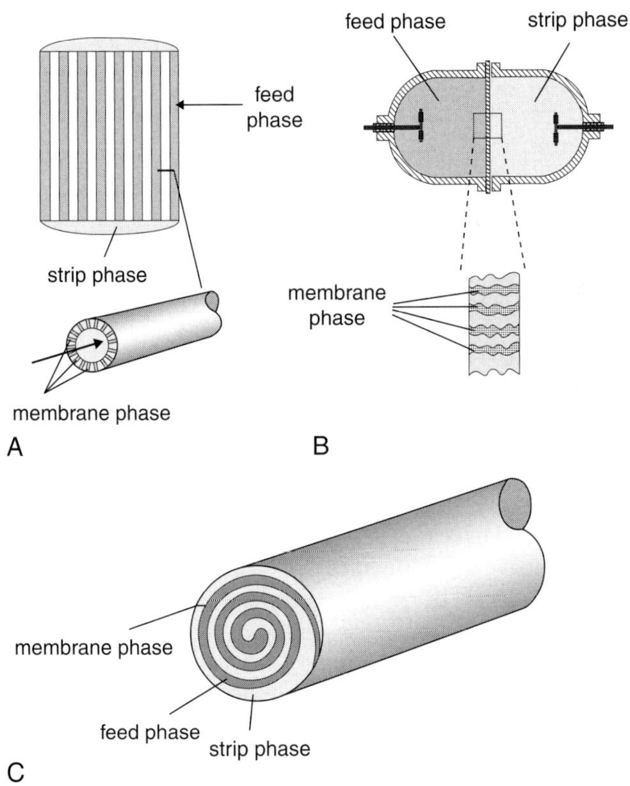

Figure 3.13 SLM configurations (designs): (A) hollow fiber, (B) flat sheet, and (C) spiral wound.

The simplest in design, the flat-sheet SLM, can be utilized on the laboratory scale, but it is difficult to scale it up for industrial applications. Usually, it is a porous polymer membrane whose pores are filled with the organic liquid and carrier, set in between the source phase and the receiving phase, which are being gently stirred (Fig. 3.13B). The design of the hollow-fiber SLM module (Fig. 3.13A) consists of an outer shell, which is a single nonporous material, through which the materials inside cannot be transported. Inside, a certain number of thin fibers are placed along the length of the shell. The source phase is pumped through the system from top to bottom, and the pores in the fibers are filled with the organic phase and the receiving phase is forced out through the sides of the shell. The use of the hollow fibers gives one of the most significant advantages, namely the surface area and membrane thickness resulting in high rate of mass transfer. Another good feature is that the source/receiving phases are easily recoverable. However, there are some drawbacks such as very hydrophobic membrane solvents are required to maintain integrity, and pore fouling often occurs due to surface effects and particles in the system.

The spiral-wound membrane is essentially a flat membrane wrapped around a perforated tube, through which the effluent streams out of the membrane. As can be seen (Fig. 3.13C) that sandwich is actually four layers: a membrane, a feed channel, another membrane, and a permeate channel, which forces all the separated material toward that perforated tube in the center. This type of membrane is an intermediate step between the flat, laboratory membrane and the hollow-fiber membrane, at least in terms of surface area per unit volume and stability.

However, the high surface area is not always the only important factor that has to be considered during SLM module design. Another is the possibility of obtaining a high feed phase/receiving phase volume ratio. Also, it is necessary to fulfill some additional conditions (e.g., sample volume, aqueous phases flow rate) resulting in the specificity of application.

A typical illustration of such an approach is the use of SLM modules in analytical chemistry for sample preparation and enrichment of the analyte. They cannot be too big, since they have to be usable in an analytical laboratory. They also should have the possibility to connect online into an analytical system. Also, they have to be suited to the volume of the sample and flow rate of the aqueous phase(s). If the sample volume is limited and concentration of the analyte is low in the sample, for example, blood or plasma, the module has to be designed to give the possibility of a very low flow rate to increase contact time between the feed and membrane phase (Fig. 3.14).

The typical modules are shown in Fig. 3.14A and B. For sample preparation use, volumes are typically in the 10–1000 μl range. By such design, it is feasible to use SLM as a very efficient sample preparation method for pharmaceutical and environmental analysis [68].

Recently, there has been more attention paid to the miniaturization of the whole analytical system [69]. The simplest way to achieve it in the case

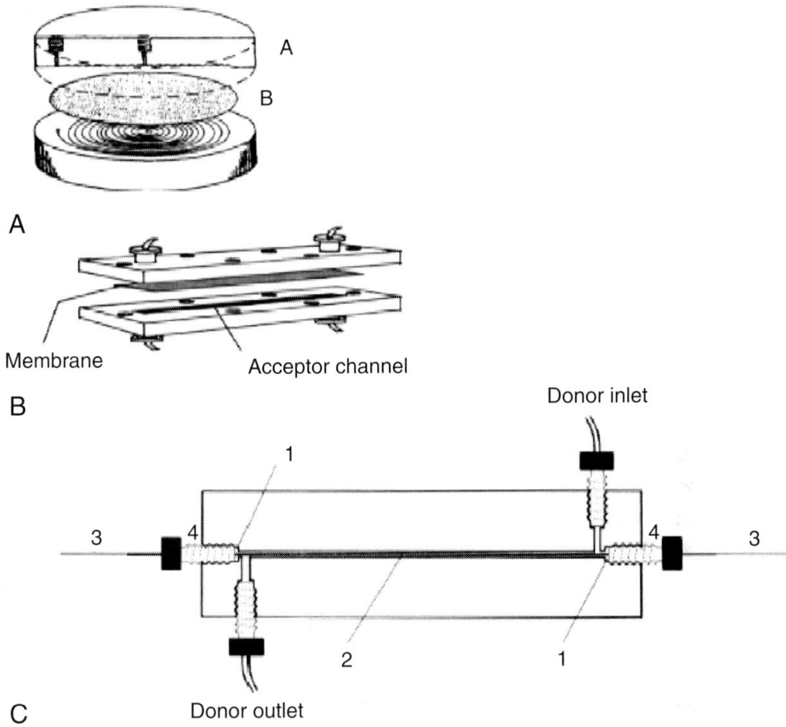

Figure 3.14 (A) Membrane unit with 1 ml channel volume (A, blocks of inert material; B, membrane). (B) Membrane unit with 10 μl channel volume. (C) Hollow-fiber membrane unit with 1.3 μl acceptor channel (lumen) volume (1, O-rings; 2, polypropylene hollow fiber; 3, fused-silica capillaries; 4, male nuts). From Ref. [150] with permission. © 2008 Elsevier.

of SLM is to use a single hollow fiber. There is a possibility to use a membrane previously soaked with liquid and filled-in lumen with a hollow-fiber receiving phase by simple immersion directly in the sample [70–78]. For this type of experiments, there is no need to design a module. However, if there is a requirement to integrate the SLM in an online system, modules have to be designed [79]. This type of unit can have channel volumes around 1 μl (Fig. 3.14C).

6. Membrane Stability

Despite many advantages, SLMs are not often used large scale in industry nowadays. The major reason for this is the membrane stability and lifetime, which are mostly too low to assure good commercial application. The stability of supported liquid membranes and mechanisms explaining SLM instability, suggestions for stability improvement are discussed in the literature [6, 80]. Since the membrane solvent in SLM is held in the pore structure solely by capillary forces, it is inevitable that during the separation process the solvent is to some extent washed or forced out of the membrane. A complete removal of the membrane phase from one of the pores enables the contact between the feed and strip phases. Once there is a continuous water path in the membrane, it will lead to solute leakage into the strip phase. In the case of carrier-mediated transport, the instability of SLMs occurs due to the loss of membrane solvent and/or carrier from the membrane phase. These effects have an influence on the flux and selectivity of the membrane [81-84]. Depending on the amounts of carrier and solvent lost from the pores of the membrane support, the solute flux might either increase, decrease, or stay almost equal (see Fig. 3.15).

Removal of solvent from the membrane to the water phases increases the carrier concentration in the membrane and therefore increases the flux (Fig. 3.15A). Carrier removal decreases their concentration in the membrane phase, so the flux decreases until the moment when the whole carrier is removed from the membrane and the solute is not transported any more (Fig. 3.15B). In the third possibility when both solvent and carriers are removed from the membrane, the carrier concentration is stable and thus the flux stays equal (Fig. 3.15C). When all of the membrane phase is lost, the membrane breaks down and a direct diffusion between the feed and strip phases takes place.

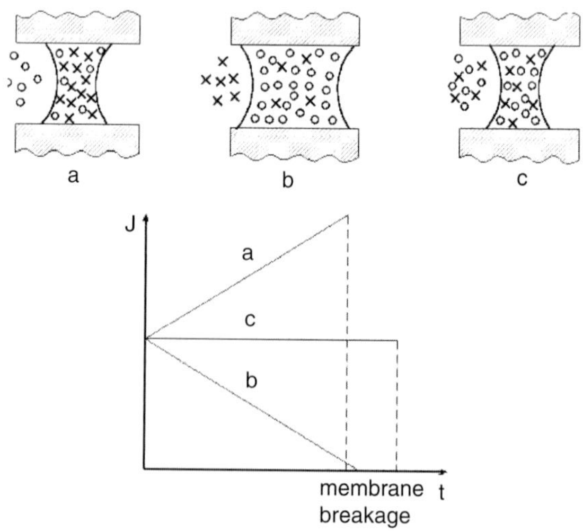

Figure 3.15 SLM degradation and its influence on the flux (*J*) during leakage of (A) organic solvent, (B) carrier, and (C) organic solvent and carrier.

The same problems occur for SLMs using ionic liquids as a membrane phase. Therefore, research studies are focused on the development of new ionic liquids with reduced water solubility [60].

6.1. Factors influencing membrane stability

Many factors, like type and characteristics of the solid polymer support, membrane solvent, and carriers as well as the operating parameters, influence the stability of an SLM. SLM lifetime seems to depend in a clear-cut way on the type of polymeric support and the nature of the liquid membrane, suggesting that solute-solvent and polymer-solvent interactions play a dominant role in membrane stability [85]. Since the membrane solvent is held within the pores by capillary forces, it is clear that the pore structure, morphology, and size influence stability. Asymmetric PTFE membranes with elliptical pores oriented with their major axis parallel to the membrane surface allow a longer lifetime than a polypropylene, polyethylene, or polycarbonate support with essentially straight-through holes [82]. The stability of SLMs strongly depends also on the pore size of the support. Stability of SLMs decreases with an increasing pore size; the smaller the pore size, the longer the observed lifetime of the membrane [86].

The physicochemical properties of organic solvents used as a membrane phase are of great importance in membrane stability. Organic liquid phases exhibit a high viscosity and high organic-water interfacial tension enhances the stability [39, 83]. Hydrophobic aromatic or aliphatic hydrocarbons and ethers are more strongly adsorbed and supported within the pores of a hydrophobic polymeric support than hydrophilic ones. SLMs with *n*-heptane or di-*n*-hexyl ether as a membrane liquid are more stable than those in which ethyl chloroacetate, nitrobenzene, or 1-octanol are used [23, 82]. It is evident that the solubility of membrane phase components in the aqueous phases causes the instability effect. The effect of membrane solvent solubility in water and volatility on membrane stability was investigated [87]. Dozol *et al.* [88] determined aqueous solubility and SLM lifetimes for a numbers of membrane solvents and concluded that a lifetime of over 200 h could be obtained when the solubility of the solvent used is lower than 12 g/l. In carrier-mediated extraction, the stability of SLMs also depends on the kind of carrier and counterion used. For ionic type of carriers, mostly used in SLM, the solubility in the aqueous feed and strip phase is very important. The stability of the membrane increases with decreasing the carrier solubility in aqueous phases. The lifetime of SLMs containing acidic carriers like DEHPA decreases when at least one of the aqueous solutions has a high pH, which is in agreement with the increased solubility of this type of compounds with increasing pH [81]. The stability of membranes in which different aliphatic amines were used as a carrier decreases in the order tertiary > secondary > primary amine, opposite to the carrier solubility [89]. The stability of

Table 3.4 Stability of the supported liquid membranes as a function of the kind of counterion and of the type of organic solvent

	Stability (days)	
Organic solvent	TrpX[a]	Trp DBSA
Di-*n*-hexyl ether	>60	>12
Butylbenzene	>50	>10
o-NPOE	>30	>6
TEHP	>30	>6

From Ref. [23] with permission. © 2008 Elsevier.
[a]X: HCl, PTSA, $HClO_4$ (PTSA, *p*-toluenesulfonic acid; DBSA, dodecylbenzylsulfonic acid).

membranes used for carrier-mediated amino acid transport strongly depends on the type of counterion (Table 3.4) [23]. Furthermore, the stability depends on the flow velocity of the aqueous phases along the membrane. The stability decreases with an increase of the flow velocity by an increasing hydrostatic pressure gradient over the membrane [83].

6.2. Degradation mechanisms

The reason for the instability of SLMs is the loss of the solvent and/or the carrier from the pores of the support, which has an influence on both flux and selectivity. The major degradation mechanisms are dissolution of carrier and membrane solvent in aqueous phases, wetting of the pores in the membrane support by aqueous phases, presence of pressure difference and osmotic pressure gradient over the membrane, emulsion formation in the liquid membrane phase, and blockage of membrane pores by precipitation of a carrier complex [6, 80].

The pressure difference over the membrane exists due to pumping of the aqueous phases through the feed and/or strip channels and has a special importance in hollow-fiber SLM configurations. When the pressure difference exceeds a certain critical value, the membrane phase is pushed out of the pores of the support.

The minimum transmembrane pressure (P_c) required to push the impregnating organic phase out of the largest pores can be calculated using the Laplace equation [84]:

$$P_c = 2\gamma \cos \theta / r, \tag{18}$$

where P_c is the pressure (N/m^2), γ is the interfacial tension between strip or feed solution and liquid membrane phase (N/m), θ is the contact angle between the membrane pores and the membrane liquid, and r is the pore radius (m).

The critical transmembrane pressure calculated using this equation is valid for cylindrical pores. Usually, the commercial hollow-fiber polymeric supports used for SLM have pores highly irregular in geometry. Therefore, especially for hydrocarbon solvents used as a membrane phase, the P_c is much larger than transmembrane pressure, which indicates that pressure difference is not the main cause of SLM degradation, but the loss of solvent and an emulsification of the membrane phase due to lateral shear forces [80].

The hypothesis for the degradation mechanism of SLMs due to emulsion formation was proposed by Neplenbroek and further developed and extended by Zha *et al.* [83, 90, 91]. The degradation of SLM has been considered to be due to the disruption of emulsion droplets caused by hydrodynamic instability at the surface. In the presence of a surface tension gradient, the interfacial hydrodynamic instabilities could be promoted or damped out, and consequently affect the liquid membrane loss. Neplenbroek observed that the more stable emulsion can be formed by organic solvent used as membrane phase with carrier acting as the emulsifier, the more unstable is the SLM impregnated with this liquid membrane. Therefore, the formation of oil-in-water (O/W) emulsion is a function of molecular structure of the carrier, the type of organic solvent, the counterion type and concentration, and salt contents in the aqueous phases. Many carriers used in SLM, such as ammonium cations (tetraoctylammonium bromide, TOMA) or phosphoric (DEHPA) and long-chain carboxylic acids, are surface-active compounds and can stabilize emulsions. The important parameter for stable O/W emulsion formation is the HLB (hydrophilic-lipophilic balance) scale of the organic phase. The HLB of an emulsifier (carrier) determines the type of emulsion that tends to be formed. The O/W emulsion is formed when the HLB value is in the 8-15 range [84]. An increase in surface area in SLM systems can be due to, for instance, local deformations of the membrane meniscus. When the liquid membrane meniscus is deformed, emulsion droplets can be formed (see Fig. 3.16).

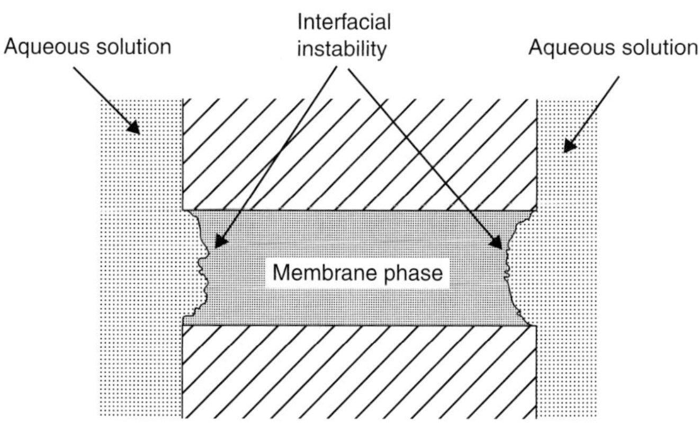

Figure 3.16 Unstable interfaces in an SLM containing microliquid necks. From Ref. [84] with permission. © 2008 Elsevier.

Additionally, the salt content and the type of counterions in the aqueous phases influence the emulsion stability and consequently SLM degradation. It was found that the amount of liquid membrane removed from SLM increases with a decrease in the salt concentration of the aqueous phases and with an increase in their flow velocities [83].

For long-term permeation, there is no one single factor determining the SLM instability. A number of phenomena appear to contribute to instability, including surface shear forces, changes in the membrane morphology, density differences, and membrane preparation protocol [92].

6.3. Improving SLM stability

Supported liquid membrane stability and lifetime limit the industrial application of this separation technique. Therefore, the stability of these membranes needs to be enhanced drastically. A proper choice of the operating and membrane composition factors might improve the lifetime of SLM systems.

The important operating conditions that influence membrane stability are membrane thickness, stirring or flow rate of aqueous phases, carrier concentration, aqueous solute concentration, and operating temperature. In general, larger membrane thickness and lower flow rate increase the stability. Membranes of higher thickness contain more liquid membrane phase and therefore it will take longer before the SLM breaks down. However, increasing membrane thickness results in the flux decreasing due to increasing the diffusion pathway. The SLM stability strongly decreased with increasing the pumping velocity of aqueous phases. This may result in a larger loss of membrane phase due to the larger transmembrane pressure differences and larger shear forces. The SLM lifetime mostly decreases with decreasing the solute and salts concentrations in aqueous phases since there is higher leaching of carrier and larger emulsification [83]. The SLM stability depends also on the operating temperature. The increasing operating temperature increases the solubility of both membrane solvent and carrier in the aqueous phases and membrane lifetime decreases; however, the flux increases due to lower viscosity of the membrane phase [93].

Considering membrane materials, we can distinguish the type of solvent for the carrier and membrane phase, the carrier itself, and the support. Increasing viscosity also increases the lifetime but the fluxes are strongly decreased [94].

Molecular structure and physicochemical properties of the carrier used are important for SLM stability, especially the lipophilicity, surface activity, and its solubility in the membrane solvent. The membrane is less stable when a more surface-active compound is used as a carrier [89]. When the carrier loss is the main reason for the SLM instability, the membrane stability can be increased by attaching the carriers to a polymer or covalently linked onto long aliphatic chains or polysiloxanes [95].

The type of polymeric microfiltration membranes used as a support influence SLM stability in several ways. In general, SLMs using a support with a lower pore size are more stable than those with larger pore size, although the surface porosity should be high enough to obtain a reasonable flux. The pore structure and morphology are also important. Membranes with asymmetric and elliptical pores oriented with their major axis parallel to the membrane surface used as a support give longer SLM lifetime than a support with essentially straight-through holes [96].

There are several other methods, which can be found in the literature, to enhance the stability of SLM, like reimpregnating of the support, stabilization by plasma polymerization, formation of barrier layers on membrane surface, or using sandwich SLM.

The SLM could be regenerated by reloading the membrane supports with fresh liquid membrane solution after they have decayed, which provides the same extraction efficiency as a newly prepared SLM [25]. The regeneration of degraded hollow-fiber liquid membranes could be done in the same way by simply reimpregnating, then pumping the fresh membrane phase at the lumen side of the support for a few minutes instead of the aqueous solution. Continuous impregnation of the membrane is also possible and is mainly applied for hollow-fiber modules. The vertical hollow-fiber module containing one single fiber with a certain amount of liquid membrane phase at the bottom was designed. The membrane liquid from this fiber was soaked into the pores of the support and moved upward through the porous network by capillary forces [97]. The continuous reimpregnation of the support was also done by adding membrane phase as an emulsion to the one of the aqueous phases. It works well, but one of the aqueous phases is still polluted with the membrane liquid [18].

One of the most important degradation mechanisms of SLM is an emulsification of the membrane phase due to lateral shear forces. Therefore, formation of barrier layers on the membrane surface by physical deposition [98] or by interfacial polymerization could prevent instability [99, 100]. A polysulfone support with *N*-methylpyrrolidone as a solvent was coated by a poly(ether ketone) layer as the outside layer and gave a specific composite membrane support. Such composite hollow-fiber membranes showed significant improvement in stability in copper ions permeation.

A plasma polymerization surface coating is another possibility to prevent the supported membrane degradation [101]. This prevention also reduces the surface membrane pores and increases mass-transfer resistance, resulting in a decreased permeability of the membrane system. The stabilization of SLM prepared by impregnating polypropylene glycol (PPG) into the pores of microporous flat-sheet polyvinylidene fluoride (PVDF) or polypropylene (PP) membranes, through crosslinking the liquid membrane phase by using γ-radiation, was reported [102, 103]. These membranes retained both their selectivity and stability over a period of more than 1 month, and the phenol mass transfer was higher than through silicon rubber tubing membrane.

6.4. Gel SLM

The stability of SLMs can be improved by gelation of the liquid membrane phase using, for example, PVC as gel-forming reagent [83]. The stabilization of SLMs by gelation could be carried out in two different ways: by a homogeneous gelation in the pores of the support and by applying a thin dense gel layer on the feed side of the membrane (see Fig. 3.17). Both gelations of liquid membranes increased the resistance against liquid displacement out of the support by effectively preventing the liquid membrane meniscus from deformation and therefore emulsion formation. However, the stability increases with PVC concentration. Only gelled liquid

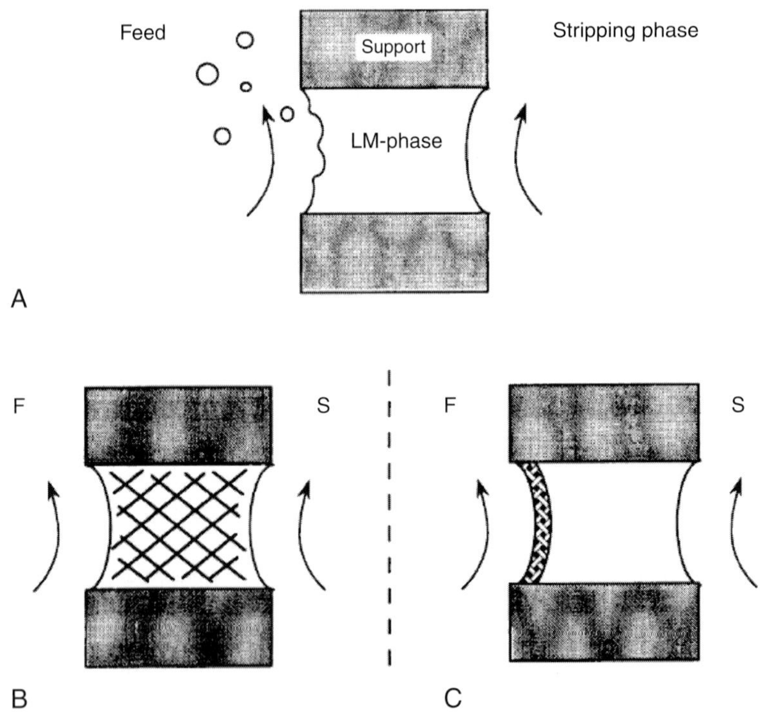

Figure 3.17 Influence of a gel network on SLM stability: (A) without gel network: SLM degradation by emulsion formation (due to local deformation of LM phase in the pores of the support); (B) homogeneous gel network in LM phase; and (C) one thin dense gel layer at the interface with the feed. From Ref. [83] with permission. © 2008 Elsevier.

membranes with low polymer concentration are interesting for practical applications because the diffusion rate of carrier molecules in a gel network decreases drastically with increasing polymer content [83]. In contrast to the result of Neplenbroek, gelation of the LM was not effective in improving the stability when trimethylammonium chloride (TOMA-Cl) and *o*-nitrophenyl were used [80].

6.5. Polymer inclusion membranes

Self-supporting membranes (SLMs) have been used for more than 30 years as polymer membrane ion-selective electrodes. This type of membranes, commonly called polymer inclusion membranes are formed by casting a solution containing an extractant, a plasticizer, and a base polymer such as CTA or PVC to form a thin stable film [104]. A number of other names are also used such as polymer liquid, gelled liquid, polymeric plasticized, fixed-site carrier, or solvent polymeric membranes. The polymeric inclusion membranes are highly resistant to carrier and plasticizer leakage and are considerably more stable than SLM. In PIMs, the carrier, plasticizer, and base membrane are well integrated into a relatively homogeneous thin film. The comparison of SLM and PIM stability has been reported in many papers. Kim *et al.* have investigated the stability of both types of membrane under similar experimental conditions. They reported no flux decline or evidence of material losses within 15 days of continuous transport experiments with PIMs containing CTA, 2-nitrophenyloctylether (2-NPOE) and macrocyclic carrier, while for SLM leakage of the organic material was observed after 48 h [105]. The better stability of PIMs compared to SLMs was also reported for a membrane using Aliquat 336 as a carrier. Under similar experimental condition, PIMs were stable for 30 days, while SLMs only for 7 days [106]. The results of several studies of PIM lifetimes are summarized in Table 3.5.

The good stability of PIMs over the various types of liquid membranes, including SLMs, and the adequate, but lower, permeability and selectivity show the potential practical applications. The main problem is the low mechanical strength of PIMs.

An excellent review providing a comprehensive summary of all aspects of those membranes was published recently [104].

6.6. Integration of SLM with other membrane processes

To overcome the problems with SLM stability, the idea of their integration with other membrane processes was also investigated. Two approaches can be distinguished. Both could lead to significant increase of the liquid membrane lifetime. One approach is to separate the liquid membrane from the feed and receiving phases. It can be achieved by placing liquid

Table 3.5 PIM lifetimes under continuous operation

Membranes (base polymer/carrier/plasticizer)	Reported lifetime and membrane performance
CTA/calix[6]arene/2-NPOE	Small flux decline after 30 days
CTA/lasalocid A/2-NPOE	No sign of flux decline or carrier and plasticizer losses after 10 days. Stable after 10 months storage in air
CTA/acyclic polyether bearing amide/2-NPOE-TBEP	Small flux decline after 15 days but no evidence of carrier and plasticizer loss
CTA/calix[4]arene/2-NPOE	Small flux decline after 20 days but no evidence of carrier and plasticizer loss
CTA/calix[4]arene/2-NPOE	Stable flux after 1 month
CTA/DC18C6/2-NPOE-TBEP	Flux decline began slowly after 100 days but no evidence of carrier and plasticizer loss
CTA/Aliquat 336/2-NPOE, DOS, DOTP, or DOP	Flux decline and carrier/plasticizer loss began after 30 days
CTA/Aliquat 336/T2EHP	Flux decline began after 18 days
CTA/*t*-BuDC18C6/2-NPOE-DNNS	Stable for several weeks

From Ref. [104] with permission. © 2008 Elsevier.
TBEP, tri(butoxyethyl)phosphate; DC18C6, dicyclohexano-18-crown-6; DOS, dioctylsebacate; DOTP, bis (2-ethylhexyl)terephthalate; DOP, dioctylphthalate; *t*-BuDC18C6, *t*-butyldicyclohexano-18-crown-6; DNNS, dinonylnaphthalenesulfonic acid.

membrane between two pieces of porous membranes. If solid membranes are not involved in the transport of the solute, but are only passively separating such a system, we have stagnant sandwich liquid membranes (SSwLMs) [107–109]. For example, such a system was used for transport of Cu(II) ions using DEHPA as a carrier and *n*-decane as the organic solvent. The sandwich membrane shows higher copper flux and longer lifetime (100 vs. 5 h) in comparison with SLM [107] (for details, see Chapter 5).

However, it is also possible that the two stagnant membranes not only are used as barriers for liquid membrane protection but also actively take part in the transport of solute. This gives an opportunity to increase membrane stability and also the selectivity of the separation. In this case, very often the term hybrid liquid membrane (HLM) is used. In most cases, such a system comprises the bulk liquid membrane placed between two solid ion-exchange membranes. Here both ion-exchange membranes are barriers, which physically separate the organic and aqueous liquid phases. Moreover, the ability of those membranes to charge species sorption from aqueous solution causes a high accumulation of reacting species at the interfaces. Many applications have been described [4, 110–123]. The typical example of a hybrid (integrated) membrane system in which liquid membrane is used was presented by Wódzki *et al.* [122]. In their study, poly (oxypropylene) bisphosphates were synthesized by modification of various polypropylene glycols and used as macroionophores of metal cations [K(I), Na(I), Ca(II), Mg(II), Zn(II), and Cu(II)]. In a comparison of the supported and hybrid liquid membranes [120], competitive transport of an equimolar mixture of cobalt(II) and nickel(II) was investigated. In both types of membranes di-2-ethylhexyl phosphoric acid (D2EHPA) as well as commercial extractants, that is, Cyanex® 272, 301, and 302 were used as ion carriers. The HLMs were composed of the cation-exchange membranes with bulk liquid membrane in the system: cation-exchange membrane-organic phase-cation-exchange membrane. After studying several factors that influence cation transport selectivity and transport effectiveness it was found that the separation of Co(II) from Ni(II) is governed by the ionic carrier used as well as by the acidity of the aqueous source phase. Interestingly, in the HLM processes, lower metal ion fluxes than in supported liquid membranes processes were observed. However, higher separation coefficients for Co(II) from Ni(II) were found for hybrid than for SLMs (for details, see Chapter 5).

The other approach to increase stability of the SLM and simultaneously alter the efficiency of the separation is to prevent the loss of liquid membrane phase by using the so-called SLM with strip dispersion. This concept was developed by Ho [124, 125]. An aqueous strip solution is dispersed in an organic membrane phase solution containing an extractant in a mixer, and the water-in-oil dispersion formed is then pumped to contact one side of a microporous support. As a result, these droplets are retained in the strip dispersion side and cannot pass through the pores to the feed solution side. The constant supply of organic membrane solution into the pores ensures a stable and continuous operation. Moreover, the direct contact between the organic and strip phases (with high-shear mixing, if necessary) provides efficient mass transfer for stripping. This process is a combination of two supported liquid membranes and emulsion liquid membrane simultaneous separation and sometimes is called strip dispersion hybrid liquid membrane (SDHLM) separation [127, 128]. For SLM with strip dispersion, the main aim is to stabilize the liquid membrane into the pores of the polymeric support by preventing emulsification of organic phase. Mostly, an SLM/strip dispersion process is used for metal separations [124–136] as well as for separation of organic substances, for example, organic acids and penicillin G [130, 137] and can be realized both in flat-sheet LM and in hollow-fiber LM. In the second case, the separation process is called [134–136] the pseudoemulsion-based hollow-fiber strip dispersion (PEHFSD) technique, but it is not different in principle from the original concept of Ho, who used a hollow-fiber polypropylene support in his pioneering study on removal and recovery of metals from wastewater and process streams.

In conclusion, there is no universal benchmark for membranes with satisfactory lifetime, permeability, selectivity, and mechanical resistance needed for industrial application. For practical purposes, a compromise between stability and efficiency is needed.

7. Supported Liquid Membranes Application

As described above, there are a variety of possibilities to design SLM processes for various purposes depending on the requirements. On the one hand, we have the simplicity of the realization of the separation/transport process and, on the other hand, its significant flexibility for adaptation. In fact, any system in which it is feasible to incorporate liquid or solution of carrier into the pores of the polymer—independent of its size, shape, or geometry—which separates two other liquid, immiscible phases, can be treated as an SLM. Moreover, it is possible to multiply such systems by proper manipulation of the manifold. Taking into account these features of SLMs, it is not surprising that there is much interest in this separation method in various fields including analytical, inorganic, and organic chemistry, hydrometallurgy, chemical engineering, biotechnology, and biomedical engineering. In fact with SLM separation, it is possible to separate various types of chemical compounds from gases or inorganic ions through many organic compounds (charged or uncharged species, hydrophobic to hydrophilic, stereoisomers, etc.) to large molecules (e.g., peptides or even proteins). Obviously, all applied approaches have their specific separation requirements depending on the nature of the separated compound, matrix composition, and also on the goal of the studies.

What follows are selected examples of applications. The aim is to show how the concept of liquid membranes is utilized in various fields where efficient separation is required. We hope that it will give the reader a better understanding how the theoretical and laboratory studies carried out to explain the transport phenomena through liquid membranes can be applied to solve a variety of separation challenges that can be encountered in real life.

7.1. Analytical applications

An essential step in the development of almost every analytical method is sample preparation. This process, no matter how it is performed, should remove potential interferences, increase the concentration of the analyte, and be reproducible independently of the sample matrix variation [138]. In recent years, several trends have emerged, such as the use of smaller initial sample sizes, small volumes or no use of organic solvents, enhancement of the specificity and selectivity, and increased automation [138]. Many techniques have been used to achieve one or all of those goals including solid-phase extraction (SPE) or liquid-liquid extraction (LLE) and also membrane techniques [138, 139]. For membrane separation, one possibility is to apply the SLM concept in achieving the goals of sample preparation. SLM extraction as a sample pretreatment technique was developed at Lund University in Sweden by Audunsson for enrichment of amines [140]. Since then SLMs have begun to be widely applied for sample cleanup, enrichment of various types of chemical substances such as metal ions, organic acids, amines, amino acids, phenolic compounds, peptides, herbicides, and drugs (Table 3.6).

Many applications have been reviewed [5, 27, 68, 151–154]. The practical and theoretical aspects of SLM extraction from an analytical chemistry point of view are extensively presented; therefore, they will not be described here. However, we would like to present some examples to show that knowledge gained from basic studies on the liquid membrane transport can be utilized to solve practical problems in sample preparation.

The first example comes from the field of environmental chemical analysis and considers SLM enrichment and determination of triazine herbicides from natural water samples [141]. It shows how simple manipulation of donor and acceptor phase pH, the simplest manner of transforming the analyte into a transportable form, can lead to an SLM system with high extraction efficiency. A porous PTFE membrane impregnated with water immiscible dihexyl ether was used as an organic solvent. The obtained detection limit of triazines ranged from 0.03 to 0.16 μg/l in natural waters with 20 min extraction time using simple UV detection.

A similar concept was used for other environmental applications, for example, phenoxy acids, sulfonureas, phenolic compounds, and other environmentally important persistent pollutants [68, 76, 141, 143, 155-166]. Also, in the same manner, several drugs were enriched and determined in body fluids such as urine [144-146, 167-172] or blood [147, 156, 157, 173, 174]. A very advanced application of SLM for analytical purposes, where transport process was based on simple diffusion with pH adjustment of aqueous phase, is the extraction of the basic drug, bambuterol, for pretreatment of plasma samples before analysis with capillary zone electrophoresis (CZE) [147]. Bambuterol was used as a model substance in a separation system, where either 6-undecanone or a mixture of di-*n*-hexyl ether (DHE) and tri-*n*-octylphosphine oxide (TOPO) was used as membrane phase. It was possible not only to achieve a very low limit of detection (~50 nmol/l) but also to ensure the removal of salts from the sample. It helped to obtain the low ionic strength of the blood plasma samples and permitted subsequent sample stacking in the capillary electrophoresis step.

In SLM extraction of metals or small multifunctional organic compounds a more difficult situation is encountered. In an aqueous environment, they are permanently charged within all pH ranges. Therefore, their affinity toward the organic phase is very low and as a result transport efficiency is

Table 3.6 Selected applications of SLM for sample preparation

Compound(s)	Sample origin	Transport mechanism, conditions	References
Trazines	Environmental	Simple diffusion, adjustment of aqueous phases	[141, 142]
Phenoxy acids	Environmental	Simple diffusion, adjustment of aqueous phases pH	[143]
Aminoglycoside antibiotics (neomycin, gentamicin, streptomycin)	Food, environmental	Simple diffusion, adjustment of aqueous phases pH	[144]
Amphetamines	Pharmaceutical	Simple diffusion, adjustment of aqueous phases pH	[145]
17β-estradiol and its metabolites	Food, environmental	Facilitated diffusion, TOPO used as carrier, adjustment of aqueous phases pH	[146]
Bambuterol	Pharmaceutical	Facilitated diffusion, TOPO used as carrier, adjustment of aqueous phases pH	[79, 147]
Metals (Cu, Cd, Co, Ni, Zn)	Environmental	Simple diffusion of the metal complex formed with 8-hydroxyquinoline in donor phase, adjustment of aqueous phases pH	[20]
Pb	Environmental	Counter-coupled carrier-mediated transport, D2EHP used as carrier, adjustment of aqueous phases pH	[148]
Peptides	Pharmaceutical	Counter-coupled carrier medieted transport, Aliquot 336 used as carrier, adjustment of aqueous phases pH	[149]
N-(phosphonomethyl) glycine (glyphosate)	Food	Counter-coupled carrier-mediated transport, Aliquot 336 used as carrier, adjustment of aqueous phases pH	[150]
Polyamines	Pharmaceutical	Counter-coupled carrier-mediated transport, D2EHP used as carrier, adjustment of aqueous phases pH	[73]

very limited. Several solutions have been proposed to overcome this. A good illustrative example of the transport of ionic species is SLM extraction of metal cations [20]. The possibility of using an SLM for the enrichment of five metals (Cu, Cd, Co, Ni, and Zn) in a flow system with offline atomic absorption spectrometry in the final analysis step has been investigated. Different approaches to transport metals across the membrane were demonstrated. In one system, 8-hydroxyquinoline was used as chelating agent in the sample (donor) solution and metal ions were extracted as a neutral complex by simple diffusion. In a second system, potassium thiocyanate was used in the donor solution and a cationic carrier (Aliquat 336) in the membrane liquid. In a third system, 8-hydroxyquinoline was used in the donor solution and Aliquat 336 in the membrane liquid. The extraction efficiency was generally better for facilitated extraction but it was dependent on the type of metal cation extracted. The approach utilizing carrier-mediated transportation of metal ions was further developed in a series of studies of Djane *et al.* [148, 175–177]. In this case, the SLM extraction of metal cations was realized by the use of anionic extractant D2EHP (di-2-ethylhexyl phosphoric acid), which served as a complexing agent incorporated into an organic liquid membrane. High extraction efficiency was strongly influenced by the acceptor and donor phase pH, which confirmed the counter-mediated carrier transport mechanism. By using D2EHP, it was feasible to determine lead in urine samples [148, 176] and other toxic metals in river water [175]. A similar approach was used for organic multifunctional compounds. As mentioned before, the simplest example is an amino acid with two functional group of opposite acid/base properties. The possibilities of realizing the amino acid transport were presented in several reports [25, 26, 178]. Similarly to metal extraction, achieving high flow rates in simple diffusion by converting the amino acid into its hydrophobic derivative and proper adjustment of the phase pH led to significant increase of extraction efficiency [178]. The use of cationic [25] or anionic [26] hydrophobic carriers and proper choice of aqueous phase pH also makes it possible to extract amino acids with high extraction efficiency. This concept was used for SLM extraction of other multifunctional substances including aminophosphonic acids, peptides, and polyamines from aqueous samples [149, 150, 179–183].

7.2. Applications of the supported liquid membrane technique in biotechnology and environmental science

SLM technology, as already mentioned, has application in many separation processes where selective recovery is one of the main requirements. Several interesting examples of SLM use for separation of various chemical species will be presented to show its flexibility and adaptability for very different purposes (Table 3.7). Additionally, the examples show the potential for the use of SLM technology in industrial processes.

Table 3.7 Examples of SLM application in biotechnology and environmental science

Substance	Application	Liquid membrane	Support	Configuration	References
Fructose	Removal from fermentation broth	Boronic acid derivative dissolved in 2-nitrophenyloctyl	PP	Flat sheet, hollow fiber	[184]
Phenol, cresols	Removal from wastewaters	Natural oil (e.g., palm) liquid membrane	PP, PTFE	Flat sheet	[185]
Chromium(VI) ions	Removal from wastewaters	A commercial amine liquid membrane	PP	Flat sheet, hollow fiber	[186]
Copper	Recovery from spent ammoniacal etching solutions	LIX54 extractant in kerosene	PP	Hollow fiber	[187]
Uranium (U), plutonium (Pu)	Removal from postnuclear plant wastewater	Tri-*n*-butyl phosphate in *n*-dodecane	PP	Hollow fiber	[188]
CO_2	Capture from industrial gases (e.g., CH_4/CO_2 mixture)	Aliphatic amine membranes	PVDF	Flat sheet	[189]

In biotechnological processes, one of the crucial problems is the recovery of the bioprocess products. For instance, during the production of ethanol from sugar cane syrup during the selective fermentation of a mixture of glucose and fructose carried out by a mutant *Saccharomyces cerevisiae* strain, a fructose may be obtained as a byproduct. In a fed-batch process, together with main product of fermentation ethanol, fructose is accumulated in the bioreactor, decreasing process performance due to the inhibition of microorganism.

Continuous removal of fructose and ethanol may prevent that inhibition. However, sugar separation is a relatively difficult and expensive task, as the most commonly used commercial method for sugar separation involves chromatographic processes. Di Luccio and coworkers [184] have investigated the feasibility of removing fructose continuously from a fermentation broth using flat-sheet and hollow-fiber SLMs. They used a liquid membrane system based on facilitated transport of fructose using boronic acid derivative dissolved in 2-nitrophenyloctylether as solvent impregnated in a porous polypropylene support. The results show that fructose removal from the fermentation broth can reduce microorganism inhibition and increase the system performance, although further improvement in membrane stability and fluxes are still necessary.

The other interesting SLM application is removal of organic compounds from wastewater, which comprises liquid waste discharged by domestic residences, businesses, industry, and/or agriculture and can encompass a wide range of potential contaminants and concentrations. In the most common usage, it refers to the municipal wastewater that contains a broad spectrum of contaminants resulting from the mixing of wastewaters from different sources. Therefore, there is a constant need to find methods that provide efficient treatment of wastewater in order to clean it up via separation and removal of the toxic substances. SLM technology is an interesting choice for selective elimination of water contaminants. For example, Venkateswaran and Palanivelu [185] investigated the transport of phenol through a flat-sheet SLM containing, interestingly, vegetable oil as liquid membrane. The results obtained show effective removal of phenol using PTFE membrane and PP as a solid support. Among the various oils tested, palm oil was chosen as the best liquid membrane with permeability of 8.5×10^{-6} m/s in acidic feed of pH 2 with 0.2 M sodium hydroxide as an effective stripping agent. They were able to transport all the phenol from the feed side to strip solution, with an initial concentration of 100 mg/l, after 6 h. A concentration factor of 5 has been achieved in the present investigation with 0.2 M sodium hydroxide as trapping reagent. They also used similar methodology to remove cresols to explore the possibility of applying this to industrial wastewater under the optimized conditions for phenol. After 14 h of the transport studies in the phenol-formaldehyde industry wastewater, phenolic concentration in the feed solution was below a detectable level (1×10^{-2} mg/l). It is an interesting application that demonstrates the use of renewable, cheap, nontoxic, naturally occurring vegetable oils as a novel, green liquid membrane for the recovery of phenol from aqueous solution in SLM.

Other important toxicological contaminants that can be found in wastewaters are metals. Toxic heavy metal ions are introduced to aquatic streams by means of various industrial activities viz. mining, refining ores, fertilizer industries, tanneries, batteries, paper industries, pesticides, etc., and possess a serious threat to the environment. The major toxic metal ions hazardous to humans as well as other forms of life are Cr, Fe, Se, V, Cu, Co, Ni, Cd, Hg, As, Pb, Zn, etc. These heavy metals are of specific concern due to their toxicity, bioaccumulation tendency, and persistency in nature [190]. The SLM technique has been widely applied for the transport and recovery of almost all important metals from various matrices; an excellent review of all aspect of metal permeation through SLM (covering both theoretical and practical considerations) is available [191]. Here, only some selected recent examples of the use of SLM for metal separation will be presented.

Chromium compounds have received considerable attention because these are used extensively in such industrial applications as electroplating, steelmaking, tanning of leather goods, and corrosion inhibition. Therefore, it is not surprising that this metal can be found in many industrial wastewaters. For example, to selectively separate and preconcentrate Cr(VI) ions, a commercial amine as the membrane liquid on the porous polypropylene support in flat-sheet configuration has been used [186]. In the first step, laboratory-scale experiments were conducted with a batch reactor made of perspex, with a membrane fixed amid the two chambers. The flux of Cr(VI) ions was maximum in very low pH (at 1) and above and below this pH it decreases. Additionally, the Cr(VI) transport through the membrane increases with rise in temperature. Tests of the efficiency of the flat-sheet SLM were conducted with higher Cr(VI) concentration (5000 ppm) for 24 h, at optimized parameters. It was observed that about one fifth of the feed Cr(VI) is left over, while the rest is transported. After the laboratory-scale experiments, the system was scaled up for a preconcentration of Cr (VI), applying the proposed SLM parameters, and using the hollow-fiber (HF) system. The highest enrichment factor (13.8) value was obtained for 50 mg/l whereby all of the metal was transported to the stripping phase and the resulting Cr concentration was 688 mg/l. This system is also a good example of how laboratory-scale experiments are useful for introducing an SLM separation method into large-scale purification.

A similar approach, but for removal of copper, was presented by Yang and Kocherginsky [192]. One of the key steps in printed circuit board production is etching of a thin copper layer. Ammoniacal etching solutions are widely used for this purpose. Earlier an SLM-based method was developed to treat wastewater containing ammonia and copper [187]. In this instance, an effective hollow-fiber supported liquid membrane (HFSLM) separation for copper recovery from spent ammoniacal etching solutions, where copper is present in much higher concentrations is described. Again, a bench-scale HFSLM system with 1.4 m^2 effective membrane surface area was first used to screen for the optimal hydrodynamic and other operational conditions. Finally, successful pilot-scale experiments were conducted on a hollow-fiber membrane contactor with a surface area of 130 m^2. The process results in copper removal by a factor of 3000 and formation of nearly saturated copper sulfate solution in the sulfuric acid, used as a stripping phase. The stability of the pilot-scale system is promising for further industrial scale-up.

The last example shows that it is also feasible to use SLMs to remove and recover efficiently radioactive metals from nuclear process effluent. By using a microporous hydrophobic polypropylene hollow-fiber supported liquid membrane (HFSLM) consisting of extractant, tri-*n*-butyl phosphate (TBP) as carrier diluted with *n*-dodecane, actinides such as uranium (U) and plutonium (Pu) were removed [188]. It was concluded after modeling and evaluation of the process conditions that it is possible to remove more than 99% of U(VI) and Pu(IV) from process effluent in the presence of fission products when stripping reagent 0.1 M hydroxylamine hydrochloride in 0.5 M HNO_3 was used.

Recently, the increase of the amount of carbon dioxide (CO_2) in the atmosphere has become more and more important. Concerns over global warming following the release of CO_2, the most important greenhouse gas, continue to grow. Therefore, carbon dioxide capture is important for both energy production and environmental preservation [193]. Each year about 28 Gt of CO_2 are released into the atmosphere (1 Gt = 1 × 10^9 metric tonnes) [194]. The exploration of capture and storage methods for CO_2 is ongoing worldwide. Also, in this field SLM technology was used for its selective capture. Al Marzouqi and coworkers [189] designed and evaluated a method to separate CO_2 from a mixture of CH_4/CO_2. They tested four different amines, the most commonly used solvents in industrial applications for carbon dioxide capture, namely diethylenetriamine (DETA), diaminoethane (DAE), diethylamine (DEYA), and bis(2-ethylhexyl)-amine (BEHA) as immobilized liquids in a facilitated transport membrane, where poly(vinylidine difluoride) (PVDF) porous membrane was used as an inert support for the amine solution. After testing various parameters influencing transport such as amine concentration, CO_2 partial pressure, and operating temperature it was observed that CO_2 permeance decreased with increasing CO_2 feed pressure, whereas the permeance of CH_4 remained constant for all tested amines. The permeance of CO_2 and the selectivity were in the order DETA > DAE > BEHA = DEYA. This order is related to the number of nitrogen atoms per amine molecule, which can be correlated to loading capacity and consequently to amine reactivity with CO_2. The main conclusion of this study was that it is possible to selectively and efficiently capture carbon dioxide from a gas mixture using amine-based SLM. This method might be a good and cheap alternative to other ways of CO_2 capture.

It can be seen that SLM technology can be applied very efficiently in many different separation processes. Despite the problem of the stability, more and more applications, also on the industrial (or at least semi-industrial) scale are being developed and investigated.

7.3. Separation of stereoisomers

Separation of stereoisomers and particularly enantiomers is a very important issue in separation science due to the relevance of the optically pure materials in the pharmaceutical industry. A detailed description of the possibilities of the realization of this process using SLMs was given earlier in this chapter. Most of the examples regarded SLM application in the fields of organic or supramolecular chemistry, for example, as a helpful tool facilitating an evaluation of the stereoselective binding properties of chiral synthetic receptors (chiral selectors, catalysts, etc.). There are many reports showing the utility of SLM for this purpose [37]. In most cases, the obtained enantioselectivities were low or moderate depending on the chiral carrier and transported compound. However, there are manners to obtain significant enantioseparation and even resolution of racemic mixtures in which SLM is the heart of a separation system (Table 3.8).

The most common manner is incorporation of a chiral species into the membrane phase, but additionally the prepared liquid membrane phase should be introduced and combined with the system, which can provide high mass transfer, for example, by maintaining large membrane surface area. However, to increase the enantioselectivity, it is also important repeatedly to introduce feed phase into contact with the membrane phase. As a result, the enantioseparation can be significantly enhanced. The easiest way to achieve such conditions is to apply hollow fibers as support for a chiral liquid membrane [195-199, 206]. Several carriers were applied as additives in hollow-fiber liquid membranes to enhance stereoselective transport of various small organic compounds. Hydrophobic derivatives or complexes of amino acids *N*-3,5-dinitrobenzoyl-L-alanine octylester [195, 196], copper(II) *N*-dodecyl-L-hydroxyproline [197] were used for the chiral separation of amino acids or organic acids. Another example is peracetylated β-CD [198] applied for resolution of drugs, for example, propranolol. Also, quinidine or quinine derivatives as chiral carriers in SLM separation for efficient resolution of *N*-blocked amino acids were reported by Maximini *et al.* [199] recently. In their studies, a continuous SLM process was developed, in which two HF-liquid membrane modules were used (each consisting of 250 individual polysulfone hollow fibers with a total membrane

Table 3.8 Stereoisomer separation utilizing SLM technique

Compound (enantiomer mixture)	Discriminating agent	Separation principle	References
Amino acids, lactic acid	*N*-3,5-dinitrobenzoyl-L-alanine octylester	Carrier-mediated transport in HFSLM	[195, 196]
Amino acids	Copper(II) *N*-dodecyl-L-hydroxyproline	Carrier-mediated transport in HFSLM	[197]
Propranolol	β-CD	Carrier-mediated transport in HFSLM	[198]
N-blocked amino acids	Quinidine or quinine derivatives	Carrier-mediated transport in HFSLM	[199]
Ofloxacin	Dibenzoyltartaric acid	Carrier-mediated transport in HFSLM	[200]
Amino acid esters	Esterases	Combination of enzymatic process with HFSLM	[201]
2-pentanol 1-phenyloethanol	*Candida antarctica* lipase B (Novozym 435)	Combination of enzymatic process with flat-sheet SLM	[202, 203]
Amino acids	Surfactant-α-chymotrypsin complex immersed in LM	Combination of enzymatic process with flat-sheet SLM	[204, 205]

surface of 0.1 m^2). Two organic liquid phases were incorporated in pores of HFs obtained by dissolution of adamantyl-carbamoyl-11-octadecylthioether quinine (module 1) and adamantyl-carbamoyl-11-octadecylthioether quinidine (module 2) with 1-decanole/pentadecane mixture. Racemic DNB leucine was chosen as a model mixture of enantiomers. After five separation steps, a 99% ee D-enantiomer and a 99% ee L-enantiomer were produced at a transmembrane flux of more than 20 mmol/m^2 h. This process led to resolution of both enantiomers with a degree of purity of 99% and in large quantities (in g). The single separation gave a moderate enantiomeric excess (60% ee). It clearly shows that multiple SLM separation of the same feed phase can give excellent enantioseparation results.

A hollow-fiber liquid membrane was used in a separation of D,L-lactic acid and D,L-alanine resolution [196]. In this case, the enantioselective transport of solutes performed in one module was facilitated by *N*-3,5-dinitrobenzoyl-L-alanine octylester chiral selector, dissolved in toluene. The maximum D,L-lactic acid separation factor achieved was 2.00 and that for the D,L-alanine was 1.75. In both cases, the D-enantiomer flux was preferred. These values correspond to the enantiomeric excess 33.5% ee and 27.2% ee, respectively, and are not as good as in the first example. However, note that in this case, only one separation step took place and feed phase was circulated in the module.

Another example of the use of HF-SLM separation concerns the resolution of racemic ofloxacin [200]. This important drug, a fluoroquinolone antibiotic with one chiral center, was separated in chiral systems by hollow-fiber liquid-supported membrane technology combining with countercurrent fractional extraction. The two chiral solutions contained L-dibenzoyltartaric acid and D-dibenzoyltartaric acid in 1-octanol, and flowed through the lumen side and the shell side of fibers, respectively. The solution which flowed through the lumen side of fibers also contained racemic ofloxacin. The wall of hollow fibers was filled with an aqueous of 0.1 mol/l Na_2HPO_4/H_3PO_4 buffer solution of pH 6.86 containing 2 mmol/l of cetyltrimethylammonium bromide for 48 h. The obtained optical purity for ofloxacin enantiomers was up to 90% when 11 hollow-fiber membrane modules of 22 cm in length in series were used.

We can also obtain an enantiopure compound from its racemic mixture with SLM by using a chemical process that transforms only one of the enantiomers. Together with the SLM separation of either the remaining unreacted enantiomer (or converted into another compound) the enantiomer's resolution was achieved. To ensure a high conversion ratio and transport, both processes should be conducted simultaneously. In most cases, an enzymatic reaction, which gives high enantioselective conversion of the racemic substrate, is used for this purpose. One interesting example was presented by Ricks and coworkers [201]. The resolution of racemic phenylalanine esters with esterases was investigated in relation to the development of a continuous process based on the use of hollow-fiber/liquid membrane (SLM) reactors. They obtained high enantioselectivity when phenylalanine isopropyl ester, whose *R*-enantiomer was converted by subtilisin Carlsberg to *S*-phenylalanine (up to 95% ee), in water at pH 7.5 and temperature 25 °C. The unreacted *R*-enantiomer of ester was removed from the reaction medium via transport of 33% *N*/*N*-diethyldodecanamide/67% dodecane organic liquid membrane through to the feed solution, where it was trapped by charging due to the low pH. The obtained enantioselectivity for ester was around 80% ee.

Another example of the enzymatic reaction-SLM separation approach is kinetic resolution of *rac*-2-pentanol. *S*-enantiomer of this compound is a chiral intermediate in the synthesis of several potential anti-Alzheimer's drugs that inhibit amyloid peptide release and/or its synthesis [202]. In this case, the immobilized enzyme *Candida antarctica* lipase B (Novozym 435) was used in the process of *rac*-2-pentanol transesterification, leading to the desired product *S*-2-pentanol. It was separated from the reaction mixture by SLM based on ionic liquid, namely 1-butyl-3-methylimidazolium tetrafluoroborate supported in nylon membrane. After testing various parameters influencing the reaction/separation process, it was possible to obtain a high enantioseparation factor, in the 60–80 range, between *S*- and *R*-enantiomer depending on the vinyl ester used as a cosubstrate of the reaction. Similarly, it was also feasible to resolve racemic 1-phenyloethanol [203], which suggests that this method can be used for other types of enzymatic/supported liquid membrane kinetic resolution of various substrates, not only those involving transesterification.

For production of enantiopure compounds, we can also use an encapsulating surfactant-enzyme complex immersed in the liquid membrane. This concept was applied by Miyako and coworkers [204, 205] to resolve a racemic mixture of amino acids and ibuprofen, a widely used drug in pain relief. Two encapsulated enzymes were used, one present in the liquid membrane which coverts one of the enantiomers into transportable form and so promotes its transport as a reaction product to the receiving phase. In this phase, the product is converted back to the original enantiomerically enriched substrate that was subjected to resolution. For example, to resolve phenylalanine [205], the encapsulated surfactant-α-chymotrypsin complex was immersed into an isooctane liquid membrane phase, and in the feed solution the racemic mixture of phenylalanine was dissolved together with methanol. The enzyme-catalyzed enantioselective esterification reaction took place at the feed-liquid membrane phase interface. The L-ester of phenylalanine produced in the course of the reaction was transported through the organic phase, while D-phenylalanine remained in the feed phase. The L-ester reached the receiving phase where it was converted into L-amino acid by the same enzyme-surfactant complex. Note that this α-chymotrypsin-facilitated SLM system achieved remarkable ee > 99% for L-phenylalanine at the end of the operation (48 h). A similar system was used (in this case utilizing lipases) for the separation of ibuprofen enantiomers using various ionic liquids as solvents in the membrane phase. The obtained resolution was up to 75% ee and depends on both the type of lipase and ionic liquid [207].

Concluding this short summary on the possibilities of the use of SLM technology for the resolution of stereoisomers, by proper choice of the method and separation conditions, it is feasible to design process that could provide very clean, enantiopure products. Moreover, SLMs can offer a high level of productivity and flexibility compared to analogous industrial-scale chiral technologies such as chiral chromatography or the diastereoisomeric crystallization method.

8. Future Perspectives

In the examples of SLM applications presented above, the possibility to separate high quantities of compounds using small volumes of organic phases shows that this method is still a very attractive choice when an efficient and selective method is necessary. Also, as a result of the development and commercialization of hydrophobic hollow-fiber membrane contactors, SLM might be applied successfully for industrial purposes. This is due to the high membrane surface per unit of volume with satisfactory liquid membrane stability and that HF-SLM technology is easily scalable. Therefore, there is much research to increase the applicability of SLM in the pharmaceutical and chemical industry, metal separation and recovery, wastewater treatment, gas separation, biotechnology, and analytical chemistry. Many of the new, interesting applications of SLM describe the use of the SLM concept. Thus, in the pharmaceutical industry, SLMs can be utilized in the production processes of fine chemicals and even drugs. One such possibility is already described in this chapter, the resolution of drug enantiomers, the production of which is a very challenging task. The other, also involving the new advance of SLM separation in biotechnology, is to use this technique for recovery of various pharmaceutically important compounds from fermentation broths. SLMs were used to separate and recover such compounds as amino acids, for example, L-valine [4, 208] and antibiotics, including β-lactam or cephalosporin antibiotics [209, 210]. Also, other important substances were separated using an SLM system with the aim of using supported liquid membranes in bioindustrial processes, for example, organic acids from fruit juices [211] and fermentation broth [4], sugars [184, 212], or even ethanol after biotechnological processing [213]. These examples show that SLM can also be used in the future for other significant compounds produced using biochemical systems or in the pharmaceutical industry. Obviously, it requires more intensive study but SLM might be a good and cheap alternative to other separation methods.

Metals separation and recovery is always of importance for industry and the environment. The theoretical and fundamental studies on metal transport through SLMs are advanced but still conducted toward implementation of laboratory-scale parameters to industrial applications [191]. This is related to the increased attention to improvement of selectivity and stability of SLM in metals separation. The increase of selectivity can be achieved by design and synthesis of new carriers [214-218]. Gain in stability of the SLM is tested by implementation of various configurations of supported liquid membranes, for example, hybrid and activated composite membranes [49] or the combined supported liquid membrane/strip dispersion process [127, 130, 133, 219]. Additionally, more attention is being paid to removal of toxic metals from wastewater effluents using mostly hollow-fiber-based SLM, which is important in environment protection [188, 218, 220-226].

An important field in which SLM separation is also employed and is still being developed is gas separation. Research is being carried out to remove greenhouse gases [61, 227, 228] and in the separation of hydrocarbons such as propylene or propane from olefin and paraffin hydrocarbon mixtures [229]. Very recently, the SLM system was even proposed as a method to remove heat and moisture [230]. In this case, SLMs were the transfer media to recover heat and moisture from exhaust air due to the high moisture diffusivity in the liquid layer. The SLM involved comprises three layers: two hydrophobic porous skin layers and a hydrophilic porous support layer where a layer of LiCl liquid solution was immobilized in the macro- and micropores as the permselective substance and acted as a moisture-capturing agent. As can be seen, there is an interesting future for SLM application in this field.

Another area in which SLMs have interesting prospects is sample preparation in analytical chemistry. Unlike in other fields where a high active area is required, the new trends are toward miniaturization, which in the case of SLM is represented by the liquid-phase microextraction (LPME) method based on porous hollow fibers [231]. The other issues being addressed are to improve automation and high throughput of the analytical methods, in which the SLM system is used as an analyte enrichment method [167, 232-234]. Additionally, the improvement of selectivity of SLM extraction is one of the main areas of interest; see the immuno-SLM method example described earlier. Several improvements in this methodology have recently been reported [35, 36, 235-237].

As can be seen from this short survey concerning current trends in SLM methodology, there is still much interest from researchers in various fields in the use of this separation method. The use of a liquid membrane immobilized in porous polymer has been applied in very surprising and unexpected ways. Even if some of the applications might look very complicated, the SLM concept itself is very easy to use in practice. This is due to its simplicity and significant flexibility. By simple variation of separation conditions, configuration of liquid membranes, use of specific carriers, or utilization of the advantages of various separation mechanisms, a very high selectivity and subsequent efficiency of the process can be achieved. This would not have been possible without the immense effort to understand, explain, and improve SLM transport. Obviously, like every separation method, SLM has several limitations (among them, one of the most important is its stability), but they can be reduced or in some cases even overcome. As a result, the use of SLMs is, and hopefully will be in the future, a promising and interesting method, which can be a good or even better alternative to other separation techniques.

REFERENCES

1. Scholander, P. F. (1960). Oxygen transport through hemoglobin solutions. Science, 131, 585-90.
2. Wittenberg, J. B. (1966). The molecular mechanism of hemoglobin-facilitated oxygen diffusion. J. Biol. Chem., 241, 104-14.
3. Li, N. N. (1968). Separating hydrocarbons with liquid membrane. US Patent 3,410,794.
4. Eyal, A. M., Bressler, E. (1993). Industrial separation of carboxylic and amino acids by liquid membranes: Applicability, process considerations, and potential advantages. Biotechnol. Bioeng., 41, 287-95.
5. Jönsson, J. Å., Mathiasson, L. (2000). Membrane-based techniques for sample enrichment. J. Chromatogr. A, 902, 205-25.
6. Kocherginsky, N. M., Yang, Q., Seelam, L. (2007). Recent advances in supported liquid membrane technology. Sep. Purif. Technol., 53, 171-77.
7. Noble, R. D., Koval, C. A., Pellegrino, J. J. (1989). Facilitated transport membrane systems. Chem. Eng. Prog., 85, 58-70.
8. Schlosser, Š., Kossaczky, E. (1980). Comparison of pertraction through liquid membranes and double liquid-liquid extraction. J. Membr. Sci., 6, 83-105.
9. Wódzki, R., Szczepanska, G., Szczepanski, P. (2004). Unsteady state pertraction and separation of cations in a liquid membrane system: Simple network and numerical model of competitive M2+/H+ counter-transport. Sep. Purif. Technol., 36, 1-16.
10. Mulder, M. (1996). *Basic principles of membrane technology*. Kluwer Academic Publishers, London.
11. Visser, H. C., Reinhoudt, D. N., De Jong, F. (1994). Carrier-mediated transport through liquid membranes. Chem. Soc. Rev., 23, 75-81.
12. Reid, R. C., Prausnitz, J. M., Sherwood, T. K. (1977). *The properties of gases and liquids*. McGraw-Hill, New York.
13. Cussler, E. L., Aris, R., Bhown, A. (1989). On the limits of facilitated diffusion. J. Membr. Sci., 43, 149-64.
14. Juang, R. S., Lee, S. H., Huang, R. H. (1998). Modeling of amine-facilitated liquid membrane transport of binary organic acids. Sep. Sci. Technol., 33, 2379-95.
15. Park, S.-W., Choi, B.-S., Kim, S.-S., Lee, J.-W. (2006). Facilitated transport of organic acid through a supported liquid membrane with a carrier. Desalination, 193, 304-12.
16. Plucinski, P., Nitsch, W. (1988). Calculation of permeation rates through supported liquid membranes based on the kinetics of liquid-liquid extraction. J. Membr. Sci., 39, 43-59.
17. Mogutov, A. V., Kocheriginsky, N. M. (1993). Macrokinetics of facilitated transport across liquid membranes: "Big carrousel". J. Membr. Sci., 79, 273-83.
18. Kocherginsky, N. M., Yang, Q. (2007). Big Carrousel mechanism of copper removal from ammoniacal wastewater through supported liquid membrane. Sep. Purif. Technol., 54, 104-16.

19. Miliotis, T., Knutsson, M., Jönsson, J. Å., Mathiasson, L. (1996). Ion-pair extraction of aromatic anionic surfactants using the supported liquid membrane technique. Int. J. Environ. Anal. Chem., 64, 35-45.
20. Papantoni, M., Djane, N. K., Ndung'u, K., Jönsson, J. Å., Mathiasson, L. (1995). Trace enrichment of metals using a supported liquid membrane technique. Analyst, 120, 1471-77.
21. Babcock, W. C., Baker, R. W., Lachapelle, E. D., Smith, K. L. (1980). Coupled transport membranes II: The mechanism of uranium transport with a tertiary amine. J. Membr. Sci., 7, 71-87.
22. Bryjak, M., Wieczorek, P., Kafarski, P., Lejczak, B. (1988). Crown-ether mediated transport of amino acids through an immobilized liquid membrane. J. Membr. Sci., 37, 287-91.
23. Wieczorek, P. (1997). Factors influencing the transport of tryptophan hydrochloride through supported liquid membranes containing macrocyclic carriers. J. Membr. Sci., 127, 87-92.
24. Wieczorek, P., Kocorek, A., Bryjak, M., Kafarski, P., Lejczak, B. (1993). Transport of dipeptides and phosphono dipeptides through an immobilized liquid membrane. Stereoselectivity of the process. J. Membr. Sci., 78, 83-91.
25. Dżygiel, P., Wieczorek, P., Mathiasson, L., Jönsson, J. Å. (1998). Enrichment of amino acids by supported liquid membrane extraction using Aliquat 336 as a carrier. Anal. Lett., 31, 1261-74.
26. Wieczorek, P., Jönsson, J. Å., Mathiasson, L. (1997). Concentration of amino acids using supported liquid membranes with di-2-ethylhexyl phosphoric acid as a carrier. Anal. Chim. Acta, 346, 191-97.
27. Jönsson, J. Å., Mathiasson, L. (1999). Liquid membrane extraction in analytical sample preparation. I. Principles. Trends Anal. Chem., 18, 318-25.
28. Lo, T. H., Baird, M. H. I., Hanson, C. (1983). *Handbook of solvent extraction.* John Wiley & Sons, New York.
29. Izatt, R. M. (1997). Review of selective ion separations at BYU using liquid membrane and solid phase extraction procedures. J. Inclusion Phenom. Mol. Recognit. Chem., 29, 197-220.
30. Shen, Y., Gronberg, L., Jönsson, J. Å. (1994). Experimental studies on the enrichment of carboxylic acids with tri-*n*-octylphosphine oxide as extractant in a supported liquid membrane. Anal. Chim. Acta, 292, 31-39.
31. Dżygiel, P., Wieczorek, P. (2001). Supported liquid membrane extraction of glyphosate metabolites. J. Sep. Sci., 24, 561-66.
32. Dżygiel, P., Wieczorek, P. (2000). Extraction of glyphosate by a supported liquid membrane technique. J. Chromatogr. A, 889, 93-98.
33. Djane, N. K., Ndung'u, K., Johnsson, C., Sartz, H., Tornstrom, T., Mathiasson, L. (1999). Chromium speciation in natural waters using serially connected supported liquid membranes. Talanta, 48, 1121-32.
34. Pichon, V., Bouzige, M., Hennion, M. C. (1998). New trends in environmental trace-analysis of organic pollutants: Class-selective immunoextraction and clean-up in one step using immunosorbents. Anal. Chim. Acta, 376, 21-35.
35. Thordarson, E., Jönsson, J. Å., Emnéus, J. (2000). Immunologic trapping in supported liquid membrane extraction. Anal. Chem., 72, 5280-84.
36. Tudorache, M., Rak, M., Wieczorek, P. P., Jönsson, J. Å., Emnéus, J. (2004). Immuno-SLM—A combined sample handling and analytical technique. J. Immunol. Methods, 284, 107-18.
37. Dżygiel, P., Wieczorek, P. (2002). Stereoselective transport of amino acids and peptides through liquid membranes. Chem. Pap., 56, 24-31.

38. Bryjak, M., Kozłowski, J., Wieczorek, P., Kafarski, P. (1993). Enantioselective transport of amino acid through supported chiral liquid membranes. J. Membr. Sci., 85, 221-8.
39. Shinbo, T., Yamaguchi, T., Yanagishita, H., Sakaki, K., Kitamoto, D., Sugiura, M. (1993). Supported liquid membranes for enantioselective transport of amino acid mediated by chiral crown ether—Effect of membrane solvent on transport rate and membrane stability. J. Membr. Sci., 84, 241-8.
40. Pietraszkiewicz, M., Kozbiał, M., Pietraszkiewicz, O. (1997). Chiral recognition of amino acids by diaza crown ethers and crowns incorporating a mannopyranoside unit, immobilized in a supported liquid membrane. Enantiomer, 2, 319-25.
41. Scrimin, P., Tonellato, U., Zanta, N. (1988). Cu(II) mediated selective transport of [alpha]-amino acids across a bulk liquid membrane using a chiral lipophilic ligand as a carrier. Tetrahedron Lett., 29, 4967-70.
42. Sessler, J. L., Andrievsky, A. (1998). Efficient transport of aromatic amino acids by sapphyrin-lasalocid conjugates. Chem. Eur. J., 4, 159-67.
43. Dżygiel, P., Wieczorek, P., Kafarski, P. (2003). Supported liquid membrane separation of amine and amino acid derivatives with chiral esters of phosphoric acids as carriers. J. Sep. Sci., 26, 1050-6.
44. Pirkle, W. H., Bowen, W. E. (1994). Preparative separation of enantiomers using hollow-fiber membrane technology. Tetrahedron Asymmetry, 5, 773-6.
45. Miyake, H., Yamashita, T., Kojima, Y., Tsukube, H. (1995). Enantioselective transport of amino acid ester salts by macrocyclic pseudopeptides containing *N,N′*-ethylene-bridged-dipeptide units. Tetrahedron Lett., 36, 7669-72.
46. Caro, J., Noack, M., Kölsch, P., Schäfer, R. (2000). Zeolite membranes—State of their development and perspective. Micropor. Mesopor. Mater., 38, 3-24.
47. Chen, Y., Xiangli, F., Jin, W., Xu, N. (2007). Organic-inorganic composite membranes prepared by self-assembly of polyelectrolyte multilayers on macroporous ceramic supports. J. Membr. Sci., 302, 78-86.
48. Cot, L., Ayral, A., Durand, J., Guizard, C., Hovnanian, N., Julbe, A., Larbot, A. (2000). Inorganic membranes and solid state sciences. Solid State Sci., 2, 313-34.
49. Resina, M., Fontàs, C., Palet, C., Muñoz, M. (2008). Comparative study of hybrid and activated composite membranes containing Aliquat 336 for the transport of Pt(IV). J. Membr. Sci., 311, 235-42.
50. McCleskey, T. M., Ehler, D. S., Young, J. S., Pesiri, G. D., Jarvinen, G. D., Sauer, N. N. (2002). Asymmetric membranes with modified gold films as selective gates for metal ion separations. J. Membr. Sci., 210, 273-8.
51. Krull, F. F., Medved, M., Melin, T. (2007). Novel supported ionic liquid membranes for simultaneous homogeneously catalyzed reaction and vapor separation. Chem. Eng. Sci., 62, 5579-85.
52. Krieg, H. M., Breytenbach, J. C., Keizer, K. (2000). Chiral resolution by β-cyclodextrin polymer-impregnated ceramic membranes. J. Membr. Sci., 164, 177-85.
53. Han, X., Armstrong, D. W. (2007). Ionic liquids in separations. Acc. Chem. Res., 40, 1079-86.
54. Dzyuba, S. V., Bartsch, R. A. (2002). Influence of structural variations in 1-alkyl (aralkyl)-3-methylimidazolium hexafluorophosphates and bis(trifluoromethylsulfonyl) imides on physical properties of the ionic liquids. ChemPhysChem, 3, 161-6.
55. Holbrey, J. D., Seddon, K. R. (1999). The phase behaviour of 1-alkyl-3-methylimidazolium tetrafluoroborates; ionic liquids and ionic liquid crystals. J. Chem. Soc., Dalton Trans., 2133-40.
56. Baker, G. A., Baker, S. N., Pandey, S., Bright, F. V. (2005). An analytical view of ionic liquids. Analyst, 130, 800-8.

57. Huang, H.-J., Ramaswamy, S., Tschirner, U. W., Ramarao, B. V. (2008). A review of separation technologies in current and future biorefineries. Sep. Purif. Technol., 62, 1-21.
58. Liu, J.-F., Jiang, G.-B., Jönsson, J. Å. (2005). Application of ionic liquids in analytical chemistry. Trends Anal. Chem., 24, 20-27.
59. Berthod, A., Ruiz-Ángel, M. J., Carda-Broch, S. (2008). Ionic liquids in separation techniques. J. Chromatogr. A, 1184, 6-18.
60. Fortunato, R., Afonso, C. A. M., Reis, M. A. M., Crespo, J. G. (2004). Supported liquid membranes using ionic liquids: Study of stability and transport mechanisms. J. Membr. Sci., 242, 197-209.
61. Hanioka, S., Maruyama, T., Sotani, T., Teramoto, M., Matsuyama, H., Nakashima, K., Hanaki, M., Kubota, F., Goto, M. (2008). CO_2 separation facilitated by task-specific ionic liquids using a supported liquid membrane. J. Membr. Sci., 314, 1-4.
62. Jiang, Y. Y., Zhou, Z., Jiao, Z., Li, L., Wu, Y. T., Zhang, Z. B. (2007). SO_2 gas separation using supported ionic liquid membranes. J. Phys. Chem. B, 111, 5058-61.
63. Matsumoto, M., Ueba, K., Kondo, K. (2006). Separation of benzene/cyclohexane mixture through supported liquid membranes with an ionic liquid. Solvent Extr. Res. Dev., 13, 51-59.
64. Matsumoto, M., Inomoto, Y., Kondo, K. (2005). Selective separation of aromatic hydrocarbons through supported liquid membranes based on ionic liquids. J. Membr. Sci., 246, 77-81.
65. Matsumoto, M., Mikami, M., Kondo, K. (2007). Selective permeation of organic sulfur and nitrogen compounds in model mixtures of petroleum fraction through supported ionic liquid membranes. J. Chem. Eng. Jpn., 40, 1007-10.
66. Branco, L. C., Crespo, J. G., Afonso, C. A. M. (2002). Studies on the selective transport of organic compounds by using ionic liquids as novel supported liquid membranes. Chem. Eur. J., 8, 3865-71.
67. Branco, L. C., Crespo, J. G., Afonso, C. A. M. (2002). Highly selective transport of organic compounds by using supported liquid membranes based on ionic liquids. Angew. Chem. Int. Ed., 41, 2771-3.
68. Jönsson, J. Å., Mathiasson, L. (1992). Supported liquid membrane techniques for sample preparation and enrichment in environmental and biological analysis. Trends Anal. Chem., 11, 106-114.
69. Luque de Castro, M., Priego Capote, F. (2008). Miniaturisation of analytical steps: Necessity and snobbism. Anal. Bioanal. Chem., 390, 67-9.
70. Raich-Montiu, J., Krogh, K. A., Granados, M., Jönsson, J. Ĺ., Halling-Sřrensen, B. (2008). Determination of ivermectin and transformation products in environmental waters using hollow fiber-supported liquid membrane extraction and liquid chromatography-mass spectrometry/mass spectrometry. J. Chromatogr. A, 1187, 275-80.
71. Piriyapittaya, M., Jayanta, S., Mitra, S., Leepipatpiboon, N. (2008). Micro-scale membrane extraction of glyphosate and aminomethylphosphonic acid in water followed by high-performance liquid chromatography and post-column derivatization with fluorescence detector. J. Chromatogr. A, 1189, 483-92.
72. Lezamiz, J., Barri, T., Jönsson, J. Å., Skog, K. (2008). A simplified hollow-fiber supported liquid membrane extraction method for quantification of 2-amino-1-methyl-6-phenylimidazo[4,5-b]pyridine (PhIP) in urine and plasma samples. Anal. Bioanal. Chem., 390, 689-96.
73. Dziarkowska, K., Jönsson, J. A., Wieczorek, P. P. (2008). Single hollow fiber SLM extraction of polyamines followed by tosyl chloride derivatization and HPLC determination. Anal. Chim. Acta, 606, 184-93.

74. Zorita, S., Mårtensson, L., Mathiasson, L. (2007). Hollow-fiber supported liquid membrane extraction for determination of fluoxetine and norfluoxetine concentration at ultra trace level in sewage samples. J. Sep. Sci., 30, 2513-21.
75. Zhao, G., Liu, J.-F., Nyman, M., Jönsson, J. Ĺ. (2007). Determination of short-chain fatty acids in serum by hollow fiber supported liquid membrane extraction coupled with gas chromatography. J. Chromatogr. B, 846, 202-8.
76. Liu, J.-F., Toräng, L., Mayer, P., Jönsson, J. Å. (2007). Passive extraction and clean-up of phenoxy acid herbicides in samples from a groundwater plume using hollow fiber supported liquid membranes. J. Chromatogr. A, 1160, 56-63.
77. Vora-adisak, N., Varanusupakul, P. (2006). A simple supported liquid hollow fiber membrane microextraction for sample preparation of trihalomethanes in water samples. J. Chromatogr. A, 1121, 236-41.
78. Bårdstu, K. F., Ho, T. S., Rasmussen, K. E., Pedersen-Bjergaard, S., Jönsson, J. Å. (2007). Supported liquid membranes in hollow fiber liquid-phase microextraction (LPME)—Practical considerations in the three-phase mode. J. Sep. Sci., 30, 1364-70.
79. Pálmarsdóttir, S., Thordarson, E., Edholm, L. E., Jönsson, J. Å., Mathiasson, L. (1997). Miniaturized supported liquid membrane device for selective on-line enrichment of basic drugs in plasma combined with capillary zone electrophoresis. Anal. Chem., 69, 1732-7.
80. Kemperman, A. J. B., Bargeman, D., Van Den Boomgaard, T., Strathmann, H. (1996). Stability of supported liquid membranes: State of the art. Sep. Sci. Technol., 31, 2733-62.
81. Danesi, P. R. (1984). Separation of metal species by supported liquid membranes. Sep. Sci. Technol., 19, 857-94.
82. Takeuchi, H., Takahashi, K., Goto, W. (1987). Some observations on the stability of supported liquid membranes. J. Membr. Sci., 34, 19-31.
83. Neplenbroek, A. M., Bargeman, D., Smolders, C. A. (1992). Supported liquid membranes: Instability effects. J. Membr. Sci., 67, 121-32.
84. Zha, F. F., Fane, A. G., Fell, C. J. D. (1995). Instability mechanisms of supported liquid membranes in phenol transport process. J. Membr. Sci., 107, 59-74.
85. Szpakowska, M., Nagy, O. B. (1997). Stability of supported liquid membranes containing Acorga P-50 as carrier. J. Membr. Sci., 129, 251-61.
86. Wieczorek, P., Tomaszewska, M. (1997). Transport of amino acids through liquid membranes supported on novel poly(vinylidenefluoride) porous flat-sheet matrix. Solvent Extr. Ion. Exc., 15, 879-94.
87. Lamb, J. D., Bruening, R. L., Izatt, R. M., Hirashima, Y., Tse, P.-K., Christensen, J. J. (1988). Characterization of a supported liquid membrane for macrocycle-mediated selective cation transport. J. Membr. Sci., 37, 13-26.
88. Dozol, J. F., Casas, J., Sastre, A. (1993). Stability of flat sheet supported liquid membranes in the transport of radionuclides from reprocessing concentrate solutions. J. Membr. Sci., 82, 237-46.
89. Chiarizia, R. (1991). Stability of supported liquid membranes containing long-chain aliphatic amines as carriers. J. Membr. Sci., 55, 65-77.
90. Zha, F. F., Fane, A. G., Fell, C. J. D. (1995). Effect of surface tension gradients on stability of supported liquid membranes. J. Membr. Sci., 107, 75-86.
91. Zha, F. F., Coster, H. G. L., Fane, A. G. (1994). A study of stability of supported liquid membranes by impedance spectroscopy. J. Membr. Sci., 93, 255-71.
92. Yang, X. J., Fane, A. G. (1999). Performance and stability of supported liquid membranes using LIX 984N for copper transport. J. Membr. Sci., 156, 251-63.
93. Saito, T. (1992). Deterioration of liquid membrane and its improvement in permeation transport of Zn(II) ion through a supported liquid membrane containing a Bathocuproine. Sep. Sci. Technol., 27, 1-9.

94. Deblay, P., Delepine, S., Minier, M., Renon, H. (1991). Selection of organic phases for optimal stability and efficiency of flat-sheet supported liquid membranes. Sep. Sci. Technol., 26, 97-116.
95. Wienk, M. M., Stolwijk, T. B., Sudholter, E. J. R., Reinhoudt, D. N. (1990). Stabilization of crown ether containing supported liquid membranes. J. Am. Chem. Soc., 112, 797-801.
96. Zha, F. F., Fane, A. G., Fell, C. J. D., Schofield, R. W. (1992). Critical displacement pressure of a supported liquid membrane. J. Membr. Sci., 75, 69-80.
97. Takeuchi, H., Takahashi, K., Nakano, M. (1990). Separation of heavy metals from aqueous solutions by hollow-fiber type supported liquid membranes in a continuous regenerating mode. Water Treatment, 5, 222-36.
98. Wijers, M. C., Jin, M., Wessling, M., Strathmann, H. (1998). Supported liquid membranes modification with sulphonated poly(ether ether ketone). Permeability, selectivity and stability. J. Membr. Sci., 147, 117-30.
99. Wang, Y., Thio, Y. S., Doyle, F. M. (1998). Formation of semi-permeable polyamide skin layers on the surface of supported liquid membranes. J. Membr. Sci., 147, 109-16.
100. Wang, Y., Doyle, F. M. (1999). Formation of epoxy skin layers on the surface of supported liquid membranes containing polyamines. J. Membr. Sci., 159, 167-75.
101. Yang, X. J., Fane, A. G., Bi, J., Griesser, H. J. (2000). Stabilization of supported liquid membranes by plasma polymerization surface coating. J. Membr. Sci., 168, 29-37.
102. Dastgir, M. G., Peeva, L. G., Livingston, A. G. (2005). The performance of composite supported polymeric liquid membranes in the Membrane Aromatic Recovery System (MARS). Chem. Eng. Sci., 60, 7034-44.
103. Dastgir, M. G., Peeva, L. G., Livingston, A. G., Morley, T. A., Steinke, J. H. G. (2005). The synthesis of polypropylene glycol based polyethers and their use in membranes for the membrane aromatic recovery system (MARS). J. Membr. Sci., 261, 87-97.
104. Nghiem, L. D., Mornane, P., Potter, I. D., Perera, J. M., Cattrall, R. W., Kolev, S. D. (2006). Extraction and transport of metal ions and small organic compounds using polymer inclusion membranes (PIMs). J. Membr. Sci., 281, 7-41.
105. Kim, J. S., Kim, S. K., Ko, J. W., Kim, E. T., Yu, S. H., Cho, M. H., Kwon, S. G., Lee, E. H. (2000). Selective transport of cesium ion in polymeric CTA membrane containing calixcrown ethers. Talanta, 52, 1143-8.
106. Scindia, Y. M., Pandey, A. K., Reddy, A. V. R. (2005). Coupled-diffusion transport of Cr(VI) across anion-exchange membranes prepared by physical and chemical immobilization methods. J. Membr. Sci., 249, 143-52.
107. Molinari, R., Argurio, P., Pirillo, F. (2005). Comparison between stagnant sandwich and supported liquid membranes in copper(II) removal from aqueous solutions: Flux, stability and model elaboration. J. Membr. Sci., 256, 158-68.
108. Molinari, R., Argurio, P., Poeiro, T., Caruso, A. (2006). Stagnant sandwich and supported liquid membrane systems for removal of pharmaceuticals in water. Desalination, 199, 529-31.
109. Molinari, R., Caruso, A., Argurio, P., Poerio, T. (2006). Diclofenac transport through stagnant sandwich and supported liquid membrane systems. Ind. Eng. Chem. Res., 45, 9115-21.
110. Teramoto, M., Tohno, N., Ohnishi, N., Matsuyama, H. (1989). Development of a spiral-type flowing liquid membrane module with high stability and its application to the recovery of chromium and zinc. Sep. Sci. Technol., 24, 981-99.
111. Schlosser, Š., Rothová, I., Frianová, H. (1993). Hollow-fiber pertractor with bulk liquid membrane. J. Membr. Sci., 80, 99-106.
112. Kedem, O., Bromberg, L. (1993). Ion exchange membranes in extraction processes. J. Membr. Sci., 78, 255-69.

113. Wódzki, R., Swiatkowski, M. (1995). Recovery and concentration of metal ions. Part 2. Multimembrane hybrid system. Sep. Sci. Technol., 30, 2763-78.
114. Kislik, V. S., Eyal, A. M. (1996). Hybrid liquid membrane (HLM) system in separation technologies. J. Membr. Sci., 111, 259-72.
115. Kislik, V. S., Eyal, A. M. (1996). Hybrid liquid membrane (HLM) and supported liquid membrane (SLM) based transport of titanium (IV). J. Membr. Sci., 111, 273-81.
116. Eyal, A., Kislik, V. (1999). Aqueous hybrid liquid membrane: A novel system for separation of solutes using water-soluble polymers as carriers. J. Membr. Sci., 161, 207-21.
117. Wódzki, R., Swiatkowski, M. (1996). Recovery and concentration of metal ions. Concentration and temperature effects in multimembrane hybrid systems. Sep. Sci. Technol., 31, 1541-53.
118. Kislik, V., Eyal, A. (2000). Aqueous hybrid liquid membrane process for metal separation. Part I. A model for transport kinetics and its experimental verification. J. Membr. Sci., 169, 119-32.
119. Kislik, V., Eyal, A. (2000). Aqueous hybrid liquid membrane process for metal separation. Part II. Selectivity of metals separation from wet-process phosphoric acid. J. Membr. Sci., 169, 133-16.
120. Gęga, J., Walkowiak, W., Gajda, B. (2001). Separation of Co(II) and Ni(II) ions by supported and hybrid liquid membranes. Sep. Purif. Technol., 22-23, 551-8.
121. Wódzki, R., Szczepanski, P. (2002). Integrated hybrid membrane systems—Membrane extraction and pertraction coupled to a pervaporation process. J. Membr. Sci., 197, 297-308.
122. Wódzki, R., Swiatkowski, M., Lapienis, G. (2005). Transport and separation properties of poly(oxypropylene) bisphosphates as macroionophores of alkali, alkaline-earth and transient metal cations in a hybrid liquid membrane system. React. Funct. Polym., 62, 195-208.
123. Wódzki, R., Szczepanski, P. (2005). Simultaneous recovery and separation of Zn^{2+} and Cu^{2+} in hybrid membrane systems. Sep. Purif. Technol., 41, 289-97.
124. Ho, W. S., Wang, B., Neumuller, T. E., Roller, J. (2001). Supported liquid membranes for removal and recovery of metals from waste waters and process streams. Environ. Prog., 20, 117-21.
125. Ho, W. S. W. (2001). Combined supported liquid membrane/strip dispersion process for the removal and recovery of radionuclides and metals. US Patent 6,328,782.
126. Ho, W. S. W., Wang, B. (2002). Strontium removal by new alkylphenylphosphonic acids in supported liquid membranes with strip dispersion. Ind. Eng. Chem. Res., 41, 381-8.
127. Gu, S., Yu, Y., He, D., Ma, M. (2006). Comparison of transport and separation of Cd (II) between strip dispersion hybrid liquid membrane (SDHLM) and supported liquid membrane (SLM) using tri-*n*-octylamine as carrier. Sep. Purif. Technol., 51, 277-84.
128. He, D., Gu, S., Ma, M. (2007). Simultaneous removal and recovery of cadmium (II) and CN^- from simulated electroplating rinse wastewater by a strip dispersion hybrid liquid membrane (SDHLM) containing double carrier. J. Membr. Sci., 305, 36-47.
129. Ho, W. S., Lee, L. T. C., Lin, K. J. (1976). Membrane hydro-metallurgical extraction process. US Patent 3,957,504.
130. Ho, W. S. W. (2003). Removal and recovery of metals and other materials by supported liquid membranes with strip dispersion. Ann. NY Acad. Sci., 984, 97-122.
131. Ho, W. S. W. (2004). Combined supported liquid membrane/strip dispersion process for the removal and recovery of radionuclides. US Patent 6,696,589.
132. Ho, W. S. W., Poddar, T. K. (2001). New membrane technology for removal and recovery of chromium from waste waters. Environ. Prog., 20, 44-52.

133. Ho, W. S. W., Poddar, T. K., Neumuller, T. E. (2002). Removal and recovery of copper and zinc by supported liquid membranes with strip dispersion. J. Chin. Inst. Chem. Eng., 33, 67-76.
134. Roy, S. C., Sonawane, J. V., Rathore, N. S., Pabby, A. K., Janardan, P., Changrani, R. D., Dey, P. K., Bharadwaj, S. R. (2008). Pseudo-emulsion based hollow fiber strip dispersion technique (PEHFSD): Optimization, modelling and application of PEHFSD for recovery of u(vi) from process effluent. Sep. Sci. Technol., 43, 3305-32.
135. Sonawane, J. V., Pabby, A. K., Sastre, A. M. (2007). Au(I) extraction by LIX-79/*n*-heptane using the pseudo-emulsion-based hollow-fiber strip dispersion (PEHFSD) technique. J. Membr. Sci., 300, 147-55.
136. Sonawane, J. V., Pabby, A. K., Sastre, A. M. (2008). Pseudo-emulsion based hollow fiber strip dispersion: A novel methodology for gold recovery. AIChE J., 54, 453-63.
137. Ho, W. S. W. (2002). Combined supported liquid membrane/strip dispersion process for the removal and recovery of penicillin and organic acids. US Patent 6,433,163.
138. Smith, R. M. (2003). Before the injection—Modern methods of sample preparation for separation techniques. J. Chromatogr. A, 1000, 3-27.
139. Ridgway, K., Lalljie, S. P. D., Smith, R. M. (2007). Sample preparation techniques for the determination of trace residues and contaminants in foods. J. Chromatogr. A, 1153, 36-53.
140. Audunsson, G. (1986). Aqueous/aqueous extraction by means of a liquid membrane for sample cleanup and preconcentration of amines in a flow system. Anal. Chem., 58, 2714-23.
141. Chimuka, L., Nindi, M. M., Jönsson, J. Å. (1997). Supported liquid membrane enrichment studies of natural water samples applied to liquid chromatographic determination of triazine herbicides. Intern. J. Environ. Anal. Chem., 68, 429-45.
142. Khrolenko, M., Dżygiel, P., Wieczorek, P. (2002). Combination of supported liquid membrane and solid-phase extraction for sample pretreatment of triazine herbicides in juice prior to capillary electrophoresis determination. J. Chromatogr. A, 975, 219-27.
143. Knutsson, M. (1992). Supported liquid membrane technique for time-integrating field sampling of acidic herbicides at sub parts per billion level in natural waters. J. Agr. Food Chem., 40, 2413-7.
144. Msagati, T. A. M., Nindi, M. M. (2005). Application of supported liquid membranes in the multi-residue extraction of aminoglycoside antibiotics in milk and urine. Bull. Chem. Soc. Jpn., 78, 2135-41.
145. Trocewicz, J. (2001). Sample preparation of amphetamine and methamphetamine by means of supported liquid membrane technique for high-performance liquid chromatography analysis. J. Sep. Sci., 24, 587-52.
146. Nindi, M. M., Msagati, T. M., Masunga, P. (2002). Supported liquid membrane extraction of 17β-estradiol and its metabolites in a variety of biological matrices. Afr. J. Biotechnol., 5, 1827-35.
147. Pálmarsdóttir, S., Lindegård, B., Deininger, P., Edholm, L. E., Mathiasson, L., Jönsson, J. Å. (1995). Supported liquid membrane technique for selective sample workup of basic drugs in plasma prior to capillary zone electrophoresis. J. Capillary Electrop., 2, 185-9.
148. Djane, N. K., Bergdahl, I. A., Ndung'u, K., Schutz, A., Johansson, G., Mathiasson, L. (1997). Supported liquid membrane enrichment combined with atomic absorption spectrometry for the determination of lead in urine. Analyst, 122, 1073-7.
149. Drapała, A., Jönsson, J. Å., Wieczorek, P. (2005). Peptides analysis in blood plasma using on-line system of supported liquid membrane and high-performance liquid chromatography. Anal. Chim. Acta, 553, 9-14.
150. Khrolenko, M. V., Wieczorek, P. P. (2005). Determination of glyphosate and its metabolite aminomethylphosphonic acid in fruit juices using supported-liquid

membrane preconcentration method with high-performance liquid chromatography and UV detection after derivatization with *p*-toluenesulphonyl chloride. J. Chromatogr. A, 1093, 111-7.
151. Jönsson, J. Å., Mathiasson, L. (1999). Liquid membrane extraction in analytical sample preparation. II. Applications. Trends Anal. Chem., 18, 325-34.
152. Jönsson, J. Å., Mathiasson, L. (2000). Membrane extraction techniques in bioanalysis. Chromatographia, 52, S8-11.
153. Jönsson, J. Å., Mathiasson, L. (2001). Membrane extraction techniques for sample preparation. Adv. Chromatogr., 41, 53-91.
154. Jönsson, J. Å., Mathiasson, L. (2001). Membrane extraction in analytical chemistry. J. Sep. Sci., 24, 495-507.
155. Nilve, G., Stebbins, R. (1991). Automated sample preparation using supported liquid membranes for liquid chromatographic determination of sulfonylurea herbicides. Chromatographia, 32, 269-77.
156. Lindegård, B., Jönsson, J. Å., Mathiasson, L. (1992). Liquid membrane work-up of blood plasma samples applied to gas chromatographic determination of aliphatic amines. J. Chromatogr.: Biomed. Appl., 573, 191-200.
157. Jönsson, J. Å., Mathiasson, L., Lindegard, B., Trocewicz, J., Olsson, A. M. (1994). Automated system for the trace analysis of organic compounds with supported liquid membranes for sample enrichment. J. Chromatogr. A, 665, 259-68.
158. Nilve, G., Knutsson, M., Jönsson, J. Å. (1994). Liquid chromatographic determination of sulfonylurea herbicides in natural waters after automated sample pretreatment using supported liquid membranes. J. Chromatogr. A, 688, 75-82.
159. Knutsson, M., Lundh, J., Mathiasson, L., Jönsson, J. Å., Sundin, P. (1996). Supported liquid membranes for the extraction of phenolic acids from circulating nutrient solutions. Anal. Lett., 29, 1619-35.
160. Knutsson, M., Mathiasson, L., Jönsson, J. Å. (1996). Supported liquid membrane work-up in combination with liquid chromatography and electrochemical detection for the determination of chlorinated phenols in natural water samples. Chromatographia, 42, 165-70.
161. Knutsson, M., Nilve, G., Mathiasson, L., Jönsson, J. Å. (1996). Supported liquid membranes for sampling and sample preparation of pesticides in water. J. Chromatogr. A, 754, 197-205.
162. Trocewicz, J. (1996). Determination of herbicides in surface water by means of a supported liquid membrane technique and high-performance liquid chromatography. J. Chromatogr. A, 725, 121-7.
163. Megersa, N., Jönsson, J. Å. (1998). Trace enrichment and sample preparation of alkylthio-*s*-triazine herbicides in environmental waters using a supported liquid membrane technique in combination with high-performance liquid chromatography. Analyst, 123, 225-31.
164. Megersa, N., Solomon, T., Jönsson, J. Å. (1999). Supported liquid membrane extraction for sample work-up and preconcentration of methoxy-*s*-triazine herbicides in a flow system. J. Chromatogr. A, 830, 203-10.
165. Megersa, N., Solomon, T., Chandravanshi, B. S., Jönsson, J. Å. (2000). Sample clean-up, enrichment and determination of *S*-triazine herbicides from southern Ethiopian lakes using supported liquid membrane extraction. Bull. Chem. Soc. Ethiopia, 14, 9-24.
166. Megersa, N., Chimuka, L., Solomon, T., Jönsson, J. Å. (2001). Automated liquid membrane extraction and trace enrichment of triazine herbicides and their metabolites in environmental and biological samples. J. Sep. Sci., 24, 567-76.
167. Jönsson, J. Å., Andersson, M., Melander, C., Norberg, J., Thordarson, E., Mathiasson, L. (2000). Automated liquid membrane extraction for high-performance liquid chromatography of Ropivacaine metabolites in urine. J. Chromatogr. A, 870, 151-7.

168. Msagati, T. A. M., Nindi, M. M. (2001). Determination of benzimidazole anthelmintic compounds by supported liquid membrane extraction and liquid chromatography. J. Sep. Sci., 24, 606-14.
169. Trocewicz, J., Suprynowicz, Z., Markowicz, J. (1996). Determination of diprivan in urine by a supported liquid membrane technique and liquid chromatography-electrochemical detection. J. Chromatogr. B, 685, 129-34.
170. Dawidowicz, A. L., Kalityński, R., Trocewicz, J., Nestorowicz, A., Fijałkowska, A., Trela-Stachurska, K. (2002). Investigation of propofol renal elimination by HPLC using supported liquid membrane procedure for sample preparation. Biomed. Chromatogr., 16, 455-8.
171. Msagati, T. A. M., Nindi, M. M. (2005). Supported liquid membrane extraction of anabolic androgenic compounds in biological matrices and detection by LC-ESI-MS. South Afr. J. Chem., 58, 67-73.
172. Trocewicz, J. (2004). Urine sample preparation of tricyclic antidepressants by means of a supported liquid membrane technique for high-performance liquid chromatographic analysis. J. Chromatogr. B, 801, 213-20.
173. Lindegård, B., Björk, H., Jönsson, J. Å., Mathiasson, L., Olsson, A. M. (1994). Automated column liquid chromatographic determination of a basic drug in blood plasma using the supported liquid membrane technique for sample pretreatment. Anal. Chem., 66, 4490-7.
174. Shen, Y., Mathiasson, L., Jönsson, J. Å. (1998). Automated capillary GC determination of local anaesthetics in plasma samples with supported liquid membranes for sample preparation. J. Microcolumn Sep., 10, 107-13.
175. Djane, N. K., Ndung'u, K., Malcus, F., Johansson, G., Mathiasson, L. (1997). Supported liquid membrane enrichment using an organophosphorus extractant for analytical trace metal determinations in river waters. Fresenius J. Anal. Chem., 358, 822-7.
176. Djane, N. K., Armalis, S., Ndung'u, K., Johansson, G., Mathiasson, L. (1998). Supported liquid membrane coupled on-line to potentiometric stripping analysis at a mercury-coated reticulated vitreous carbon electrode for trace metal determinations in urine. Analyst, 123, 393-6.
177. Ndung'u, K., Djane, N. K., Mathiasson, L. (1998). Determination of trace metal ions by ion-pair chromatography after enrichment using supported liquid membrane. J. Chromatogr. A, 826, 103-108.
178. Wieczorek, P., Jönsson, J. Å., Mathiasson, L. (1997). Extraction of dansylated amino acids using the supported liquid membrane technique. Anal. Chim. Acta, 337, 183-9.
179. Dziarkowska, K., Koprek, K., Wieczorek, P. P. (2008). Studies of polyamines transport through liquid membranes with D2EHPA as a carrier. J. Sep. Sci., 31, 372-9.
180. Poliwoda, A., Ilczuk, N., Wieczorek, P. P. (2007). Transport mechanism of peptides through supported liquid membranes. Sep. Purif. Technol., 57, 444-9.
181. Drapała, A., Wieczorek, P. (2004). Facilitated SLM extraction of peptides with D2EHPA as a carrier. Desalination, 163, 47-53.
182. Drapala, A., Wieczorek, P. (2002). Extraction of short peptides using supported liquid membranes. Desalination, 148, 235-9.
183. Drapała, A., Dżygiel, P., Jönsson, J. Å., Wieczorek, P. (2001). Supported liquid membrane extraction of peptides. Acta Biochim. Pol., 48, 1113-6.
184. Di Luccio, M., Smith, B. D., Kida, T., Alves, T. L. M., Borges, C. P. (2002). Evaluation of flat sheet and hollow fiber supported liquid membranes for fructose pertraction from a mixture of sugars. Desalination, 148, 213-20.
185. Venkateswaran, P., Palanivelu, K. (2006). Recovery of phenol from aqueous solution by supported liquid membrane using vegetable oils as liquid membrane. J. Hazard. Mater., 131, 146-52.

186. Ashraf, M. W., Mian, A. (2006). Selective separation and preconcentration studies of chromium(VI) with Alamine 336 supported liquid membrane. Toxicol. Environ. Chem., 88, 187-96.
187. Kocherginsky, N. M., Grishchenko, A. (2003). Method for metal recovery from aqueous solutions. US Patent 6,521,117.
188. Rathore, N. S., Sonawane, J. V., Gupta, S. K., Pabby, A. K., Venugopalan, A. K., Changrani, R. D., Dey, P. K. (2004). Separation of uranium and plutonium from aqueous acidic wastes using a hollow fiber supported liquid membrane. Sep. Sci. Technol., 39, 1295-319.
189. Al Marzouqi, M. H., Abdulkarim, M. A., Marzouk, S. A., El-Naas, M. H., Hasanain, H. M. (2005). Facilitated transport of CO_2 through immobilized liquid membrane. Ind. Eng. Chem. Res., 44, 9273-8.
190. Sud, D., Mahajan, G., Kaur, M. P. (2008). Agricultural waste material as potential adsorbent for sequestering heavy metal ions from aqueous solutions—A review. Bioresource Technol., 99, 6017-27.
191. De Gyves, J., De San Miguel, E. R. (1999). Metal ion separations by supported liquid membranes. Ind. Eng. Chem. Res., 38, 2182-202.
192. Yang, Q., Kocherginsky, N. M. (2006). Copper recovery and spent ammoniacal etchant regeneration based on hollow fiber supported liquid membrane technology: From bench-scale to pilot-scale tests. J. Membr. Sci., 286, 301-9.
193. Stott, P. A., Stone, D. A., Allen, M. R. (2004). Human contribution to the European heatwave of 2003. Nature, 432, 610-4.
194. International Energy Agency (2002). World Energy Outlook.
195. Hadik, P., Kotsis, L., Eniszné-Bódogh, M., Szabó, L. P., Nagy, E. (2005). Lactic acid enantioseparation by means of porous ceramic disc and hollow fiber organic membrane. Sep. Purif. Technol., 41, 299-304.
196. Hadik, P., Szabó, L. P., Nagy, E. (2002). D,L-Lactic acid and D,L-alanine enantioseparation by membrane process. Desalination, 148, 193-8.
197. Huang, D., Huang, K., Chen, S., Liu, S., Yu, J. (2008). Enantioseparation of racemic α-cyclohexyl-mandelic acid across hollow fiber supported liquid membrane. J. Brazilian Chem. Soc., 19, 557-62.
198. Ferreira, Q., Coelhoso, I. M., Ramalhete, N., Marques, H. M. C. (2006). Resolution of racemic propranolol in liquid membranes containing TA-β-cyclodextrin. Sep. Sci. Technol., 41, 3553-68.
199. Maximini, A., Chmiel, H., Holdik, H., Maier, N. W. (2006). Development of a supported liquid membrane process for separating enantiomers of *N*-protected amino acid derivatives. J. Membr. Sci., 276, 221-31.
200. Kewen, T., Chunshan, Z., Xinyu, J. (2003). Racemic ofloxacin separation by supported-liquid membrane extraction with two organic phases. Sci. China Ser. B, 46, 96-103.
201. Ricks, E. E., Estrada-Valdes, M. C., McLean, T. L., Lacobucci, G. A. (1992). Highly enantioselective hydrolysis of (*R*,*S*)-phenylalanine isopropyl ester by subtilisin carlsberg. Continuous synthesis of (,*S*)-phenylalanine in a hollow fiber/liquid membrane reactor. Biotechnol. Progr., 8, 197-203.
202. Hernández-Fernández, F. J., de los Ríos, A. P., Tomás-Alonso, F., Gómez, D., Víllora, G. (2008). On the development of an integrated membrane process with ionic liquids for the kinetic resolution of rac-2-pentanol. J. Membr. Sci., 314, 238-246.
203. Hernández-Fernández, F. J., de los Ríos, A. P., Tomás-Alonso, F., Gómez, D., Rubio, M., Víllora, G. (2007). Integrated reaction/separation processes for the kinetic resolution of rac-1-phenylethanol using supported liquid membranes based on ionic liquids. Chem. Eng. Process., 46, 818-24.

204. Miyako, E., Maruyama, T., Kubota, F., Kamiya, N., Goto, M. (2005). Optical resolution of various amino acids using a supported liquid membrane encapsulating a surfactant-protease complex. Langmuir, 21, 4674-79.
205. Miyako, E., Maruyama, T., Kamiya, N., Goto, M. (2004). Highly enantioselective separation using a supported liquid membrane encapsulating surfactant-enzyme complex. J. Am. Chem. Soc., 126, 8622-3.
206. Jiao, F. P., Huang, K. L., Peng, X. H., Zhao, X. H., Yu, J. G. (2006). Hollow fiber liquid-supported membrane technology for enantioseparation of racemic salbutamol by combinatorial chiral selectors. J. Central South Univ. Technol., 13, 39-43.
207. Miyako, E., Maruyama, T., Kamiya, N., Goto, M. (2003). Enzyme-facilitated enantioselective transport of (*S*)-ibuprofen through a supported liquid membrane based on ionic liquids. Chem. Commun., 2926-7.
208. Deblay, P., Minier, M., Renon, H. (1990). Separation of L-valine from fermentation broths using a supported liquid membrane. Biotechnol. Bioeng., 35, 123-31.
209. Ghosh, A. C., Bora, M. M., Dutta, N. N. (1996). Developments in liquid membrane separation of beta-lactam antibiotics. Bioseparation, 6, 91-105.
210. Sahoo, G., Dutta, N. (2002). Perspectives in liquid membrane extraction of cephalosporin antibiotics. In *History and trends in bioprocessing and biotransformation* (T. Scheper, ed.), pp. 209-42. Springer, Berlin.
211. Schäfer, A., Hossain, M. M. (1996). Extraction of organic acids from kiwifruit juice using a supported liquid membrane process. Bioprocess Biosyst. Eng., 16, 25-33.
212. Di Luccio, M., Smith, B. D., Kida, T., Borges, C. P., Alves, T. L. M. (2000). Separation of fructose from a mixture of sugars using supported liquid membranes. J. Membr. Sci., 174, 217-24.
213. Christen, P., Minier, M., Renon, H. (1990). Ethanol extraction by supported liquid membrane during fermentation. Biotechnol. Bioeng., 36, 116-23.
214. Alpoguz, H. K., Kaya, A., Memon, S., Yilmaz, M. (2007). Facilitated supported liquid membrane transport of Hg^{2+} using calix[4]arene derivatives. J. Macromol. Sci. A, 44, 17-20.
215. Fontás, C., Anticó, E., Vocanson, F., Lamartine, R., Seta, P. (2007). Efficient thiacalix[4]arenes for the extraction and separation of Au(III), Pd(II) and Pt(IV) metal ions from acidic media incorporated in membranes and solid phases. Sep. Purif. Technol., 54, 322-8.
216. Raut, D. R., Mohapatra, P. K., Ansari, S. A., Manchanda, V. K. (2008). Evaluation of a calix[4]-bis-crown-6 ionophore-based supported liquid membrane system for selective ^{137}Cs transport from acidic solutions. J. Membr. Sci., 310, 229-36.
217. Zaghbani, A., Tayeb, R., Dhahbi, M., Hidalgo, M., Vocanson, F., Bonnamour, I., Seta, P., Fontàs, C. (2007). Selective thiacalix[4]arene bearing three amide groups as ionophore of binary Pd(II) and Au(III) extraction by a supported liquid membrane system. Sep. Purif. Technol., 57, 374-9.
218. Belkhouche, N. E., Didi, M. A., Romero, R., Jönsson, J. Å., Villemin, D. (2006). Study of new organophosphorus derivates carriers on the selective recovery of M(II) and M(III) metals, using supported liquid membrane extraction. J. Membr. Sci., 284, 398-405.
219. He, D., Luo, X., Yang, C., Ma, M., Wan, Y. (2006). Study of transport and separation of Zn(II) by a combined supported liquid membrane/strip dispersion process containing D2EHPA in kerosene as the carrier. Desalination, 194, 40-51.
220. Venkateswaran, P., Gopalakrishnan, A. N., Palanivelu, K. (2007). Di(2-ethylhexyl) phosphoric acid-coconut oil supported liquid membrane for the separation of copper ions from copper plating wastewater. J. Environ. Sci., 19, 1446-53.
221. Van De Voorde, I., Pinoy, L., De Ketelaere, R. F. (2004). Recovery of nickel ions by supported liquid membrane (SLM) extraction. J. Membr. Sci., 234, 11-21.

222. Stankovic, V. (2007). Metal removal from effluents by electrowinning and a new design concept in wastewater purification technology. Chem. Biochem. Eng. Q., 21, 33-45.
223. Muthuraman, G., Palanivelu, K. (2006). Removal of CI Reactive Yellow 125, CI Reactive Red 158 and CI Reactive Red 159 dyes from aqueous solution with a supported liquid membrane containing tributylphosphate as carrier. J. Text. Inst., 97, 341-7.
224. Iyer, R. H. (2003). Separation and recovery of radioactive and non-radioactive toxic trace elements from aqueous industrial effluents. Indian J. Exp. Biol., 41, 1002-11.
225. Chaudry, M. A., Bukhari, N., Mazhar, M., Abbasi, W. (2007). Coupled transport of chromium(III) ions across triethanolamine/cyclohexanone based supported liquid membranes for tannery waste treatment. Sep. Purif. Technol., 55, 292-9.
226. Ashraf, W., Bukhari, A. (2007). Separation of organic species from wastewater using a polyol supported liquid membrane. Pol. J. Chem., 81, 1621-8.
227. Ilconich, J. B., Luebke, D. R., Myers, C., Pennline, H. W. (2006). Carbon dioxide separation through supported ionic liquids membranes in polymeric matrixes. In *23rd Annual International Pittsburgh Coal Conference, PCC—Coal-Energy, Environment and Sustainable Development*, Pittsburgh.
228. Ilconich, J., Myers, C., Pennline, H., Luebke, D. (2007). Experimental investigation of the permeability and selectivity of supported ionic liquid membranes for CO_2/He separation at temperatures up to 125 °C. J. Membr. Sci., 298, 41-7.
229. Duan, S., Ito, A., Ohkawa, A. (2003). Separation of propylene/propane mixture by a supported liquid membrane containing triethylene glycol and a silver salt. J. Membr. Sci., 215, 53-60.
230. Zhang, L. Z., Xiao, F. (2008). Simultaneous heat and moisture transfer through a composite supported liquid membrane. Int. J. Heat Mass Transfer, 51, 2179-89.
231. Pedersen-Bjergaard, S., Rasmussen, K. E. (2008). Liquid-phase microextraction with porous hollow fibers, a miniaturized and highly flexible format for liquid-liquid extraction. J. Chromatogr. A, 1184, 132-42.
232. Varanusupakul, P., Vora-adisak, N., Pulpoka, B. (2007). In situ derivatization and hollow fiber membrane microextraction for gas chromatographic determination of haloacetic acids in water. Anal. Chim. Acta, 598, 82-6.
233. Sandahl, M., Mathiasson, L., Jönsson, J. Å. (2002). On-line automated sample preparation for liquid chromatography using parallel supported liquid membrane extraction and microporous membrane liquid-liquid extraction. J. Chromatogr. A, 975, 211-7.
234. Liu, J. F., Liang, X., Jiang, G. B., Cai, Y. Q., Zhou, Q. X., Liu, G. G. (2003). Evaluation of an on-line coupled continuous flow liquid membrane extraction and precolumn system as trace enrichment technique by liquid chromatographic determination of bisphenol A. Talanta, 60, 1155-61.
235. Tudorache, M., Zdrojewska, I. A., Emnéus, J. (2006). Evaluation of progesterone content in saliva using magnetic particle-based immuno supported liquid membrane assay (μ-ISLMA). Biosens. Bioelectron., 22, 241-6.
236. Tudorache, M., Emnéus, J. (2006). A micro-immuno supported liquid membrane assay (μ-ISLMA). Biosens. Bioelectron., 21, 1513-20.
237. Tudorache, M., Emnéus, J. (2005). Selective immuno-supported liquid membrane (ISLM) extraction, enrichment and analysis of 2,4,6-trichlorophenol. J. Membr. Sci., 256, 143-9.

CHAPTER 4

Emulsion Liquid Membranes: Definitions and Classification, Theories, Module Design, Applications, New Directions and Perspectives

Mousumi Chakraborty,* Chiranjib Bhattacharya,† *and* Siddhartha Datta†

1. Introduction and Definitions

Membrane processes for separation of chemical species from a mixture are gaining in importance and are emerging as a viable alternative to conventional separation processes. The interest in the mass transfer through membranes can be attributed to membrane processes being technically simple and having low energy consumption.

However, the use of permeable and semipermeable membranes in microfilters, ultrafilters, osmosis, reverse osmosis, dialysis (which are comparatively newer methods of separation) has problems like high capital costs, low mass transfer rate, low selectivity, and large equipment size.

Liquid membrane (LM) separation provides a potentially powerful technique for effecting diverse separation operations. Compared to conventional processes, emulsion liquid membrane (ELM) and liquid surfactant membrane (LSM) processes have some attractive features, for example, simple operation, high efficiency, extraction and stripping in one stage, larger interfacial area, scope of continuous operation. The ELM technique has great potential for recovery and removal of different metal ions and hydrocarbons from wastewater where conventional methods provide lower separation efficiency.

*Department of Chemical Engineering, S.V. National Institute of Technology, Surat 395007, India
†Department of Chemical Engineering, Jadavpur University, Kolkata 700032, India

Liquid Membranes: Principles and Applications in Chemical Separations and Wastewater Treatment
DOI: 10.1016/B978-0-444-53218-3.00004-0

The objective of this chapter is to provide comprehensive knowledge-based information by critical analysis, classification, model description, and applications of a variety of ELM separation processes. New perspectives and directions of development in these fields are also discussed.

1.1. Description of liquid membranes

Liquid membrane processes are those involving a selective liquid membrane phase in which simultaneous extraction/stripping occurs. Separation is achieved by permeation of solute through this liquid phase from a feed phase to a receiving phase. The feed and receiving phases are normally miscible while the membrane phase is immiscible in both.

Liquid membranes can be of three types—bulk liquid membrane, immobilized on a solid supported liquid membrane, and liquid membrane as double emulsions. Of these three types, ELMs can achieve much higher mass transfer area than the other two membranes. ELMs were first used by Li [1] for separation of hydrocarbons. Since then, considerable work has been done to demonstrate qualitatively the feasibility of performing separations with specific formulations.

The system in the form of double emulsions may be of two types: water-in-oil emulsion dispersed in an external aqueous phase and oil-in-water emulsion dispersed in an outer organic phase. The membrane phase in the water-in-oil-in-water (W/O/W) type is the immiscible oil phase separating the aqueous phases, while in the O/W/O type the immiscible water phase separating the two organic phases acts as the membrane. Hence, the liquid membrane serves a dual purpose of permitting selective transfer of one or more components through it from external phase to internal droplets and vice versa and preventing mixing of external and internal phases.

2. Mechanisms of ELM Transport

The principal rate-determining step in solute permeation through a liquid membrane is solute diffusion through the membrane. However, separation can be enhanced by the use of additives, specific carriers, chemical reagents, or external electric or photoelectric impulses. The various separation mechanisms are broadly classified into two types—simple and facilitated—that are discussed below in detail.

2.1. Simple permeation mechanism

Since an LSM is a thin film of liquid (oil or aqueous) composed of surfactants and their solvents between a feed and a receiving phase, any immiscible liquid can serve as a membrane between two liquid or gas phases containing a solute at different concentrations [2]. If the solute is soluble in the membrane phase and has a reasonable diffusivity through the membrane, then its selective transport through the membrane from higher to lower concentration can be achieved. This type of permeation is not of much technical importance and is suitable only for studies on emulsion stability.

2.2. Facilitated transport mechanism

In this type, the effectiveness of separation through a liquid membrane is improved by maximizing the flux through the membrane phase and the capacity for the diffusing species in the receiving phase. This may be achieved by two mechanisms:

Type I. In this case the mass transfer rate through the membrane phase is increased by incorporating a stripping agent in the internal phase which reacts with the solute yielding a membrane insoluble product. Examples of this system are extraction of weak acids or bases from wastewater such as phenol removal by NaOH solution as the internal phase [3-8] and removal of ammonia by H_2SO_4 solution as the internal phase [9-10].

Type II. In this case a reactive component or carrier is incorporated in the membrane phase to carry the diffusing species across the membrane to the internal phase thus promoting its transfer through the membrane. This method is applicable to separation of metal ions from wastewater or mine leaching solutions. At the membrane-external phase interface the carrier, which is soluble only in the membrane phase, forms a membrane-soluble compound by reversible reaction with the solute to be transported. The reaction product diffuses through the membrane to the membrane-internal phase interface and dissociates, discharging the solute to the internal phase. The unchanged carrier then diffuses back to the membrane-external phase interface. In this way each carrier molecule is able to transport solute molecules as many times as necessary so that only a small amount of carrier is required in the membrane phase even for achieving a high degree of separation (see Fig. 4.1). If the solute is insoluble in the membrane phase, the only means of transport of the solute through the membrane phase is in the form of a carrier-solute complex. In that case, the concentration gradient of the complex across the membrane phase can be maximized by reaction with a stripping agent at the membrane-internal phase interface (carrier transport with chemical reaction [11]). In some cases of facilitated transport with chemical reaction, for example, in carrier-mediated metal ion transport, two different species are transported at the same time. This phenomenon is called coupled mass transport. The carriers used in coupled metal ion transport may be acidic carriers like –COOH, $-SO_3H$ or chelating groups (LIX agents) as well as basic carriers like amines or quaternary ammonium

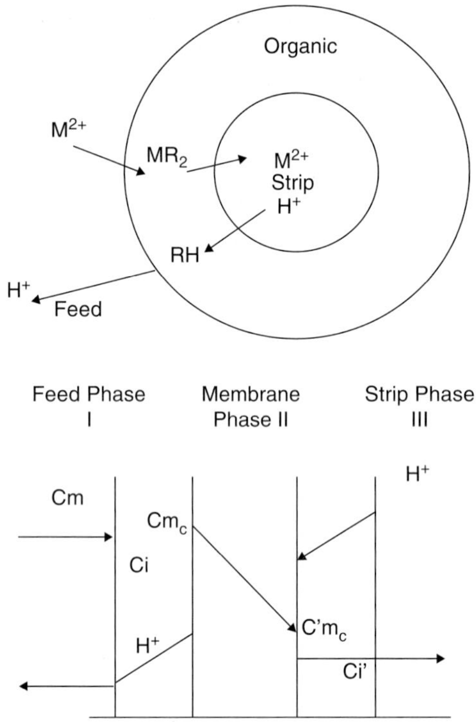

Figure 4.1 A schematic representation of the liquid membrane globule and concentration profile of solute through ELM.

salts. If the transports of these two different species occur in the same direction it is called cotransport [12], while transport in opposite direction is called counter transport.

It is easy to make the feed and strip solutions with different pH so that ion exchange processes on the two membrane surfaces will be shifted in opposite directions. This creates a concentration gradient of different forms of the carrier (with and without ions) in the membrane, and results in the directed ion flux through the membrane. Evidently the process leads to the transport of targeted ionic species across the membrane against their concentration gradient. This type of so-called "uphill" or active transport will continue until one driving factor (difference of chemical potentials) is balanced by the difference between chemical potentials of another transported ion. If the carrier selectively extracts one species (A) in the presence of another species (B), A in the feed mixture will be finally separated from B, and then purified and concentrated in the strip solution. It is important that the process uses only chemical energy as a driving factor and does not need transmembrane pressure or voltage. The most common examples of counter transport include copper recovery by LIX agents [13] and chromium extraction with quaternary ammonium salts [14].

3. Modeling of Liquid Membranes

A number of studies have been reported in the literature for the development and testing of mathematical descriptions for solute transport through liquid membranes. The existing models can be broadly classified into (i) the membrane film model in which the entire resistance to mass transfer is assumed to be concentrated in a membrane film of constant thickness and (ii) distributed resistance model which considers the mass transfer resistance to be distributed throughout the emulsion drop.

3.1. Film models for liquid membrane separations

Depending upon the membrane geometry, film models can be of two types: (a) Uniform flat sheet model, which assumes the membrane to be a planar film and (b) Spherical shell model in which an emulsion globule is characterized as a double shell with the membrane around a single internal phase droplet. Planar geometry models have been used [15-17]. Kremesec and Slattery considered the overall mass transfer resistance as a sum of the resistance through continuous, membrane, and internal phases [17].

The spherical shell model was first proposed by Cahn and Li [3]. They suggested that the mass transfer rate through the film is directly related to the solute concentration difference, ΔC across the film. The rate equation then becomes:

$$-\frac{\mathrm{d}C_c}{\mathrm{d}t} = DA(\Delta C/\delta) \tag{1}$$

where C_c is the solute concentration in the external continuous phase, D is the diffusivity of the solute through the membrane phase, and A is the mass transfer area per unit volume of feed solution.

Since A and δ are difficult to measure for a liquid membrane system, DA/δ can be replaced by $D'(V_e/V_c)$ where D' is an effective diffusivity and V_e/V_c is the treat ratio or holdup ratio (volume ratio of emulsion to external phase). In the case of type-I facilitated transport mechanism, where the solute is removed by reaction with internal stripping reagent, the solute concentration in the internal phase can be considered to be zero and hence Eq. (1) becomes

$$\ln\frac{C_{c0}}{C_c} = D'(V_e/V_c)t \tag{2}$$

Equation (2) indicates the external phase concentration as a fraction of the initial concentration. Cahn and Li [3] also used Eq. (2) to analyze data from batch experiments of phenol removal from wastewater with NaOH as the internal reagent. They assumed a constant mass transfer resistance with time and also neglected solute accumulation in the membrane as well as in the internal phase. These authors also applied this model for analysis of non-facilitated transport processes such as separation of hydrocarbons using an aqueous medium as the membrane phase separating the organic feed and receiving phases [18]. Boyadzhiev *et al.* [19] used the membrane film model suggested by Cahn and Li for carrier-mediated transport. However their analysis does not account for the internal reagent consumption. Hochchauser and Cussler [20] also analyzed carrier-mediated transport of chromium with tridodecylamine as the mobile carrier with the help of a spherical shell membrane film model. Similarly Lee *et al.* [21] and Volkel *et al.* [22] used the spherical shell model for studying copper extraction using LIX-64N and LIX-65N as carriers. Kondo *et al.* [23] used benzoylacetone as carrier for copper transport through the membrane film. All these investigators used various approximations regarding chemistry of the process and specific reaction rates within the membrane or at its surface while developing their models. Krishna *et al.* [24] also proposed a steady-state membrane film model for hydrocarbon separations by considering nonselective leakage. Ulbricht *et al.* [25] and Gupta *et al.* [26] used membrane film models to describe liquid membrane hydrocarbon transfer. Matulevicius and Li [27] attempted to improve upon this model by considering the unsteady-state mass transfer of solute through the membrane with the mass transfer resistance through the membrane film as the controlling resistance. The model equation is:

$$\frac{\partial C_{\mathrm{m}}}{\partial t} = D\frac{1}{r^2}\frac{\partial}{\partial r}\left(r^2\frac{\partial C_{\mathrm{m}}}{\partial r}\right) \tag{3}$$

with initial and boundary conditions:

$$\begin{aligned} C_{\mathrm{m}} &= 0 && \text{at } t = 0(r < R) \\ C_{\mathrm{m}} &= \alpha C_{\mathrm{c}} && \text{at } r = R(t > 0) \\ C_{\mathrm{m}} &= 0 && \text{at } r = Rj(t > 0) \end{aligned}$$

where C_{m} is the membrane phase solute concentration and C_{c} is the continuous phase solute concentration.

Matulevicius and Li [27] obtained a good agreement between their experimental results and the model predictions. On this basis, they concluded that the amount of internal circulation within the drops was negligible compared to the overall rate process.

3.2. Distributed resistance models for liquid membrane separations

The distributed resistance models describe adequately the emulsion homogeneity resulting from the presence of the droplets of internal reagent dispersed in the emulsion globules. Two different approaches have been reported based on the nature of reaction with the internal reagent.

3.2.1. Advancing front model

In advancing reaction front model, the stripping reaction is assumed to be instantaneous and irreversible. Therefore in this model the solute is unable to penetrate into the globule beyond those droplets that are completely depleted of internal reagent as it is immediately removed by reaction with the reagent. Thus, there exists a sharp boundary or reaction front at which the reaction takes place separating the inner region containing no solute from the outer region containing no stripping reagent. The reaction products are immobilized and hence are incapable of back diffusion.

In their first model of simple permeation of solute through emulsion globules, Kopp *et al.* [28] considered a reversible reaction between the solute and the internal reagent. The model equations were

$$\frac{\partial C_{\mathrm{m}}}{\partial t} = D_{\mathrm{eff}} \frac{1}{r^2} \frac{\partial}{\partial r} \left(\frac{r^2 \partial C_{\mathrm{m}}}{\partial r} \right) - \frac{\partial C_{\mathrm{d}}}{\partial t} \tag{4}$$

$$\frac{\partial C_{\mathrm{i}}}{\partial t} = K_{\mathrm{f}} C_{\mathrm{m}} - K_{\mathrm{b}} C_{\mathrm{d}} \tag{5}$$

where C_{m} and C_{d} are the permeate concentrations in the membrane phase and encapsulated droplets respectively, C_{i} is the internal reagent concentration, D_{eff} is an effective molecular diffusivity, and K_{f} and K_{b} are the rate constants of the forward and backward reactions, respectively.

The advancing front model of Kopp *et al.* [28] was improved upon by Ho *et al.* [29]. They assumed a diffusional controlled mass transfer mechanism for emulsion drops of uniform size and considered the curvature effects and change in external phase concentration with time. Ho *et al.* [29] considered the extraction to take place in well-agitated systems where the continuous phase mass transfer resistance is negligible. Stroeve and Varanasi [30] overcame this shortcoming of the model of Ho *et al.* by incorporating the external phase mass transfer resistance with pseudo-steady-state assumption in their advancing reaction front model. The model equations are as given below.

The mass flux from the continuous phase to the surface of the emulsion drop of radius R is given by:

$$N = K_{\mathrm{e}} 4\pi R^2 (C_{\mathrm{e}}' - C_{\mathrm{S}}') \tag{6}$$

where K_e is the continuous phase mass transfer coefficient, C_e' is the solute concentration in external continuous phase, and C_S' is the solute concentration in the external continuous phase at the surface of the double emulsion drops.

The flux of solute arriving at the reaction front is given by:

$$D_{\text{eff}} \frac{\mathrm{d}C}{\mathrm{d}r} = \frac{N}{4\pi r^2} \tag{7}$$

where D_{eff} is the effective solute diffusivity in the double emulsion drops and C is the solute concentration in the reacted region of the double emulsions.

The rate of consumption of the internal reagent is given by:

$$-\frac{\mathrm{d}}{\mathrm{d}t}\left[\frac{4\pi R_f^3}{3}\frac{V_i}{(V_i+V_m)}C_{io}\right] = N - \frac{\mathrm{d}}{\mathrm{d}t}\left[\int_{R_f}^{R} C(4\pi r^2)\mathrm{d}r\right] \tag{8}$$

where R_f is the reaction front position, V_i is the volume of internal reagent phase, V_m is the total membrane phase volume, and C_{io} is the initial solute concentration in the external continuous phase. The second term in the right-hand side accounts for the accumulation of unreacted solute present in the region depleted of internal reagent that is a partial correction for ignoring the solute accumulation.

Fales and Stroeve [31] investigated the effect of the continuous phase mass transfer resistance on solute extraction with double emulsion in a batch reactor. They presented an extension of the perturbation analysis technique to give a solution of the model equations of Ho *et al.* [29] taking external phase mass transfer resistance into account. Kim *et al.* [5] also developed an unsteady-state advancing reaction front model considering an additional thin outer liquid membrane layer and neglecting the continuous phase resistance.

Dutta *et al.* [32] modified the pseudo-steady-state advancing reaction front model of Stroeve and Varanasi [30] by considering the polydispersity of the emulsion globules and the external phase mass transfer resistance. They also included the outer membrane film resistance in their model [5]. Their results were in good agreement with experimental data for phenol extraction.

An advancing reaction front model considering emulsion globules of various sizes with same sauter mean diameter had been proposed by Chakraborty *et al.* to determine the nickel(II) extraction rate [33]. A diagram of the model is given in Fig. 4.2. Sauter mean diameter was taken as an average diameter of different sized emulsion globules. The effect of polydispersity on ELM had been taken into account in this model and the polydisperse effect was found to have very little influence on mass transfer behavior. The material balance for the solute in the reacted region of the globules of *j*th drop size ($R_{0j} < r < R_{fj}$) in the membrane phase was

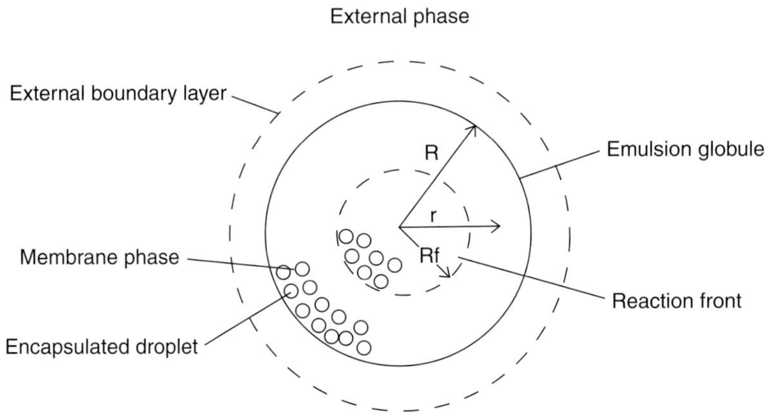

Figure 4.2 Schematic diagram of the model. From Ref. [33] with permission.

$$\frac{\partial C}{\partial t} = \frac{D_e}{(1-\varphi)r^2}\frac{\partial}{\partial r}\left(r^2\frac{\partial C}{\partial r}\right) \tag{9}$$

where

$$t = 0, \quad C = 0, \quad (r \leq R_{0j}) \tag{10}$$

$$r = R_{fj}(t), C = 0, \quad (t \geq 0) \tag{11}$$

$$\text{at } r = R_j, \quad C = C_D C_e, \quad (t \geq 0) \tag{12}$$

The material balance for the solute in the external phase was

$$-\frac{dC_e}{dt} = \frac{3V}{V_e}\sum_{j=1}^{k}\frac{D_e}{R_j}f_j\frac{\partial C}{\partial r}\bigg|_{r=R_j} \tag{13}$$

$$t = 0, \quad C_e = C_{eo} \tag{14}$$

The material balance for the solute at the reaction front is

$$-\frac{d}{dt}\left(\frac{4}{3}\pi R_{fj}^3\varphi C_i\right) = 4\pi R_{fj}^2 D_e\frac{\partial C}{\partial r}\bigg|_{r=R_{fj}(t)} \tag{15}$$

Rate of change of internal reagent mass within the reactive core = mass flow rate of solute (nickel) at the reaction front.

$$t = 0, \quad R_{fj} = R_j \tag{16}$$

The above equations transformed into dimensionless form by defining

$$\eta = \frac{r}{R}, \quad \chi = \frac{R_\mathrm{f}}{R}, \quad \tau = \frac{D_\mathrm{e} t}{R^2},$$
$$g = \frac{C}{C_\mathrm{eo}}, \quad h = \frac{C_\mathrm{e}}{C_\mathrm{eo}}, \quad m = \frac{C_\mathrm{i}}{C_\mathrm{eo}}, \tag{17}$$
$$E = \frac{3V}{V_\mathrm{e}}, \quad R0 = (R_0/R)^2$$

Then the diffusion equation became

$$\frac{\partial g}{\partial \tau} = \frac{R0_j}{(1-\varphi)\eta_j^2} \frac{\partial}{\partial \eta_j}\left(\eta^2 \frac{\partial g}{\partial \eta_j}\right) \tag{18}$$

$$\tau = 0, \quad g = 0, \quad (\eta_j \leq 1) \tag{19}$$

$$\eta_j = \chi_j, \quad g = 0, \quad (\tau \geq 0) \tag{20}$$

$$\eta_j = 1 \quad \text{then}\, g = C_\mathrm{D} h \quad \text{at}\, \tau \geq 0 \tag{21}$$

The material balance equation in the external phase was

$$\frac{\mathrm{d}h}{\mathrm{d}\tau} = -E \sum_{j=1}^{k} R0_{jfj} \frac{\partial g}{\partial \eta_j}\bigg|_{\eta_j = 1} \tag{22}$$

$$\tau = 0, \quad h = 1 \tag{23}$$

The material balance equation at the reaction front was as follows:

$$-\frac{\mathrm{d}\chi_j}{\mathrm{d}\tau} = \frac{R0_j}{\varphi m} \frac{\partial g}{\partial \eta_j}\bigg|_{\eta_j = \chi_j} \tag{24}$$

$$\tau = 0, \quad \chi_j = 1 \tag{25}$$

The coupled equations had been solved by numerical computation using an implicit finite difference technique [34]. While solving the above equations, the emulsion globule size d_{32} (sauter mean diameter) was calculated by using the following correlation [35]:

$$d_{32} = 0.12(We)^{-0.5} d_\mathrm{I} \tag{26}$$

The diffusivity of di-(2-ethylhexyl) phosphoric acid (D2EHPA)-nickel(II) complex in the membrane phase was determined by the Willke-Chang correlation [36]):

$$D_\mathrm{m} = \frac{(117.3 \times 10^{-18})(\psi M)^{0.5} T}{\mu_\mathrm{m} v_\mathrm{c}^{0.6}} \mathrm{m}^2/\mathrm{s} \tag{27}$$

The effective diffusivity of the complexes in the emulsion was obtained by the Jefferson-Witzell-Sibbitt correlation [37],

$$D_{\mathrm{eff}} = 10^4 D_{\mathrm{m}} \left[\frac{4(1+2p)^2 - \pi}{4(1+2p)^2} \right] \mathrm{cm}^2/\mathrm{s} \tag{28}$$

where

$$p = 0.403(\varphi)^{-1/3} - 0.5 \tag{29}$$

The polydisperse effect had very little influence on mass transfer behavior (see Fig. 4.3).

The model showed that for larger drops the reaction front penetration was small, while for smaller drops the movement of the reaction front was very fast and most of the nickel is consumed in the early stage (see Fig. 4.4). Afterward there was virtually simple physical absorption of nickel and these smaller drops were no longer effective in extracting the solute, resulting in a loss of effective interfacial area for solute transport.

Chakraborty *et al.* [38] also used the advancing front model to analyze simultaneous extraction of copper(II) and nickel(II) using D2EHPA as extractant and hydrochloric acid as stripping membrane phase

$$\frac{\partial C}{\partial t} = \frac{D_{\mathrm{e}}}{(1-\varphi)r^2} \frac{\partial}{\partial r} \left(r^2 \frac{\partial C}{\partial r} \right) \tag{30}$$

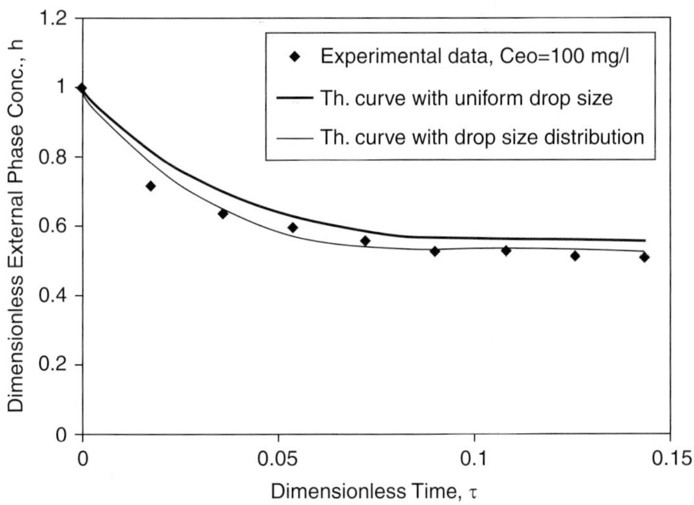

Figure 4.3 Effect of drop size distribution on external phase nickel concentration. From Ref. [33] with permission.

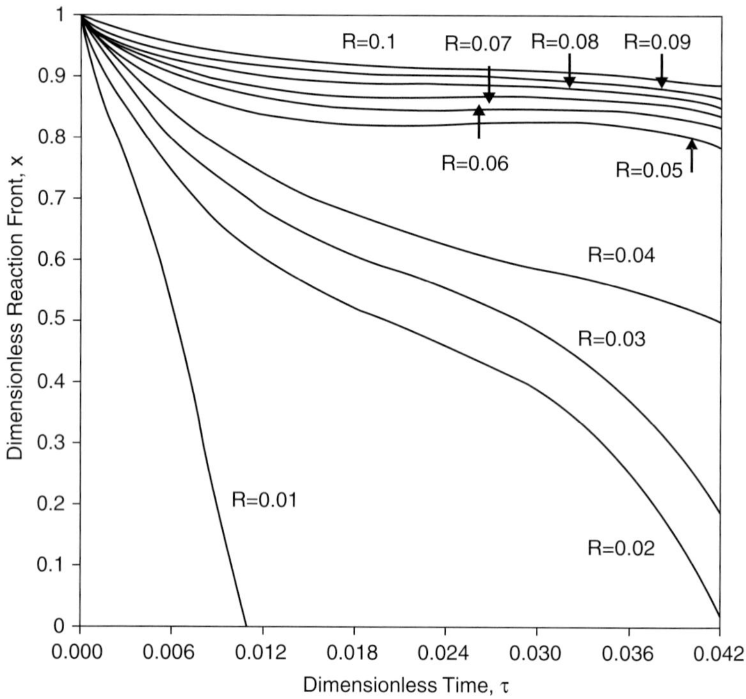

Figure 4.4 Reaction front progress inside emulsion globules of various sizes with same sauter mean diameter of 0.065 cm. From Ref. [33] with permission.

$$\frac{\partial C'}{\partial t} = \frac{D'_{\mathrm{e}}}{(1-\varphi)r^2}\frac{\partial}{\partial r}\left(r^2\frac{\partial C'}{\partial r}\right) \tag{31}$$

where

$$R_{\mathrm{f}}(t) < r < R, \quad t > 0 \tag{32}$$

$$t = 0, \quad C = 0, \quad C' = 0 \quad (r \leq R) \tag{33}$$

$$r = R_{\mathrm{f}}(t), \quad C = 0, \quad C' = 0 \quad (t \geq 0) \tag{34}$$

$$r = R, \quad C = C_{\mathrm{D}}C_{\mathrm{e}}, \quad C' = C'_{\mathrm{D}}C'_{\mathrm{e}} \quad (t \geq 0) \tag{35}$$

The material balances for the solutes in the external phase were

$$-V_{\mathrm{e}}\frac{\mathrm{d}C_{\mathrm{e}}}{\mathrm{d}t} = \frac{3}{R}VD_{\mathrm{e}}\frac{\partial C}{\partial r}\bigg|_{r=R} \tag{36}$$

$$-V_{\mathrm{e}}\frac{\mathrm{d}C'_{\mathrm{e}}}{\mathrm{d}t} = \frac{3}{R}VD'_{\mathrm{e}}\frac{\partial C'}{\partial r}\bigg|_{r=R} \tag{37}$$

$$t = 0, \quad C_e = C_{eo}, \quad C_e^{'} = C_{eo}^{'} \tag{38}$$

The material balances of the solutes at the reaction front were

$$-\frac{d}{dt}\left(\frac{4}{3}\pi R_f^3 \varphi C_i\right) = 4\pi R_f^2 D_e \frac{\partial C}{\partial r}\bigg|_{r = R_f(t)} + 4\pi R_f^2 D_e^{'} \frac{\partial C^{'}}{\partial r}\bigg|_{r = R_f(t)} \tag{39}$$

$$t = 0, \quad R_f = R \tag{40}$$

The above equations transformed into dimensionless form by defining

$$\begin{aligned} &\eta = \frac{r}{R}, \quad \chi = \frac{R_f}{R}, \quad E = \frac{3V}{V_e}, \quad \tau = \frac{D_e t}{R^2}, \quad \beta = \frac{D_e^{'}}{D_e} \\ &g = \frac{C}{C_{eo}}, \quad g^{'} = \frac{C^{'}}{C_{eo}^{'}} \\ &h = \frac{C_e}{C_{eo}}, \quad h^{'} = \frac{C_e^{'}}{C_{eo}^{'}} \\ &m = \frac{C_i}{C_{eo}}, \quad m^{'} = \frac{C_i}{C_{eo}^{'}} \end{aligned} \tag{41}$$

Then the diffusion equations became

$$\frac{\partial g}{\partial \tau} = \frac{1}{(1-\varphi)\eta^2}\frac{\partial}{\partial \eta}\left(\eta^2 \frac{\partial g}{\partial \eta}\right) \tag{42}$$

$$\frac{\partial g^{'}}{\partial \tau} = \frac{\beta}{(1-\varphi)\eta^2}\frac{\partial}{\partial \eta}\left(\eta^2 \frac{\partial g^{'}}{\partial \eta}\right) \quad \text{at}\, \chi < \eta < 1, \quad \tau > 0 \tag{43}$$

$$\tau = 0, \quad g = 0, \quad g^{'} = 0 \quad (\eta \leq 1) \tag{44}$$

$$\eta = \chi, \quad g = 0, \quad g^{'} = 0 \quad (\tau \geq 0) \tag{45}$$

$$\eta = 1 \quad \text{then}\, g = C_D h, \quad g^{'} = C_D^{'} h^{'} \quad \text{at}\, \tau \geq 0 \tag{46}$$

The material balance equations in the external phase were

$$\frac{dh}{d\tau} = -E\frac{\partial g}{\partial \eta}\bigg|_{\eta = 1} \tag{47}$$

$$\frac{dh^{'}}{d\tau} = -E\beta\frac{\partial g^{'}}{\partial \eta}\bigg|_{\eta = 1} \tag{48}$$

$$\tau = 0, \quad h = 1, \quad h^{'} = 1 \tag{49}$$

The material balance equation at the reaction front is as follows:

$$-\frac{d\chi}{d\tau} = \frac{1}{\varphi m}\frac{\partial g}{\partial \eta}\bigg|_{\eta=\chi} + \frac{\beta}{\varphi m'}\frac{\partial g'}{\partial \eta}\bigg|_{\eta=\chi} \quad (50)$$

$$\tau = 0, \quad \chi = 1 \quad (51)$$

where

C, metal ion concentration in saturated zone of emulsion globule
C_{i0}, initial internal reagent conc. in internal phase
C_i, internal reagent conc. in internal phase
C_e, metal ions conc. in external phase
C_{eo}, initial metal ion conc. in external phase
D_e, effective diffusivity of metal ions in saturated zone of emulsion globules
f_j, volume fraction of emulsion globules of jth size
C_D, distribution coefficients of the solutes between membrane and external phase
C_S, equilibrium solutes concentrations in the aqueous phase
r, radial coordinate in emulsion globules
R, radius of emulsion globules
R_f, reaction front position
R_0, sauter mean emulsion globule radius
m, $\sum n_j R_j^3 / n_j R_j^2$
n, number of emulsion globules
t, time
V, total volume of emulsion phase
V_e, volume of external phase
We, Weber number, $\rho N^2 d_I^3/\sigma$
d_{32}, sauter mean globule diameter
ρ, density of the aqueous external phase
N, stirring rate
σ, interfacial tension between membrane and external phase
d_I, impeller diameter
D_m, diffusivity of D2EHPA-nickel(II) complex in the membrane phase
E, $3V/V_e$
φ, volume fraction of internal aqueous phase in the emulsion
θ_i, dimensionless internal reagent conc. $= C_i/C_{i0}$
j, jth drop size

The coupled equations had been solved by numerical computation using an implicit finite difference technique [34]. It was found from Figs. 4.5 and 4.6 that the fraction of solutes [copper(II) and nickel(II)] extracted were higher with a lower initial external phase solutes concentration. This was due to a higher distribution coefficient for a lower initial external phase solutes concentration.

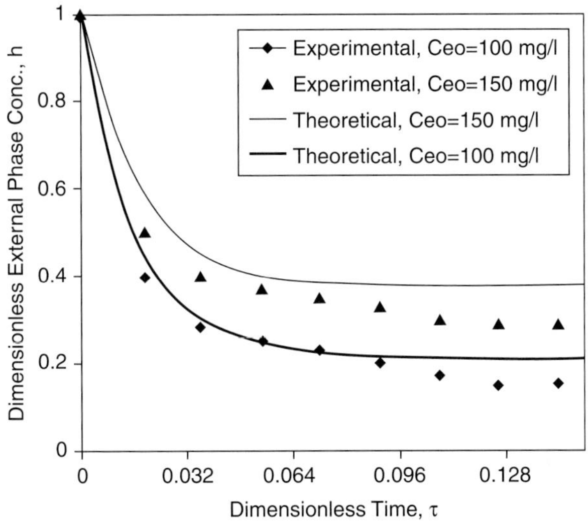

Figure 4.5 Variation of external phase copper ion concentration with initial external phase copper concentration. From Ref. [38] with permission.

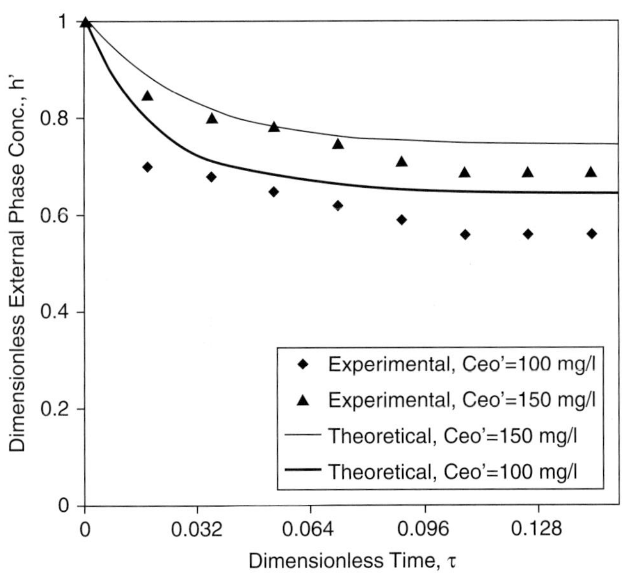

Figure 4.6 Variation of external phase nickel ion concentration with initial external phase nickel concentration. From Ref. [38] with permission.

One shortcoming of the advancing reaction front approach is the assumption of reaction irreversibility, which when combined with instantaneous kinetics requires that the reagent concentration be identically zero in the reacted core. This condition is not satisfied with systems having lower distribution coefficients and low solute concentrations. This has led to the development of the reversible reaction model in which there are no separate reacted and unreacted regions. Thus, it is conceivable that the solute can reach the center of the globule either without contacting the internal reagent or undergoing a series of forward and reverse reactions.

3.2.2. Reversible reaction model

The reversible reaction model removes the restriction of irreversibility of the stripping reaction. The reaction reversibility precludes the reaction advancing front since there is no separate reacted and unreacted region. Solute diffusing into the emulsion globule reacts with the internal reagent or distributes itself between the two phases. The product formed inside the encapsulated droplets may give rise to reverse reaction producing unreacted solute that can diffuse back into the membrane. Thus, it is possible that the solute diffuses to the center of the globule without coming into contact with the internal reagent or through a series of forward and reverse reaction steps. The earliest model allowing reaction reversibility was developed by Teramoto *et al.* [6, 39]. Because of low stirring rates and moderately high diffusion coefficients in their studies, the contribution of the external phase mass transfer resistance to extraction rates was significant. They also incorporated an additional membrane film resistance as proposed by Kim *et al.* [5] that is meant to simulate the observation that internal water droplets cannot reach the surface of the emulsion globule. However, Teramoto *et al.* [39] observed that the mass transfer resistance offered by this peripheral oil layer is negligible for their system of kerosene, amines, and HCl.

Bunge and Noble [8] proposed a way to approximate the effect of enhanced internal phase diffusivity using individual diffusion coefficients for the solute in the internal and the membrane phases and by introducing an enhancement parameter involving the initial concentration of the reagent and the volume fraction of the internal phase in the globule. The material balance equations describing the solute concentration in the membrane portion of the globule, C_{Am} and in the external phase, C_{Ab} are as follows:

$$\frac{\partial C_{\mathrm{Am}}}{\partial t} = \frac{D_{\mathrm{eff}}}{r^2}\frac{\partial}{\partial r}\left(\frac{r^2 \partial C_{\mathrm{Am}}}{\partial r}\right) - \left(\frac{1-f_{\mathrm{m}}}{f_{\mathrm{m}}}\right)\left(\frac{\partial C_{\mathrm{Ai}}}{\partial t} + \frac{\partial C_{\mathrm{Pi}}}{\partial t}\right) \tag{52}$$

with initial and boundary conditions

$$t = 0 \quad C_{\mathrm{Am}} = 0 \quad (R > r \geq 0)$$

$$r = R \quad C_{\mathrm{Am}} = k_{\mathrm{bm}} C_{\mathrm{Ab}} \quad (t \geq 0)$$

$$r = 0 \quad \frac{\partial C_{\mathrm{Am}}}{\partial r} = 0 \quad (\text{for all } t)$$

$$\frac{\partial C_{\mathrm{Ab}}}{\partial t} = -\frac{3D_{\mathrm{eff}}}{R f_{\mathrm{b}}}(1 - f_{\mathrm{b}}) f_{\mathrm{m}} \frac{\partial C_{\mathrm{Am}}}{\partial r}\bigg|_{r=R} \tag{53}$$

where C_{Ai} and C_{Pi} are the concentrations of unreacted solute and reaction product, respectively, in the internal phase, R is the mean globule radius, k_{bm} is the solute partition coefficient between bulk and membrane phases, f_{m} and f_{b} are the volume fractions of membrane and bulk phases, respectively, and D_{eff} is the mean effective diffusivity based on the membrane phase driving force, including diffusion of both the reacted and unreacted solutes through the internal phase. If local phase and reaction equilibria are established between the internal phase droplets and the membrane phase, then

$$\frac{\partial C_{\mathrm{Ai}}}{\partial t} + \frac{\partial C_{\mathrm{Pi}}}{\partial t} = \frac{1}{k_{\mathrm{im}}} \left[1 + \frac{K C_{\mathrm{Bi}}^{0}}{\left(1 + \dfrac{K C_{\mathrm{Am}}}{k_{\mathrm{im}}}\right)^{2}} \right] \frac{\partial C_{\mathrm{Am}}}{\partial t} \tag{54}$$

where K is the equilibrium constant of the reversible reaction and k_{im} is the solute partition coefficient between the internal and membrane phases.

Baird *et al.* [40] extended the reversible reaction model of Bunge and Noble to include the effects of multicomponent mixtures on the extraction rates of the individual components. Lin and Long [41] also developed a model for carrier-mediated batch extraction of nitric acid based on the reversible reaction model. Casamatta *et al.* [42] proposed a model for permeation of hydrocarbons through an aqueous membrane phase that was a combination of the distributed resistance and membrane film models. Yan *et al.* [43] developed a diffusion as well as reaction controlled model for type-I facilitated transport of acetic acid through a LSM with NaOH as the internal reagent. The mass transfer, both inside and outside the emulsion globules, as well as the reaction between the solute and the internal reagent were taken into consideration in their model. Since far excess reagent was used, pseudo-first-order reaction between the solute and the internal reagent was assumed. The mass transfer equations proposed in this model are as follows:

Globules ($0 < r < R$, $t > 0$)

$$\frac{D_{\mathrm{e}}}{r^2} \frac{\partial}{\partial r}\left(r^2 \frac{\partial C}{\partial r}\right) = \frac{\partial C}{\partial t} + \frac{\varphi_{\mathrm{i}} k_1 C}{\alpha} \tag{55}$$

with initial and boundary conditions

$$C = 0 \quad (0 \le r \le R,\ t = 0)$$

$$\frac{\partial C}{\partial r} = 0 \quad (r = 0,\ t \ge 0)$$

$$D_e \frac{dC}{dr} = K(C_e - C/\alpha), \quad r = R, \quad t \ge 0 \tag{56}$$

External phase:

$$V_e \frac{dC_e}{dt} = \frac{3}{R}(V_i + V_m) D_e \frac{\partial C}{\partial r}\Big|_{r=R}, \text{ with initial condition } C_e = C_{e0}(t = 0) \tag{57}$$

where C is the solute concentration inside the emulsion globules averaged locally over the membrane and internal phases, D_e is the diffusivity in the emulsion globules, R is the sauter mean emulsion globule radius, φ_i is the volume fraction of the internal reagent, k_l is the reaction rate constant, K is the external phase mass transfer coefficient, α is the solute distribution coefficient between the external and emulsion phases at equilibrium, C_e is the external phase solute concentration, V_e, V_m, and V_i are the volumes of the external, membrane, and internal phases, respectively. Guoyu *et al.* [44] developed a generalized mass transfer model applicable to both type-I and type-II facilitated transport. Sahoo and Dutta [45] also developed a reaction and diffusion controlled model and tested it with batch permeation experiments of cephalexin. Bhowal and Dutta [46] also proposed a pseudo-steady-state model for type-I facilitated transport through LSMs that may be considered to be a combination of advancing reaction front and reversible reaction models. The model has been found to predict satisfactorily the experimental results of the extraction of weak acids and weak bases in a batch separation system.

Advancing front model and three reversible reaction models were applied to describe 2-chlorophenol permeation from aqueous solutions [47]. The numerical implementation seemed more stable in the Bunge and Noble [8] model than in reversible reaction models that allowed changes of effective diffusivity with solute concentration in membrane phase, although results were quite similar for the three models. Kargari *et al.* [48] studied the selective separation of gold(III) ions from acidic aqueous solutions, using MIBK as carrier and LK-80 as emulsifier. They found only Au^{3+} ions is transported across the liquid membrane and nearly all (Pd^{2+}, Pt^{4+}, Cu^{2+}, and Fe^{3+}) of other ions remained in the external solution without change in their concentration. Reversible reaction model of the system shows a good compatibility with experimental results for gold(III) ions transport.

3.3. Equilibrium extraction correlation

Experiments were conducted to investigate the extraction of 2-chlorophenol (2-CP) from aqueous solution by liquid membrane [49]. Depending on the pH of the aqueous solution, molecular 2-CP and its ionic form are known to exist in equilibrium according to the following equation:

$$\mathrm{PhClOH} = \mathrm{PhClO}^- + \mathrm{H}^+$$

On the basis of the above equilibrium, an equation was established to relate the 2-CP concentration in the external aqueous phase to that in the W/O emulsion. The empirical Langmuir isotherm is first adopted for this purpose. This isotherm is represented by

$$Q_e = \frac{abC_e}{1 + bC_e} \tag{58}$$

in which Q_e is the equilibrium extraction capacity of W/O emulsion (in mg 2-CP removed per ml emulsion), C_e is the 2-CP equilibrium concentration in the aqueous phase (mg/l), and a and b are the constant parameters. To determine the two parameters, Eq. (57) is rewritten as

$$\frac{1}{Q_e} = \frac{1}{a} + \frac{1}{abC_e} \tag{59}$$

The 2-CP equilibrium concentration (C_e) in the aqueous phase was measured in a test run and the equilibrium extraction capacity (Q_e) was readily calculated using the measured data and the known initial conditions. According to Eq. (58), a plot of $1/Q_e$ versus $1/C_e$ would yield a straight line and the model parameters (a and b) are obtained from the slope and intercept. The model involved only single empirical parameter, which is easy to establish from the experimental data.

3.4. Advanced stripping model

The assumptions of the model are:

1. The chemical reactions in phases and interfaces are immiscible
2. The stripping reactions at the interface are irreversible
3. Sauter mean diameter was used for characterization of globules and internal droplets described by advanced front model and modeling of facilitated transport
4. Loss of cation in feed = loss of cation in membrane + loss of cation in strip
5. Loss of cation in membrane phase is negligible
6. The complete ionization of stripping phase, the completely extraction of cations

Assuming that the steady state of diffusion into the globule exits and the reaction of the front moves slowly in comparison to solute diffusion, a quasi-steady-state periphery of the emulsion globule is stated. The flux of the solute can be obtained by

$$J = \left[\frac{1}{K} + \left(\frac{1}{\alpha D_{\mathrm{m}}}\right)\left(\frac{R}{R_{\mathrm{f}}}\right)(R - R_{\mathrm{f}}) + \frac{1}{\alpha\varphi k_1}\right]^{-1} C_{\mathrm{e}} t = 0, \quad R_{\mathrm{f}} = R, \quad C_{\mathrm{e}} = C_{\mathrm{e0}} \tag{60}$$

where R_{f} is the radius of unreacted core in emulsion, $1/D_{\mathrm{m}}$ approximately represents the membrane phase diffusion resistance within emulsion globule; $1/k_1$ is the interfacial reaction, and $1/K$ is the combined overall resistance. Therefore, the overall mass transfer resistance can accurately be the mass transfer capacity of the chosen ELM system under the operating conditions. Modeling of transport of Cs (137) in xylene promoted by ephedrine hydrochloride in stripping phase was described by Said *et al.* [50]. The proposed model satisfactorily predicts the experimental results.

3.5. Models for continuous operations

Despite a continuing strong interest in ELM operations, most of the studies conducted so far have been devoted to batch extraction operations and little attention has been devoted to continuous operation, which is necessary for successful scale-up design procedures of these processes. Some of the models developed for continuous liquid membrane operations are discussed below.

3.5.1. Multistage mixer settler operations

The first attempt in sizing of continuous multistage mixer-settler systems for liquid membrane processes was made by Cahn and Li [3]. Their membrane film model when extended to multistage systems with each stage being of equal volume, the approximate volume per mixer is

$$V_{\mathrm{M}} = K\left[\sqrt[N]{\frac{C_0}{C_1}} - 1\right] \tag{61}$$

where

$$K = V_{\mathrm{w}}(1 + 1/\varphi_{\mathrm{e}})/D' \tag{62}$$

C_0 and C_1 are the feed and desired effluent concentration, respectively, N is the number of stages, V_{w} is the water phase flow rate, D' is the permeation rate constant, φ_{e} is the holdup ratio (volume ratio of emulsion to external phase in the mixer). While this approach is simple, there are nevertheless certain drawbacks to its use. The permeation coefficient D' or equivalently the membrane film thickness varies depending on conditions. Hatton and

Wardius [51] extended the advancing front model to be employed to multistage mixer-settler systems for liquid membrane operations. They presented a zero order solution to the perturbation equations based on the model developed by Ho *et al.* [29]. The emulsion globule residence time distribution in each mixer was assumed to be exponential and the fractional utilization of internal reagent was given by

$$f = \frac{v_{\mathrm{w}}(C_0 - C_1)}{v_{\mathrm{e}} C_{\mathrm{IR}}} = 3\int_0^1 \chi^2 \exp\left[-\frac{\theta}{6}(1 - 3\chi^2 + 2\chi^3)\right] \mathrm{d}\chi \tag{63}$$

where

$$\theta = \frac{C_{\mathrm{IR}}}{C_1} \frac{\frac{R^2}{\alpha D_{\mathrm{e}}}}{\frac{V_{\mathrm{E}}}{v_{\mathrm{e}}}} \tag{64}$$

C_0 and C_1 are the continuous phase solute concentrations in feed and inside the mixer, respectively, v_{w} and v_{e} are the feed rates of the continuous and emulsion phases, respectively, C_{IR} is the internal reagent concentration (based on volume of emulsion), R is the emulsion globule radius, V_{E} is the emulsion phase holdup volume in the mixer, α is the partition coefficient for solute between external phase and emulsion, D_{e} is the effective solute diffusivity in the emulsion, and χ is the dimensionless reaction front position. Hatton and Wardius [51] also extended their analysis to develop simple graphical and numerical procedures for the prediction of multistage extraction performance of mixer-settler trains operating either cocurrently or countercurrently without any external recycle over individual stages. For a typical stage i in a cocurrent mixer-settler, they defined the parameter θ as

$$\theta_i = \frac{C_{\mathrm{IR}i-1}}{C_i} \frac{\frac{R^2}{\alpha D_{\mathrm{e}}}}{\frac{V_{\mathrm{E}}}{v_{\mathrm{e}}}} \tag{65}$$

It is an unique function of the fractional utilization f_i for that stage which is given by

$$f_i = \frac{v_{\mathrm{w}}}{v_{\mathrm{e}}} \frac{(C_{i-1} - C_{\mathrm{i}})}{C_{\mathrm{IR}i-1}} \tag{66}$$

Combining Eqs. (65) and (66), the concentration ratio γ_{i-1} is given by

$$\gamma_{i-1} = \frac{v_{\mathrm{e}} C_{\mathrm{IR}i-1}}{v_{\mathrm{w}} C_{i-1}} = \frac{\theta_i}{\varphi + \theta_i f_i} \tag{67}$$

Combining all equations they got

$$\gamma_{i-1} = \frac{\theta_i}{(\varphi + \theta_i f_i)}(1 - f_i) \tag{68}$$

Since θ_i and f_i are uniquely related for countercurrent as well as cocurrent systems, both γ_i and γ_{i-1} can be plotted as functions of θ_i with φ as parameter. Therefore, from the known values of φ and γ_0 (the feed condition), the number of stages required to effect a given separation can be stepped off graphically in a manner directly analogous to that used in the McCabe-Thiele method for equilibrium stage operations. In a later study Wardius and Hatton [52] analyzed and compared different cascaded mixer systems with and without inter stage settling using the advancing reaction front model.

3.5.2. Column type operations

Like in continuous mixer-settler systems, mathematical developments for the characterization of the mass transport processes in continuous columns are also rather limited. One of the main parameters in modeling of column type operations is axial mixing and therefore dispersion effects within the column should be taken into consideration. For the continuous phase, this is generally considered as a diffusional process superimposed on plug flow while for the dispersed phase, it is found that the distribution in drop residence times caused by variations in drop rise velocities depending on their relative sizes is far more significant than the dispersion arising due to usual turbulent and entrainment back mixing mechanisms of Levenspiel and Fitzgerald [53].

In order to develop a continuous separation process, Kataoka *et al.* [54] simulated permeation of metal ion in continuous countercurrent column. They developed the material balance equation considering back mixing only in the continuous phase and steady-state diffusion in the dispersed emulsion drops which is similar to the liquid extraction situation. Bart *et al.* [55] also modeled the extraction of copper in a continuous countercurrent column. They considered only the continuous phase back mixing in the model and assumed that the reaction between copper ions and carrier is slow, so that the differential mass balance equation for external phase in their model is

$$D\frac{\partial^2 C}{\partial Z^2} - V_c\frac{\partial C}{\partial Z} - K\left(C - \frac{1}{m}\right) = 0 \tag{69}$$

where C is the external phase solute concentration, Z is the axial distance, V_c is the velocity of the external phase, K is the mass transfer coefficient in the external phase. The above equation was solved analytically to obtain the height of the column.

Lorbach and Hatton [56] analyzed the polydispersity and back mixing effects in terms of the advancing reaction front model by assuming pseudo-steady-state diffusion within the macrodrop so that the zero order solution to the perturbation expansion could be used. Mok *et al.* [57] proposed a mathematical model to describe in a continuous countercurrent column. They employed the advancing reaction front model for deriving the overall mass transfer coefficient in the emulsion globule while the axial dispersion model was applied to the external continuous phase. Their predicted results were consistent with experimental data obtained for carrier-mediated extraction of penicillin G.

4. ELM Design Considerations

4.1. Operational aspects in emulsion liquid membranes

The operations encountered in a ELM separation process are as follows (see Fig. 4.7):

(1) Emulsification of the membrane and internal phases
(2) Emulsion-continuous phase contacting
(3) Settling of the emulsion and external phases after extraction
(4) De-emulsification to recover the membrane phase

4.2. Preparation of emulsion liquid membranes

The important part in liquid membrane process is its formulation, which includes the selection of carrier, strip agent, surfactant, and diluents. The choice of those components and its formulation often decides the success of

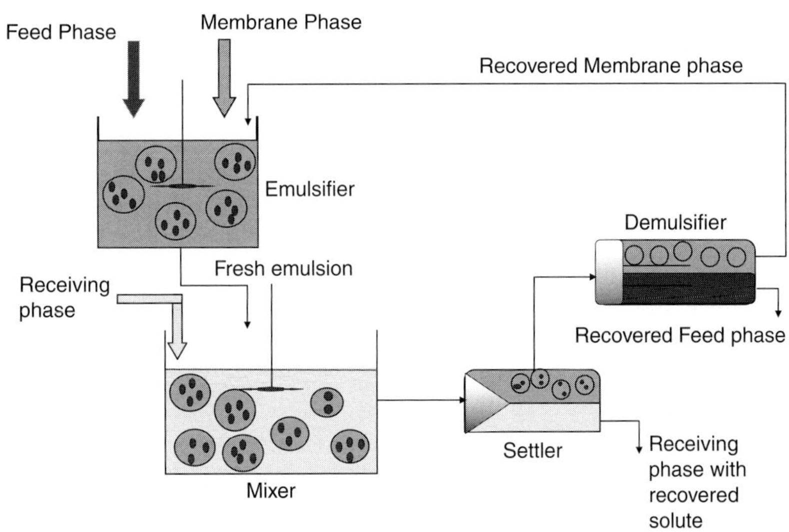

Figure 4.7 Flow diagram of the batch mixer-settler operation.

the process. The carrier must be very selective to the target metal ions in both external and internal aqueous phases while strip agent and type of surfactant must be properly chosen in minimizing the cotransport of water during extraction process. Various types of surfactants have been tested, only a few are suitable, like Span 80 and ECA 4360. There are no special requirements on the choice of the diluents except it should provide high extractant solubility, high-boiling point, and low solubility in feed and stripping phases, nontoxic and cheap.

For the preparation of emulsions, laboratory-scale high-speed mixers or homogenizers are being used by most investigators. Chaudhuri and Pyle [58] used a compressed air driven high-speed stirrer for emulsification. Goto *et al.* [59] performed agitation under ultrasonic irradiation in a test tube. Groeneweg *et al.* [60] used a thermostated vessel provided with a wire stirrer for emulsification. Marr *et al.* [61] used a continuous static homogenization device in industrial scale plant. For emulsion-continuous phase contacting, almost all the researchers used a baffled mixing vessel stirred by a turbine type impeller. In some cases the vessel was thermostated to maintain a constant temperature throughout the extraction process. Some studies were also conducted with continuous column type mixers. Mok *et al.* [57] used a continuous countercurrent-mixing column stirred at 5.5 rps. Jeong *et al.* [62] used pilot scale column contactors for hydrocarbon separation. Goswami *et al.* [63] used a continuous Oldshue-Rushton type stirred column. Designs of some settlers for separation of emulsion and continuous phases have been reported [64].

4.3. Emulsification and surfactants

The term surfactant is a blend of "surface acting agent." Surfactants are usually organic compounds that are amphipathic, meaning they contain both hydrophobic groups (their "tails") and hydrophilic groups (their "heads"). Therefore, they are soluble in both organic solvents and water. A micelle—the lipophilic ends of the surfactant molecules dissolve in the oil, while the hydrophilic charged ends remain outside, shielding the rest of the hydrophobic micelle. Surfactants reduce the surface tension of water by adsorbing at the liquid-gas interface. They also reduce the interfacial tension between oil and water by adsorbing at the liquid-liquid interface. Surfactants are also often classified into four primary groups; anionic, cationic, nonionic, and zwitterionic (dual charge). Some commonly encountered surfactants of each type include:

- *Anionic* (based on *sulfate*, *sulfonate*, or *carboxylate* anions): Sodium dodecyl sulfate (SDS), ammonium lauryl sulfate, fatty acid salts, and so on.
- *Cationic* (based on *quaternary ammonium cations*): Cetyl trimethylammonium bromide (CTAB) and so on.
- *Zwitterionic* (*amphoteric*): Dodecyl betaine, dodecyl dimethylamine oxide, cocamidopropyl betaine, coco ampho glycinate, and so on.

- *Nonionic*: Alkyl poly (ethylene oxide), copolymers of poly (ethylene oxide) and poly(propylene oxide), alkyl polyglucosides, and so on.

HLB stands for hydrophile-lipophile balance. Surfactants with a low HLB are more lipid loving and thus tend to make a water-in-oil emulsion while those with a high HLB are more hydrophilic and tend to make an oil-in-water emulsion. The HLB number of a mixture composed of $x\%$ of surfactants of HLB A and $y\%$ of surfactants of HLB B is obtained by the following formula

$$\mathrm{HLB}(A+B) = (Ax+By)/(x+y) \tag{70}$$

HLB in the range of 1-10 is more soluble in oil than water. Those in the HLB range of 10-20 are more soluble in water than oil.

- HLB ca. 1-3.5: Antifoams
- HLB ca. 3.5-8: Water-in-oil emulsifiers
- HLB ca. 7-9: Wetting and spreading agents
- HLB ca. 8-16: Oil-in-water emulsifiers
- HLB ca. 13-16: Detergents
- HLB ca. 15-40: Solubilizers

4.4. Stripping agents

Acid or base can be used as an internal phase or stripping phase for ELM process depending upon solute to be extracted, for example, Cahn and Li [3] used NaOH solution as the internal phase for phenol removal from wastewater. H_2SO_4 solution was used as the internal phase for removal of ammonia [4-8]. The solute extraction rate also increases with an increase in the amount of internal reagent present in the emulsion.

4.5. Extractant agents

Extractant is present in membrane phase and known as carrier. Carrier is promoting solute transfer through the membrane (facilitated transport), for example, the carriers used in coupled metal ion transport may be acidic carriers like –COOH, $-SO_3H$, or chelating groups (LIX agents) as well as basic carriers like amines or quaternary ammonium salts. Table 4.1 shows some previous studies on metal extraction using different types of extractant.

4.6. De-emulsification

For breaking W/O emulsion in wastewater treatment, electrostatic de-emulsification techniques are generally used [11, 14, 66-69]. Other methods of de-emulsification have been tried to include heat treatment, phase dilution, and high shear [70-72].

Table 4.1 Extractant used in metals extraction using ELM processes Othman *et al.* [65]

Type of extractant	Metal ions	Feed solution	Surfactant	Stripping solution	Diluent
Group 1					
Cyanex 272	Cu	$CuSO_4$	ECA5025 DNP-8	6N H_2SO_4	Tetradecane
LIX 63/LIX 64N		Cu salt	Span 80	HCl	Kerosene
LIX 860		$CuSO_4$	ECA 5025 DNP-8	0.25-2.5M H_2SO_4	Kerosene
Acorga P17/ Acorga P50		Cu salt	Span 80	H_2SO_4	Tetradecane
KELEX 100/ SME 529					
Cyanex 272	Ni	$NiNO_3$	ECA5025 DNP-8	6N H_2SO_4	Tetradecane
D2EHPA		$NiCl_2$		HNO_3	Kerosene
PC-88A		$NiSO_4$	Span 80	dil H_2SO_4	*n*-Heptane
Cyanex 272/ DEHPA	Zn	$ZnSO_4$	ECA5025 DNP-8	6N H_2SO_4	Tetradecane
D2EHPA		$ZnCl_2$	Span 80	HNO_3	Kerosene
DEHMTPA		$ZnSO_4$	ECA5025 Span 80	Thiourea	*n*-Dodecane
D2EHPA	Ag	$AgNO_3$	Span 80	HNO_3	Toluene
D2EHPA	Pb	$Pb(NO_3)_2$	ECA5025	HCl	Toluene
PC-88A	Co	$CoSO_4$	PX 100	H_2SO_4	Paraffin oil
MSP-8	Pd	Simulated Waste	ECA4360 Span 80	H_2SO_4	*n*-Heptane
TOA	Hg	$HgCl_2$	Span 80	NaOH	Toluene
Group 2					
Adogen 364	Cd	Pure Cd	Span 80	NaOH	Dimethyl benzene
Primene JMT	Ag	Ag salt	Not mentioned	H_2SO_4	Tetradecane
Aliquat 336	Mo	Na-Mo salt	Monesan	NaOH	Kerosene, heptane
Aliquat 336	Cr	Cr(IV)	Span 80	NaOH	Kerosene
Group 3					
Calix[4]arene					
Carboxyl/ *p*-tert-octylcalix[*n*] arene (1,4,6)	Rare earth	Lanthanide chloride	$2C_{18}\Delta^9GE$	H_2SO_4	Toluene

Group 1, Ion formation extractant; Group 2, Ion association extractant; Group 3, Solvating extractant.

4.7. Various parameters affecting extraction rate/permeability

The effects of various operating conditions, such as nature of the membrane and its composition, stirrer speed, initial concentration of the solute in the external phase, pH of the external phase, and the temperature on the extraction rate, have been studied by several researchers.

4.7.1. Membrane thickness and its composition

The membrane composition should be such that the membrane should be stable and at the same time should have reasonable solute extraction capability. Studies by a number of researchers [10, 45, 58] reveal that the initial extraction rate increases with increase in surfactant concentration. This is due to the fact that addition of more surfactant lowers the surface tension and results in smaller droplet size of the W/O emulsion, which gives a larger mass transfer area and thus more efficient solute extraction. With low surfactant concentration, it was observed in certain cases that after large contact times, the external phase solute concentration slightly increases. Lee and Chan [10] observed that a large amount of surfactant increases the viscosity of the membrane phase and lowers the diffusivity of the solute through the membrane thereby decreasing solute extraction rate. Hence the surfactant concentration cannot be increased indefinitely and there exists an optimum value beyond which if it is increased, the extraction rate may be adversely affected. Chaudhuri and Pyle [58] found that for extraction of lactic acid with Span 80 as surfactant, with 1-2% of the surfactant concentration, the external phase lactic acid concentration increased after about 3-4 min contact time. Lee and Chan [10] found this optimum value to be 6%.

Chakraborty *et al.* [73] found that the variation of membrane thickness, within the range studied 3-10% surfactant (Span 80) concentration, has almost negligible effect in the nickel(II) extraction rate (see Fig. 4.8). This is presumably because, when the concentration of surfactant is less, the continuous phase resistance is found to be the rate-controlling factor. Extraction was found to be favored by increase in the carrier concentration. This is because of the fact that the carrier concentration in the membrane phase phenomenologically increases the interfacial solute concentration and hence the driving force for extraction providing an increased extraction rate. Also, due to the equilibrium of the stripping reaction, an increase in the carrier concentration increases the equilibrium solute concentration in the internal phase and hence reduces the external phase solute concentration. However, in practice, if the carrier concentration in the membrane phase is increased beyond a certain limit, there is a decrease in the extraction rate. Thien and Hatton [74] suggested this to be about 10% (v/v) of the membrane phase. Hano *et al.* [75] reported that the limiting carrier concentration for lactic acid extraction to be 5% (v/v). Sahoo and Dutta [45] showed that

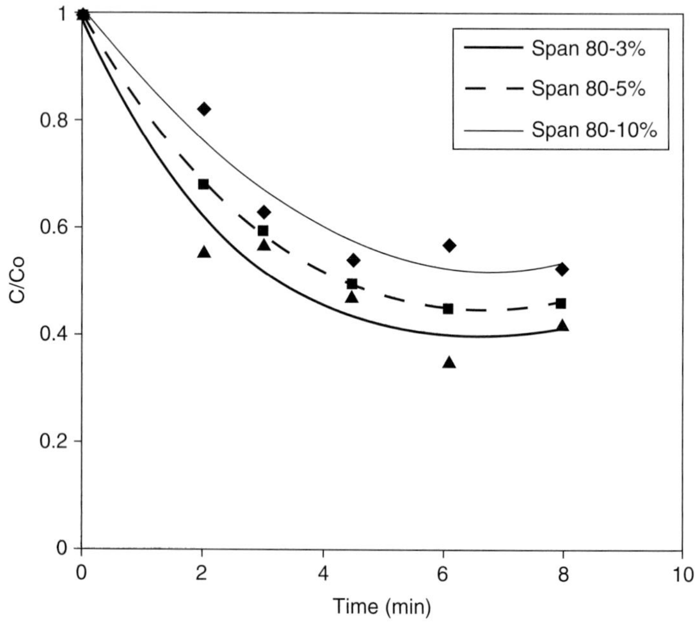

Figure 4.8 Variation of external phase nickel ion concentration with surfactant concentration. From Ref. [73] with permission.

the stability of the emulsion depends on the ratio of surfactant to extractant concentration. They found that a 2:1 ratio of Span 80 and Aliquat 336 generates a stable emulsion.

Chakraborty *et al.* [76] also found that degree of chromium(VI) extraction is a function of time in acidic pH with the extractants Aliquat 336, tri-*n*-octyl amine and their mixtures (see Fig. 4.9). Furthermore, a slight synergistic effect was observed when the two extractants were mixed. Valenzuela *et al.* [77] studied that the initial copper extraction rate was proportional to the concentration of oximic carrier in the organic phase and to the metal content in the acid mine drainage.

Recently Othman *et al.* [78] improved the emulsion stability with the increase in surfactant concentration. They found that increasing concentration of surfactant from 1 to 7% (w/v) increases the stability of the liquid membrane which leads to the decrease in the breakup rate, hence the extraction degree of silver was also increased. Extraction of silver is enhanced by increasing the concentration of Cyanex 302 from 0.03 to 0.05 M, while the emulsion breakup decreased and further increase of carrier concentration does not affect the extraction performance very much. Sabry *et al.* [79] observed that the emulsion stability improves by increasing the surfactant concentration till 8% v/v then the emulsion

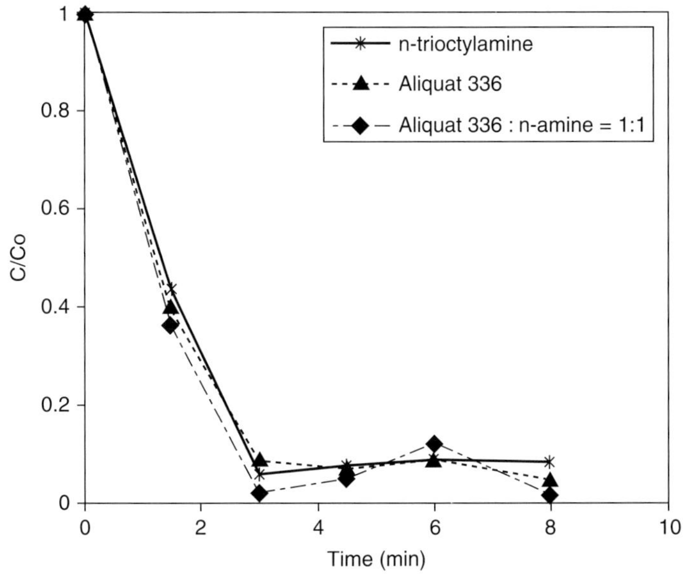

Figure 4.9 Variation of external phase chromium ion concentration with different types of carriers. From Ref. [76] with permission.

stability becomes constant. On the other hand, an increase in the surfactant concentration decreases the removal efficiency of lead due to mass transfer resistance caused by the surfactant film.

4.7.2. Stirring rate

Another parameter affecting extraction to a large extent was found to be the stirring rate. Experiments by various researchers [10, 41, 45, 80] reveal that the efficiency of ELM extraction increases with increase in stirring speed. This is due to the fact that with increase of stirring rate during extraction of solute, the sizes of the emulsion droplets become smaller providing more mass transfer area. However, as the stirring rate is increased, the emulsion droplets become more unstable and more is the leakage of the internal phase that adversely affects the solute extraction rate at larger extraction times. This theory has been validated by the experimental results of Lee and Chan [10], which show that as the stirring rate exceeds 400 rpm, the external phase concentration falls to a minimum and then bounces back and increases. This indicates that there are two competing processes involved in the solute transport, that is, the diffusion of the solute through the emulsion membranes into the internal phase and the leakage of the internal solution due to breaking of some emulsion globules. Kargari *et al.* [81] found that by increasing agitation speed, the shear forces which acts on the emulsion globules increased and this makes the globules smaller, resulted more gold extraction rate. Othman *et al.* [78] found that the speed of agitation plays a major role in the mass transfer rate of silver through the liquid membrane. It was observed that increasing the speed of agitation from 3.33 to 4.166 Hz (200 to 250 rpm) increased the rate of extraction but further increase in the speed of agitation to 6.66 Hz (400 rpm) resulted in reduction in the degree of extraction from 95 to 50%. Actually an increase in the level of agitation would increase the interfacial area and the mass transfer coefficient. However, this is true up to certain level of agitation beyond which an increase in the level is likely to break the emulsion droplets thereby reducing overall enrichment and the degree of extraction.

4.7.3. Feed phase solute concentration

The feed phase solute concentration also influences the solute extraction and degree of extraction increases with decrease in the solute concentration. Hence it may be concluded that ELM method is more suitable for solute removal from dilute aqueous solutions. As the solute concentration is increased, the transport rate may decrease largely due to the reduced capacity of the internal phase to strip the transported solute. Also at a high solute concentration, the internal droplets in the peripheral region are more readily saturated with the solute. Therefore, the complex must diffuse through the membrane phase to a region deeper inside the emulsion globule to release the solute in the receiving phase such that the internal mass transfer resistance is significant.

When the solute concentration is low, the external mass transfer is controlling and the extraction rate is higher than that at high solute concentration. This observation of high solute transport rate and recovery at low solute concentration is important from the process design point of view. It is possible to achieve the required degree of separation with small solute concentration in the feed using a multistage train of mixer-settler devices in a practical recovery process. It was observed that with increase in the initial concentration of nickel(II), rate of separation will be slower as the separation progresses although the initial rates of separation were same [73]. With increase in the initial concentration of nickel(II), the percentage of extraction decreases as shown in the Fig. 4.10. By increasing the percentage of the initial concentration, extraction decreases to almost 7.5%. This is due to high metal ion concentration, which gets limited extractant for complex formation. While by decreasing initial concentration from 150 to 100 ppm, percentage of extraction increases to 15%. This is due to low metal ion concentration compared to large amount of extractant.

4.7.4. Feed phase pH

The pH of the external phase has a very important role in solute extraction particularly in case of carrier-mediated transport. This may be due to the fact that in many cases, the equilibrium constant of the reaction between solute

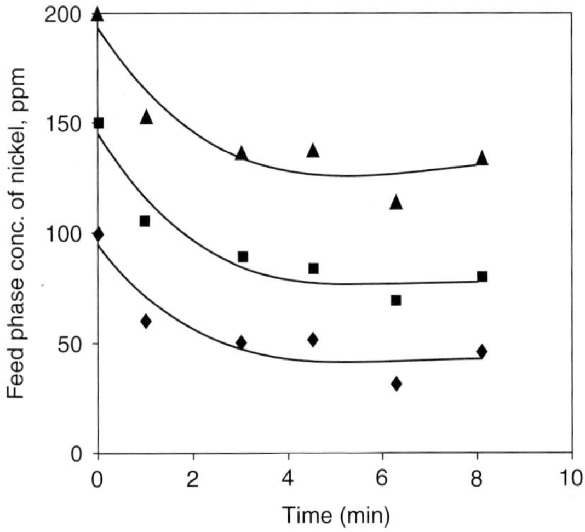

Figure 4.10 Variation of external phase nickel ion concentration with initial external phase nickel concentration. From Ref. [73] with permission.

and the carrier at the external surface of the emulsion globules is sensitive to the external phase pH.

Strzelbicki *et al.* [82] showed that chromium(VI) extraction rate decreases sharply after a pH value of 6.0. In case of phenol extraction, Wan *et al.* [11] observed that low values of pH favored extraction as the undissociated form of phenol which is the only form soluble in oil is predominant only at pH < 4.0. Liu and Zhang [83] showed that the extraction rate of Samarium (Sm^{3+}) in low pH range is inversely proportional to the H^+ concentration. Sahoo and Dutta [45] noted that the solute distribution coefficient for cephalexin extraction increased with increase in pH up to a pH value of 9.5 beyond which the distribution coefficient appears to decrease marginally. Chakraborty *et al.* [76] found similar result for extraction of chromium (see Fig. 4.11).

If pH in the external phase is kept low, then the extraction will be faster and high in percentage. The chromate ion may exist in aqueous phase in different ionic forms ($HCrO_4^-$, CrO_4^{2-}, $Cr_2O_7^{2-}$, $HCr_2O_7^-$). Strzelbicki *et al.* [81] reported that the distribution coefficient decreases sharply when the equilibrium pH exceeds 6 for the extraction of chromium(VI) oxyanions with Aliquat 336. This is because the quaternary ammonium salts exhibit reasonable extraction abilities only toward monovalent oxyanions, that is, $HCrO_4^-$, while CrO_4^{2-} is the dominant species at that pH. This anion is predominant only at low pH and hence the faster depletion of solute occurs if the external phase is acidic. Chakraborty *et al.* [84] also found that in case of competitive transport of copper(II) and nickel(II) different extraction mechanism controls the removal of metals from the

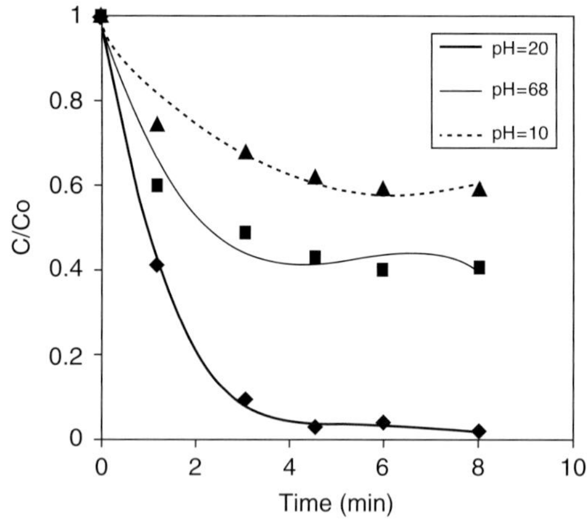

Figure 4.11 Effect of external phase pH on extraction of chromium. From Ref. [76] with permission.

external aqueous solution under differing feed solution acidity conditions. When feed pH is around 3.5, as nickel's permeation rate decreases, value of separation factor of is around 2.0 and it is found that 90% recovery (maximum) of copper(II) is possible along with 45% recovery of nickel(II) at pH 3.5 (see Fig. 4.12).Valenzuela *et al.* [77] found that the initial copper extraction rate was inversely proportional to the hydrogen ion concentration in the external aqueous feed solution.

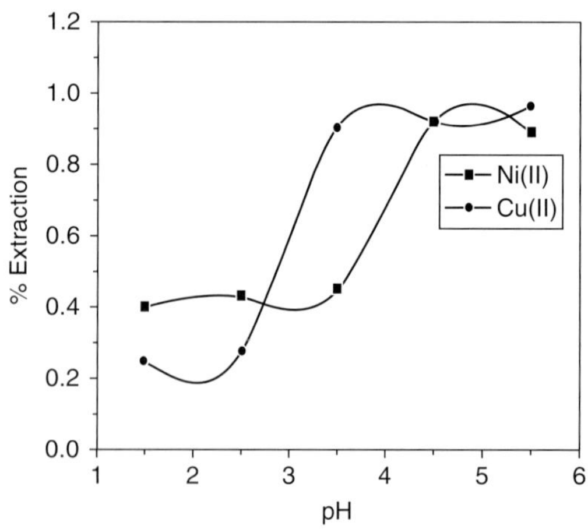

Figure 4.12 Extraction profile of Cu(II) and Ni(II) from binary mixture. From Ref. [84] with permission.

4.7.5. Volume ratio of emulsion to external phase (treat ratio)

The volume ratio of emulsion to external phase when increased tends to increase the extraction rate due to increase in the capacity of the membrane and the internal phase for enhanced permeation and stripping of the solute. Furthermore, an increase in this ratio also leads to an improvement in the dispersibility of the emulsion system. Using silver ion facilitated O/W/O membrane experiments have been carried out to separate toluene from *n*-heptane [85]. It has been found that more than 50% extraction of toluene is possible within 20 min and maximum separation factor is 12 at 3 min of operation. The value of R (ratio of the volume of the surfactant solution to that of the feed in emulsion) is fixed at 0.3 and K (ratio of the volume of the solvent to that of the emulsions) varies from 2 to 4. For low values of K, the emulsion aggregates tend to coalesce and the mass transfer area is small. Besides, the concentration driving force for low values of K will be small because the concentration of the permeated feed in the solvent phase is relatively higher than for high values of K, even when the amount of feed permeated into the solvent phase is the same. With gradual increase in K, the separation factor as well as % extraction of toluene (see Figs. 4.13 and 4.14) tends to increase because the possibility of emulsion coalescence

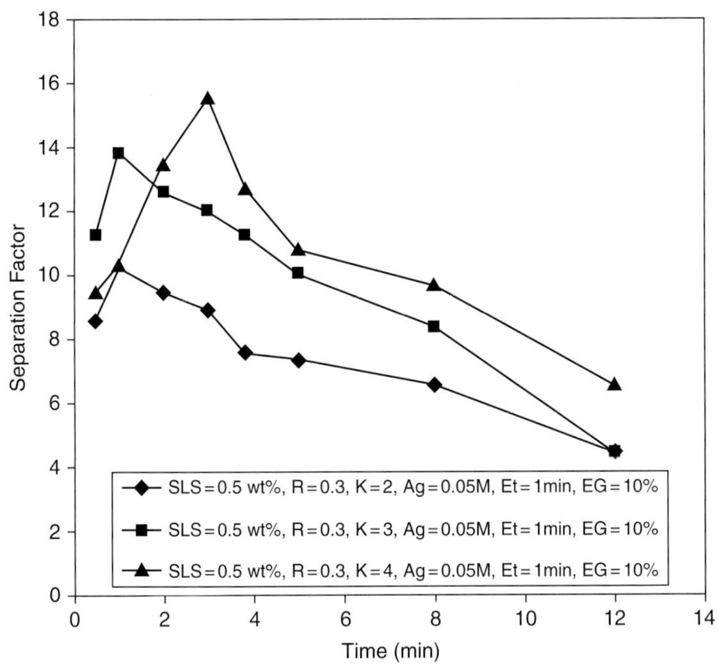

Figure 4.13 Effects of the relative amount of solvent on separation factor. From Ref. [85] with permission.

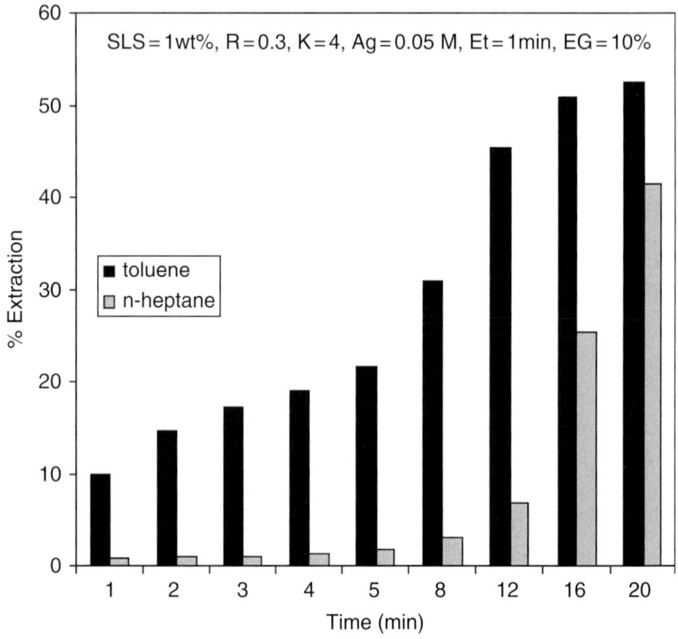

Figure 4.14 Effects of the relative amount of solvent on % extraction. From Ref. [85] with permission.

decreases and the concentration driving force begins to increase. However, large emulsion volumes may also increase leakage of the internal solution into the external phase, somewhat retarding the extraction [10].

4.7.6. Internal stripping reagent concentration and the volume fraction of the internal phase

The solute extraction rate also increases with an increase in the amount of internal reagent present in the emulsion. The amount of internal reagent can be changed by changing the internal reagent concentration and the volume fraction of the internal phase. In case of competitive transport of copper(II) and nickel(II) keeping feed phase pH at 3.5, if internal strip phase acid concentration is varied from 1.0 to 2.0N recovery of copper(II) will increase up to 99% but at same time recovery of nickel(II) will also increase. Figure 4.15 illustrates the change of separation factor along with the internal strip phase acid concentration. Maximum value of separation factor (β) is found to be 2.10 at internal strip phase acid concentration 1.4N [84].

Valenzuela *et al.* [77] showed the influence of sulfuric acid concentration as stripping agent on the extent of copper extraction. Extraction is practically constant regardless of acid content in the strip liquor, indicating the reduced effect of the back-extraction step on the overall process rate.

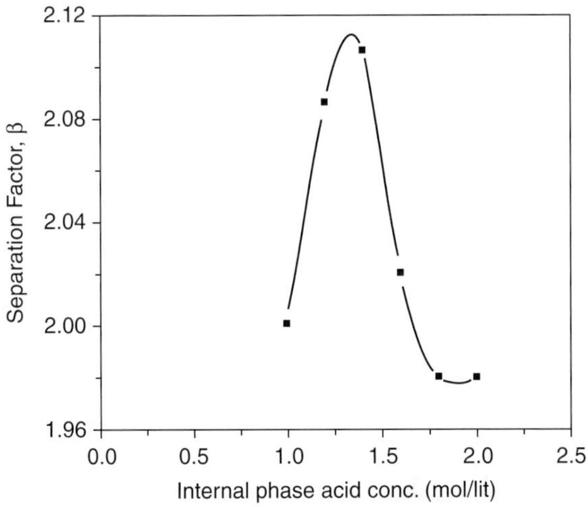

Figure 4.15 Effect of internal acid concentration on separation factors of Cu(II) and Ni(II) ions. From Ref. [84] with permission.

Sabry *et al.* [79] showed that the removal efficiency of lead is increased by increasing the acidity in the stripping phase. The internal phase volume fraction, if increased beyond an optimum value, may also result in phase inversion. However, the internal reagent concentration as well as the internal phase volume fraction cannot be increased indefinitely. If the concentration of the internal reagent is beyond some limiting value, it will be harmful to the stability of the membrane and will increase water permeability in the membrane leading to obvious swelling due to high osmotic pressure gradient between the internal and external phases.

4.7.7. Temperature

Another factor that affects the solute extraction is the temperature. This is significant in case of reaction-controlled processes such as thiocyanate extraction by trimethyl ammonium chloride [86]. Thiocyanate ions are rapidly transported into the membrane with increase in temperature. This is due to the strong influence of temperature on the reaction rate constants.

By summarizing the results of studies by various researchers as mentioned above, it can be concluded that for successful use of ELMs for practical purposes, the different operating parameters should be carefully chosen so as to maximize the separation yield.

4.8. Hydrodynamics of liquid membranes

The diameters of emulsion drops as well as the internal droplets are important particularly in the modeling of mass transfer through ELMs. While most workers have used experimentally determined values by direct photographic techniques, the designer often calls for predictive models to estimate drop sizes in stirred vessels related to vessel geometry and physical properties of the dispersion system as reported in the literature. Most of them are based on the expression derived by Hinze [87], which compared the restoring stress due to interfacial tension with the inertial stress due to turbulence and is described as

$$\frac{d_{\max}}{d_1} \propto We^{(0.6)} \tag{71}$$

where d_1 is the impeller diameter.

Legisetty *et al.* [88] and Calabrese *et al.* [89] considered the effect of dispersed phase viscosity on drop breakage. In their models, attempts have been made to incorporate the effect of dispersed phase holdup on drop breakage by considering the turbulent velocity fluctuations to be damped by the dispersed phase drops as

$$\overline{u^2}(d)\mid_{\varphi} = \overline{u^2}(d)\mid_{\varphi=0}(1 + a\varphi)^{-2} \tag{72}$$

Although the models are capable of predicting the diameter of the internal droplets reasonably well, they underpredict the drop diameter of the liquid membrane emulsion drops in batch stirred vessels. Kataoka and Nishiki [90] proposed the following correlation for size of W/O/W macrodrops based on analysis of their experimental data.

$$d_{32} = 6.6 \times 10^{-2} \left[\frac{We \frac{\Delta\rho}{\rho_C} \mu_C}{\mu_M} \right]^{-0.75} d_I \tag{73}$$

Ohtake *et al.* [35] obtained the following correlation:

$$\mathrm{d}_{32} = 0.12 We^{-0.5} \quad \text{for } \mu_d < 160\,\mathrm{m\,Pa\,S} \tag{74}$$

$$d_{32} = 0.002 \mu_d^{0.8} We^{-0.5} \quad \text{for } \mu_d > 160\,\mathrm{m\,Pa\,S} \tag{75}$$

Rautenbach and Machhamer [91] developed a correlation for prediction of drop sizes of emulsion globules in stirred column contactors as

$$d_{32} = 0.024 \left[\frac{n^2 d_I^3 \rho_C}{\sigma} \right]^{-1.2} \left[\frac{n d_I^2}{\nu_C} \right]^{0.6} d_I \tag{76}$$

which is similar to the correlation used for liquid-liquid extractors. Most of the researchers conducting studies on drop size estimation either by experimental methods or by empirical correlations, considered the steady-state sauter mean drop size, d_{32}.

Very few studies have been conducted so far to determine the transient drop size distributions used to elucidate the dynamic processes related with breakage and coalescence of the dispersed phase. Bajpai *et al.* [92] proposed a method for the measurement of the unsteady-state drop size distributions by a detergent stabilization technique. Hong and Lee [93, 94] measured photographically the transient drop size distributions for a number of different systems. Chatzi *et al.* [95] used a laser diffraction technique for the online measurement of the drop size distribution in an agitated vessel and investigated the effect of agitation rate and temperature on the transient drop size distribution.

4.9. Leakage and stability in emulsion liquid membranes

ELMs since their development in 1968 have shown tremendous promise in a wide variety of industrial separations. But the problem that inhibits the application of this technology in industrial equipment is the loss of extraction efficiencies that often occur in these systems due to lack of stability of the emulsion globules. Stability of double emulsions is generally understood as the resistance of the individual globules against coalescence. The breakdown of W/O/W type dispersions is described through several possible mechanisms [96] which include (i) coalescence of the internal aqueous droplets into larger internal droplets (ii) coalescence of the emulsion globules suspended in the external continuous phase (iii) expulsion of the internal droplets following rupture of the thin membrane film during interaction of the internal and external continuous phases, and (iv) swelling or contraction due to water permeation through the oil membrane by diffusion in case of W/O/W emulsion.

The first study on stability was performed experimentally by Hochchauser and Cussler [20]. They used 0.1 M sodium dichromate solution as the internal phase and water as the external bulk phase with Span 80 as the surfactant dissolved in an organic membrane phase. The solubility that of dichromate found in the external water phase was due to the rupture of the membranes. The stirring rate and the concentration of sodium dichromate were found to have no effect on breakup. They also observed a rapid breakage during an initial period with no further breakage at longer times. Martin and Davies [97] along with their mass transfer studies on the extraction of copper from an aqueous solution also performed stability study using sulfuric acid as the encapsulated phase. The organic phase used by them consisted of a commercial chelating agent (LIX 64N), an organic diluent and a nonionic surfactant. Their study revealed that the emulsion breakup was dependent on factors like the operating conditions in the mixing device, size of the subdrops, type of surfactant used and impeller speed. Breakage was also found to increase linearly with time. This is however in contrast to the findings of Kita *et al.* [98], Kondo *et al.* [99], Takahasi *et al.* [100], all of whom measured the stability of double emulsions using a tracer technique. Release of the tracer from the subdrops was a direct measure of the breakage of emulsion globules as the tracer transport through the membrane was negligible. The effects of variables like tracer or salt concentration, agitation time, size of the emulsion drops, and concentration of the emulsifying agent and pH of the internal droplets on breakage were studied in the above investigations. Shere and Cheung [101] measured the stability of LSMs as a function of time using NaOH as a tracer and Span 80 as the surfactant. Stroeve *et al.* [102] used this approach too on the 1934 study of Taylor [103] who was the first to study both experimentally and theoretically the deformation and breakup of a drop of a newtonian liquid suspended in another immiscible newtonian liquid undergoing simple shear and plane hyperbolic flows. Similar studies have been conducted experimentally [104–106] using the tracer technique to study emulsion stability and proposed a breakup rate constant K_b, which is defined by

$$\ln(1-\varepsilon) = -K_b t \tag{77}$$

$$\varepsilon = \frac{V_e C_e}{V_i^0 C_i^0} \tag{78}$$

where V_e and C_e are the volume and concentration, respectively, of the external phase and V_i^0 and C_i^0 are the initial volume and concentration, respectively, of the encapsulated phase. They observed that the breakage constant K_b decreases with the increase in surfactant concentration up to a critical value, beyond which it remains practically constant. Wan *et al.* [11] studied the effect of carrier concentration on the stability of ELMs for extraction of phenol. The carrier used was N_{503}, that is, *N,N*-di(l-methylheptyl)acetamide. Stability was slightly affected when the carrier concentration varied from 2.0 to 5.0%. However on further increasing the concentration, particularly beyond 10%, the stability was affected significantly. Chakraborty *et al.* [107] found that the leakage phenomena, which reflect the stability, are influenced by pH of the feed phase, speed of agitation, emulsion drops size per unit specific interfacial area, surfactant concentration, pH in inner aqueous phase, and the presence of different tracers. They observed that percentage breakage of emulsion drop is different with different tracers (copper sulfate, sodium thiocyanate, and sodium chloride) as internal phase (see Fig. 4.16).

4.10. Internal droplet size distribution

In the separation process using ELMs, the dispersed drop sizes as well as internal droplet sizes are important in determining efficiency of extraction, stability of the liquid membranes and evaluating the interfacial contact area.

Chakraborty *et al.* [108] studied separation of toluene from *n*-heptane using ELMs and examined photomicrographs of the emulsions prepared using 4000 and 8000 rpm immediately after preparation. In both cases, the diameters of the dispersed emulsion droplets were in the 10–120 μm range and those of the internal oil droplets were in the 0.001–0.12 μm range.

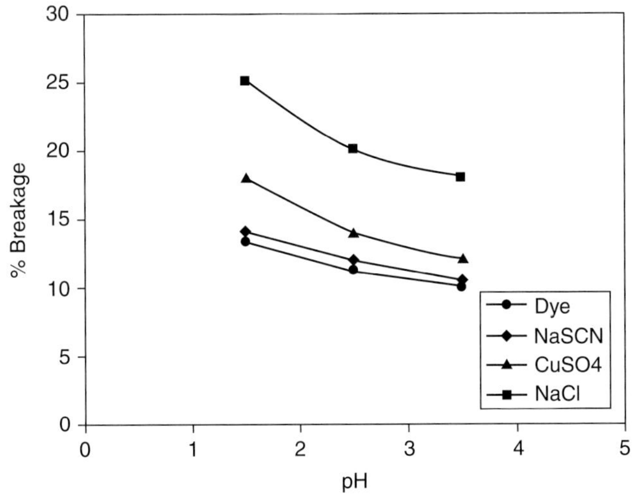

Figure 4.16 Effect of tracers' concentration on membrane breakage. From Ref. [107] with permission.

However, the dispersed emulsion drops prepared at 4000 rpm contain relatively less number of internal droplets (see Fig. 4.17A) than those prepared at 8000 rpm (Fig. 4.17B).

From Fig. 4.18, which showed log normal as well as cumulative internal droplet size distribution, it can be observed that emulsion prepared using 4000 rpm impeller speed resulted in a narrow size distribution with internal droplets mean diameter 0.024 μm. In contrast, emulsion prepared using 8000 rpm impeller speed yields broader droplet size distribution with lower mean diameter of 0.015 μm.

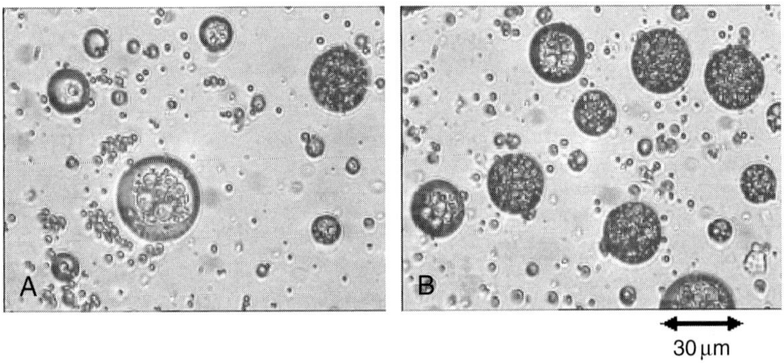

Figure 4.17 Photomicrograph of emulsion prepared at (A) 4000 rpm (B) 8000 rpm. From Ref. [108] with permission.

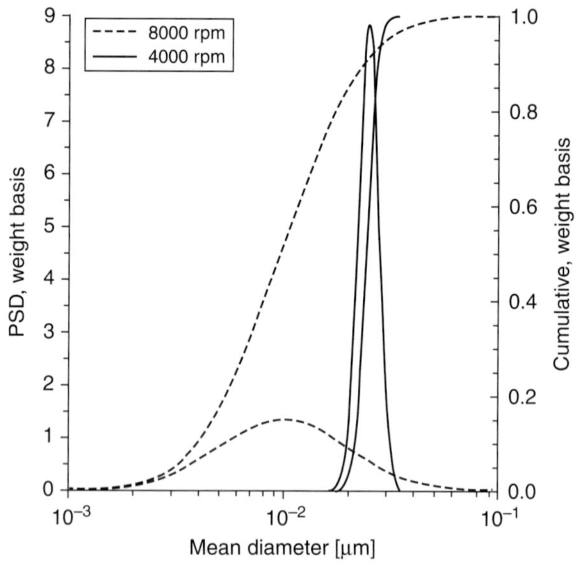

Figure 4.18 Effect of stirring speed on internal droplet size distribution. From Ref. [108] with permission.

By selecting optimum impeller speed during emulsification, surfactant concentration, volume ratio of surfactant solution, carrier concentration, and a suitable feed phase composition, uniformly distributed stable emulsion could be obtained to provide a high separation factor and higher mass transfer rate.

5. Applications of ELM Technology

The extraction capabilities of liquid membranes have been used successfully in many areas. Since 1968 efforts have been made for successful industrial application of liquid membrane technology. Emphasis has been on facilitated transport of LSMs. Some of the possible commercial applications are discussed below.

5.1. Metal ion extraction

Studies have been made on the mechanism of separation, process kinetics, mass transfer modeling, and engineering evaluation with metals like copper, zinc, cadmium, cobalt, nickel, mercury, uranium, chromium, rhenium, and several others, including noble metals like gold and silver, lanthanides and rare earths. Table 4.2 presents some examples.

Table 4.2 Different metals extraction using ELM process

S. No.	References	Metal ion recovered and carriers used	Parameters/Modeling studied
1	Martin and Davies [97]	Cu from dilute aqueous solutions	Membrane composition, pH, acid content of the feed, internal phase, agitation speed
2	Lee *et al.* [21]	Cu transport with LIX-64N and LIX-65N	Competitive transport of copper and nickel
3	Kondo *et al.* [23]	Separation of Cu using benzoylacetone	The rates of complex formation and the effects of process variables on the mass transfer rate
4	Strzelbicki *et al.* [109]	Separation of Cu, Co, and Ni using LIX-70 and Kelex-100	The multistage separation with fresh and reused liquid surfactant membrane
5	Bock and Valint [110]	Recovery of U from wet process phosphoric acid using TOPO	The liquid membrane process is superior to solvent extraction systems
6	Mar *et al.* [61]	Zn removal using (D2EHPA)	Effect of surfactant on the rate of interfacial mass transfer
7	Bart *et al.* [111]	Recovery of Cu using LIX and Acorga	A pseudo-homogeneous reaction model was assumed to calculate column height
8	Hirato *et al.* [112]	U recovery using TNOA	Two-stage counter current liquid membrane process is very much effective
9	Eroglu *et al.* [113]	Extraction of Sr ions using D2EHPA	Extraction increases with volume of emulsion, stirring rate, D2EHPA concentration and decreases with pH of the inner phase
10	Chakraborty and Datta [114]	Te^{4+} transport using D2EHPA	Stripping phase acid concentration, ratio of volume of emulsion to feed, Biot number
11	Chakravarti *et al.* [115]	Extraction of Cr(VI) by Aliquat 336 through coupled countertransport	Oil-membrane constituents and different types of chemical reactions at the interface

(continued)

Table 4.2 (*continued*)

S. No.	References	Metal ion recovered and carriers used	Parameters/Modeling studied
12	Li *et al.* [116]	Transport of Hg(II) ions using TOA	Variiation of concentrations of HCl, KCl, TOA, Span 80, and NaOH
13	Li *et al.* [117, 118]	Separation of Cd using TOA	Effect of concentrations of HCl, KI, TOA, Span 80, and NaOH
14	Longquan *et al.* [119]	Separation of Co and Ni in presence of EDTA with P-204	EDTA acts as a masking agent for Ni form a stable chelate in the pH range of 4.4-5.2
15	Xingrong and Xiujuan [120]	Extraction of Sm^{3+} using D2EHPA	Two simplified mass transfer equations are derived and simulated
16	Kasaini *et al.* [106]	Selective recovery of Co and Ni with PC-88A	Higher permeation rate of Co compared to Ni at pH higher than 5.5
17	Katsushi *et al.* [121]	Synergistic extraction of Ni with LIX63-DOLPA and LIX63-D2EHPA	LIX63-DOLPA has higher synergistic effect than LIX63-D2EHPA mixture
18	Kondo and Matsumoto [122]	Permeation of In using DISPA	Control of the permeation process by the interfacial reaction
19	Kulkarni *et al.* [123]	Recovery of Ni using D2EHPA	Membrane viscosity, volume ratios of emulsion to feed, carrier, nickel, and strip acid concentration
20	Sznejer and Marmur [124]	Removal of Cd by D2EHPA	Separation is independent of the water to oil phase ratio of the inner emulsion
21	Ye *et al.* [125]	Transport of Na^{+} ions by *p*-tert-butylcalix[6]arene and K^{+} ions by *p*-tert-butylcalix[8]arene	The driving force for the transport is the pH gradient between the source and receiving phases
22	Serga *et al.* [126]	Extraction of Ni(II) using D2EHPA	Direct current applied to the process contributes to the complete extraction
23	Chakraborty *et al.* [127]	Extraction of Cu from wastewater	Optimization of the process parameters

24	Urtiaga *et al.* [128]	Cd removal from wet phosphoric acid using Cyanex 302	Rate varied in the order NDSX > ELM > SLM
25	Kulkarni *et al.* [129]	Recovery of Ni(II) using D2EHPA	Stability with hydrochloric, sulfuric, nitric, and methane sulfonic acid as strippant
26	Uddin and Kathiresan [130]	Extraction from ternary (copper, nickel, and cobalt) mix using polyethylene glycol as bifunctional surfactant	Characterization of the reactions (stoichiometry and reaction order) through equilibrium and kinetic studies
27	Said *et al.* [50]	Permeation of Cs with crown ether (18C6).	Analysis and testing of pertraction rates on a new model.
28	Kulkarni and Mahajani [131]	Extraction of Mo(VI) with Aliquat 336	Residence time, membrane viscosity, extractant, strip phase concentration, rate of agitation, phase ratio
29	Kulkarni *et al.* [132]	Separation a of U(VI) using TOPO	Strip phase concentration and feed phase pH are two critical parameters
30	Sayed [133]	Separation of U using Aliquat 336	Permeation rate and percent permeation of uranium
31	Valenzuela *et al.* [77]	Cu removal from an acidic mine drainage using a salicylaldoxime	Carrier concentration, pH, and metal content in the mine water
32	Gurel *et al.* [134]	Pb removal from storage battery industry wastewater	Amount of organic solvents, external aqueous phase pH, concentration of surfactant, carrier, internal stripping phase, and ligand
33	Kargari *et al.* [135]	Separation of gold(III) using MIBK	Only Au^{3+} ions transported from Au^{3+}, Pd^{2+}, Pt^{4+}, Cu^{2+}, and Fe^{3+} mixture
34	Othman *et al.* [78]	Separation of Ag using Cyanex 302	Agitation speed, surfactant, carrier concentration, treat ratio, and types of diluents, stripping solutions
35	Sabrya *et al.* [79]	Removal of Pb(II) using D2EHPA	Removal efficiency 99-99.5% at the optimum operating conditions
36	Gasser *et al.* [136]	Co(II) from thiocyanate medium by CYANEX 923	Recovery 95% in the inner phase
37	Fouad and Bart [137]	Extraction of Zn using D2EHPA	Hollow-fiber extractor eliminated leakage and swelling of emulsion

5.2. Removal of weak acids/bases

Weak acids like phenol and cresol and weak bases like ammonium and amines have been successfully removed from wastewater. Among them, the separation-concentration of phenol has been intensively investigated. In most of these systems phenol extraction through the membranes was based on the solubility difference of phenol between the aqueous and organic phases. Phenol, being somewhat oil soluble, was transferred into the membrane phase and then diffused across the membrane into the internal aqueous caustic phase where it was immediately neutralized by the caustic and tied up as phenolate that is insoluble in oil and consequently could not diffuse back again. As a result, a high phenol concentration gradient was maintained across the liquid membrane and thus the phenol was readily removed from the external aqueous phase.

Phenol extraction by ELMs has been studied [3-5, 7, 29, 41]. Teramoto *et al.* [6] removed phenol and cresol from wastewater. Wan *et al.* [11] increased the solubility of phenol in the membrane phase by incorporating a mobile carrier [*N*,*N*-di(1-methylheptane)acetamide] and thus improved the efficiency of phenol removal. Cahn and Li [138] studied removal of ammonium sulfide from wastewater. Amine extractions have been carried out [39, 40].

Dzygiel and Wieczorek [139] used ELMs for separation of amino acids. Separation of binary amino acids from aqueous solutions by solvent extraction and LSM was compared by using D2EHPA as an active carrier [140]. The distribution ratios were measured by batch SX experiments. Important variables affecting LSM permeation experiments were systematically studied. Experiments on liquid membrane extraction of 2-chlorophenol examined the extraction efficiencies under various operating conditions [141]. Emphasis was placed on identifying the optimal conditions of operating variables in the extraction process and in the demulsification of exhausted water-in-oil (W/O) emulsion. Correia and Carvalho [47] and Lin *et al.* [49] studied 2-chlorophenol extractions by LSMs. Devulapalli and Jones [72] separated aniline from dilute aqueous solution. Up to 99.5% of the aniline is removed from feed solutions. Leakage is minimal and swelling is only about 3% after 5 min of processing. Datta *et al.* [142] also studied removal of aniline from aqueous solution in a mixed flow reactor. The maximum removal of aniline obtained in this study was 98.53%. Taylor-Couette flow was applied to treat model industrial wastewaters containing phenols and selected substituted phenols (hydroquinone, three chlorophenols, and two nitrophenols) [143].

5.3. Separation of inorganic species

Apart from ammonia, some other inorganic species extracted by liquid membranes are strong acids like nitric acid and thiocyanate ions from aqueous solutions using carrier-mediated coupled transport process.

Lin and Long [41] applied ELM to effectively separate nitrate ions (94% efficiency) from water with tri-*n*-octyl amine as the extractant and sodium carbonate as the internal phase. H^+ ions were also transported along with NO_3^- from external to the internal phase (cotransport). Kobya *et al.* [86] studied the kinetics of thiocyanate ion removal from aqueous potassium thiocyanate solution by counter-transport process using quaternary ammonium salt of hexadecyl trimethyl ammonium chloride as the carrier and sodium chloride solution as the stripping phase.

5.4. Hydrocarbon separations

Liquid membrane technology has been applied to a great extent for separation of mixtures of saturated and aromatic hydrocarbons. Investigations reveal that the LSM process offers potential for dearomatization of petroleum streams like naphtha and kerosene to meet product specifications for naphtha cracker feedstock and aviation kerosene, respectively [25, 63, 85, 144-146]. The separation is based on a simple permeation technique and occurs due to the difference in solubility and diffusivity of permeating species through the membrane. Kato and Kawasaki [70] conducted studies on the enhancement of hydrocarbon permeation by the use of a polar additive like sulfolane or triethyl glycol. Sharma *et al.* [147] enhanced the selectivity of the membrane by several orders with the addition of a carrier. Chakraborty *et al.* [85] used cyclodextrins to enhance the separation factor and removal efficiency of aromatic compound.

5.5. Biochemical and biomedical applications

ELM has promise in the fields of biotechnology and biomedicine and has found application in the separation of organic acids, extraction of fatty acids and amino acids, purification of antibiotics, enzyme catalyzed reactions, and detoxification of blood, and so on. Table 4.3 shows some examples of these studies.

Table 4.3 Various biomedical and biochemical applications

Sr. No.	References	Compound recovered
1	Yan *et al.* [148]	Separation of acetic acid from wastewater
2	Thien and Hatton [73], Itoh *et al.* [149], and Teramoto *et al.* [150]	Separation of amino acids using D2EHP as carrier and industrial surfactants as stabilizers
3	Yan *et al.* [151]	Batch extraction of lactic acid from wastewater using NaOH as the internal reagent
4	Eyal and Bressler [152]	Separation of carboxylic acids and amino acids

(continued)

Table 4.3 (*continued*)

Sr. No.	References	Compound recovered
5	Scholler *et al.* [153], Chaudhuri and Pyle [58], and Juang *et al.* [154]	Amine-mediated extraction of lactic acid with Na_2CO_3 as the stripping reagent
6	Reisinger and Marr [155] and Juang *et al.* [156]	Separation of citric and lactic acids from multiacid mixtures
7	Hano *et al.* [157]	Separation of penicillin G using ECA 4360J (polyamine type) as a surfactant
8	Nakano *et al.* [158]	Formation and extraction of Ag-π complex of polyunsaturated fatty acid
9	Mok *et al.* [57]	Extraction of penicillin G using ECA 4360J and amberlite LA2
10	Lee *et al.* [159]	Penicillin G extraction with surfactants Span 80/PARABAR
11	Sang *et al.* [160]	Penicillin G separation using Oldshue-Rushton type countercurrent extraction column
12	Sahoo and Dutta [45]	Extraction of cephalexin using 7-aminodeacetoxy cephalosporic acid with Aliquat 336 as carriers
13	Sahoo *et al.* [161]	Extraction of cephalosporin-C (CPC) using Aliquat 336
14	Lee [162]	Continuous extraction of penicillin G using an optimal surfactant composition
15	Pal *et al.* [163, 164]	Immobilization of multienzymes (α-glucosidase and glucose oxidase) in ELMs to use as EELM reactors
16	Bayraktar [165]	Optimization of the separation of DL-tryptophan from aqueous solution
17	Habaki *et al.* [166]	Simulation of permeation of erythromycin, macrolide antibiotics, into multilayer liquid membrane
18	Ishizu *et al.* [167]	Separation of compactin (ML-236B), a highly effective pharmaceutical compound
19	Lee [168]	Improvement of emulsion stability by a dilute polymer solution and extraction of penicillin G
20	Lee and Lee [169]	Extraction of neutral species by a mixture of an organoboronic acid and a quaternary ammonium as carriers
21	Kaghazchi *et al.* [170]	Improvement of extraction efficiency of L-lysine by an increase in the pH of the feed phase, the concentration of $[H^+]$ in the internal phase and carrier

5.6. Preparation of fine particles using emulsion liquid membrane

Size-(submicrometer-sized) and morphology-(spherical) controlled composite Gd–Eu oxalate particles were prepared by Hirai *et al.* [171] using W/O/W system. The oxalate particles thus prepared were calcined in air to obtain $Gd_2O_3 : Eu^{3+}$ phosphor particles and in sulfur atmosphere to obtain $Gd_2O_2S : Eu^{3+}$ phosphor particles. These submicrometer-sized spherical phosphor particles showed photoluminescence properties with emission peak at 614 nm for $Gd_2O_3 : Eu^{3+}$ and 628 nm for $Gd_2O_2S : Eu^{3+}$.

Hirai *et al.* [172] had prepared Y_2O_3 nanoparticulate thin films using an emulsion liquid membrane (water-in-oil-in-water (W/O/W) emulsion) system, consisting of Span 83 (sorbitan sesquioleate) as a surfactant and VA-10 (2-methyl-2-ethylheptanoic acid) as an extractant (cation carrier). Y_2O_3 nanoparticulate thin film was prepared by casting the W/O emulsion, separated from the external phase and containing the Y oxalate nanoparticles, on a Si substrate, followed by calcination in air. Sun and Deng [173] synthesized $CaCO_3$ nanoparticles particles via mass restriction method in an emulsion liquid membrane process. Instrumental analysis, such as Fourier transform infrared (FTIR), wide X-ray diffraction (WXRD), scanning electron microscopy (SEM), and energy dispersive X-ray (EDX) analysis, confirmed a predominant calcite form of the final product via this process. Jarudilokkul *et al.* [174] also synthesized HAp nanoparticles using ELM consisting of Span 20 and Tween 80 as a mixture of biodegradable surfactant and caproic acid as an extractant. specific surface area reduced from 227 to 58 m^2/g after thermal treatment.

6. Liquid Membrane Industrial Plant

6.1. Zinc removal

Zinc removal from wastewater in the viscous fiber industry was first commercialized in 1986 at Lenzing, AG, Austria, having a capacity of 75 m^3/h. The plant is capable of removing 500 ppm zinc selectively in a wastewater containing 5-8 gm/l H_2SO_4 and 25 gm/l Na_2SO_4, some long chain amine modifiers and some solid particles. Zinc could be reduced down to 3 ppm in the treated water and concentrated to 50-60 gm/l in the inner phase. A 1.5 m diameter Oldshue-Rushton column with 10 m active height has been used together with a premixer for emulsification. De-emulsification has been carried out in an electrostatic splitter. Energy consumption is around 4 Kwh/m^3 emulsion. Bis(2-ethyl hexyl) di-thiophosphoric acid has been used as carrier as it is more selective with respect to Ca^+.

To date, there have been three more industrial plants installed for zinc recovery: the 700 m^3/h capacity plant at Glanzstoff, AG, Austria, the 200 m^3/h capacity plant at CFK Schwarza, Germany, and the 200 m^3/h capacity plant at AKZO Iede, Netherlands.

6.2. Phenol removal

Phenol removal from wastewater was first commercialized around 1986 at the Nachung Plastic Factory in Guangzhou, China. In this, phenol can be removed from about 1000 ppm to 0.5 ppm with an extraction efficiency of greater than 99.95%. The internal reagent NaOH converts the phenol to sodium phenolate and the phenolate is trapped into the internal phase.

6.3. Cyanide removal

Cyanide removal from wastewater in gold processing is being commercialized at the Hnang-hua Mountain Gold plant, near Tian-Fin, China. Cyanide can be reduced from about 130 ppm to 0.5 ppm with an extraction efficiency of 99.6%.

7. Summary

7.1. Advantages

Liquid membrane separation processes have received considerable attention because of their potential advantages over other separation processes, both conventional (distillation, solvent extraction) and newly developed (solid membranes).

Some of the advantages of liquid membranes are:

- They have high surface area to volume ratio (100-200 m^2/m^3 for solid supported and 3000 m^2/m^3 for ELMs) compared to solid membranes (10 m^2/m^3) [3].
- The diffusivity of most molecules through liquids is much higher than through polymer membranes, which are so low that exceedingly thin membranes must be constructed to produce industrially acceptable fluxes, which may lead to fabrication and mechanical integrity problems.
- High solute transfer flux and selectivity through the membrane can be achieved by incorporation of a chemical component in the membrane phase which enhances the transport of one component through it.
- The liquid membrane processes can be operated continuously and are scaleable. Because of its close analogy with liquid-liquid extraction

operation, existing design procedures for liquid-liquid extraction can be adopted for scale-up of permeaters [73]).
- The fraction of membrane phase to external phase can be very high (up to 1:40) [175]. Since membrane fluid corresponds to solvent of solvent extraction, this indicates a significantly smaller solvent requirement in liquid membrane processes, particularly at lower solute concentrations.
- Since both extraction and stripping take place in a single stage in liquid membrane separation processes, there is a substantial saving in contacting equipment volume: separate contactors for extraction and stripping processes are not required.
- The volume of the stripping phase is much smaller than the external phase that enables concentration of the solute along with extraction.
- In facilitated transport, unlike solvent extraction and other equilibrium stage wise processes, the overall mass transfer rate is not governed by the usual equilibrium considerations alone. Instead, the solute transport process is controlled by a combination of the diffusion rate and the complexation reaction rate and in case of coupled transport; the solute can be transported against its concentration gradient thus opening up the possibilities of separation from even very dilute solute solutions.

7.2. Disadvantages

Although liquid membranes have received considerable attention in several diverse fields they are associated with some significant drawbacks:

- *De-emulsification*: one of the stumbling blocks in the construction of a continuous liquid membrane process plant is de-emulsification and recovery of the solvent. De-emulsification involves coalescence of dispersed droplets into larger droplets with subsequent phase separation by gravity. The most popular method to augment this process is by application of an electric field. This indicates that liquid membrane plants will be energy intensive.
- *Stability of emulsions*: Another obstacle to the application of ELMs to industrial separations is the stability of emulsion globules. Stability of emulsion is affected by two phenomena—globule rupture and osmotic swell [176]. Interfacial shear between the continuous and membrane phase causes the liquid membrane to thin and, in some cases, rupture, thereby affecting separation. Emulsion swelling occurs when water in the continuous phase diffuses through the organic membrane and swells the inner aqueous droplet phase. The increased volume of the internal phase due to emulsion swell leads to increased breakage and dilution of the concentrated droplet phase.

8. Future Prospects

As an emerging technology, liquid membrane separations have been extensively examined for potential application in many fields such as metal recovery, gas separation, organic compound removal, pollutant removal, and bioseparations. The difficulties in the application of these processes did not consist in sophisticated equipment or installation but in the adequate choice of reagent to allow the selective extraction of solute in required quantity. The widespread use of the ELM process has been limited due to the instability of emulsion globules against fluid shear. Numerous studies have attempted to enhance the stability of ELMs. Examples include adding more surfactants into the membrane phase and increasing the membrane viscosity. However, in most reported attempts increased stability has been unfortunately accompanied by loss in extraction efficiency and rate. A unique contacting device, a Taylor-Couette column, provides a relatively low and uniform fluid shear that helps maintain the stability of emulsion without compromising the extraction efficiency of a target compound. Current research has also been directed to minimizing membrane leakage or rupture through the use of bifunctional surfactants, which act as both emulsifier and extractant, and on additives (e.g., polymer) to impact elasticity of membrane. Research is going on to integrate ELM, reverse micelle and affinity separation processes. Such integration is expected to provide high selectivity, purification, and concentration at the same time.

REFERENCES

1. Li, N.N. (1998). Separating hydrocarbons with liquid membranes, US Patent 3,410,794.
2. Way, J.D., Noble, R.D., Flynn, T.M. and Sloan, E.D. (1982). Liquid membrane transport: A survey. J. Membr. Sci., 12, 239-59.
3. Cahn, R.P. and Li, N.N. (1974). Separation of phenol from wastewater by the liquid membrane. Sep. Sci. Technol., 9, 505-19.
4. Terry, R.E., Li, N.N. and Ho, W.S. (1982). Extraction of phenolic-compounds and organic-acid by liquid membranes. J. Membr. Sci., 10, 305-23.
5. Kim, K.S., Choi, S.J. and Ihm, S.K. (1983). Simulation of phenol removal from wastewater by liquid membrane emulsion. Ind. Eng. Chem. Fundam., 22, 167-72.
6. Teramoto, M., Takihana, H., Shibutani, M., Yuasa, T. and Hara, N. (1983). Extraction of phenol and cresol by liquid surfactant membrane separation. Sep. Sci. Technol., 18, 397-419.
7. Boyadzhiev, L., Bezenshek, E and Lazarova, Z. (1984). Removal of phenol from waste water by double emulsion membranes and creeping film pertraction. J. Membr. Sci., 21, 137-44.
8. Bunge, A.L. and Noble, R.D. (1984). A diffusion model for reversible consumption in emulsion liquid membranes. J. Membr. Sci., 21, 55-71.

9. Chan, C.C. and Lee, C.J. (1987). A mass transfer model for the extraction of weak acids/bases in emulsion liquid-membrane systems. Chem. Eng. Sci., 42, 83-95.
10. Lee, C. J. and Chan, C.C. (1990). Extraction of ammonia from a dilute aqueous solution by emulsion liquid membranes. 1. Experimental studies in a batch system. Ind. Eng. Chem. Res., 29, 96-100.
11. Wan, Y.H., Wang, X.D. and Zhang, X.J. (1997).Treatment of high concentration phenolic waste water by liquid membrane with N503 as mobile carrier. J. Membr. Sci., 135, 263-70.
12. Sastre, A., Madi, M., Cortina, J.L. and Miralles, N. (1998). Modelling of mass transfer in facilitated supported liquid membrane transport of gold(III) using phospholene derivatives as carriers. J. Membr. Sci., 139, 57-65.
13. Bhavani, R., Neena, T. and John, W. (1994). Emulsion liquid membranes for waste-water treatment: Equilibrium models for some typical metal-extractant systems. Environ. Sci. Technol., 28, 1090-8.
14. Salazar, E., Ortiz, M.I., Uritaga, A.M. and Irabien, J. A. (1992). Kinetics of the separation-concentration of chromium(VI) with emulsion liquid membranes. Ind. Eng. Chem. Res., 31, 1523-9.
15. Cussler, E.L. (1971). Membranes which pump. AIChE J., 17, 1300-3.
16. Kremesec, V.J. (1981). Modeling of dispersed-emulsion separation systems. Sep. Purif. Methods, 10, 117-57.
17. Kremesec, V.J. and Slattery, J.C. (1982). Analysis of batch, dispersed-emulsion separation systems. AIChE J., 28, 492-500.
18. Cahn, R.P. and Li, N.N. (1976). Separations of organic compounds by liquid membrane processes. J. Membr. Sci., 1, 129-42.
19. Boyadzhiev, L., Sapundzhiev, T. and Bezenshek, E. (1977). Modelling of carrier-mediated extraction. Sep. Sci. Technol., 12, 541-51.
20. Hochhauser, A.M. and Cussler, E.L. (1975). Concentrating chromium with liquid surfactant membranes. AIChE Symp. Ser., 71, 136-42.
21. Lee, K.H., Evans, D.F. and Cussler, E.L. (1978). Selective copper recovery with two types of liquid membranes. AIChE J., 24, 860-8.
22. Volkel, W., Halwachs, W. and Schugerl, K. (1980).Copper extraction by means of a liquid surfactant membrane process. J. Membr. Sci., 6, 19-31.
23. Kondo, K., Kita, K., Koida, I., Irie, J. and Nakashio, F. (1979). Extraction of copper with liquid surfactant membranes containing benzoylacetone. J. Chem. Eng. Jpn., 12, 203-9.
24. Krishna, R., Goswami, A.N. and Sharma, A. (1987). Effect of emulsion breakage on selectivity in the separation of hydrocarbon mixtures using aqueous surfactant membranes. J. Membr. Sci., 34, 141-5.
25. Ulbricht, M., Marr, R. and Draxler, J. (1991). Selective separation of organic solutes by aqueous liquid surfactant membranes. J. Membr. Sci., 59, 189-203.
26. Gupta, T.C.S.M., Goswami, A.N. and Rawat, B.S. (1990). Mass transfer studies in liquid membrane hydrocarbon separations. J. Membr. Sci., 54, 119-30.
27. Matulevicius, E.S. and Li, N.N. (1975).Facilitated transport through liquid membranes. Sep. Purif. Methods, 4, 73-96.
28. Kopp, A.G., Marr, R.J. and Moser, F.E. (1978).A new concept for mass transfer in liquid surfactant membranes without carriers and with carriers that pump. Int. Chem. Eng. Symp. Ser., 54, 279-90.
29. Ho, W.S., Hatton, T.A., Lightfoot E.N. and Li, N.N. (1982). Liquid surfactant membranes: A diffusion controlled model, AIChE J., 28, 662-70.
30. Stroeve, P. and Varanasi, P.P. (1984). Extraction with double emulsions in a batch reactor: Effect of continuous phase resistance. AIChE J., 30, 1007-9.

31. Fales, J.L. and Stroeve, P. (1984). A perturbation solution for batch extraction with double emulsions: Role of continuous phase mass transfer resistance. J. Membr. Sci., 21, 35-53.
32. Datta, S., Mukhopadhyay, A. and Sanyal, S.K. (1993). Facilitated transport through a liquid surfactant membrane with continuous phase resistance: Role of drop-size distribution. Sep. Sci. Technol., 28, 1327-40.
33. Chakraborty, M., Bhattacharya, C. and Datta, S. (2003). Effect of drop size distribution on mass transfer analysis of the extraction of nickel(II) by emulsion liquid membrane. Colloids Surf. A, 224, 65-74.
34. Carnahan, B., Luther, N.A. and Wilkes, J.O. (1969). *Applied Numerical Methods*. Wiley, New Jersey.
35. Ohtake, T., Hano, T., Takagi, K. (1987). Effects of viscosity on drop diameter of w/o emulsion dispersed in a stirred tank. J. Chem. Eng. Jpn., 20, 443-7.
36. Reid, R.C., Prausnitz, J.M. and Sherwood, T.K. (1988). *The Properties of Gases and Liquids*, 4th ed. McGrawHill, New York.
37. Jefferson, T.B., Witzell, O.W. and Sibbitt, W.L. (1958). Thermal conductivity of graphite-silicone oil and graphite-water suspensions. Ind. Eng. Chem., 50, 1589-92.
38. Chakraborty, M., Bhattacharya, C. and Datta, S. (2003). Mathematical modeling of simultaneous copper (II) and nickel (II) extraction from wastewater by emulsion liquid membranes. Sep. Sci. Technol., 38, 2081-106.
39. Teramoto, M., Takihana, H., Shibutani, M., Yuasa, T., Miyake, Y. and Teranishi, H. (1981). Extraction of amine by w/o/w emulsion system. J. Chem. Eng. Jpn., 14, 122-8.
40. Braid, R.S., Bunge, A.L. and Noble, R.D. (1987). Batch extraction of amines using emulsion liquid membranes importance of reaction reversibility. AIChE J., 33, 43-53.
41. Lin, C.C. and Long, R.L. (1997). Removal of nitric acid by emulsion liquid membrane: Experimental results and model prediction. J. Membr. Sci., 134, 33-45.
42. Casamatta, G., Chavarie, C. and Angelino, H. (1978). Hydrocarbon separation through a liquid water membrane: Modelling of permeation in an emulsion drop. AIChE J., 24, 945-9.
43. Yan, N., Shi, Y. and Su, Y.F. (1992). A mass transfer model for type I facilitated transport in liquid membranes. Chem. Eng. Sci., 47, 4365-71.
44. Guoyu, M., Yuanli, J. and Chang, K.S. (1997). A general mass transfer model for liquid surfactant membranes. Chem. Eng. Sci., 52, 433-41.
45. Sahoo, G.C. and Dutta, N.N. (1998). Studies on emulsion liquid membrane extraction of cephalexin. J. Membr. Sci., 145, 15-26.
46. Bhowal, A. and Datta, S. (1998). A mathematical model for the analysis of extraction data of weak acids/bases in liquid surfactant membranes. J. Membr. Sci., 139, 103-8.
47. Correia, P.F.M.M. and de Carvalho, J.M.R.A. (2001). Comparison of models for 2-chlorophenol recovery from aqueous solutions by emulsion liquid membranes. Chem. Eng. Sci., 56, 5317-25.
48. Kargari, A., Kaghazchi, T., Sohrabi, M. and Soleimani, M. (2004). Batch extraction of gold(III) ions from aqueous solutions using emulsion liquid membrane via facilitated carrier transport. J. Membr. Sci., 233, 1-10.
49. Lin, S.H., Pan, C.L. and Leu, H.G. (2002). Equilibrium and mass transfer characteristics of 2-chlorophenol removal from aqueous solution by liquid membrane. Chem. Eng. J., 87, 163-9.
50. Said, N.E., Sheref, E.E. and Borai, E. (2003). Modeling of transport of Cs (137) by emulsion liquid membrane (18C6) in xylene promoted by ephedrine hydrochloride in stripping phase. J. Membr. Sci., 211, 183-91.
51. Hatton, T.A. and Wardius, D.S. (1984). Analysis of staged liquid surfactant membrane operations. AIChE J., 30, 934-44.

52. Wardius, D.S. and Hatton, T.A. (1985). A model for liquid membrane extraction with instantaneous reaction in cascaded mixers. Chem. Eng. Commun. 37, 159-71.
53. Levenspiel, O. and Fitzgerald, T.J. (1983). A warning on the misuse of the dispersion model. Chem. Eng. Sci., 38, 489-91.
54. Kataoka, T., Nishiki, T., Yamauchi, M. and Zhong, Y. (1987). A simulation for liquid surfactant membrane permeation in a continuous countercurrent column. J. Chem. Eng. Jpn., 20, 410-5.
55. Bart, H.J., Wachter, R., Marr, R. and Muller, H.J. (1988). Mass transfer in a permeation column. J. Membr. Sci., 36, 413-23.
56. Lorbach, D.M. and Hatton, T.A. (1988). Polydispersity and backmixing effects in diffusion controlled mass transfer with irreversible chemical reaction and analysis of liquid emulsion membrane processes. Chem. Eng. Sci., 43, 405-18.
57. Mok, Y.S., Lee, W.K. and Lee, Y.K. (1997). Modeling of liquid emulsion membranes facilitated by two carriers. Chem. Eng. J., 66, 11-20.
58. Chaudhuri, J.B. and Pyle, D.L. (1992). Emulsion liquid membrane extraction of organic acids. I: A theoretical model for lactic acid extraction with emulsion swelling. Chem. Eng. Sci., 47, 41-8.
59. Goto, M., Kakoi, T., Yoshii, N., Kondo, K. and Nakashio, F. (1993). Effect of synthesized surfactants in the separation of rare earth metals by liquid surfactant membranes. Ind. Eng. Chem. Res., 32, 1681-5.
60. Groeneweg, F., Voorst Vader, F.V. and Agterof, W.G.M. (1993). The dynamic stability of w/o emulsions prepared from vegetable oil. Chem. Eng. Sci., 48, 229-38.
61. Marr, R., Bart, H.J. and Draxler, J. (1990). Liquid membrane permeation. Chem. Eng. Process., 27, 59-64.
62. Jeong, Y., Ihm, S. and Won, Y. (1987). Experimental investigation of the oil/water/oil liquid-membrane separation of toluene and n-heptane – II. Continuous test of a countercurrent permeator. J. Membr. Sci., 32, 47-57.
63. Goswami, A.N., Gupta, T.C.S.M., Sharma, S.K., Sharma, A. and Krishna, R. (1993). Unsteady-state modeling and analysis for liquid surfactant membrane hydrocarbon separation processes. Ind. Eng. Chem. Res., 32, 634-40.
64. Lackner, H., Bart, H.J., and Marr, R. (1986). Hydrodynamics in mixer-settlers with liquid membranes. In: *Proceedings of ISEC '86*, München, Vol. 3, pp. 339-44.
65. Othman, N., Goto, M. and Mat, H. (2003). Extractant selectivity of metal extraction from liquid photographic waste. In: *17th Symposium of Malaysian Chemical Engineers (SOMChE 2003)*, Penang, Malaysia, December 29-30.
66. Nakashio, F., Goto, M., Matsumoto, M., Irie, J. and Kondo, K. (1988). Extractant selectivity of metal extraction from liquid photographic waste. J. Membr. Sci., 38, 249-60.
67. Goto, M., Irie, J., Kondo, K. and Nakashio, F. (1989). Electrical demulsification of W/O emulsion by continuous tubular coalescer. J. Chem. Eng. Jpn., 22, 401-6.
68. Kataoka, T. and Nishiki, T. (1990). Development of a continuous electric coalescer of w/o emulsions in liquid surfactant membrane process. Sep. Sci. Technol., 25, 171-85.
69. Lu, G., Lu, Q. and Li, P. (1997). Break-down of liquid membrane emulsion under high electric field. J. Membr. Sci., 128, 1-6.
70. Kato, S. and Kawasaki, J. (1987). Enhancement of hydrocarbon permeation by polar additives in liquid emulsion membranes. J. Chem. Eng. Jpn., 20, 585-90.
71. Kawasaki, J., Kato, S. and Egashira, R. (1988). Separation of hydrocarbons by emulsified liquid membrane – Improvement in selectivity of separation. In: *Proceedings of The First Sino-Japanese Symposium on Liquid Membranes*, Hangzhou. pp. 23-5.
72. Devulapalli, R. and Jones, F. (1999). Separation of aniline from aqueous solutions using emulsion liquid membranes. J. Haz. Mat., 70, 157-70.

73. Chakraborty, M., Bhattacharya, C. and Datta, S. (2002). Studies on transport mechanism of nickel (II) from an acidic solution using emulsion liquid membranes. J. Energy Heat Mass Transf., 24, 75-88.
74. Thien, M.P. and Hatton, T.A. (1998). Liquid emulsion membranes and their applications in biochemical processing. Sep. Sci. Technol., 23, 819-53.
75. Hano, T., Ohtake, T., Matsumoto, M., Kitayama, D., Hori, F. and Nakashio, F. (1991). Extraction equilibrium of amino acids with quaternary ammonium salt. J. Chem. Eng. Jpn., 24, 20-4.
76. Chakraborty, M., Murthy, Z.V.P., Bhattacharya, C. and Datta, S. (2005). Process intensification: Extraction of chromium (VI) by emulsion liquid membrane. Sep. Sci. Technol., 40, 2353-64.
77. Valenzuela, F., Cabrera, J., Basualto, C. and Hagar, J. S. (2005). Kinetics of copper removal from acidic mine drainage by a liquid emulsion membrane. Miner. Eng., 18, 1224-32.
78. Othman, N., Mat, H. and Goto, M. (2006). Separation of silver from photographic wastes by emulsion liquid membrane system. J. Membr. Sci., 282, 171-7.
79. Sabrya, R., Hafeza, A., Khedra, M. and Hassaninb, A.E. (2007). Removal of lead by an emulsion liquid membrane. Desalination, 212, 165-75.
80. Nakano, K., Kato, S., Noritomi, H. and Nagahama, K. (1996). Extraction of polyunsaturated fatty acid ethyl esters from sardine oil using Ag^+-containing o/w/o emulsion liquid membranes. J. Membr. Sci., 110, 219-27.
81. Kargari, A., Kaghazchi, T., Sohrabi, M. and Soleimani, M. (2004). Batch extraction of gold(III) ions from aqueous solutions using emulsion liquid membrane via facilitated carrier transport. J. Membr. Sci., 233, 1-10.
82. Strzelbicki, J., Charewicz, W.A. and Mackiewicz, A. (1984). Permeation of chromium (VI) and rhenium(VII) oxyanions through liquid organic membranes facilititated by quaternary ammonium chlorides. Sep. Sci. Technol., 19, 321-36.
83. Liu, X. and Zhang, X. (1997). Simplified model for extraction of rare-earth ions using emulsion liquid membrane. J. Membr. Sci., 128, 223-9.
84. Chakraborty, M., Bhattacharya, C. and Datta, S. (2004). Study of the extraction of copper (II) and nickel (II) from wastewater by emulsion liquid membranes. J. Inst. Eng. (India) Chem. Eng. Div., 85, 1-6.
85. Chakraborty, M., Mitseva, P.I. and Bart, H.J. (2006). Selective separation of toluene from n-heptane via emulsion liquid membranes containing substituted cyclodextrins as carriers. Sep. Sci. Technol., 41, 3539-52.
86. Kobya, M., Topcu, N. and Demircioglu, N. (1997). Kinetic analysis of coupled transport of thiocyanate ions through liquid membranes at different temperatures. J. Membr. Sci., 130, 7-15.
87. Hinze, J.O. (1955). Fundamentals of the hydrodynamic mechanism of splitting in dispersion processes. AIChE J., 1, 289-95.
88. Lagisetty, J., Das, P., Kumar, R. and Gandhi, K.S. (1986). Breakage of viscous and non-newtonian drops in stirred dispersions. Chem. Eng. Sci., 41, 65-72.
89. Calabrese, R.V., Wang, C.Y. and Bryner, N.P. (1986). Drop breakup in turbulent stirred-tank contactors part III: Correlations for mean size and drop size distribution. AIChE J., 32, 677-81.
90. Kataoka, T. and Nishiki, T. (1986). Dispersed mean drop sizes of (w/o)/w emulsions in a stirred tank. J. Chem. Eng. Jpn., 19, 408-12.
91. Rautenbach, R. and Machhamer, O. (1988). Modeling of liquid membrane separation processes. J. Membr. Sci., 36, 425-44.
92. Bajpai, R.K., Ramkrishna, D. and Prokop, A. (1975). Coalescence frequencies in waldhof-agitated systems. Biotechnol. Bioeng., 17, 1697-703.

93. Hong, P.O. and Lee, J.M. (1983).Unsteady-state liquid-liquid dispersions in agitated vessels. Ind. Eng. Chem. Proc. Des. Dev., 22, 130-5.
94. Hong, P.O. and Lee, J.M. (1985). Changes of the average drop sizes during the initial period of liquid-liquid dispersions in agitated vessels. Ind. Eng. Chem. Proc. Des. Dev., 24, 868-72.
95. Chatzi, E.G., Boutris, C.J. and Kiparissides, C. (1991). Online monitoring of drop size distributions in agitated vessels. 1. Effects of stabilizer concentration. Ind. Eng. Chem. Res., 30, 1307-13.
96. Hou, W. and Papadopoulos, K.D. (1996). Stability of water-in-oil-in-water type globules. Chem. Eng. Sci., 51, 5043-51.
97. Martin, T.P. and Davies, G.A. (1977). The extraction of copper from dilute aqueous solutions using a liquid membrane process, Hydrometallurgy, 2, 315-20.
98. Kita, Y.S., Matsumoto, S., and Yonezawa, D. (1977). *Nippon Kagaku Kaishi (in Japanese)*, p. 748.
99. Kondo, K., Kita, K., Koida, I., Irie, J. and Nakashio, F. (1979). Extraction of copper with liquid surfactant membranes containing benzoylacetone. J. Chem. Eng. Jpn., 12, 203-9.
100. Takahasi, K., Ohtsubo, F. and Takeuchi, H. (1981). A study of the stability of (w/o)/w-type emulsions using a tracer technique. J. Chem. Eng. Jpn., 14, 416-8.
101. Shere, A.J. and Cheung, H.M. (1988). Effect of preparation parameters on leakage in liquid surfactant membrane systems. Sep. Sci. Technol., 23, 687-701.
102. Stroeve, P., Prabodh, P.V., Elias, T. and Ulbrecht, J.J. (1981). Stability of double emulsion droplets in shear flow. In: *53rd Annual Meeting of the Society of Rheology*, Louisville, KY, October, 11-15.
103. Taylor, G.I. (1934). The formation of emulsions in definable fields of flow. Proc. R. Soc. Lond., Ser. A, 146, 501-23.
104. Grace, H.P. (1982). Dispersion phenomena in high-viscosity immiscible fluid systems and application of static mixers as dispersion devices in such systems. Chem. Eng. Commun., 14, 225-77.
105. Goto, M., Matsumoto, M., Kondo, K. and Nakashio, F. (1987). Development of new surfactant for liquid surfactant membrane process. J. Chem. Eng. Jpn., 20, 157-64.
106. Kasaini, H., Nakashio, F. and Goto, M. (1998). Application of emulsion liquid membranes to recover cobalt ions from a dual-component sulphate solution containing nickel ions. J. Membr. Sci., 146, 159-68.
107. Chakraborty, M., Bhattacharya, C. and Datta, S. (2004). Study of the stability of (w/o)/w-type emulsion during the extraction of nickel (II) via emulsion liquid membrane. Sep. Sci. Technol., 39, 1-17.
108. Chakraborty, M. and Bart, H.J. (2006). Emulsion liquid membranes: Role of internal droplet size distribution on toluene/n-heptane separation. Colloids Surf. A, 272, 15-21.
109. Strzelbicki, J. and Charewicz, W. (1980). The liquid surfactant membrane separation of copper, cobalt and nickel from multicomponent aqueous solutions. Hydrometallurgy, 5, 243-54.
110. Bock, J. and Valint, P.L. (1982). Uranium extraction from wet process phosphoric acid. A liquid membrane approach. Ind. Eng. Chem. Fundam., 21, 417-22.
111. Bart, H.J., Draxler, J. and Marr, R. (1988). Residence time selection in liquid membrane permeation for recovery. Hydrometallurgy, 19, 351-60.
112. Hirato, T., Kishigami, I., Awakura, Y. and Majima, H. (1991). Concentration of uranyl sulfate solution by an emulsion-type liquid membrane process. Hydrometallurgy, 26, 19-33.
113. Eroglu, I., Kalpakci, R. and Gunduz, G. (1993). Extraction of strontium ions with emulsion liquid membrane technique. J. Membr. Sci., 80, 319-25.

114. Chakraborty, R. and Datta, S. (1994). Transport of Te^{4+} through liquid surfactant membrane using D2EHPA as the carrier. J. Membr. Sci., 96, 233-40.
115. Chakravarti, A.K., Chowdhury, S.B., Chakrabarty, S., Chakrabarty, T and Mukherjee, D.C. (1995). Liquid membrane multiple emulsion process of chromium(VI) separation from waste waters. Colloids Surf. A, 103, 59-71.
116. Li, Q., Liu, Q., Li, K. and Tong, S. (1997). Separation study of cadmium through an emulsion liquid membrane. Talanta, 44, 657-62.
117. Li, Q.M., Lu, Q., Zhang, Q.F., Wei, X.J. and Guo, J.Z. (1998). Separation study of cadmium through an emulsion liquid membrane using triisooctylamine as mobile carrier. Talanta, 46, 927-32.
118. Longquan, L., Cheng, W. and Yadong, L. (1997). Separation of cobalt and nickel by emulsion liquid membrane with the use of EDTA as masking reagent. J. Membr. Sci., 135, 173-7.
119. Xingrong, L. and Xiujuan, Z. (1997). Simplified model for extraction of rare-earth ions using emulsion liquid membrane. J. Membr. Sci., 128, 223-9.
120. Li, Q., Liu, Q. and Wei, X. (1998). Separation study of mercury through an emulsion liquid membrane. Talanta, 43, 1837-42.
121. Katsushi, Y., Takashiko, K., Hiroshi, K. and Masahiro, G. (1998). Synergistic extraction of nickel by liquid surfactant membranes. Sep. Sci. Technol., 33, 369-85.
122. Kondo, K. and Matsumoto, M. (1998). Separation and concentration of indium (III) by an emulsion liquid membrane containing diisostearylphosphoric acid as a mobile carrier. Sep. Purif. Technol., 13, 109-15.
123. Kulkarni, P.S., Tiwari, K.K. and Mahajani, V.V. (1999). Studies in extraction of nickel by liquid emulsion membrane process. Indian J. Chem. Technol., 6, 329-35.
124. Sznejer, G. and Marmur, A. (1999). Cadmium removal from aqueous solutions by an emulsion liquid membrane. The effect of resistance to mass transfer at the outer oil–water interface. Colloids Surf. A, 151, 77-83.
125. Ye, Z.F., Wang, Y.P., Liu, Y.S., Jiang, Z.L., Shen, X., Zhu, L.G. and Shi, X.F. (1999). Transport of Na^{+}, K^{+} ions in a bubbling pseudo-emulsion liquid membrane system with *p*-tert-butyl calixarenes as carriers. J. Membr.Sci., 163, 367-72.
126. Serga, V.E., Kulikova, L.D. and Purin, B.A. (2000). Extraction of nickel by liquid membranes in an electric field. Sep. Sci. Technol., 35, 299-313.
127. Chakravarti, A.K., Chowdhury, S.B. and Mukherjee, D.C. (2000). Liquid membrane multiple emulsion process of separation of copper(II) from waste waters. Colloids Surf. A, 166, 7-25.
128. Urtiaga, A.M., Alonso, A., Ortiz, I., Daoud, J.A., Reefy, S.A.E., Ortiz, S.P.D. and Gallego, T. (2000). Comparison of liquid membrane processes for the removal of cadmium from wet phosphoric acid. J. Membr. Sci., 164, 229-40.
129. Kulkarni, P.S., Tiwari, K.K. and Mahajani, V.V. (2001). Recovery of nickel via liquid emulsion membrane process using methane sulfonic acid as a strippant. Sep. Sci. Technol., 36, 639-56.
130. Uddin, M.S. and Kathiresan, M. (2000). Extraction of metal ions by emulsion liquid membrane using bi-functional surfactant: Equilibrium and kinetic studies. Sep. Purif. Technol., 19, 3-9.
131. Kulkarni, P.S. and Mahajani, V.V. (2002). Application of liquid emulsion membrane (LEM) process for enrichment of molybdenum from aqueous solutions. J. Membr. Sci., 201, 123-35.
132. Kulkarni, P.S., Mukhopadhyay, S., Bellary, M.P. and Ghosh, S.K. (2002). Studies on membrane stability and recovery of uranium (VI) from aqueous solutions using a liquid emulsion membrane process. Hydrometallurgy, 64, 49-58.
133. Sayed, M.S.E. (2003). Uranium extraction from gattar sulfate leach liquor using aliquat-336 in a liquid emulsion membrane process. Hydrometallurgy, 68, 51-6.

134. Gurel, L., Altas, L. and Buyukgungor, H. (2005). Removal of lead from wastewater using emulsion liquid membrane technique. Environ. Eng. Sci., 22, 411-20.
135. Kargari, A., Kaghazchi, T. and Soleimani, M. (2006). Experimental and modeling of selective separation of gold(III) ions from aqueous solutions by emulsion liquid membrane system. J. Membr. Sci., 279, 389-93.
136. Gasser, M.S., Hefny, N.E.E. and Daoud, J.A. (2008). Extraction of Co(II) from aqueous solution using emulsion liquid membrane. J. Hazard. Mater., 151, 610-5.
137. Fouad, E.A. and Bart, H.J. (2008). Emulsion liquid membrane extraction of zinc by a hollow-fiber contactor. J. Membr. Sci., 307, 156-68.
138. Cahn, R.P., Li, N.N. and Minday, R.M. (1978). Removal of ammonium sulfide from wastewater by liquid membrane process. Environ. Eng. Sci., 12, 1051-5.
139. Dzygiel, P. and Wieczorek, P. (2000). Extraction of amino acids with emulsion liquid membranes using industrial surfactants and lecithin as stabilizers. J. Membr. Sci., 172, 223-32.
140. Juang, R.S. and Wang, Y.Y. (2002). Amino acid separation with D2EHPA by solvent extraction and liquid surfactant membranes. J. Membr. Sci., 207, 241-52.
141. Lin, S.H., Pan, C.L. and Leu, H.G. (1999). Liquid membrane extraction of 2-chlorophenol from aqueous solution. J. Hazard. Mater., 65, 289-304.
142. Datta, S., Bhattacharya, P.K. and Verma, N. (2003). Removal of aniline from aqueous solution in a mixed flow reactor using emulsion liquid membrane. J. Membr. Sci., 226, 185-201.
143. Park, Y., Skelland, A.H.P., Forney, L.J. and Kim, J.H. (2006). Removal of phenol and substituted phenols by newly developed emulsion liquid membrane process. Water Res., 40, 1763-72.
144. Sharma, A., Goswami, A.N., Rawat, B.S. and Krishna, R. (1987). Effect of surfactant type on the selectivity for separation of 1-methyl naphthalene and dodecane using liquid membranes. J. Membr. Sci., 42, 19-30.
145. Garti, N. and Kovacs, A. (1991). Facilitated emulsion liquid membrane separation of complex hydrocarbon mixtures. J. Membr. Sci., 56, 239-46.
146. Ghosh, A.C., Patil, G.S. and Dutta, N.N. (1994). Liquid membrane permeation of aromatic hydrocarbons in LPG condensate. Fuel Process. Technol., 38, 17-30.
147. Sharma, A., Goswami, A.N. and Krishna, R. (1989). Use of additives to enhance the selectivity of liquid surfactant membranes. J. Membr. Sci., 40, 329-42.
148. Yan, N., Huang, S. and Shi, Y. (1987). Removal of acetic acid from wastewater with liquid surfactant membranes: An external boundary layer and membrane diffusion controlled model. Sep. Sci. Technol., 22, 801-18.
149. Itoh, H., Thien, M.P., Hatton, T.A. and Wang, D.I.C. (1990). A liquid emulsion membrane process for the separation of amino acids. Biotechnol. Bioeng., 35, 853-60.
150. Teramoto, M., Yamashiro, T., Inoue, A., Yamamoto, A., Matsuyama, H. and Miyake, Y. (1991). Extraction of amino acids by emulsion liquid membranes containing di(2-ethylhexyl)phosphoric acid as a carrier biotechnology; coupled, facilitated transport; diffusion. J. Membr. Sci., 58, 11-32.
151. Yan, N., Shi, Y. and Su, Y.F. (1992). A mass transfer model for type I facilitated transport in liquid membranes. Chem. Eng. Sci., 47, 4365-71.
152. Eyal, A.E. and Bressler, E. (1993). Mini-review industrial separation of carboxylic acid and amino acids by liquid membranes: Applicability, process considerations and potential advantages. Biotechnol. Bioeng., 41, 287-95.
153. Scholler, C., Chaudhuri, J.B. and Pyle, D.L. (1993). Emulsion liquid membrane extraction of lactic acid from aqueous solutions and fermentation broth. Biotechnol. Bioeng., 42, 50-8.

154. Juang, R.S., Lee, S.H. and Shiau, R.C. (1997). Mass-transfer modeling of permeation of lactic acid across amine-mediated supported liquid membranes. J. Membr. Sci., 137, 231-9.
155. Reisinger, H. and Marr, R. (1992). Multicomponent-liquid-membrane permeation of organic acids. Chem. Eng. Technol., 15, 363-70.
156. Juang, R.S., Huang, R.H. and Wu, R.T. (1997). Separation of citric and lactic acids in aqueous solutions by solvent extraction and liquid membrane processes. J. Membr. Sci., 136, 89-99.
157. Hano, T., Ohtake, T., Matsumoto, M., Ogawa, S. and Hari, F. (1990). Extraction of penicillins with liquid surfactant membrane. J. Chem. Eng. Jpn., 23, 771-5.
158. Nakano, K., Kato, S., Noritomi, H. and Nagahama, K. (1996). Extraction of polyunsaturated fatty acid ethyl esters from sardine oil using Ag^+-containing o/w/o emulsion liquid membranes. J. Membr. Sci., 110, 219-27.
159. Lee, S.C., Lee, K.H., Hyun, G.H. and Lee, W.K. (1997). Continuous extraction of penicillin G by an emulsion liquid membrane in a countercurrent extraction column. J. Membr. Sci., 124, 43-51.
160. Sang, C.L., Kwi, H.L., Gun, H.H. and Won, K.L. (1997). Continuous extraction of penicillin G by an emulsion liquid membrane in a countercurrent extraction column. J. Membr. Sci., 124, 43-51.
161. Sahoo, G.C., Dutta, N.N. and Das, N.N. (1999). Liquid membrane extraction of cephalosporin-c from fermentation broth. J. Membr. Sci., 157, 251-61.
162. Lee, S.C. (2000). Continuous extraction of penicillin G by emulsion liquid membranes with optimal surfactant compositions. Chem. Eng. J., 79, 61-7.
163. Pal, P., Dutta, S. and Bhattacharya, P. (2000). Studies on the modeling and simulation of a sequential bienzymatic reaction system immobilized in emulsion liquid membrane. Biochem. Eng. J., 5, 89-100.
164. Pal, P., Dutta, S. and Bhattacharya, P. (2002). Multi-enzyme immobilization in eco-friendly emulsion liquid membrane reactor—A new approach to membrane formulation. Sep. Purif. Technol., 27, 145-54.
165. Bayraktar, E. (2001). Response surface optimization of the separation of dl-tryptophan using an emulsion liquid membrane. Process Biochem., 37, 169-75.
166. Habaki, H., Egashira, R., Stevens, G.W. and Kawasaki, J. (2002). A novel method improving low separation performance for W/O/W ELM permeation of erythromycin. J. Membr. Sci., 208, 89-103.
167. Ishizu, H., Habaki, H. and Kawasaki, J. (2003). Permeation and concentration of compactin by a liquid membrane technique. J. Membr. Sci., 213, 209-19.
168. Lee, S.C. (2004). Comparison of extraction efficiencies of penicillin G at different w/o ratios in the emulsion liquid membrane systems with dilute polymer solutions. J. Membr. Sci., 237, 225-32.
169. Lee, S.C. and Lee, H.K. (2005). Development of an emulsion liquid membrane system for extraction of a neutral species. J. Membr. Sci., 264, 13-9.
170. Kaghazchi, T., Kargari, A., Yegani, R. and Zare, A. (2006). Emulsion liquid membrane pertraction of L-lysine from dilute aqueous solutions by D2EHPA mobile carrier. Desalination, 190, 161-71.
171. Hirai, T., Hirano, T. and Komasawa, I. (2002). Preparation of Gd2O3: $Eu3^+$ and Gd2O2S: $Eu3^+$ phosphor fine particles using an emulsion liquid membrane system. J. Colloid Interface Sci., 253, 62-9.
172. Hirai, T., Kawamura, Y. and Komasawa, I. (2004). Preparation of Y_2O_3 nanoparticulate thin films using an emulsion liquid membrane system. J. Colloid Interface Sci., 275, 508-13.

173. Sun, Q. and Deng, Y. (2004). Synthesis of micron to nanometer CaCO3 particles via a mass restriction process in emulsion liquid membrane. J. Colloid Interface Sci., 278, 376-82.
174. Jarudilokkul, S., Tanthapanichakoon, W. and Boonamnuayvittaya, V. (2007). Synthesis of hydroxyapatite nanoparticles using an emulsion liquid membrane system. Colloids Surf. A, 296, 149-53.
175. Marr, R. and Kopp, A. (1982). Liquid membrane technology – A survey of phenomena, mechanisms, and models. Int. Chem. Eng., 22, 44-60.
176. Yan, J. and Pal, R. (2001). Osmotic swelling behaviour of globules of W/O/W emulsion liquid membranes. J. Membr. Sci., 190, 79-91.

CHAPTER 5

Bulk Hybrid Liquid Membrane with Organic Water-Immiscible Carriers: Application to Chemical, Biochemical, Pharmaceutical, and Gas Separations

Vladimir S. Kislik

Nomenclature

6-APA	6-aminopenicillanic acid
7-ACA	7-aminocephalosporanic acid
AMP	2-amino-2-methyl-1-propanol
Cyanex 302	di-(2,4,4-trimethylpentyl) monothiophosphinic acid, a product of CYTEX Canada, Inc. (nowadays Cognis)
Cyanex 471X or TIBPS	triisobutylphosphine sulfide.
DB18C6	dibenzo-18-crown-6
DEA	diethanolamine
DEHPA	di-(2-ethylhexyl) phosphate
DMCCA	dimethyl(cyclopropane)carboxylic acid
EHPNA	2-ethylhexyl phosphonic acid, mono-2-ethylhexyl ester (PC-88A), Daihachi Chemical Industries. Co., Ltd
Hostarex A 327	(*n*-octyl(*n*-decyl)amine
MAs	mandelic acids
MDEA	methyldiethanolamine
MEA	monoethanolamine
MEHPA	(2-ethylhexyl) phosphate ($C_8H_{19}O_4P$), containing 45% of mono-(2-ethylhexyl) phosphate on molar basis
MPOEP	methoxy-poly(oxyethylene) phosphate (1)
PAA	phenylacetic acid
PABA	*p*-aminobenzoic acid
PEG 4000 and 6000	poly(ethylene glycol)s
PEGP	poly(ethylene glycol) phosphate

Institute of Applied Chemistry, The Hebrew University of Jerusalem, Campus Givat Ram, Jerusalem 91904, Israel

Liquid Membranes: Principles and Applications in Chemical Separations and Wastewater Treatment
DOI: 10.1016/B978-0-444-53218-3.00005-2

POEBMP	poly(oxyethylene) Me ether phosphate
POEBP	poly(oxyethylene) bisphosphate (2), poly(oxyethylene) bis(di-Me phosphate) (3)
POEP	poly(oxyethylene) phosphate
PPG, PPO	poly(propylene glycol)
PPOPP	poly[poly(oxypropylene) phosphate]s (Mn = 5800, 8100, 10,400), with different POP units (400, 1200, 2000)
Star-shaped polymer	$[CH_3O(CH_2CH_2O)_n]_x$-[DGEEG]-$[(CH_2CH_2O)_mP(O)(ONa)_2]_x$, (MPOE-[DGEEG]-POEP), (DGEEG is ethylene glycol diglycidyl ether)
TBP	tri-*n*-butylphosphate
THP	tri-hexyl phosphate
TOA	tri-*n*-octylamine.
TOMAC	tri-(octylmethyl) ammonium chloride
TOPO	tri-*n*-octylphosphinic oxide

1. Introduction and Definitions

Technological problems with stability of emulsion LM and supported LM have led researchers in recent years to look for alternative bulk LM or bulk water-immiscible LM. The term bulk organic hybrid liquid membrane (BOHLM) includes all bulk liquid membrane processes incorporating liquid-liquid extraction (LLX) and membrane separation in one continuously working module. BOHLM utilizes an extracting reagent (carrier) solution, immiscible with water, circulating or flowing between membranes as barriers. The concept of BOHLM transport is quite simple (see Fig. 5.1): a solution of an extracting reagent (carrier phase, E) flows between two membranes, which separate the carrier phase from the feed (F) and receiving (R) phases. A solute of interest diffuses to the F/E interface, are extracted from the feed phase by a LM as a result of the thermodynamic conditions at the F/E interface; the solute-carrier complex diffuses to the E/R interface and is simultaneously stripped by the receiving phase due to the different thermodynamic conditions at the E/R interface. The membranes are permeable to solutes, but block transfer of the carrier solution to the feed and to the strip solutions. Blocking is accomplished through membrane hydrophilic or hydrophobic or ion-exchange properties, or through their pore size.

The BOHLM term includes several similar LM systems, developed by different research groups, such as hybrid liquid membrane (HLM) by Kislik *et al.* [1-3]; hollow-fiber contained liquid membrane (HFCLM) by Sirkar *et al.* [4–8] and Schlosser *et al.* (HFLM) [9-11]; pertraction

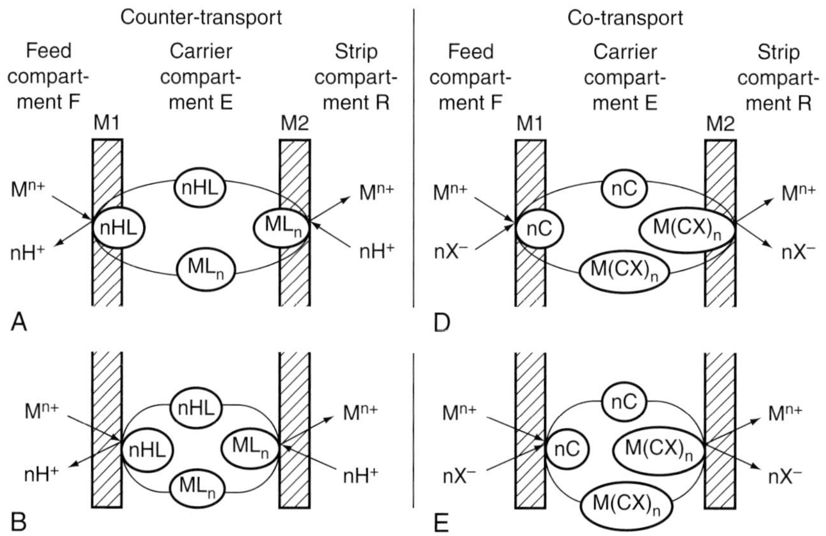

Figure 5.1 Schematic transport models of a bulk organic hybrid liquid membrane (BOHLM) system with (A, D) hydrophobic membranes and (B, E) hydrophilic or ion-exchange membranes. From Ref. [2] with permission.

by Boyadzhiev *et al.* [12-22], Schlosser *et al.* [23-26], and Wodzki *et al.* [27-29]; flowing liquid membranes (FLMs) by Teramoto *et al.* [30-37]; membrane-based solvent extraction and stripping by Sirkar *et al.* [38], Schlosser *et al.* [39-41], Kedem *et al.* [42-44], and Eyal *et al.* [45, 46]; and multimembrane hybrid system (MHS) by Wodzki *et al.* [47-52]. All these systems are based on the membrane-based nondispersive (as the means for blocking the organic reagent from mixing with the aqueous feed and strip solutions) selective LLX coupled to permselective diffusion of solute-extractant complexes and selective stripping of the solute in one continuous dynamic process. The large number of terms for similar bulk LM processes can cause confusion. Some of these terms vary by membrane type used (hollow fiber [4-11], flat neutral, ion-exchange sheets [12-30]), some are by module design [4-8, 12-14, 23-25, 39-41, 47-54]. In some articles, the definitions of the process are not correct, for example, the term membrane-based (or nondispersive) solvent extraction [39-46] (for details see Section 3.4.9). Therefore, all above-mentioned bulk LM processes may be unified under the term bulk organic hybrid liquid membrane systems.

In this chapter, the basic BOHLM principles such as interaction mechanisms and theories of transport, and design considerations are analyzed by comparison of the modifications by different research groups. The application of BOHLM systems to solutes separation in the last decade is also presented.

The scope of this chapter is bulk liquid membrane processes driven mainly by the concentration gradient. Application of a pressure difference, an electric field, or temperature considerably intensifies the process, but these special methods are beyond the scope of this overview.

2. Theory: Mass-Transfer Mechanisms and Kinetics

No universal model exists for all these types of transport, and the available knowledge concerning the specific interfacial processes should be taken into account in the description of a real membrane process. Most models published are very sophisticated because they assume many possible types of control, nonlinear equilibria, phase interactions, etc.

A number of reviews and articles have appeared on related subjects. Theoretical aspects of the hollow-fiber liquid membrane (HFCLM) selective separation processes have been covered in the reviews and articles by Sirkar *et al.* [4-7, 55, 56], and with new modifications by Schlosser *et al.* [9-11, 25]. Theory for separations by the FLM is presented in the work by Teramoto *et al.* [31, 34-37]; theoretical considerations for rotating film pertraction systems have been described by Boyadzhiev [18, 57-60], and for the multi-membrane hybrid systems by Wodzki *et al.* [27, 29, 47, 48, 50, 52].

Examples of the model considerations presented below can be regarded as simplified examples only. The processes have been developed using analytical [1] and numerical [27] solutions for both the local steady-state [1, 2, 61-65] and dynamic unsteady [27, 66, 67] conditions, respectively. Both models, HLM [1] and MHS [27] with some modifications, may be used for theoretical analysis of all BOHLM systems. Considerations for hollow-fiber systems are presented also in short.

2.1. Model for the BOHLM system [1-3]

2.1.1. Mass-transfer mechanisms and kinetics

The theory for BOHLM is developed for flat thin uncharged symmetric membranes without variation in porosity and pore sizes across the membrane thickness. To develop a three-phase system model [1, 2], the transport model simplification analysis, developed by Hu [68] for the two-phase system, is used. Titanium(IV) was chosen, as an example for transport model verification, because of the extensive experimental data available on liquid-liquid extraction and membrane separation [1, 2, 64, 65] and for its extraction double-maximum acidity dependence phenomenon [63]. The last was observed for most extractant families: basic (anion exchangers), neutral (complexants), and acidic (cation exchangers). So, it is possible to study both counter- and cotransport mechanisms at pH ≥ 0.5 and $[H]^+ \geq 7$ mol kg^{-1} feed solution acidities, respectively, using neutral (hydrophobic, hydrophilic) and ion-exchange membranes.

In general, the mass-transfer rate (or flux) of any solute passing through the barrier (membrane) is a function of distance and time: $J = f(x, t)$. First, let us consider the flux as a function of distance.

Three liquids, having bulk solute i concentration $[C_i]_F$, $[C_i]_E$, and $[C_i]_R$, constant volumes V_F, V_E, and V_R, respectively, are separated by two membranes with the same working area S. Stirring of bulk liquids is effective in such a way that the aqueous (h_{fe}, h_{re}) and the organic (h_{ef}, h_{er}) boundary layers become sufficiently thin and constant. Concentration profiles with hydrophobic membranes are demonstrated in Fig. 5.2, while those containing hydrophilic or ion-exchange membranes are in Fig. 5.3. Using the concept of the one-dimensional series of diffusion resistances, and

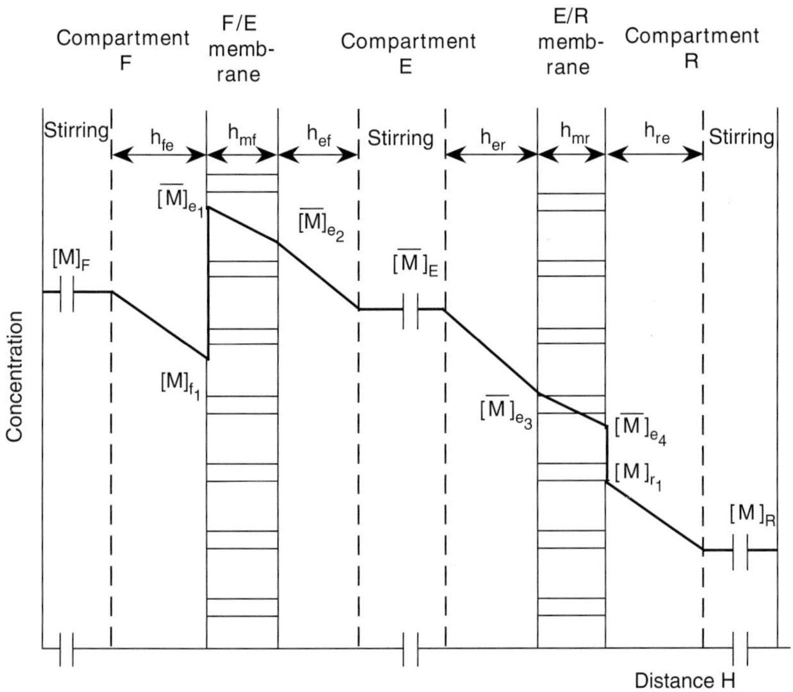

Figure 5.2 Schematic concentration profile of each Ti(IV) chemical species, transported through the BOHLM system with hydrophobic membranes. Layers controlling the permeation rate are h_{fe}—feed-side aqueous boundary layer; h_{mf}—feed-side microporous membrane, immobilized by membrane solution; h_{ef}—feed-side boundary layer of the membrane solution; h_{er}—strip-side boundary layer of the membrane solution; h_{mr}—strip-side microporous membrane, immobilized by membrane solution; and h_{re}—strip-side aqueous boundary layer. From Ref. [1] with permission.

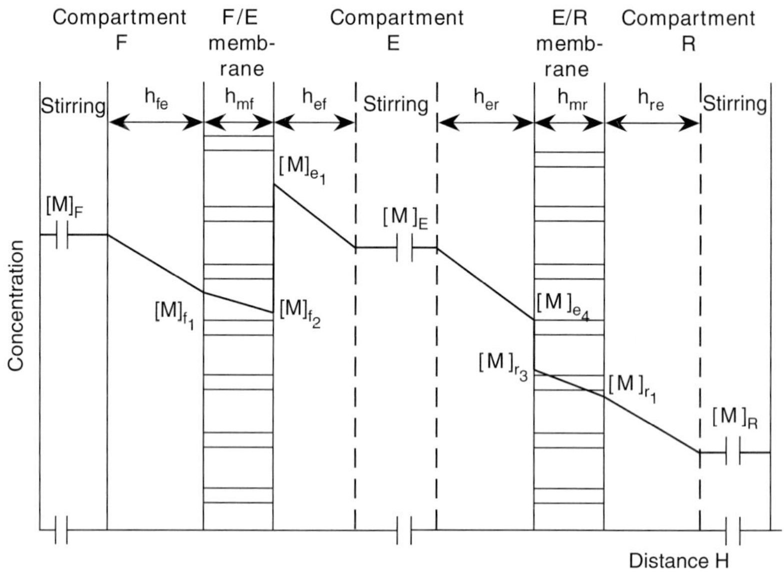

Figure 5.3 Schematic concentration profile of each Ti(IV) chemical species, transported through the BOHLM with hydrophilic or ion-exchange membranes. Layers controlling the permeation rate are the same as those in Fig. 5.2, except for h_{mf}—feed-side microporous membrane is immobilized by feed aqueous solution and h_{mr}—strip-side microporous membrane is immobilized by strip aqueous solution. From Ref. [1] with permission.

regarding the principle of resistance additivity [1-41, 47-54], the overall mass-transfer coefficients (or permeability coefficients [59, 62]) $K_{F/E}$ on the feed side and $K_{E/R}$ on the strip side are related to the individual films (including membrane barrier) mass-transfer coefficients, k:

For hydrophobic membranes,

$$K_{F/E} = \frac{k_{fe}k_{fm}k_{ef}E_F}{k_{fe}k_{ef} + k_{fe}k_{mf} + k_{mf}k_{ef}E_F}, \tag{1}$$

$$K_{E/R} = \frac{k_{re}k_{mr}k_{er}}{k_{mr}k_{er}E_R + k_{mr}k_{re} + k_{re}k_{er}}. \tag{2}$$

For hydrophilic or ion-exchange membranes,

$$K_{F/E} = \frac{k_{fe}k_{mf}k_{ef}E_F}{k_{fe}k_{mf} + k_{fe}k_{ef}E_F + k_{mf}k_{ef}E_F}, \tag{1'}$$

$$K_{E/R} = \frac{k_{re}k_{mr}k_{er}}{k_{mr}k_{er}E_R + k_{re}k_{er}E_R + k_{mr}k_{re}}. \tag{2'}$$

where E_F and E_R are distribution coefficients of solutes between membrane and aqueous feed, and strip phases, respectively, at local equilibrium.

Individual film mass-transfer coefficients may be determined by the following considerations. According to postulates of nonequilibrium thermodynamics [69], the general equation that relates the flux, J, of the solute to its concentration, C, and its derivative is [70]

$$J = UC - D\frac{dC}{dx}, \tag{3}$$

where U is the flow rate and D is the sum [71] of all "effective" diffusion coefficients.

Referring to the equation continuity, as x approaches zero, the steady-state zones or layers are formed next to the phase's interface (it is not for the bulk LM, where $x \gg 0$). Separation then occurs by differential displacement permeation through the interface. Efficiency of the separation hinges directly on the distribution of solute in the steady-state layers.

Using Giddings' analysis [72] of such a system, we obtain

$$\ln\left[\frac{C - \frac{J_o}{U}}{C_o - \frac{J_o}{U}}\right] = U\frac{h}{D}. \tag{4}$$

According to Refs [4-41, 47-54, 64, 73, 74], the mass-transfer coefficient of every layer, k_i, is

$$k_i = \frac{D_i}{h_i}. \tag{5}$$

where D_i is the free diffusion coefficient of the solute in the layer and h_i is the thickness of the layer [63].

Replacing Eq. (4) with (5), we obtain

$$k_i = U\left\{\ln\left[\frac{C_i - \frac{J_{ss}}{U}}{C_{i-1} - \frac{J_{ss}}{U}}\right]\right\}^{-1}, \tag{6}$$

where C_i is the concentration of the solute in the bulk solution at time of sampling t_i; C_{i-1} is the concentration of the solute in the same solution at time of previous sampling t_{i-1}.

According to Sirkar's [38] assumption, the diffusion mass-transfer rate through a membrane having a solvent-filled pore (hydrophobic), or an aqueous solution-filled pore (hydrophilic or ion exchange), may be expressed through the diffusion coefficients of the solute in the respective interface layers.

For hydrophobic membranes,

$$k_{mf} = \frac{D_{ef}\epsilon_m}{h_{mf}\tau_m}, \tag{7}$$

$$k_{mr} = \frac{D_{er}\epsilon_m}{h_{mr}\tau_m}. \tag{8}$$

For hydrophilic or ion-exchange membranes,

$$k_{\mathrm{mf}} = \frac{D_{\mathrm{fe}}\epsilon_{\mathrm{m}}}{h_{\mathrm{mf}}\tau_{\mathrm{m}}}, \tag{7'}$$

$$k_{\mathrm{mr}} = \frac{D_{\mathrm{re}}\epsilon_{\mathrm{m}}}{h_{\mathrm{mr}}\tau_{\mathrm{m}}}, \tag{8'}$$

where ϵ_{m} is the membrane porosity and τ_{m} is the membrane tortuosity.

These expressions are valid when the following assumptions are held:

- There is unhindered diffusion of the solute (solute dimensions $\times\ 10^2 \leq$ pore dimensions).
- The membrane is symmetric and completely wetted by the designated phase.
- No two-dimensional effects occur.

The applicability of these expressions for charged membranes is doubtful because of hindered diffusion of solute in their pores, but it may be evaluated experimentally. One more serious simplification: the influence of the flow velocities extends to inside the membrane pores.

Another way of evaluating mass-transfer coefficients through the membranes may be proposed, providing the membranes possess the same properties (diffusion resistance, porosity, tortuosity, etc.). They may, however, be of different thicknesses. It is clear from Eq. (4) that the diffusion coefficient (exactly U/D, where U is known) of the solute through the feed-side and strip-side membranes may be evaluated as a slope of the plots:

$$-\ln\left(\frac{C}{C_0}\right)F = f(h_{\mathrm{mf}}), \tag{9}$$

$$\ln\left(\frac{C}{C_0}\right)R = f(h_{\mathrm{mr}}). \tag{10}$$

Individual mass-transfer coefficients of the solute (e.g., titanium compound) can be experimentally determined in every layer and membrane. Thus, overall mass-transfer coefficients of the feed (K_{C_F}) and strip (K_{C_R}) sides of the BOHLM system may be calculated according to Eqs (1), (2) or (1′), (2′).

Now, the concentration profiles in the system can be analyzed as a function of time.

Introducing the assumptions as:

- Linear concentration gradients
- Concentration of the solute permeating species is lower than that of the carrier in the membrane phase
- Instantaneous interfacial chemical reactions and local reaction equilibria at the interfaces [1, 4–8, 30–38, 47–52, 75, 76]

we can derive the transport rate and distribution relation equations for any solute [M] permeating species:

$$V_F\left(\frac{d[M]_F}{dt}\right) = -SK_F([M]_F V_F - [\overline{M}]_E V_E), \tag{11}$$

$$V_E\left(\frac{d[\overline{M}]_E}{dt}\right) = SK_F([M]_F V_F - [\overline{M}]_E V_E) - SK_R([\overline{M}]_E V_E - [M]_R V_R). \tag{12}$$

From the overall material balance of the system,

$$[M]_R = \frac{Q^0}{V_R} - \frac{V_F}{V_R}[M]_F - \frac{V_E}{V_R}[\overline{M}]_E, \tag{13}$$

where $[M]_F^0$, $[\overline{M}]_E^0$, and $[M]_R^0$ are initial concentrations of a solute species in the feed, membrane, and strip phases, respectively; Q^0 is the overall initial quantity of the solute in the BOHLM system.

The system described with Eqs (11)-(13) results in an analytical solution under the assumption that the mass-transfer coefficients are constant[1] (at constant flow or stirring rates). The complete set of the model equations follows

$$[M]_F = \frac{\gamma}{\beta} + C_1 e^{\gamma_1 t} + C_2 e^{\gamma_2 t}, \tag{14}$$

$$[\overline{M}]_E = \left[\frac{\gamma}{\beta} + C_1 e^{\gamma_1 t}\left(1 + \frac{\gamma_1}{K_F S}\right) + C_2 e^{\gamma_2 t}\left(1 + \frac{\gamma_2}{K_F S}\right)\right]\frac{V_F}{V_E}, \tag{15}$$

$$[M]_R = \frac{Q^0}{V_R} - [M]_F \frac{V_F}{V_R} - [\overline{M}]_E \frac{V_E}{V_R}, \tag{16}$$

$$\alpha = 2S(K_F + K_R), \tag{17}$$

$$\beta = 3S^2 K_F K_R, \tag{18}$$

$$\gamma = K_F K_R \frac{S^2}{V_F} Q^0, \tag{19}$$

$$\frac{\gamma}{\beta} = \frac{Q^0}{3V_F} = \frac{[M]_F^0 V_F + [\overline{M}]_E^0 V_E + [M]_R^0 V_R}{3V_F}, \tag{20}$$

$$\gamma_1 = -\frac{\alpha}{2} + \sqrt{\frac{\alpha^2}{4} - \beta}, \tag{21}$$

[1]In the real continuous BOHLM system, mass-transfer coefficients $K_{F/E}$ and $K_{E/R}$ may be close to constant at the stabilized conditions of the process (acidity or pH, temperature, flow rates, etc.).

$$\gamma_2 = -\frac{\alpha}{2} - \sqrt{\frac{\alpha^2}{4} - \beta}, \tag{22}$$

$$C_1 = \frac{\gamma_2\left([\mathrm{M}]_\mathrm{F}^0 - \frac{\gamma}{\beta}\right) - SK_\mathrm{F}\left([\overline{\mathrm{M}}]_\mathrm{E}^0 \frac{V_\mathrm{E}}{V_\mathrm{F}} - [\mathrm{M}]_\mathrm{F}^0\right)}{\gamma_2 - \gamma_1}, \tag{23}$$

$$C_2 = -\frac{\gamma_1\left([\mathrm{M}]_\mathrm{F}^0 - \frac{\gamma}{\beta}\right) - SK_\mathrm{F}\left([\overline{\mathrm{M}}]_\mathrm{E}^0 \frac{V_\mathrm{E}}{V_\mathrm{F}} - [\mathrm{M}]_\mathrm{F}^0\right)}{\gamma_2 - \gamma_1}. \tag{24}$$

2.1.2. Driving forces

The transport of titanium(IV) species through BOHLM can be formally described as a simultaneous combination of diffusion, extraction, and stripping operations occurring in nonequilibrium conditions. These systems are very complicated to analyze and therefore some assumptions are needed for simplification. Kinetics of extraction and stripping processes are much faster than their diffusion for most metal ions, and many researchers have adopted "the local extraction equilibrium" at feed-extractant and extractant-strip membrane interfaces. So, the chemistry of the BOHLM system at equilibrium conditions has to be analyzed.

The chemical reactions responsible for the transport can be schematized [62-65]. At the feed solution-membrane (carrier) solution interface, for feed:

At low acidity (pH region),

$$[\mathrm{Ti} \star 2\mathrm{H_2O}]_\mathrm{F}^{4+} + 4\overline{\mathrm{HL}}_\mathrm{E} \longrightarrow [\overline{\mathrm{TiL_4} \star 2\mathrm{H_2O}}]_\mathrm{E} + 4\mathrm{H}_\mathrm{F}^+. \tag{25}$$

At high (>7 mol kg^{-1}) acidity,

$$[\mathrm{Ti} \star 2\mathrm{HCl}]_\mathrm{F}^{4+} + 4\overline{\mathrm{HL}}_\mathrm{E} \longrightarrow [\overline{\mathrm{TiL_4} \star 2\mathrm{HCl}}]_\mathrm{E} + 4\mathrm{H}_\mathrm{F}^+. \tag{26}$$

At the membrane solution-strip solution interface:

At low acidity (pH region),

$$[\overline{\mathrm{TiL_4} \star 2\mathrm{H_2O}}]_\mathrm{E} + 2\mathrm{H}_\mathrm{R}^+ \longrightarrow [\mathrm{TiO}]_\mathrm{R}^{2+} + 4\overline{\mathrm{HL}}_\mathrm{E} + \mathrm{H_2O}_\mathrm{R}. \tag{27}$$

At high acidity,

$$[\overline{\mathrm{TiL_4} \star 2\mathrm{HCl}}]_\mathrm{E} + \mathrm{H_2O}_\mathrm{R} \longrightarrow [\mathrm{TiO}]_\mathrm{R}^{2+} + 4\overline{\mathrm{HL}}_\mathrm{E} + 2\mathrm{Cl}_\mathrm{R}^-. \tag{28}$$

These reactions are characterized by equilibrium constants:

$$K_\mathrm{F/E} = \frac{[\overline{\mathrm{TiL_4} \star 2\mathrm{H_2O}}]_\mathrm{E} \star [\mathrm{H}]_\mathrm{F}^4}{[\overline{\mathrm{HL}}]_\mathrm{E}^4 \star [\mathrm{Ti} \star 2\mathrm{H_2O}]_\mathrm{F}} = E_\mathrm{F} \star \left(\frac{[\mathrm{H}]_\mathrm{F}}{[\overline{\mathrm{HL}}]_\mathrm{E}}\right)^4 \tag{29}$$

or

$$K'_{F/E'} = \frac{[\overline{TiL_4 \star 2HCl}]_E \star [H]_F^4}{[\overline{HL}]_E^4 \star [Ti \star 2HCl]_F} = E_{F'} \star \left(\frac{[H]_F}{[\overline{HL}]_E}\right)^4; \qquad (29')$$

$$K_{E/R} = \frac{[TiO]_R \star [\overline{HL}]_E^4 \star [H_2O]}{[\overline{TiL_4 \star 2H_2O}]_E \star [H]_R^2} = \frac{[\overline{HL}]_E^4}{[H]_R^2 \star E_R}, \qquad (30)$$

or

$$K'_{E/R'} = \frac{[TiO]_R \star [\overline{HL}]_E^4 \star [CL]_R^2}{[TiL_4 \star 2HCl] \star [H_2O]} = \frac{[\overline{HL}]_E^4 \star [Cl]_R^2}{E_R}, \qquad (30')$$

where E_F and E_R are distribution coefficients of titanium ions between membrane and aqueous feed, and strip phases, respectively, and concentration of $[H_2O]$ is neglected:

$$K = K_{F/E}K_{E/R} = \left(\frac{E_F}{E_R}\right)\left(\frac{[H]_F^4}{[H]_R^2}\right) \qquad (31)$$

or

$$K' = K'_{F/E}K'_{E/R} = \left(\frac{E_F}{E_R}\right)([H]_F^4 \star [Cl]_R^2), \qquad (31')$$

K (or K') is denoted as a Driving force coefficient of the BOHLM system.

Consider

$$K = K_c K_d, \quad \text{where} \quad K_c = \frac{E_F}{E_R} \quad \text{and} \quad K_d = \frac{[H]_F^4}{[H]_R^2} \qquad (32)$$

or

$$K' = K'_c K'_d, \quad \text{where} \quad K'_c = \frac{E_F}{E_R} \quad \text{and} \quad K'_d = [H]_F^4 \star [Cl]_R^2, \qquad (32')$$

K_c is denoted as an internal [77] (carrier) driving force coefficient, derived from the extraction distribution ratio between liquid membrane phase and feed, and receiving phases. K_d (or $K_{d'}$) is denoted as an external driving force coefficient, derived from the coupling effect of the BOHLM system.

These initial parameters are easily accessible experimentally by equilibrium extraction experiments [77]. For example, extraction of titanium(IV) from 0.1 mol kg^{-1} Ti(IV) hydrochloric acid solutions at 0.45 mol kg^{-1} (pH 0.65), 2 mol kg^{-1}, and at 7 mol kg^{-1} HCl by 1 mol kg^{-1} DEHPA in benzene, at aqueous phase/organic phase = 5/1, have shown the initial distribution coefficients (for details, see [65]).

At pH 0.65, $E_{F_1} = 20\text{-}25$; at 7.0 mol kg^{-1} HCl, $E_{F_2} = 200\text{-}220$; and at 2.0 mol kg^{-1} HCl, $E_{F_3} = 1.5\text{-}2.0$.

The initial distribution ratios $K_c = E_F/E_R$ (internal or liquid membrane driving force coefficients) are

$$K_c = 10 - 15, \quad K_c^{'} = 100 - 150.$$

Of course, during the transport process in the closed BOHLM system, the distribution ratio will change in accordance with changing feed and strip phase conditions (acidities, titanium concentration, etc.). At $K_c = 1$ ($E_F = E_R$) concentration of titanium(IV) in the carrier solution should be $[\text{Ti}]_R \star E_R$ (the system at equilibrium). Therefore, E_R may be denoted as an irreversibility coefficient of the BOHLM for both closed and open [69] systems (flowing feed, strip streams, buffered acidities, etc.). K_c is an "uphill pumping" border of the BOHLM system.

2.2. Numerical model of competitive M^{2+}/H^+ countertransport [27, 78-81]

A mathematical model to be solved numerically has been developed and used to predict the separation effects caused by nonstationary conditions for a bulk liquid membrane transport. Numerical calculations compute such pertraction[2] characteristics as input and output membrane selectivity (ratio of respective fluxes), concentration profiles for cations bound by a carrier in a liquid membrane phase, and the overall separation factors all being dependent on time. The computations of fluxes and separation factors as dependent on time have revealed high separation efficiency of unsteady-state pertraction as compared with steady or near-steady-state process (with reactions near equilibrium).

The mechanism of the competitive pertraction system (CPS) is presented schematically in Fig. 5.4 together with the compartmental model necessary for constructing the reaction-diffusion network. The simple flat-layered bulk liquid membrane of the thickness L_m and interface area S separates the two reservoirs (f, feed and s, stripping) containing transported divalent cations A^{2+} and B^{2+} (most frequently Zn^{2+} and Cu^{2+} or Ca^{2+} and Mg^{2+}) and/or antiported univalent cations H^+. At any time of pertraction t, their concentrations are $[A]_f$, $[B]_f$, and $[H]_f$ and $[A]_s$, $[B]_s$, and $[H]_s$, for the feed and stripping solution, respectively. The hydrophobic liquid membrane contains a carrier of total concentration [C]. Its main property is the ability to react reversibly with cations at respective reaction zone and to diffuse throughout the liquid membrane phase.

[2] The term "pertraction," used by the authors [27] did not change in the following text (for further explanation, see Section 1).

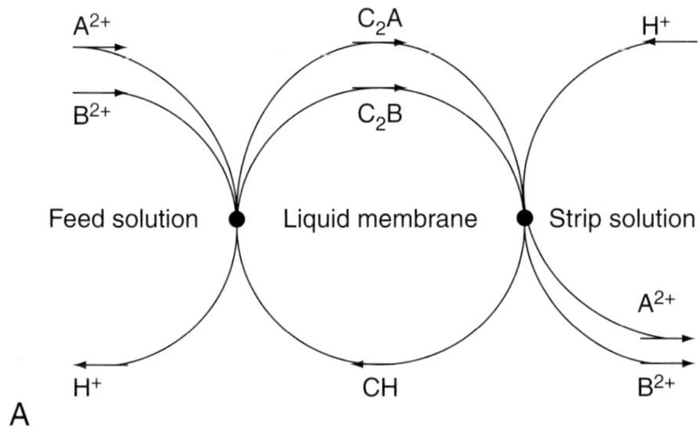

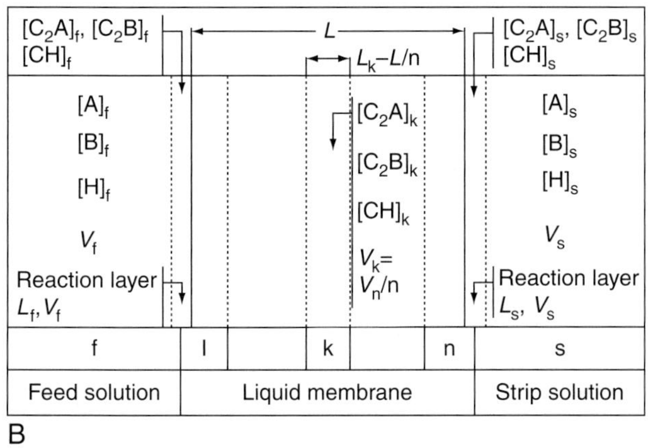

Figure 5.4 (A) Scheme of A^{2+}, B^{2+}/H^{+} competitive countertransport mediated by an ionic carrier CH and (B) compartmental model of pertraction system. From Ref. [27] with permission.

It is also assumed that ionic species or their ionic pairs cannot diffuse in the membrane phase without intervention by the carrier. The membrane is formally divided into n homogeneous compartments which contain the carrier in the form C_2A, C_2B, and CH the local concentrations of which are equal to $[C_2A]_k$, $[C_2B]_k$, and $[CH]_k$, with $k = 1, 2,\ldots,n$. In the case of bulk agitated or flowing liquid membranes, the model should be modified by adding a large central compartment (layer).

Assumptions of the model:

- The concentrations of ionic species within interfacial reaction layers of thickness L_f and L_s are be the same as in the feed or stripping solution, respectively.
- The concentrations of cations bound by the carrier at respective interfacial layer, that is, $[C_2A]_i$, $[C_2B]_i$, and $[CH]_i$ with i = f or s, are the same as the concentrations in the membrane compartments indexed by $k = 1$ or n.

- Interfacial carrier molecules are present as hydrophilic fragments in adjacent aqueous solutions and as hydrophobic parts in an organic phase of a liquid membrane. Thus, the reaction can be treated as a homogeneous type occurring in an adjacent reaction zone instead of a heterogeneous adsorption-type reaction process.
- Initial concentration of the carrier is the same as in a bulk liquid membrane.
- Concentration of free cations is the same as in the bulk aqueous phase.
- Thermodynamic couplings between transported species are negligible.

Under these assumptions and simplifications, the thermodynamic network analysis (TNA) [82] can be applied to analyze LM transport. Certainly, in the case of a real specific system, the detailed mechanism of reaction-diffusion interfacial phenomena should be taken into account as far as possible. The model presented can be regarded as a simplified example only. When describing a real membrane process, the accessible knowledge concerning the specific interfacial processes should be taken into account.

According to the scheme presented in Fig. 5.4, two competitive cation-exchange processes occur at each of the interfaces.

At the feed interface,

$$A^{2+}_{(f)} + 2CH_{(f)} \underset{k_{-fA}}{\overset{k_{fA}}{\rightleftarrows}} C_2A_{(f)} + 2H^{+}_{(f)} \quad (R_{fA}), \tag{33}$$

$$B^{2+}_{(f)} + 2CH_{(f)} \underset{k_{-fB}}{\overset{k_{fB}}{\rightleftarrows}} C_2B_{(f)} + 2H^{+}_{(f)} \quad (R_{fB}). \tag{34}$$

At the strip interface,

$$C_2A_{(s)} + 2H^{+}_{(s)} \underset{k_{-sA}}{\overset{k_{sA}}{\rightleftarrows}} 2CH_{(s)} + A^{2+}_{(s)} \quad (R_{sA}), \tag{35}$$

$$C_2B_{(s)} + 2H^{+}_{(s)} \underset{k_{-sB}}{\overset{k_{sB}}{\rightleftarrows}} 2CH_{(s)} + B^{2+}_{(sB)} \quad (R_{sB}). \tag{36}$$

The reaction rates between the carrier and the two cations are represented by coupled differential equations resulting from networks in Fig. 5.5.

The symbols $k_{i,M}$ and $k_{-i,M}$ (i = f, s; M = A, B) denote the kinetic constants for the rates of reversible reactions given by Eqs (37)-(40):

$$J_{f,A} = k_{f,A}[A]_f[CH]_f^2 - k_{-f,A}[C_2A]_f[H]_f^2, \tag{37}$$

$$J_{f,B} = k_{f,B}[B]_f[CH]_f^2 - k_{-f,B}[C_2B]_f[H]_f^2, \tag{38}$$

$$J_{s,A} = k_{s,A}[C_2A]_s[H]_s^2 - k_{-s,A}[CH]_s^2[A]_s, \tag{39}$$

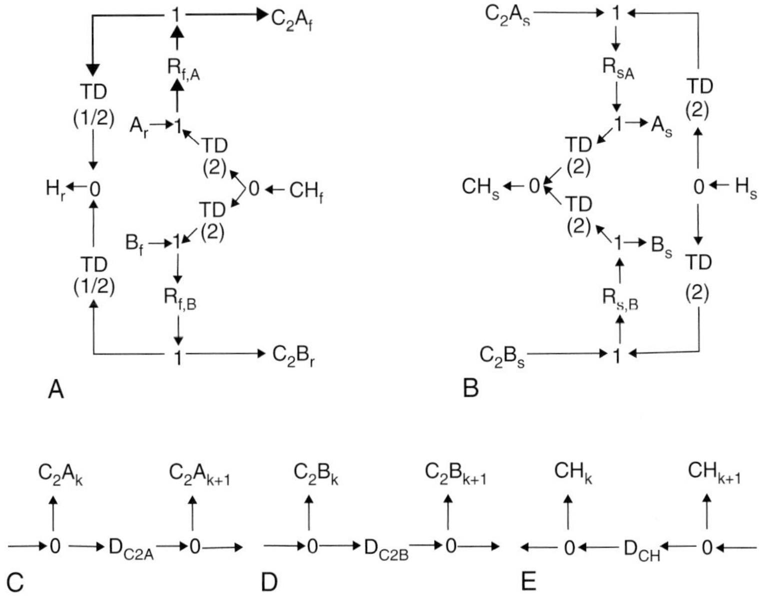

Figure 5.5 Network for feed (A), strip (B) competitive cation-exchange reactions, and local diffusion processes: C_2A (C), C_2B (D), and CH (E) species. From Ref. [27] with permission.

$$J_{s,B} = k_{s,B}[C_2B]_s[H]_s^2 - k_{-s,B}[CH]_s^2[B]_s. \tag{40}$$

According to the scheme of the compartments in Fig. 5.4B, all local diffusion fluxes of M species from k to $k+1$ compartment can be defined by a phenomenological Eq. (41) corresponding with Fick's first law for diffusion:

$$N_{j(k,k+1)} = P_j\Delta[X]_{j(k,k+1)}. \tag{41}$$

where P_j denotes the permeability coefficient of species (e.g., carrier in different forms) between the k and $k+1$ compartment:

$$P_j = D_jS/(L_m/n). \tag{42}$$

In Eqs (37)-(41), $\Delta[X]_{j(k,k+1)}$ denotes the concentration difference between the compartment k and $k+1$ for the carrier transporting cation j (j = A^{2+}, B^{2+}, H^+) with the diffusion coefficient D_j. Consequently, the diffusion coefficients of a carrier transporting different cations are different but constant throughout the liquid membrane (independent of their local concentration).

The reaction-diffusion network, as presented in Fig. 5.5, consists of four coupled loops representing the pertraction of A^{2+}, B^{2+}, and H^+ cations. All model equations used further in numerical calculations can be deduced with the help of Kirchhoff's law for a "0" junction (KCL):

$$\sum_k \nu_k N_k = 0, \tag{43}$$

where k is the index of the flux entering or leaving a given "0" junction, υ is $+1$ or -1 valued depending on the bond direction. Note that all the reactions and diffusion rates should be expressed in mol s^{-1} to maintain the compatibility of units.

The mathematical model derived from the network in Fig. 5.5, together with the specification of all the local fluxes, time-dependent variables, and parameters, is presented in Table 5.1. The model consists of ordinary differential equations (ODE) describing the evolution of all capacitances (local concentrations) characteristic for a membrane and external solutions.

A step-by-step simulation of the system can be carried out by numerical calculations. The calculations have been performed using the data from Table 5.1 after assuming $n = 4$. However, the increase of the number of layers, that is, increasing n, will always result in more precise calculations and predictions. The $n = 4$ should be treated as the lowest limit required for obtaining quantitative data sufficient for the interpretation of the separation effects. The Madonna Berkeley (ver.3) solver of ODE, the set of parameters, and initial conditions describing a liquid membrane system (listed in Table 5.2) were applied to numerically investigate the properties of CPS. The main relations resulting from the computations were time-dependent values of (i) the concentration profiles of transported and antiported species in a liquid membrane, (ii) the concentration of A^{2+} and B^{2+} in the external aqueous solutions, and (iii) input and output fluxes of A^{2+} and B^{2+}. To characterize the membrane selectivity towards the target A^{2+} cations, the following quantities are calculated and discussed.

The selectivity ($S_{N,\mathrm{i}}$) calculated as the ratio of A^{2+} and B^{2+} fluxes (N) for feed (f) and strip (s) interface, henceforth called as input or output flux selectivity:

$$S_\mathrm{i} = N_{\mathrm{A,i}}/N_{\mathrm{B,i}}, \quad \mathrm{i} = \mathrm{f} \quad \text{or} \quad \mathrm{s}. \tag{44}$$

The overall concentration of divalent cations bound by the carrier in a liquid membrane phase and reaction zones ($[A]_\mathrm{m}$, $[B]_\mathrm{m}$), and their ratio (S_m) which characterizes the competitive accumulation of one of the cations in a liquid membrane phase:

$$[\mathrm{A}]_\mathrm{m} = \left([\mathrm{C_2A}]_{\mathrm{f},z} V_{\mathrm{f},z} + \sum_{k=1}^{n} [\mathrm{C_2A}]_k V_k + [\mathrm{C_2A}]_{\mathrm{s},z} V_{\mathrm{s},z} \right) / (V_\mathrm{m} + V_{\mathrm{f},z} + V_{z,\mathrm{s}}), \tag{45}$$

$$[\mathrm{B}]_\mathrm{m} = \left([\mathrm{C_2B}]_{\mathrm{f},z} V_{\mathrm{f},z} + \sum_{k=1}^{n} [\mathrm{C_2B}]_k V_k + [\mathrm{C_2B}]_{\mathrm{s},z} V_{\mathrm{s},z} \right) / (V_\mathrm{m} + V_{\mathrm{f},z} + V_{z,\mathrm{s}}), \tag{46}$$

Table 5.1 Mathematical description of competitive pertraction of A^{2+}, B^{2+} cation mediated by an ionic carrier CH

Time-dependent variables—concentrations (mol cm^{-3})	
Feed (f) and strip (s) phase	$[A]_i[B]_i[H]_i$, i = f or s
Interfaces	$[C_2A]_i[C_2B]_i[CH]_i$, i = f or s
Membrane	$[C_2A]_k[C_2B]_k[CH]_k$, $k = 1, 2,\ldots,n$
Feed solution	Stripping solution
Capacitance fluxes (mol cm^{-3} s^{-1})	
$d[A]_f/dt = -(V_{f,r}/V_f)J_{f,A}$	$d[A]_s/dt = -(V_{s,r}/V_s)J_{s,A}$
$d[B]_f/dt = -(V_{f,r}/V_f)J_{f,B}$	$d[B]_s/dt = -(V_{s,r}/V_s)J_{s,B}$
$d[H]_f/dt = 2(V_{f,r}/V_f)(J_{f,A}+J_{f,B})$	$d[H]_s/dt = 2(V_{s,r}/V_s)(J_{s,A}+J_{s,B})$
Accumulation of A^{2+}	Accumulation of B^{2+}
Membrane and interfaces	
$d[C_2A]_{f,r}/dt = J_{f,A} - N_{C_2A_{(f,1)}}/V_{f,r}$	$d[C_2B]_{f,r}/dt = J_{f,B} - N_{C_2B_{(f,1)}}/V_{f,r}$
$d[C_2A]_1/dt = [N_{C_2A_{(f,1)}} - N_{C_2A_{(1,2)}}]/V_2$	$d[C_2B]_1/dt = [N_{C_2B_{(f,1)}} - N_{C_2B_{(1,2)}}]/V_1$
$d[C_2A]_k/dt = [N_{C_2A_{(k-1,k)}} - N_{C_2A_{(k-2,k)}}]/V_k$	$d[C_2B]_k/dt = [N_{C_2B_{(k-1,k)}} - N_{C_2B_{(k,k+1)}}]/V_k$
$d[C_2A]_n/dt = [N_{C_2A_{(n-1,n)}} - N_{C_2A_{(n-2,n)}}]/V_n$	$d[C_2B]_n/dt = [N_{C_2B_{(n-1,n)}} - N_{C_2B_{(n-2,n)}}]/V_n$
$d[C_2A]_s/dt = N_{C_2A_{(n,s)}}/V_{s,r} - J_{s,A}$	$d[C_2B]_s/dt = N_{C_2B_{(n,s)}}/V_{s,r} - J_{s,B}$
Accumulation of H^+	
$d[CH]_{f,r}/dt = N_{CH_{(f,1)}}/V_{f,r} - 2(J_{f,A}V) = SL_f$	
$d[CH]_1/dt = [N_{CH_{(2,1)}} - N_{CH(1,f)}]/V_{s,r} = SL_s$	
$d[CH]_k/dt = [N_{CH_{(k+1,k)}} - N_{CH_{(k,k-1)}}]$	
$d[CH]_n/dt = [N_{CH_{(s,n)}} - N_{CH_{(n,n-1)}}]$	
$d[CH]_{s,r}/dt = 2(J_{A,s} - J_{B,s}) - N_{CH_{(s,n)}}/V_{s,r}$	
Reaction rates (J mol^{-1} cm^{-3} s^{-i}) and local flows (N mol s^{-1})	
$J_{f,A} = k_{f,A}[A]_f[CH]_f^2 - k_{-f,A}[C_2A]_f[H]_f^2$	$J_{f,B} = k_{f,B}[B]_f[CH]_f^2 - k_{-f,B}[C_2B]_f[H]_f^2$
$J_{s,A} = k_{s,A}[C_2A]_s[H]_s^2 - k_{-s,A}[A]_s[CH]_s^2$	$J_{s,B} = k_{s,B}[C_2B]_s[H]_s^2 - k_{-s,B}[CH]_s^2$
$N_{C_2A_{(f,1)}} = 2P_{C_2A}([C_2A]_f - [C_2A]_1)$	$N_{C_2B_{(f,1)}} = 2P_{C_2B}([C_2B]_f - [C_2B]_1)$

(continued)

Table 5.1 (*continued*)

Time-dependent variables—concentrations (mol cm^{-3})	
$N_{C_2A_{(k,k+1)}} = 2P_{C_2A}([C_2A]_k - [C_2A]_{k+1})$	$N_{C_2B_{(k,k+1)}} = 2P_{C_2B}([C_2B]_k - [C_2B]_{k+1})$
$N_{C_2A_{(n,s)}} = 2P_{C_2A}([C_2A]_n - [C_2A]_s)$	$N_{C_2B_{(n,s)}} = 2P_{C_2B}([C_2B]_n - [C_2B]_s)$
$N_{CH_{(1,f)}} = 2P_{CH}([CH]_1 - [CH]_f)$	$P_{C_2A} = nD_{C_2A}\ S/L_m$
$N_{CH_{(k,l+k)}} = P_{CH}([CH]_{k+1} - [CH]_k)$	$P_{C_2B} = nD_{C_2B}\ S/L_m$
$N_{CH_{(s,n)}} = 2P_{CH}([CH]_s - [CH]_n)$	$P_{CH} = nD_{CH}\ S/L_m$
Parameters	
Kinetic constants	$k_{f,A}$, $k_{-f,A}$, $k_{f,B}$, $k_{-f,B}$, $k_{s,A}$, $k_{-s,A}$, $k_{s,B}$, $k_{-s,B}$ mol^{-2} s^{-1} (cm^3)
Diffusion coefficients	D_{C_2A}, D_{C_2B}, D_{CH} (cm^2 s^{-1})
Membrane area and thickness	S (cm^2): L_m (cm)
Thickness of reaction compartments	L_f, L_s (cm)
Volume of feed and stripping solutions and membrane	V_f, V_s, V_m (cm^3)
Number of membrane compartments	n

Source: From Ref. [27] with permission.

$$S_m = [A]_m/[B]_m. \tag{47}$$

The overall separation factor (α_B^A):

$$\alpha_B^A = \frac{[A]_s[B]_f}{[B]_s[A]_f}. \tag{48}$$

To emphasize the kinetic source of separations studied, all computations have been carried out after assuming the same equilibrium constants for cation-exchange reactions appearing in the pertraction system, that is,

$$K_{f,A} = K_{f,B} = k_{f,A}/k_{-f,A} = k_{f,B}/k_{-f,B}, \tag{49}$$

$$K_{s,A} = K_{s,B} = k_{s,A}/k_{-s,A} = k_{s,B}/k_{-s,B}. \tag{50}$$

The conditions (49) and (50) impose no selectivity for the CPS under stationary conditions when the reactions attain their equilibrium state and the diffusion coefficients D_{C_2A} and D_{C_2B} are equal.

The bond-graph network of liquid membrane process can be successfully exploited for modeling the separation and transport ability of complex reaction-diffusion phenomena. However, such models involving appropriate mathematical formulations are especially useful in predicting the system's response to the changes in operating conditions and specific characteristics of the liquid membrane components. In general, such models are not

Table 5.2 Parameters and operational conditions for simulation of competitive transport process

Parameter variables	Symbol	Value
(1) Initial concentrations (mol cm^{-3})		
Feed phase	$[A]_{f,o}$	1×10^{-5}
	$[B]_{f,o}$	1×10^{-5}
	$[H]_{f,o}$	1×10^{-10}
Stripping phase	$[A]_{s,o}$	0
	$[B]_{s,o}$	0
	$[H]_{s,o}$	3×10^{-3}
Membrane	$[C_2A]_{k,o}$	0
	$[C_2B]_{k,o}$	0
	$[CH]_{k,o}$	5×10^{-4}
Reaction zones	$[C_2A]_{i,o}$	0
	$[C_2B]_{i,o}$	0
	$[CH]_{i,o}$	5×10^{-4}
(2) Diffusion coefficients (cm^2 s^{-1})		
	D_{C_2A}	1×10^{-5}-1×10^{-7}
	D_{C_2B}	1×10^{-5}-1×10^{-7}
	D_{CH}	1×10^{-5}-1×10^{-7}
(3) Kinetic constants (cm^6 mol^{-2} s^{-1})		
	$k_{f,A}$	1×10^6-1×10^9
	$k_{-f,A}$	1×10^6-1×10^9
	$k_{s,A}$	1×10^6-1×10^9
	$k_{-s,A}$	1×10^6-1×10^9
	$k_{f,B}$	1×10^6-1×10^9
	$k_{-f,B}$	1×10^6-1×10^9
	$k_{s,B}$	1×10^6-1×10^9
	$k_{-s,B}$	1×10^6-1×10^9
(4) Dimensions		
	V_f (cm^3)	100
	V_s (cm^3)	10
	L_m (cm)	2.5×10^{-3}
	L_f (cm)	5×10^{-5}
	L_s (cm)	5×10^{-5}
	S_m (cm^2)	1

Source: From Ref. [27] with permission.

applicable for the assessment of unknown parameters. The idea of network modeling needs as much information as possible to be introduced into a model. Thus, at the initial step no factor can be ignored. Consequently, the related calculations should be treated as experiments aimed at searching for new quantitative differences in transport and separation systems.

2.3. The theory of hollow-fiber liquid membrane transport [5, 9-11, 39, 55, 56, 83, 84]

The theory of a HFCLM transport was developed by Sirkar *et al.* and described in detail [56, 85-87]. Its modified model, HFLM, developed by Schlosser *et al.* [5, 9-11, 25, 39, 83] is presented below.

Concentration profiles of a solute in three-phase system transport through LM in a hollow-fiber module are similar to that presented in Fig. 5.3.

Following assumptions are held [11, 25]:

- Steady-state flux
- Consider only diffusional resistances, R_{i}. Reactions on both interfaces are fast and negligible accumulation in boundary layers is assumed
- A high stoichiometric excess of reactant in the stripping solution
- Fibers are identical and the number of fibers in both bundles (for feed and strip solutions) is the same:

$$S_{\mathrm{wF}} = S_{\mathrm{wR}} = S_{\mathrm{w}}; \quad \varepsilon_{\mathrm{F}} = \varepsilon_{\mathrm{R}} = \varepsilon; \quad \tau_{\mathrm{F}} = \tau_{\mathrm{R}} = \tau; \quad \text{and} \quad k_{\mathrm{wF}} = k_{\mathrm{wR}}, \tag{51}$$

where S_{wF} and S_{wR} are surface area (m^2) in the feed and strip; ϵ is the porosity of the fiber; τ is the tortuosity of the pores; k_i are the individual mass-transfer coefficients (m s^{-1}).

This results in concentration of solute in the feed referring to its concentration in the strip, C_{FRM}, equals zero:

$$C_{\mathrm{FRM}} = C_{\mathrm{R}} D_{\mathrm{R}} / D_{\mathrm{F}} = 0 \tag{52}$$

and

$$R_{\mathrm{R}} = 0, \tag{53}$$

where C_i are the molar solute concentrations; D_i are the distribution coefficients at equilibrium at feed-membrane and membrane-strip interfaces; and R_i are resistivities.

The relation for the overall mass-transfer resistance can be derived [11, 25, 83]:

$$\frac{1}{K_{\mathrm{p}}} = \frac{\varepsilon_{\mathrm{F}}}{k_{\mathrm{F}}} + \frac{S_{\mathrm{F}}}{DS_{\mathrm{wF}} k_{\mathrm{wF}}} + \frac{S_{\mathrm{F}} \epsilon_{\mathrm{F}}}{DS_{\mathrm{M}} k_{\mathrm{M}}} + \frac{S_{\mathrm{F}} \varepsilon_{\mathrm{F}}}{DS_{\mathrm{wR}} \varepsilon_{\mathrm{R}} k_{\mathrm{wR}}} + \frac{S_{\mathrm{F}} \varepsilon_{\mathrm{F}}}{S_{\mathrm{R}} k_{\mathrm{R}}}, \tag{54}$$

$$R = R_{\mathrm{F}} + R_{\mathrm{wF}} + R_{\mathrm{M}} + R_{\mathrm{wR}} + R_{\mathrm{R}}. \tag{55}$$

Identification of the individual mass-transfer coefficients depends on the construction of the module (material and structure of fibers in both bundles) and on its mode of operation. For the laminar flow inside hollow fibers, the following correlation can be used [84]:

$$\frac{k_{\mathrm{F}} d_i}{D_{\mathrm{f}}} = 1.5 \left(\frac{d_i^2 u_{\mathrm{F}}}{L D_{\mathrm{f}}} \right)^{1/3}, \tag{56}$$

where d_i is the fiber diameter (m); D_f is the diffusion coefficient (m s^{-1}); u_i are the linear velocities of flow (m s^{-1}); and L is the effective length of the fibers (m).

Individual mass-transfer coefficients in the pores are estimated by

$$k_w = k_{wF} = k_{wR} = \frac{2Df_M}{\tau(d_0 - d_i)} \tag{57}$$

and

$$R_{wF} - R_{wR} = R_w. \tag{58}$$

The resistance in the bulk liquid membrane is calculated from

$$R_M = R - (R_F + 2R_w). \tag{59}$$

Differential mass balance of solute in the feed stream along the length of the HF contactor for constant volumetric flow rates of the solutions is

$$-V_F\frac{dc_F}{dz} = K_{ploc}(\pi d_{iF} N_F \varepsilon_F)(c_F - c_{FRM}), \tag{60}$$

where V_i are volumes of the solution passed; K_{ploc} is defined for effective surface area of the feed/LM interface. When concentration c_{FRM} is close to zero, after rearranging of Eq. (60):

$$k_{ploc}dz = -\frac{V_F}{\pi d_{iF} N_F \varepsilon_F}\frac{dc_F}{c_F}. \tag{61}$$

Integrating Eq. (60) for initial and boundary conditions $z = 0$, $c_F = c_{F_1} = c_{F_0}$, and for $z = L$, $c_F = c_{F_2}$, the following relation is obtained:

$$L = \frac{V_{F_0}}{k_p \pi d_{iF} N_F \varepsilon_F}\ln\frac{c_{F_0}}{c_{F_2}} \tag{62}$$

and the concentration of the solute in the strip, c_{F_2} can be calculated by equation

$$\frac{c_{F_2}}{c_{F_0}} = \exp[-(\pi d_i N_F / V_F)(K_p L)], \tag{63}$$

where K_p is the integral value of the overall mass-transfer coefficient of the module.

These considerations and the set of equations for transport of the solute through the HF modules are modified and simplified in comparison to the HFCLM [85] theoretical considerations. The models presented in Sections 2.1 and 2.2 consider different diffusion parameters at the F/M and M/R interfaces, and different kinetics of chemical reactions. So, they are more identical to the real transport processes. These models, modified for a hollow-fiber module, may be used for the solute transport through hollow-fiber permeator also.

3. Module Design for Separation Systems

3.1. Preliminary design and optimization of the module

3.1.1. Determination and optimization of the transport rate parameters

According to the theoretical model for transport kinetics (see above) most preliminary parameters, needed for the BOHLM process design and optimization, may be obtained by a number of known or experimentally obtained data. Individual mass-transfer coefficients of solute species in the feed, carrier, and strip interfacial boundary layers are determined experimentally by feed, carrier, and strip flow rate variations:

$$k_{c(\mathrm{fe})_i} = U_{\mathrm{F}_j}\left\{-\ln\left[\left([\mathrm{C}_{\mathrm{F}_i}] - \frac{J_{\mathrm{F}_{ss}}}{U_{\mathrm{F}_i}}\right)\Big/\left(\left[\mathrm{C}_{\mathrm{F}_{(i-1)}}\right] - \frac{J_{\mathrm{F}_{ss}}}{U_{\mathrm{F}_i}}\right)\right]\right\}^{-1} \text{(for feed layer)}, \tag{64}$$

$$\frac{1}{2}k_{c(\mathrm{ef})_i}\ \left(\text{or } k_{c(\mathrm{er})_i}\right) = U_{\mathrm{E}_j}\left\{\pm\ln\left[\left([\mathrm{C}]_{\mathrm{E}_i} - \frac{J_{\mathrm{E}_{ss}}}{U_{\mathrm{E}_j}}\right)\Big/\left([\mathrm{C}]_{\mathrm{E}_{(i-1)}} - \frac{J_{\mathrm{E}_{ss}}}{U_{\mathrm{E}_j}}\right)\right]\right\}^{-1} \text{(for carrier layers)}, \tag{65}$$

where "plus" (positive) is at increasing concentrations versus time, and "minus" (negative) is at decreasing ones:

$$k_{c(\mathrm{re})_i} = U_{\mathrm{R}_j}\left\{\ln\left[\left([\mathrm{C}]_{\mathrm{R}_i} - \frac{J_{\mathrm{R}_{ss}}}{U_{\mathrm{R}_j}}\right)\Big/\left([\mathrm{C}]_{\mathrm{R}_{(i-1)}} - \frac{J_{\mathrm{R}_{ss}}}{U_{\mathrm{R}_j}}\right)\right]\right\}^{-1} \text{(for strip layer)}, \tag{66}$$

where U_{F_j}, U_{E_j}, and U_{R_j} are jth flow velocities; $J_{\mathrm{F}_{ss}}$, $J_{\mathrm{E}_{ss}}$, and $J_{\mathrm{R}_{ss}}$ are fluxes at steady state; and $[\mathrm{C}]_{\mathrm{F}_i}$, $[\mathrm{C}]_{\mathrm{E}_i}$, and $[\mathrm{C}]_{\mathrm{R}_i}$ are concentrations of solute species, sampled at time i, in the feed, carrier, and strip solutions, respectively.

The correlation factor between $[\mathrm{C}]_i$ and J_{ss}/U_i in Eqs (64)-(66) has been checked in experiments with different metal ions transport [62] in the range $U = 0.1$-$1.0\ \mathrm{cm^3\ s^{-1}}$. Results showed that $[\mathrm{C}] > J_{ss}/U$ is regularly more than 1.5 orders in the feed, carrier and strip solutions. It means that the J_{ss}/U ratio may be excluded from Eqs (64)-(66). Therefore, we obtain

$$k_{c(\mathrm{fe})_i} = U_{\mathrm{F}_i}\left\{-\ln\left([\mathrm{C}]_{\mathrm{F}_i}\right)\Big/\left([\mathrm{C}]_{\mathrm{F}_{(i-1)}}\right)\right\}^{-1}$$

or

$$k_{c(fe)_i} = U_{F_i}\left\{\ln\left([C]_{F_{(i-1)}}\right)/\left([C]_{F_i}\right)\right\}^{-1}, \tag{64'}$$

$$\frac{1}{2}k_{c(ef)_i}\ \left(\text{or } k_{c(er)_i}\right) = U_{E_j}\left\{\pm\ln\left[\left([C]_{E_i}\right)/\left([C]_{E_{(i-1)}}\right)\right]\right\}^{-1}, \tag{65'}$$

$$k_{c(re)_i} = U_{R_i}\left\{\ln\left([C]_{R_i}\right)/\left([C]_{R_{(i-1)}}\right)\right\}^{-1}. \tag{66'}$$

Equations (64′)-(66′) are equal to those used by other researchers [3-20, 28, 35-37]. These authors obtained Eqs (64′)-(66′) by considering the basic Stokes-Einstein equation. We obtained the same equations as a particular case from Eq. (6), based on kinetics of irreversible processes (nonequilibrium thermodynamics). In Table 5.3 are presented examples of individual and overall mass-transfer coefficients, obtained at titanium(IV) ion transport through the BOHLM system [1].

Concentration profiles of titanium(IV) species in the feed, carrier, and strip solutions were calculated using model equations. Comparison of the experimental and the simulated data shows that:

- Diffusion of titanium(IV) species through the feed aqueous boundary layer does not control the transport rate. Thus, variations of the feed flow-rate have little effect on the titanium transport performance.
- The magnitudes of individual titanium mass-transfer coefficients are similar at carrier or strip flow rates variations. Resistance to diffusion in the carrier solution layers and membrane pores is not a rate-controlling step, since the overall mass-transfer coefficients on the strip side of the BOHLM are two orders less than that on the feed side. Thus, rate-controlling steps could act as resistance of the strip solution layer, or the interfacial backward extraction reaction rate.
- Dependence of titanium transport rate to the strip on strip flow rate is much stronger at low acidity feed than at high acidity feed. Thus, we can conclude that the interfacial backward extraction reaction rate is a rate-controlling step of titanium transport in the BOHLM system.
- At high acidity feed solutions, discrepancies between experimentally obtained and theoretically calculated data for titanium concentration in the strip phase are 15-30 times or 15-230 times at feed flow or strip flow rate variations, respectively.
- At low acidity feed solutions, the discrepancies are about 4.5 times at feed flow rate variations (independent of it), or 1-112 times at strip flow rate variations.

These differences between the experimental and the simulated data have the following explanation. According to the model, mass transfer of titanium from the feed through the carrier to the strip solutions is dependent on the resistances: boundary layer resistances on the feed and strip sides, resistances

Table 5.3 Individual and overall mass-transfer coefficients, obtained using only internal (carrier) driving force motivated transport equations

No.	Feed phase acidity (mol kg^{-1})	Flow velocity (cm^3 s^{-1})	Ti flux from feed, J_F ($\times10^{-8}$ mol cm^{-2} s^{-1})	Ti flux to carrier phase, J_E ($\times10^{-8}$ mol cm^{-2} s^{-1})	Ti flux to strip, J_R ($\times10^{-10}$ mol cm^{-2} s^{-1})	$k_{c(fe)}$ (cm s^{-1})	$k_{c(ef)}$ = ($\times10^{-2}$ cm s^{-1})	$k_{c(re)}$ ($\times10^{-2}$ cm s^{-1})	D_E ($\times10^{-2}$ cm^2 s^{-1})	$k_{c(mf)}$ = ($\times10^{-2}$ cm s^{-1})	K_{c_F} ($\times10^{-3}$ m s^{-1})	K_{c_R} ($\times10^{-5}$ m s^{-1})
	1	2	3	4	5	6	7	8	9	10	11	12
1	7.0	0.1	1.91	1.04	0.69	0.86	1.2	1.5			3.78	2.5
2		0.2	2.78	2.77	1.56	1.1	2.2	2.7			4.55	3.2
3		0.4	3.47	4.86	6.08	1.8	4.2	4.5	6.2	0.47	5.78	3.7
4		0.6	3.99	6.94	12.15	2.3	6.1	6.3			6.33	4.0
5		1.0	4.34	7.99	15.63	3.5	10	10			7.15	4.2
6	0.45	0.1	3.30	3.82	0.35	47	1.1	1.3			124	43
7		0.2	5.21	5.21	3.13	56	2.1	2.0			168	68
8		0.4	7.29	7.29	14.76	73	4.0	3.5	0.3	2.8	227	97
9	pH	0.6	8.33	8.33	38.19	92	5.9	4.9	6		269	120
10	0.65	1.0	9.55	9.38	112.85	130	9.8	7.5			326	150

Notes. (1) Initial titanium(IV) concentration in the feed phase 0.1 mol kg^{-1}. (2) Results represented in columns 3 and 6 were obtained at various feed flow velocities (column 2) and fixed ($U = 0.2$ cm^3 s^{-1}) carrier and strip solutions' flow velocities. Results in the columns 4 and 7 were obtained at various carrier solution flow velocities and fixed ($U = 0.2$ cm^3 s^{-1}) feed and strip solutions' flow velocities. Results in the columns 5 and 8 were obtained at various strip solution flow velocities and fixed ($U = 0.2$ cm^3 s^{-1}) feed and carrier solutions' flow velocities. (3) Coefficient DE in the column 9 is defined as an "effective" diffusion coefficient (for details and determination, see in the text). (4) *Source*: From Ref. [1] with permission.

of the free carrier and titanium-carrier complex through the carrier solution boundary layers, including those in the pores of the membrane, and resistances due to interfacial reactions at the feed- and strip-side interfaces. In the model equations, we took into consideration only mass-transfer relations, motivated by internal driving force (forward extraction-backward extraction distribution ratio, K_c). Mass-transfer relations, motivated by external driving force between feed and strip phases, indicated by K_d coefficient, were not considered. Thus, resistances to the titanium transport, due to diffusion kinetics through the feed and strip boundary layers of protons or protons and Cl-anions, generated at the feed-membrane and strip-membrane interfaces (see Eqs (25)-(28)), resistance to free carrier molecules diffusion through the boundary layers of the carrier solution and through the membrane pores opposite to titanium direction, were not taken into account.

There are two ways to evaluate individual mass-transfer coefficients of these processes:

(1) By sampling and determining proton and Cl-anion concentrations in the feed and strip phases, and free carrier concentration in the carrier solution during experiments with flow rate variations
(2) By comparing the experimentally obtained titanium concentration profiles with model predicted ones, which were calculated using only mass-transfer coefficients, K_c. We used the second, simpler way

Individual external mass-transfer coefficients, k_d (see Table 5.4), were evaluated using Eqs (64′)-(66′), where $[\mathrm{Ti}]_{\mathrm{F}_i}$, $[\mathrm{Ti}]_{\mathrm{E}_i}$, and $[\mathrm{Ti}]_{\mathrm{R}_i}$ were taken from model-calculated data and $[\mathrm{Ti}]_{\mathrm{F}_{(i-1)}}$, $[\mathrm{Ti}]_{\mathrm{E}_{(i-1)}}$, and $[\mathrm{Ti}]_{\mathrm{R}_{(i-1)}}$ were experimentally obtained data under the same conditions and sampling time, t_i.

Overall mass-transfer coefficients K_{dF} on the feed side and K_{dR} on the strip side were calculated by the following equations:

$$K_{\mathrm{dF}} = \frac{k_{\mathrm{d(fe)}} \times k_{\mathrm{d(mf+ef)}}}{k_{\mathrm{d(fe)}} + k_{\mathrm{d(mf+ef)}}} \tag{67}$$

and

$$K_{\mathrm{dR}} = \frac{k_{\mathrm{d(re)}} \times k_{\mathrm{d(mr+er)}}}{k_{\mathrm{d(re)}} + k_{\mathrm{d(mr+er)}}}. \tag{68}$$

The overall mass-transfer coefficients of the BOHLM system (see Eqs (32) and (32′)) were calculated by equations:

$$K_{\mathrm{F}} = K_{\mathrm{cF}} K_{\mathrm{dF}} \tag{69}$$

and

$$K_{\mathrm{R}} = K_{\mathrm{cR}} K_{\mathrm{dR}}. \tag{70}$$

Comparing the simulated results with the experimental data, it appears that at higher flow rates, where boundary layers resistance becomes less

Table 5.4 Individual and overall mass-transfer coefficients, accounting for coupling effects of the titanium transport (external driving force): combined overall mass-transfer coefficients

No.	Feed phase acidity (mol kg^{-1})	Flow velocity (cm^3 s^{-1})	$k_{d(fe)}$ (m s^{-1})	$k_{d(mf+ef)} = k_{d(mr+er)}$ (m s^{-1})	$k_{d(re)}$ (m s^{-1})	K_{d_F} (m s^{-1})	K_{d_R} (m s^{-1})	$K_{F/E}$ (×10^{-2} m s^{-1})	$K_{E/R}$ (×10^{-4} m s^{-1})
	1	2	3	4	5	6	7	8	9
1	7.0	0.1	0.802	−0.60	0.625	2.38	15.2	0.90	3.80
2		0.2	0.798	−0.63	0.647	2.99	23.1	1.36	7.40
3		0.4	0.895	−0.69	0.697	3.01	70.3	1.74	26.00
4		0.6	0.980	−0.76	0.765	3.38	125.0	2.14	50.00
5		1.0	1.10	−0.83	0.835	3.38	147.6	2.41	62.00
6	0.45	0.1	0.190	0.465	−0.785	0.135	1.14	1.67	1.00
7		0.2	0.227	1.12	2.69	0.189	1.09	3.18	7.40
8		0.4	0.295	6.25	4.94	0.282	2.76	6.4	26.80
9	pH 0.65	0.6	0.412	60.0	6.01	0.409	5.46	11.0	65.50
10		1.0	0.524	100.0	16.73	0.521	14.33	17.0	215.00

Notes. (1) For an explanation of the results, represented in columns 3-5, see note 1 of Table 5.3. (2) The meaning of the negative mass-transfer coefficients in columns 4 and 5 is not a real case, but was used for feasibility of calculations (see Eqs (64′)-(66′) and (67), (68)). We stated that the model-calculated concentrations of the solute will be in the numerator, and the experimentally obtained ones in the denominator. Therefore, if the magnitude of experimentally obtained concentration prevails the model-calculated one, the mass-transfer coefficient will be "negative." (3) *Source*: From Ref. [1] with permission.

important, membrane-strip interfacial reaction kinetics dominates as a rate-controlling step for titanium transport. It is evident that in both cotransport (high acidity feed) and countertransport (low acidity feed) the rate-controlling step of titanium transport is the backward extraction reaction rate at the membrane-strip interface.

In the case of high acidity feed (cotransport), mass-transfer coefficient values approach a constant when the feed or strip flow rates are above 0.6 $cm^3\ s^{-1}$. Below this border the dependence may be evaluated by the following equations:

$$K_{F/E} = 6.04e^{-3} + 1.9e^{-2U_F}, \tag{71}$$

$$K_{E/R} = -2.17e^{-4} + 6.9e^{-3U_R}. \tag{72}$$

In the case of low acidity feed (countertransport), the dependence is evident in all ranges of flow rates studied and may be evaluated by

$$K_{F/E} = -1.53e^{-3} + 0.174e^{-2U_F}, \tag{73}$$

$$K_{E/R} = -4.59e^{-4} + 2.37e^{-2U_R}. \tag{74}$$

Model analysis shows that the solute concentration enrichment cannot exceed the value of $[C]_E/E_R$ (solute concentration in the carrier phase/backward extraction distribution coefficient), or $[C]_F \times E_F/E_R$. Thus, extraction distribution parameters control the enrichment ability of the BOHLM.

Many researchers [3-20] propose the application of processes, based on the steady state of the system. Experimental and model simulation data show much higher mass-transfer rates through the BOHLM, with solute concentration in the carrier solution, reaching its maximum. At this stage, both the internal (extraction-backward extraction distribution ratio) and the external (coupling) driving forces motivate the solute transport in an optimal way. At a steady-state transport, solute permeation is motivated mostly by an external driving force and the fluxes are about an order lower (in the case with titanium). A much more effective BOHLM module, with continuously flowing feed (open system), can be designed if the feed-side membrane area S_F and the feed flow rate U_F enable us to obtain a fixed solute feed outlet concentration at a contact time t_{max}, close to that at the maximum on the simulated concentration profile C_{Emax} of the carrier solution.

The values of overall mass-transfer coefficients govern the location of C_{Emax} and t_{max}. This may be found by differentiating Eq. (15) and setting $dC_E/dt = 0$:

$$t_{max} = \frac{\ln(K_R/K_F)}{K_R - K_F}, \tag{75}$$

where t_{max} is the processing time at which maximum quantity (C_{Emax}) of the metal species occurs in the LM solution. The C_{Emax} may be found by combining Eqs (15), (17)-(19), and (67):

$$C_{\mathrm{Emax}} = C_{\mathrm{F}}^{0}(K_{\mathrm{F}}/K_{\mathrm{R}})^{K_{\mathrm{R}}(K_{\mathrm{R}}-K_{\mathrm{F}})}. \tag{76}$$

The greatest transport rate of the metal species to the strip solution occurs at $t_{\max}$ and C_{Emax} (inflection point on the curve C_{R} vs. time).

Let us consider two extreme cases for the C_{R} curve versus time. From Eqs (16)-(19), we obtain

(1) At $K_{\mathrm{R}} \gg K_{\mathrm{F}}$,

$$C_{\mathrm{R}} = C_{\mathrm{F}}^{0}(1 - \mathrm{e}^{-SK_{\mathrm{F}}t}). \tag{77}$$

So, the rate-controlling step of the BOHLM transport of the solute species to the strip phase is determined by the K_{F} overall mass-transfer coefficient.

(2) At $K_{\mathrm{F}} \gg K_{\mathrm{R}}$,

$$C_{\mathrm{R}} = C_{\mathrm{F}}^{0}(1 - \mathrm{e}^{-SK_{\mathrm{R}}t}). \tag{78}$$

In this case, the rate-controlling step of the BOHLM transport of the solute to the strip phase is determined by the K_{R} overall mass-transfer coefficient.

One more advantage of the BOHLM system is realized from theoretical simulations. At feed-side resistances not controlling the solute transport (as in the case with titanium) the fluxes of the solute to the strip are approaching these in the SLM systems, but without many drawbacks of the latter.

3.1.2. Determination of the selectivity parameters

Selectivity of the BOHLM system is an important parameter that should be used in designing a process (see also Chapter 2). According to the transport model equations, the selectivity of two solutes (e.g., two metal species, $S_{\mathrm{M_1/M_2}}$) is determined by the relation

$$S_{\mathrm{M_1/M_2}} = \frac{K_{\mathrm{M_1}}(C_{\mathrm{F}}^{0})_{\mathrm{M_1}}}{K_{\mathrm{M_2}}(C_{\mathrm{F}}^{0})_{\mathrm{M_2}}}, \tag{79}$$

where subscripts $\mathrm{M_1}$ and $\mathrm{M_2}$ refer to the two metal species; $K_{\mathrm{M_1}}$ and $K_{\mathrm{M_2}}$ are the total overall mass-transfer coefficients; $(C_{\mathrm{F}}^{0})_{\mathrm{M_1}}$ and $(C_{\mathrm{F}}^{0})_{\mathrm{M_2}}$ are the initial concentrations of two solutes (metal ions) in the treated feed phase.

Introducing a separation factor, A, defined as a ratio of the total overall mass-transfer coefficients of the solutes (metal species)

$$A_{\mathrm{M_1/M_2}} = K_{\mathrm{M_1}}/K_{\mathrm{M_2}}, \tag{80}$$

an equation for the BOHLM system selectivity is obtained:

$$S_{\mathrm{M_1/M_2}} = A_{\mathrm{M_1/M_2}} \frac{[\mathrm{C_F^0}]_{\mathrm{M_1}}}{[\mathrm{C_F^0}]_{\mathrm{M_2}}}. \tag{81}$$

Based on the principle of resistance additivity, the total overall mass-transfer coefficient, K_{M}, of every solute passing through the BOHLM, is related to the overall mass-transfer coefficients on the feed and strip sides as follows:

$$K_M = \frac{[K_F K_R]_M}{[K_F + K_R]_M}. \tag{82}$$

Overall mass-transfer coefficients on the feed and strip sides are calculated using Eqs (1), (1′) and (2), (2′). E_F and E_R are determined experimentally through distribution coefficients at membrane-based equilibrium forward and backward extraction (in detail, see Section 2.1).

Thus, the separation factor of the two metal species is

$$A_{M_1/M_2} = \frac{[K_F K_R]_{M_1} [K_F + K_R]_{M_2}}{[K_F K_R]_{M_2} [K_F + K_R]_{M_1}}. \tag{83}$$

For preliminary evaluation of the selectivity of two metal species separation, we can assume that in the same solution environment (water) the diffusion coefficients of these metal ions with the same charge have similar values and the diffusion coefficients of the metal-carrier complexes have similar values. Thus, substituting Eqs (1) and (2) in Eq. (83), we can represent separation factor as dependent only on the distribution coefficients at forward and backward extraction:

$$A_{M_1/M_2} \approx \frac{[E_F E_R]_{M_1}}{[E_F E_R]_{M_2}}. \tag{84}$$

The distribution coefficient may be expressed as a function of the metal association (stability) constants in the LM solution, the association constants of metal ions with solvent environment in the feed and in the strip solutions, and partition coefficients of the carrier and metal ion. In this case, the separation factor can be determined by the stability constants of the metal complexes, formed with functional groups of carrier, if we assume that the metal ions are predominantly present (a) as free ions in the acid solution, so that complex concentrations can be disregarded, and (b) as complexes in the LM solutions, so that free ion concentrations can be disregarded:

$$A_{M_1/M_2} \approx \frac{[\beta_F \beta_R]_{M_1}}{[\beta_F \beta_R]_{M_2}}, \tag{85}$$

where β_F and β_R are stability constants of the metal-carrier complexes in equilibrium with the feed and strip solvent environment (as a rule acids for metal ions), respectively. Therefore, preliminary selectivity data of metal species separation by BOHLM may be evaluated without experimentation, if stability constants data are available in the literature.

From Eqs (84) and (85), it follows that high separation factors are favored when approaching conditions $[E_F E_R]_{M_1} \gg [E_F E_R]_{M_2}$ or $[\beta_F \beta_R]_{M_1} \gg [\beta_F \beta_R]_{M_2}$. On the other hand, the system loses its selectivity when distribution parameters of both metal species are either extremely high or extremely low [75]. Selectivity can be increased by choosing a selective carrier with intermediate distribution data values, and adjusting its concentration, its volume, the acidity of the feed and strip solutions in such a way as to approach the above conditions.

3.2. Membrane types used as barriers

Most membranes, used as a wall between phases in the BOHLM systems, are polymeric although inorganic membranes have also become available. Membranes can be classified according to their structure as homogeneous, asymmetric, or composite; according to their geometry as flat sheet, hollow fiber, or tubular; according to their pore surface as neutral hydrophobic or hydrophilic and surface charged (ion exchange). Hydrophobic membranes, flat sheet, or hollow fibers (see Tables 5.5 and 5.6), are readily wettable by most organic solvents, but water does not wet it. An organic solvent, brought on one side of the membrane, immediately wets the membrane and appears on the other side, or immobilizes the interface. Aqueous solution to be contacted with the organic solution of a carrier is brought at a pressure higher than that of organic solution to prevent the last from coming through the membrane, but not so high to be able to displace organic phase from inside the pores. The two phases contact each other at a mouth of the pores or inside the membrane (see Figs 5.1A and D and 5.2).

Contrarily, hydrophilic and ion-exchange membranes (see Tables 5.5-5.7) are readily wettable by aqueous solutions, but immiscible by organic phases. So, contact between aqueous and organic phases takes place on the opposite side to the bulk aqueous phase interface of the membrane (see Figs 5.1B and E and 5.3).

Although a number of inorganic ion-exchange materials (zeolites, bentonites) are now available, most membranes used are polymeric. The properties of ion-exchange membranes are closely related to those of ion-exchange resins. The properties important for the separation reasons are permselectivity, long-term electrical (resistance) and chemical stability and ion-exchange capacity (see Table 5.7). Flat sheet, hollow fiber, and tubular ion-exchange membranes are also available.

The limits of pore dimensions in BOHLM membranes are purposely chosen to minimize convective transport. To achieve selective mass transfer of the solutes that have to be separated, a combination of solute and membrane properties is taken into account. Two solutes can be separated if their diffusion coefficients differ by at least an order of magnitude. This diffusivity is modified by the matrix structure and pore size of the membrane. Knowledge of the relative solute diffusivities in the membrane gives us the knowledge of retention abilities of the membrane. The sieving properties of the membranes allow them to retain some macromolecules while selectively removing solutes of middle molecular weight and less selective removal of solutes with low molecular weight.

Table 5.5 Examples of flat-sheet polymer membranes, used in the referenced works, and their properties

Trademark	Supplier	Type	Porosity (%)	Pore size (μm)	Thickness (μm)
Celgard® 2400	Hoechst-Celanese Co., USA	Hydrophobic	38	0.05–0.125	25
Celgard® 2402			38		50
Celgard® 4400			38		175
Celgard® 3400		Hydrophilic	38		25
Celgard® 5401			38		175
Duragard® 2500	Polyplastics Co., Ltd, Japan	Hydrophobic	47	0.04 × 0.4	25
Duragard® 2502			45	0.04 × 0.4	50
Fluoropore® FP-010	Sumitomo Co., Ltd, Japan		67	0.1	60
Mesh KDJ-6	Dainippon Plastics Co., Ltd, Japan		87	[a]	800
Spacer KDN-4			73	[a]	900
Hydrophilic Spacer		Hydrophilic	70	[a]	800–1000
Nylon-6 (polyamide 6)	ENKA America, Inc., Ltd		70	0.2	110
Cuprophan 150 PM (polyamide 150)			59	30–50 Å[b]	22

Note. Celgard® 3400 is made hydrophilic by coating a surfactant onto Celgard® 2400.
[a]Pore size is not available in manufacturer's catalog.
[b]The Cuprophan® membranes were designed for an artificial kidney and therefore have the 30–50 Å pore diameters.

Table 5.6 Examples of hollow-fiber membranes, used in the referenced works, and their properties

Trademark	Supplier	Type	ID (μm)	CD (μm)	Pore size (μm)	Porosity (%)
Celgard® X-10	Hoechst-Celanese Co., USA	Hydrophobic	100	150	0.03	20
Celgard® X-20			240	290	0.03	40
Cuprophan C1	AKZO (Asheville, NC, USA)	Hydrophilic	140	200	30–50 Å	55
Cuprophan D4			220	270	30–50 Å	60
Nylon-6			600	1000	0.2	75
Capillary ultrafiltration membranes	Daicel Chemical Industries, Ltd	Hydrophobic	800	1300	150,000[a]	[b]

[a]Molecular weight cutoff.
[b]Data are not available in manufacturer's catalog.

Table 5.7 Examples of ion-exchange membranes, used in the referenced works, and their properties (see also Table 6.6 in Chapter 6)

Supplier	Trademark	Type	Gel water (%)	Thickness (mm)	Permselectivity (%)	Capacity (meq g^{-1})
Tokuyama Soda Co., Japan	Neosepta, CMX	Cation-exchange, low electric resist.	24	0.13-0.16	90	2.0-2.5
	Neosepta, AFN-7	Anion exchange; antifouling	45	0.16	[a]	3.2
	Neosepta, BP-1	Cation-anion exchange	20	0.3	85	2.0
Du Pont De Nemours, USA	Nafion-117	Cation exchange	18.5	0.22	90	3.9
	Nafion-120		44	0.26	96	4.3
FumaTech GmbH, Germany	FKS	Cation exchange	40	0.6	79	1.8

[a] Not available in manufacturer's catalog.

The knowledge of the difference between solute diffusivities in the membrane, immobilized by organic phase (hydrophobic), vis-à-vis the membrane, immobilized by aqueous phase (hydrophilic, ion exchange), would aid in the choice of membrane type. In all cases, the membrane material would have to be completely long-term resistant to the solvents of either phase.

The first choice would be to use as thin a membrane as possible, with high porosity and small pore size, while still providing the mechanical stability for potting and pumping pressure.

3.3. Carrier types used

Solute transport through the LM enhances by water-immiscible species, dissolved in the solvent (water immiscible also) and selectively interacted with solutes. These species are named carriers and transport is facilitated, or membrane-mediated, or carrier-enhanced, or... (authors use different terms for the same phenomenon). Most LM carriers are originally extractants developed for solvent extraction processes. The reader can find their descriptions in the Solvent Extraction Handbooks, for example [88, 89]. Many new carriers are developed specially for LM (see Section 4). The use of LM with carriers offers an alternative to solvent extraction for selective separation and concentration of metal ions from dilute solutions.

Facilitated LM transport involves reversible complexation/decomplexation reactions in addition to dissolution and diffusion.

Performance of LM is strongly related to the characteristics of a carrier. The main parameters of carriers in LM transport are high selectivity, high capacity, high complexation constant, E_F, and high decomplexation, or stripping constant, E_R, rapid kinetics of complexation and destraction of the complex on membrane interfaces, rapid kinetics of diffusion of the carrier-solute complexes, etc. (for details, see Chapter 2).

Different mechanisms of diffusion take place in LM: diffusion of the carrier-solute complex, diffusion of the uncomplexed carrier in the opposite direction, and diffusion of the uncomplexed solutes. The last transport mechanism is not accessible to solutes that do not react with the carrier species. It is the complexation reaction that makes facilitated LM transport highly selective.

A great variety of carriers are used in the BOHLM transport (see Tables 5.8-5.13). They may be divided into cation, anion exchangers and neutral ligands. The first group comprises the large number of organic acids and their derivatives and related proton donors (see Table 5.8). For example, some commercially available extractants: di(2-ethylhexyl)phosphoric acid (DEHPA), bis(2,2,4-trimethylpentyl)phosphinic acid (Cyanex 272), some hydroxyoximes (LIX or Acorga series), oligoamide compounds, containing 8-hydroxyquinolyl groups (Kelex or LIX 26 series). Crown ethers and related macrocyclic multidentate ligands have a pronounced selectivity towards cations (metal ions). Their ability to selectively and reversibly bind

Table 5.8 Selected references on metal ion separations by BOHLM techniques

Metal ions	Liquid membrane	Polymer membrane support	References
Hybrid liquid membrane (HLM) technique			
Ti(IV)	DEHPA	Celgard 2400, 2402, 3400 Neosepta CM-1, ACH-45	[1, 2]
Cu(II), Cd(II), Zn(II)	MEHPA, DEHPA Cyanex 302	Neosepta CM-1, CM-2 Celgard 2400, 2402, 3400	[69, 84, 115–119]
Cd(II)	Tri-*n*-octylamine (TNOA)	Celgard 2400	[120]
Strip dispersion hybrid liquid membrane (SDHLM) technique			
Cd(II)	Tri-*n*-octylamine	Microp. polyprop. 0.02	[121, 122]
Cu(II)	N 902, N 530	Microp. polyprop. 0.02	[123, 124]
Multimembrane hybrid system (MHS) with (or without) pervaporation (PV) technique			
K(I), Na(I), Mg(II), Ca(II), Cu(II), Zn(II)	Poly(oxypropylene) bisphosphates (POPBP), Poly[poly(oxypropylene)phosphate]s (PPOPP)	Nafion-117, Neosepta CMX	[125, 126]
Cu(II), Zn(II)	5-nonylsalicylaldoxime (Acorga P-50), DEHPA	Nafion-117, Neosepta CMX	[127]
Cd(II)	Tri-*n*-octylamine (TNOA) + octyl alcohol	Microp. polyprop. 0.02 μm	[128]
K(I), Na(I), Mg(II), Ca (II), Cu(II), Zn(II)	Star-shaped polymers, PEG	Nafion-117, Neosepta CMX, AFN-7, Pervap 1000, 2201	[28, 129–133]
K(I), Na(I), Zn(II), Cu (II), Mg(II), Ca(II)	DEHPA, Acorga P-50, Cyanex 302	Nafion-120, FKS, Pervap 1000	[29, 47, 74, 86–88, 105, 132]
Zn(II), Mn(II), Cu(II), Co(II), Ni(II), Cr (III) Fe(III)	DEHPA, Acorga P-50	Nafion-117, 120, Neosepta CMX, BP-1	[49–51, 85, 134]

(continued)

Table 5.8 (*continued*)

Metal ions	Liquid membrane	Polymer membrane support	References
Rotating film module			
Cu(II), Zn(II), Fe(II), Ni(II)	Kelex 100, TOA, LIX 54, 65, 65N, 860, Hydroxyoximes, Acorga P-5100	Hydrophilic coating with viscose	[12, 14, 16, 17]
		Hydrophilic support	
Ag(I)	TIBPS	Hydrophilic support	[21, 96-98]
Creeping film module			
Cu(II)	Acorga P-5100, 5300	Hydrophilic support of viscose	[19, 22, 57]
Jodine	DB18C6	Hydrophilic support	[102]
Hollow-fiber module			
Cu(II), Cr(III), Cr (VI), Hg(II)	LIX 84, TOA	Celgard X-10	[8, 135, 136]
Nd, Ho	DEHPA	Celgard X-10	[9, 24, 39, 137, 138]
U(VI)	di-*n*-hexyl octanamide (DHOA)	Microporous hydrophobic polypropylene capillaries	[139]
Pb, Cd	DEHPA	Microporous hydrophobic HF	[140, 141]
Zn, Fe (III)	TBP	Microporous hydrophobic HF	[142, 143]
Au(I), Ag(I)	LIX 79, mixt. LIX 79-TOPO	Microporous hydrophobic HF	[144, 145]
Spiral-type flowing module			
Cr(VI), Zn	TOA, TOMAC, EHPNA	Duragard 2500, 2502	[34]
Co, Ni, Zn	DEHPA, Cyanex 302	Duragard 2500, 2502, Mesh spacer KDJ-6	[35-37]

Layered (without polymer membrane walls) bulk liquid membrane module		
Li, Ag, Na, K, Sr	Crown 18C6, Cyanex 301, Calixcrown oligomers (amide, ether, ester, siloxan)	[146-153]
Zn, Co	DEHPA	[25, 42, 53, 154-156]
Ag	Cyanex 471X, *N,N*-diethyl-*N*′-camphanyl thiourea	[157, 158]
Cd, Cu, Zn	Calixcrown oligomers, Janus Green, LIX 860,	[149, 159-172]
Toxic heavy metals	5-nonylsalicylaldoxime	
Pd, Au	Crowns 18C6, hexadecylpyridinium bromide (HDPB)	[173-177]
Bi	Cyanex 301	[148]
Hg(II)	Crown 18C6, 4′-nitrobenzo[15-crown-5], calix oligomer deriv., bis-calix[4]arene ester derivative	[178-186]
Er, Nd	HEH(EHP)	[187-189]
U	TBP, calixarenes	[190-192]
$[UO_2]^{2+}$	Thenoyltrifluoroacetone (TTA), DEHPA, Aliquat 336	
Cr(III)	Dinonylnaphthalene sulfonic acid (DNNSA)	[193]
Cr(VI)	*p*-tert-Bu calix[4]arene 3-morpholino Pr diamide	[194-196]
	Triphenylphosphine (TPP)	
Co(II), Ni(II), Cu(II), Zn(II), As(I), Cd (II), Pb(II)	*N,N*-dibutyl-*N*′-benzoylthiourea	[197-200]
	Ligand surfactants and alkyl pyridin-2-yl ketoximes (chelat. agent)	
Th(IV)	2-Thenoyltrifluoroacetone (HTTA), DEHPA	[201, 202]

(*continued*)

Table 5.8 (*continued*)

Metal ions	Liquid membrane	Polymer membrane support	References
Pb(II)	Mixt. of benzylaza-12-crown-4 and oleic acid, dibenzopyridino-18-crown-6		[203, 204]
Cd	Ketoconazole and oleic acid		[205]
Mo	8-Hydroxyquinoline		[206]
Fe(III)	*N*,*N*′-dihydroxy-*N*,*N*′-dimethyldecanediamide (H2L8) + *β*-alanine, or L-glutamic acid, or glycine, or L-lysine hydroxamic acids		[207]
	Rhodotorulic acid (H2LRA) or alcaligin (H2LAG), and *N*,*N*′-dihydroxy-*N*,*N*′-dimethyldecanediamide (H2L8)		[208]
Cu(II), Cd(II), Hg(II), Pb(II)	2,2′-bis(*p*-octyloxybenzyl) diethylenetriamine (bis-pODET)		[209]
In(III)	DEHPA		[210]
Cs(I), Ba(II)	5′-(Tert-butyldimethylsilyl)-2′,3′-*O*-isopropylidene isoguanosine (isoG 1)		[211]
Au(III)	Tetrathia-12-crown-4		[212]

Table 5.9 Selected references on carboxylic and amino acids separations by BOHLM techniques

Acids	Liquid membrane	Polymer membrane walls	References
Hybrid liquid membrane (HLM) technique			
Lactic, acetic, citric, propionic	Alamine 336 Alamine 304	Celgard 2400, 3400 Neosepta AM-3, AMX	[78, 213]
Multimembrane hybrid system (MHS) with (or without) pervaporation (PV) technique			
Acetic, formic, lactic, propionic, oxalic, citric, tartaric	TOPO, TBP, MIBK, TOA, octanol, hydrocarbons C_6-C_{10}	Neosepta AFN-7 Nafion-120	[29, 46, 48, 214-217]
Rotating film module			
Phenol, L-lysine, butyric	DEHPA	Hydrophilic support	[21, 23, 100, 103, 104, 218]
Hollow-fiber module			
Phenol, acetic, oxalic, succinic, citric acids	TOA, MIBK, Hostarex A327, Decanol, Xylene	Celgard X-10, X-20, Nylon-6, Cuprophan	[6, 11, 26, 40, 101, 219-221]
Butyric acid, DMCCA, 5-methyl-2-pyrazinecarboxylic acid, heterocyclic carboxylic acid, PAA, MAs 6-APA	TOA, Hostarex A327, Cyanex 471, lipase in octanol	Celgard X-10, X-20, Cuprophan	[39, 222, 223]
Butyric acid, 5-methyl-2-pyrazinecarboxylic acid (MPCA)	TOA, Cyanex 471	Celgard, Liqui-Cel 10×	[224, 225]

(continued)

Table 5.9 (*continued*)

Acids	Liquid membrane	Polymer membrane walls	References
Spiral-type flowing module			
Fatty acids	DEHPA, silver nitrate	Duragard 2500, 2502, Mesh spacer KDJ-6	[32, 226]
Eicosapentaenoic, docosahexaenoic acids	Silver nitrate, DEHPA	Fluoropore FP-010, Mesh spacer KDN-4	[227]
Layered (without polymer membrane walls) bulk liquid membrane module			
Butyric acid, 5-methyl-2-pyrazinecarboxylic acid	TOA, Hostarex A327, THP		[43, 44, 222]
Amino acids, L-isoleucine, 7-ACA	DEHPA, DB18C6, Aliquat 336		[228-230]
Aminocephalosporanic, *p*-aminobenzoic acid, phenylalanine	Carboxylated poly (styrene), octanoic, lauric, decanoic acids		[231-233]
Dihydrogenphosphate anion	Oxomolybdenum (V) tetraphenylporphyrin complex		[234]
α-Amino acids, phenylalanine (Phe), histidine (His), tryptophan (Trp)	Chiral lipophilic ligand *N*-dodecyl-ephedrinium		[235, 236]
Amino acid picrates	Isosteviol and derivatives		[237]
Leucine, valine, and glycine	di-, tri-, tetraethylene glycols and their derivatives		[238]
L-Phenylalanine, L-tryptophan, l-leucine, and l-isoleucine	DEHPA		[239]
Phenol	Cyanex 923, Amberlite LA 2 and trioctylamine (TOA)		[240]

Separated compounds	Liquid membrane	Polymer membrane walls	References
Hybrid liquid membrane (HLM) technique			
Compactin (ML-236B)	Et acetate	Polytetrafluoroethylene (PTFE), hydrophilic	[241]
Rutaecarpine	Kerosene	Hydrophobic, flat support	[242]
Rotating film module			
Erythromycin, tylosin, L-lysine	DEHPA	Hydrophilic coating, hydrophilic support	[15, 243]
Phenylalanine	DEHPA	Hydrophilic support	[244]
Hollow-fiber module			
Diltiazem, drugs	Octanol, Decanol	Celgard X-10, X-20, Cuprophan	[245-250]
Fructose	Alcohols		[251]
Antibiotics	1-Decanol, *n*-decanol, *n*-octanol		[252-254]
Bioproducts		Celgard X-10, X-20	[255]
Phenylalanine	DEHPA	Hydrophobic HF	[256]
Layered (without polymer membrane walls) bulk liquid membrane module			
Indole alkaloid vincamine	Trichloroethylene		[257]
Tryptophan, serotonin	M-^{t}Bu$_4$-salphen-3*n*-cr-*n* complexes (M = Mn, Fe)		[258]
Proteins	Reversed micellar solution		[96, 259]
Enantiomers	Aerosol reversed micelles with Ba^{2+}, β-cyclodextrin		[260-263]
Cephalosporin-C, cephalexin	Aliquat 336,		[230, 264, 265]
Racemic clenbuterol (chiral resolution)	D-(+)-dibenzoyltartaric acid		[257]
Layered (without polymer membrane walls) bulk liquid membrane module			
Lysozyme	Riv. micelles of aerosol-OT		[258]
Tropane alkaloids	Diisopropyl ether		[259]

Table 5.11 Selected references on organic compounds separations by BOHLM techniques

Organic compounds	Liquid membrane	Polymer membrane walls	References
Rotating film module			
Aromatic amines	C_{11}-C_{13} normal paraffins	Disk with hydrophilic support	[13, 21, 260]
Spiral-type flowing module			
Ethylene/ethane	Silver nitrate	Duragard 2500, 2502, Mesh spacer KDN-4	[30, 261]
Benzene	Silver nitrate	Fluoropore FP-010, Mesh spacer KDN-4	[33]
Hollow-fiber module			
Volatile organic compounds (VOC)	1-Octanol, vacuum	Silicone-coated HF	[262-266]
2-Propanol/*n*-heptane	Water, sulfolane	Silicone-coated HF	[267, 268]
2-Phenylethanol (PE), benzaldehyde (BA), ethyl butyrate (EB), dimethyltrisulfide (DMTS)	Hexane	Liqui-Cel X-30 or X-40 module	[269, 270]
Layered (without polymer membrane walls) bulk liquid membrane module			
Olefin/paraffin	Silver nitrate		[271, 272]
Phenol	Dialkylamine, trioctylamine, Cyanex 923		[23]
Textile anionic dyes	Tetrabutylammonium bromide (TBAB)		[273]
Et® 4-chloro-3-hydroxybutyrate (CHBE) from racemic mixt.	Cinchonidine, cyclodextrin		[274]
D-Ribose, D-xylose, and D-glucose	Amphiphilic cyclodextrin		[275]
Toluene	Ag^+		[276]

Table 5.12 Selected references on gas separations by BOHLM techniques

Gases	Liquid membrane	Polymer membrane walls	References
Spiral-type FLM module			
CO_2 (mixt. with CH_4)	Water, sol. $K_2CO_3/KHCO_3$, diethanolamine (DEA)	Duragard 2500, 2502	[277]
		Fluoropore FP-010	
CO_2 (mixt. with O_2, N_2)		Spacers: KDJ-6, KDN-4	
Capillary ultrafiltration membrane module			
CO_2 (mixt. with N_2)	Aqueous sol. of 2-(butylamino)ethanol, 2-(methylamino-1-propanol) (AMP), monoethanolamine (MEA), diethanolamine (DEA)	Polyethersulfone capillary ultrafiltration membranes	[31, 111-113, 278]
Hollow-fiber module			
CO_2 from flue gas	Monoethanolamine (MEA), triethanolamine (TEA)	Poly(vinylidene fluoride) (PVDF) hollow fiber	[279-283]
CO_2-N_2 mixt.		Celgard 2400, Saint-Gobain R128-10	[284]
		Poly(1-trimethylsilyl-1-propyne) (PTMSP)	
CO_2	Polyamidoamine (PAMAM) dendrimer, carbonic anhydrase	Celgard polypropylene (PP) hollow fibers	[285-288]
Volatile organic compounds (VOC)	Inert, nontoxic solv., silicone oil	Perfluorodimethyldioxole-tetrafluoroethylene (PDD-TFE) hollow fibers	[289-294]
H_2S	Aqueous sol. of alkali carbonate, MDEA	Hydrophobic or hydrophilic microporous membranes	[295-297]
CO_2, O_2	Vacuum	Perfluorodimethyldioxole-tetrafluoroethylene (PDD-TFE) hollow fibers	[7, 298, 299]
Hollow-fiber module			
Spiral-type FLM module	Fe^{2+} + EDTA or Fe^{3+} + EDTA aq. sol.	Polypropylene, Celgard X-10, X-30	[300]
CO_2-N_2 mixt.	Water, diethanolamine (DEA)	Polypropylene, Celgard X-10, X-30	[301-309]
CO_2-CH_4 mixt.			

Table 5.13 Selected references on reactor conversion-separations by BOHLM techniques

Solutions treated	Solutes treated	Liquid membrane	Polymer membrane walls	References
Wastewater treatment	Reviews	Enzymes, fermentors, biocatalysts, catalysts	MF, UF membranes	[310–320]
	Metals, hydrogen		UF flat sheet (spiral bound), HF and capillary membrane	[321–324]
	Anions			[325–329]
	Organic acids			[330–335]
Enzymatic conversion	Organic compounds, hydrolysis	Catalysts, enzymes	UF flat sheet (spiral bound), HF and capillary membrane	[336–342]
				[343–345]
Electrokinetic conversion	Lactic acid	Fermentors	Ultrafiltration HF membranes	[346]

metal ions may enable a LM to perform difficult separations. A wide variety of macrocyclic carriers exist and are being developed in recent years. The complex stability, donor abilities of functional atoms (O, N, S, P, etc.), and/or groups of macrocyclic carriers are intensively studied to develop some rules for predicting their carrier properties.

The second group comprises water-immiscible primary, secondary, tertiary amines and their derivatives, quaternary amine salts and other proton acceptors (see Tables 5.9 and 5.10). Considerable effort has been devoted to the development of anion transporting agents, which play an important role in biological and biochemical processes. Several amines, their derivatives, surfactants, lipophilic metal complexes, macrocycles with positively charged subunits are known and are being developed as anion carriers.

The third group are water-immiscible organic species with electron donor or acceptor properties (see Tables 5.10-5.13) or solvating carriers. They include carbon-oxygen compounds (amides, ethers, ketones); phosphorus-oxygen compounds [tri-*n*-butylphosphate (TBP), dibutyl-phosphate (DBP), or dibutyl-phosphonate (DBBP)]; phosphine oxides [tri-*n*-octylphosphine oxide (TOPO)]; phosphine sulfides (Cyanex 471); alkyl sulfides (dihexyl, diheptyl sulfides); nitrogen-containing compounds (CLX 50), etc. All of them are known as selective extractants, but few have been tested as carriers in LM processes.

Some LM carriers tested show one of the serious problems: a low rate of decomplexation which becomes a rate-controlling step of all LM transport. To overcome this disadvantage is one of the directions for future investigations.

3.4. Examples of BOHLM systems

A large number of BOHLM separation systems are described in the literature. They can be divided according to type of membrane walls (barriers) used: planar, spiral wound, and hollow fiber, and to hydrophobic/hydrophilic properties of the membrane. The last are hydrophobic membranes, immobilized by LM organic solution, and hydrophilic or ion-exchange membranes, immobilized by aqueous feed and strip solutions.

Hydrophilic (or ion-exchange) membranes were used for designing rotating disk, creeping film, hybrid liquid membrane, and multimembrane hybrid membrane systems, hollow-fiber LM modules. Hydrophobic membranes were used for designing hybrid liquid membrane, multimembrane hybrid system, flowing LM, hollow-fiber contained LM, capillary liquid membrane modules (or contactors). Below, some of these systems are referenced and described shortly.

3.4.1. Layered bulk liquid membrane modules [90, 91]

The simplest U- or H-shaped layered BLM modules, without membrane walls, are used mainly in transport studies. Examples of BLM layered module with mixing of all three solutions are presented in Fig. 5.6.

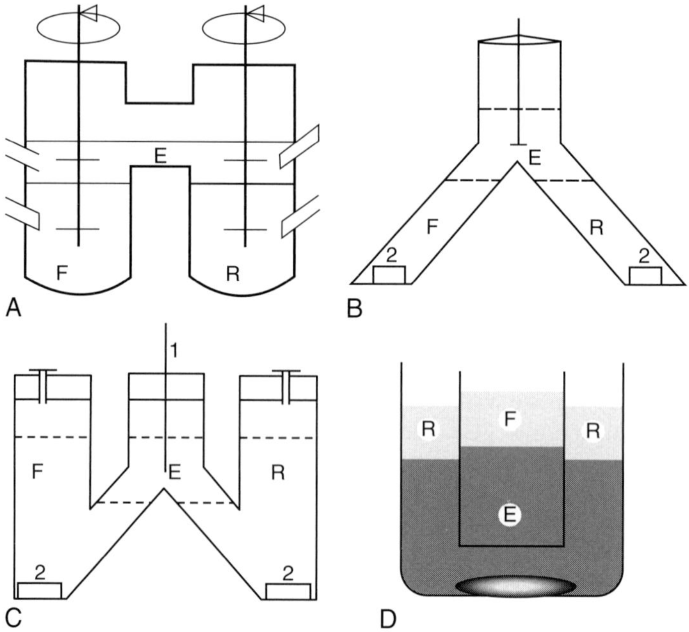

Figure 5.6 Examples of layered bulk liquid membrane modules.

3.4.2. Rotating disk modules [12, 14, 16, 18, 20, 92-100]

As can be seen in Fig. 5.7, hydrophilic membrane disks are fixed on a horizontal shaft and their lower parts immersed to compartments, with disks, in which aqueous phases (feed or strip) are immersed, due to rotation,

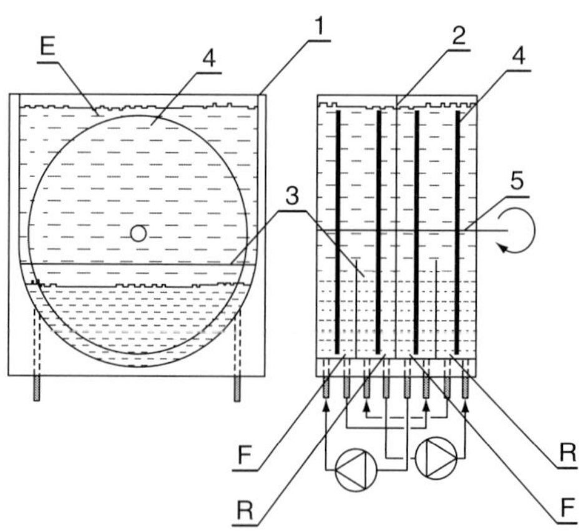

Figure 5.7 Scheme of a rotating film contactor: (1) body, (2) stage wall, (3) feed/stripping solution separating walls, (4) rotating disks, and (5) common shaft. From Ref. [12] with permission.

are contacting with the LM phase. Mass transfer of the solute from the feed into the strip solutions occurs through the bulk LM phase.

A contactor with two parallel shafts with disks, one for the feed and one for the strip solutions, has been presented [97].

3.4.3. Creeping film modules [19, 21, 22, 86, 101-104]

Creeping film process (CFP) is a liquid membrane technique for simultaneous removal and concentration of dissolved species from their diluted aqueous solutions. CFP contactor is presented schematically in Fig. 5.8. Feed and strip solutions flow down the vertical hydrophilic porous membrane sheets. A mobile organic LM interposed between two creeping aqueous films. CFP is a continuous mass-transfer process in which eddy diffusion controls the mass fluxes in all three liquid films.

3.4.4. Hybrid liquid membrane modules [1-3, 40, 43, 44]

A HLM contactor is shown schematically in Fig. 5.1. A liquid membrane solution flows (or circulates) between two membranes, which separate the LM phase (E) from the feed (F) and receiving (R) phases. A solute (or solutes) diffuses to the F/E interface, is extracted from the feed phase by a carrier; as a result of the thermodynamic conditions at the F/E interface, the solute-carrier complex diffuses to the E/R interface and is simultaneously stripped by the receiving phase due to the different thermodynamic conditions at the E/R interface. The membranes may be hydrophobic, immersed by LM, or hydrophilic (or ion exchange), immersed by feed and strip aqueous phases.

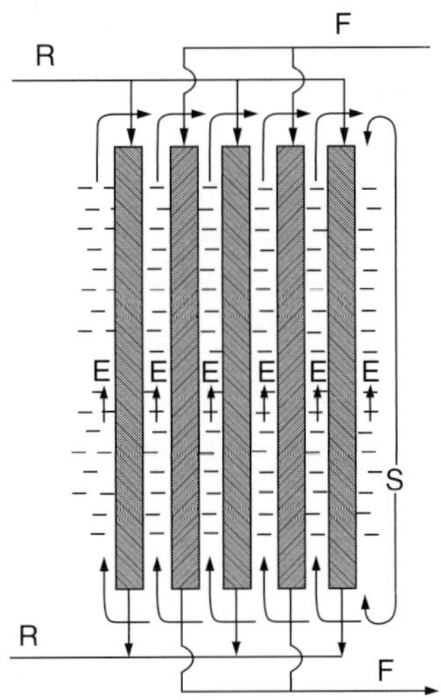

Figure 5.8 Scheme of a creeping film contactor: F and R are the inlet and outlet of the feed and receiving solutions, respectively; E is the flowing liquid membrane organic solution. From Ref. [19] with permission.

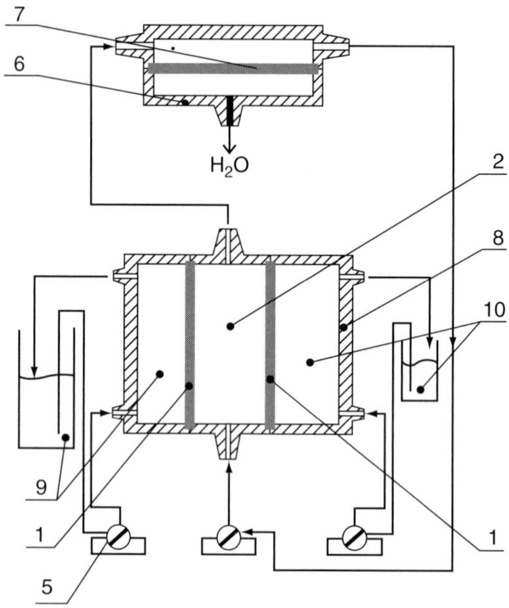

Figure 5.9 Scheme of the MHS[FLM-PV] multimembrane hybrid system: (1) cation-exchange membranes, (2) flowing liquid membrane, (5) vacuum system, (6) pervaporation unit, (7) pervaporation membranes, (8) contactor, (9) feed solution, and (10) stripping solution. From Ref. [29] with permission.

3.4.5. Multimembrane hybrid systems [29, 47, 51, 105, 106]

MHS with pervaporation of water from LM (MHS-PV) is presented in Fig. 5.9. Contrary to the simple MHS with an agitated bulk liquid membrane, separated from the feed and strip solutions by flat hydrophobic or hydrophilic or ion-exchange membranes, the MHS-PV system exploits a liquid membrane continuously flowing between the two flat cation-exchange and two pervaporation membranes. To couple the separation and pervaporation processes, the LM is simultaneously pumped through the MHS and PV modules. The pervaporation membranes are placed on stainless steel porous supports. Aqueous feed and strip solutions are intensively agitated.

The MHS (without pervaporation section) and HLM are very similar systems.

3.4.6. Flowing liquid membrane modules [19, 30, 33-36, 107]

In FLM, the LM organic solution flows in a thin channel between two hydrophobic microporous membranes separating the LM phase from an aqueous feed and strip solutions. The FLM differs from the HLM and MHS modules with hydrophobic membranes by application of a spiral-type module. A schematic diagram of the spiral-type FLM module is shown in Fig. 5.10. The microporous membrane films (see Table 5.5) and one mesh spacers were

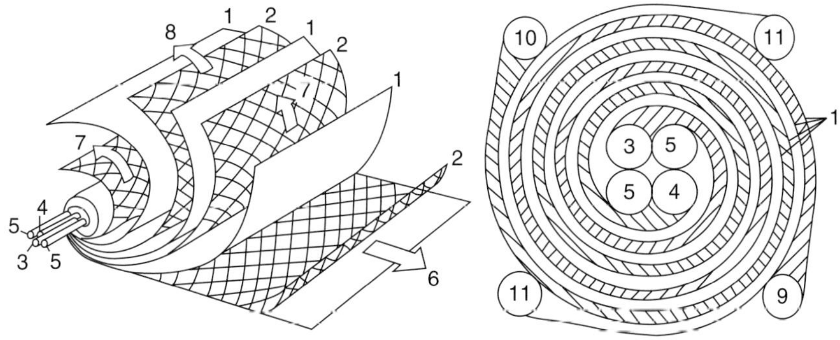

Figure 5.10 Schematic diagram of a spiral-type flowing liquid membrane module: (1) microporous hydrophobic membrane, (2) mesh spacer, (3) inlet and (9) outlet pipe of feed, (4) inlet and (10) outlet pipe of strip, (5) inlet and (11) outlet tube of organic membrane solution, (6) feed solution, (7) organic LM solution, and (8) strip solution. From Ref. [34] with permission.

spirally wound around acrylic resin pipes through which the feed, strip and organic LM solutions were supplied to the module [34, 36]. The outer surface and the top and bottom ends of the module were sealed with an adhesive.

3.4.7. Hollow-fiber liquid membrane modules [4-6, 9-11, 38, 101]

Several types of three-phase hollow-fiber (HF) modules have been described. They can be classified into two groups. The first one with parallel phase flow is developed by Sirkar group, termed as a "hollow-fiber contained liquid membrane" and described in detail in Ref. [4]. Another types of HF module, developed by Schlosser's group, "hollow-fibers in tube pertractor" with parallel flow [10, 11, 23-26, 90, 101] consists of one or two intermixed U-shaped bundles of microporous polypropylene fibers, inserted into polysulfon (or glass) tubes (see Fig. 5.11). For better mixing, pulsation of the LM phase along the fibers is used. It increases the transport rate by 35-60%.

The second type of the HF modules have been developed by Schlosser group also [25, 38-40, 85, 101]. The module enables a crossflow of the liquid membrane phase or its pulsation perpendicularly to fibers, as shown in Fig. 5.12. The planar HF elements can be assembled for contacting four or more phases.

Some modified plate and frame HF modules [108-110] were proposed. Both ends of fibers were fixed on opposite sides of the frame forming elements of the stack. The transversal flow of one phase was possible. The module with "fiber in fiber" units and crossflow of one phase is described in Ref. [9].

3.4.8. Capillary liquid membrane systems [111-113]

Facilitated transport for gas separation using a capillary membrane module was developed by the Teramoto group [111, 112]. The concept of a capillary membrane apparatus is shown in Fig. 5.13. Both a feed gas and a LM solution

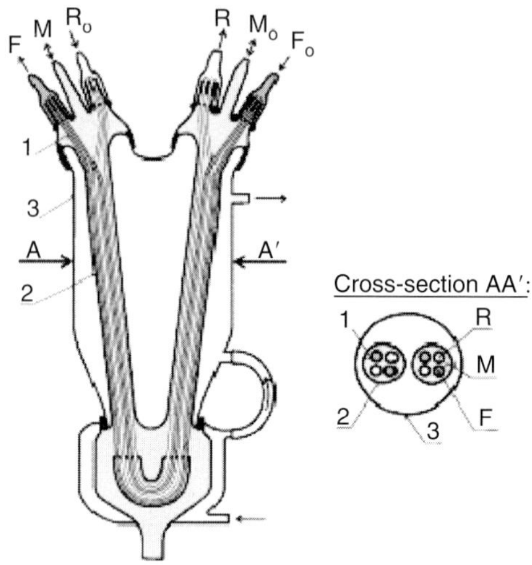

Figure 5.11 Scheme of the hollow-fiber contactor with parallel flow of phases, distributed U-shaped bundles of fibers and with separated inlet and outlet end chambers [F: feed (donor phase), E: liquid membrane phase, R: stripping solution, 1: hollow fiber for the feed, 2: glass tubulet, 3: glass tempering jacket of the contactor]. From Ref. [9] with permission.

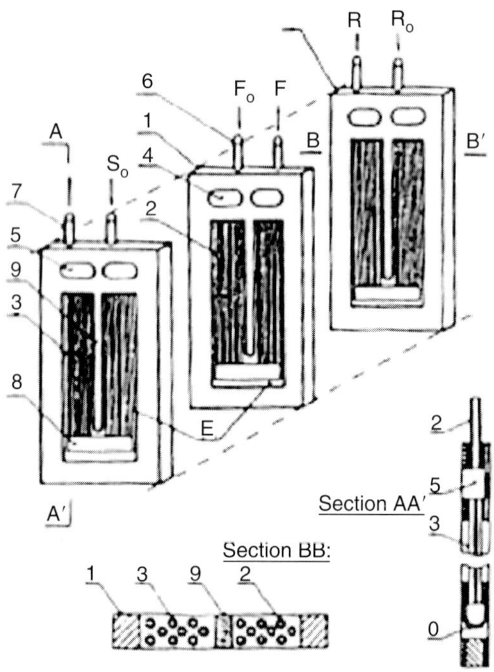

Figure 5.12 Scheme of the hollow-fiber contactor of three phases with crossflow of LM phase: (1) body of element, (2, 3) hollow fiber in downstream and upstream part, (4, 5) inlet and outlet chamber, (6,7) inlet and outlet tube, (8) flowing head, (9) central baffle, F: feed, E: LM phase, R: strip solution. From Ref. [25] with permission.

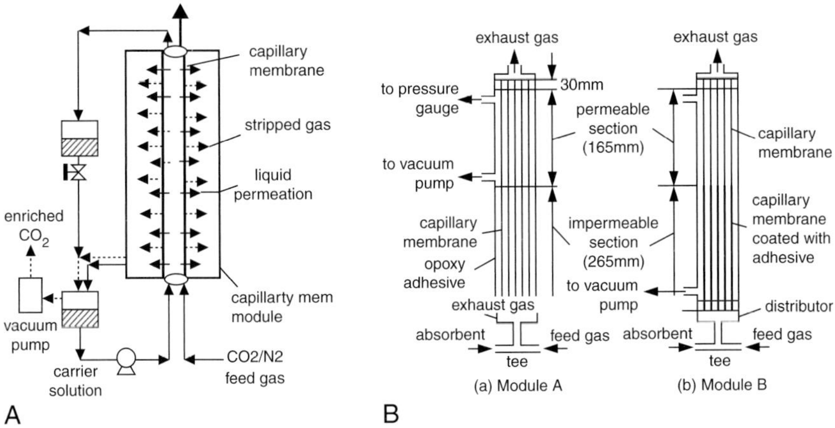

Figure 5.13 Schematic diagrams of facilitated transport of gas using capillary membrane module for removal and enrichment of carbon dioxide: (A) experimental capillary membrane apparatus and (B) capillary membrane modules with permeation of carrier solution. From Ref. [112] with permission.

are supplied to the lumen side (high-pressure feed side) of the capillary ultrafiltration membrane module and flow upward. The LM solution, which contains dissolved solute gas (CO_2 in the present case), flows to the permeate side (low-pressure shell side), where the solution liberates dissolved gas to become a lean solution. The lean solution is returned to the lumen of the capillary module by a pump. In the shell side, dissolved gas is stripped from the liquid flowing down on the outer surface of the capillary, and discharged from a vacuum pump through a liquid reservoir.

3.4.9. Membrane-based or nondispersive solvent extraction systems

Inclusion of this technique to the BOHLM has to be explained. Solvent extraction or partition of the solute between two immiscible phases is an equilibrium-based separation process. So, the membrane-based or nondispersive solvent extraction process has to be equilibrium based also. "Liquid membrane separation is a rate process and the separation occurs due to a chemical potential gradient, not by equilibrium between phases" [114]. According to these definitions, many authors who refer to their works as membrane-based or nondispersive solvent extraction processes are not correct.

The theory of membrane-based solvent extraction suggests that overall mass transfer of treated solute consists of several steps: diffusion of the solute through the aqueous layer from the bulk source aqueous solution to the phases' interface (nonequilibrium process), interaction of the solute with extractant and formation of the solute-extractant complex (as a rule, the process is rapid and reaches equilibrium), diffusion of the solute-extractant complex through the membrane support itself (nonequilibrium process), and diffusion of the solute-extractant complex through the organic layer to the bulk organic solution (nonequilibrium process). Here, we can see that only one of the components may be equilibrium based. So, we can conclude that all systems may be equilibrium based in two cases: (1) if the kinetics of solute-extractant interaction is a rate-controlling process, which is not true in the majority of separations published in the literature, and (2) if the overall mass transfer of the solute from the bulk source solution to the bulk membrane solution comes to equilibrium.

The authors of very many works on so-called membrane-based or nondispersive solvent extraction could not prove that the process reaches equilibrium. Therefore, we cannot confirm the processes, published in these works, as membrane-based solvent extraction, but can confirm them as liquid membrane processes. Liquid membrane separations are dynamic nonequilibrium processes, in which only local equilibrium at immiscible phases' interface may be suggested.

The works, in which the authors prove the overall equilibrium state, are not referenced here. The works, in which the equilibrium state is not proved, or cannot be reached (e.g., the process where the feed-membrane-strip flows are arranged in one continuously operating module) are referenced in the next section as hybrid liquid membrane or hollow-fiber liquid membrane processes.

4. Selected Applications

Applications of the BOHLM processes are mainly in metal separation, wastewater treatment, biotechnologies, drugs recovery-separation, organic compounds, and gas separation. In recent years, integrated hybrid systems incorporating two or more functions in one module, for example biotransformation and separation, become of great interest to researchers. The recent applications of different types of the BOHLM systems are summarized below.

4.1. Metal separation-concentration

Metal ions separation for hydrometallurgical and wastewater treatment applications has attracted considerable interest. Selective separations of alkali, alkali earth, rare earth, heavy metal ions, precious metals, etc., are studied by many authors using all above described techniques. Referenced works in metal separations, classified according to the BOHLM techniques with types of membrane walls and carriers used, are listed in Table 5.8.

4.2. Biotechnological products recovery-separation

Recovery and separation of carboxylic and amino acids from fermentation broth have been tested using layered BLM, rotating, creeping, spiral-type FLM, HFLM, HLM, and MHS-PV techniques of the BOHLM processes. The research from the last 15 years in this field, classified according to the BOHLM techniques with types of membrane walls and carriers used, are referenced in Table 5.9.

4.3. Pharmaceutical products recovery-separation

Selected papers from the last 15 years, presenting results of separation of drugs, antibiotics, enantiomers by layered BLM, hollow-fiber LM, rotating film techniques, are referenced in Table 5.10. Hollow-fiber contactors, developed from modular elements with optimum configured stacks for individual applications, may be an interesting direction for research.

4.4. Organic compounds separation and organic pollutants recovery from wastewaters

Separation of ethylene, benzene, propanol, olefin, aromatic amines from organic liquid mixtures, of volatile organic compounds (VOC) and phenol from wastewater, were investigated (Table 5.11), using a rotating film module, spiral-type FLM, hollow-fiber and layered LM techniques. High separation factors (>1000) in pilot- and industrial-scale experiments were found.

4.5. Fermentation or enzymatic conversion-recovery-separation (bioreactors)

The BOHLM systems, integrating reaction, separation, and concentration functions in one apparatus (bioreactor), attracted great interest in the last few years. Bioreactors combine the use of specific biocatalyst for the desired chemical reactions, with repeated or continuous application of it under very specific conditions. Such techniques were termed hybrid membrane reactors. In biotechnology and pharmacology, these applications are termed hybrid membrane bioreactors or simply bioreactors (see Table 5.13). An example of an experimental setup of the bioreactor system is shown schematically in Figs 5.14 and 5.15.

4.6. Analytical applications

In the last two decades, BOHLM techniques were intensively used in analytical chemistry for separation and preconcentration of metals, organic acids, organic, and pharmaceutical compounds. Some membrane-based

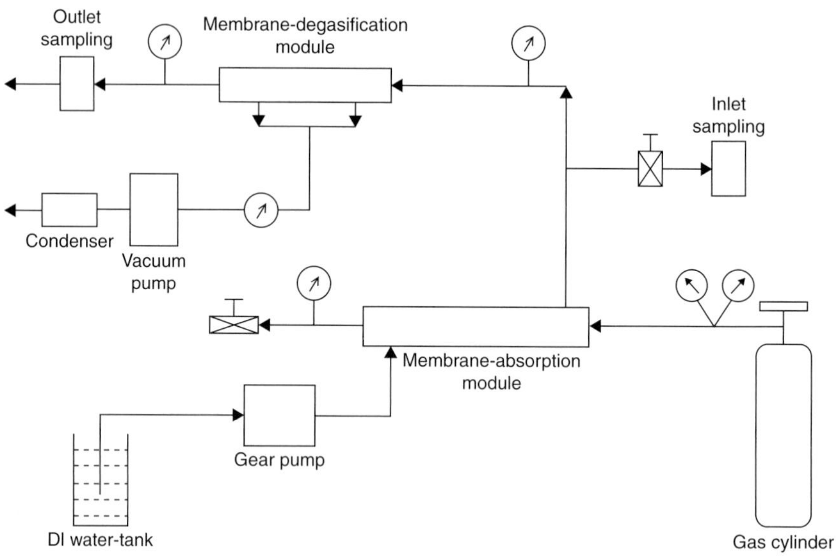

Figure 5.14 Experimental setup: gas absorption and subsequent degasification using membrane modules. From Ref. [7] with permission.

techniques for analytical sample preparations are described in detail in the reviews [347-355], some are referenced in Table 5.14.

Preparation of samples for analytical purposes seems to be one of the prospective directions in the application of the BOHLM processes.

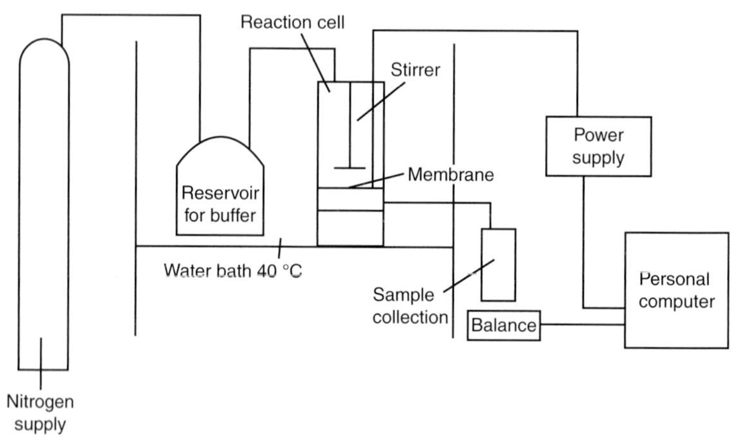

Figure 5.15 Schematic flow diagram of the integrated membrane reactor system. From Ref. [341] with permission.

Table 5.14 Selected references on analytical separations and preconcentration by BOHLM techniques

Organic compounds	BOHLM technique	Liquid membrane	References
Metals			
Heavy metals, Cu			[356-359]
Trace metal enrichment	HF cartridges	8-Quinolinol chelat. agent	[360]
Organic acids			
	Membrane extraction	Amine and their derivatives	[361]
Organic and pharmaceutical species			
Dichlorophenol, clenbuterol, ambroxol, ropivacaine metabolites, triazine herbicides, pesticides	HF membrane extraction		[362-370]

5. Summary Remarks

In comparison with liquid-liquid extraction (LLX), supported liquid membrane (SLM), and emulsion liquid membrane (ELM), BOHLM has the potential to provide many economic and operational advantages, such as: low carrier losses, long membrane lifetime, "once through" continuous operation, compact equipment, application of different driving forces (chemical potential, pressure, temperature gradients between different compartments, electric field, etc.), no need of surface formation, impregnation, gravity gradients, high membrane capacity, etc. Commercially available membrane modules and equipment may be used in the BOHLM.

One more advantage of the BOHLM system is realized from theoretical simulations. At feed-side resistances not controlling the solute transport (as in the case with titanium) the fluxes of the solute to the strip are approaching these in the SLM systems, but without many drawbacks of the latter.

Currently, the only commercial application of the LM technologies is waste treatment (water and degasification), where low concentration solutes are removed from large volumes of effluents. Solvent extraction and ion exchange are often not economically convenient in these cases. Due to the complexation reaction and low quantity of complexing agent-carrier required, the BOHLM technologies are suited for high recoveries of dilute solutes.

Despite promising technological performance, few BOHLM techniques have been commercialized. This is due to a number of economic and technical factors. The main drawbacks of the BOHLM systems are the poisoning of the carrier by irreversible reactions, low diffusion rate of the large organic molecules, losses of the LM and contamination of the treated and product aqueous solutions by the LM organic components, the membrane walls blinding or fouling due to the emulsion, and gel formation on their surfaces. Long-term stability of complexing carriers in the LM, stability of membrane walls, their long-term permeability are problems that limit the commercialization of the BOHLM technologies.

To optimize the BOHLM processes and improve separation and transport properties, it is necessary to develop complexation chemistry and new selective carriers, or to improve existing ones. Such improvements and reducing the price of membrane film or hollow-fiber production will definitely speed industrial application of the BOHLM technologies, especially for gas, pharmaceutical and bioreactor applications.

REFERENCES

1. Kislik V, Eyal A. Hybrid liquid membrane (HLM) system in separation technologies. J Membr Sci 1996; 111: 259-272.
2. Kislik V, Eyal A. Hybrid liquid membrane (HLM) and supported liquid membrane (SLM) based transport of titanium(IV). J Membr Sci 1996; 111: 273-281.
3. Gega J, Walkowiak W, Gajda B. Separation of Co(II) and Ni(II) ions by supported and hybrid liquid membranes. Sep Purif Technol 2001; 22-23: 551-558.
4. Majumdar S, Sirkar KK. Hollow-fiber contained liquid membrane. In: Ho WSW, Sirkar KK, eds. *Membrane Handbook*. Van Nostrand Reinhold, New York, NY, 1992: 764-808.
5. Qin Y, Cabral J. Theoretical analysis on design of hollow fiber modules and modules cascades for the separation of diluted species. J Membr Sci 1998; 143: 197-205.
6. Dai X-P, Yang Z-F, Luo RG, Sirkar KK. Lipase-facilitated separation of organic acids in a hollow fiber contained liquid membrane module. J Membr Sci 2000; 171: 183-196.
7. Bhaumik D, Majumdar S, Fan O, Sirkar KK. Hollow fiber membrane degassing in ultrapure water and microbiocontamination. J Membr Sci 2004; 235: 31-41.
8. Sengupta A, Basu R, Sirkar KK. Separation of solutes from aqueous solutions by contained liquid membranes. AIChE J 1988; 34: 1698-1713.
9. Schlosser S, Sabolova E. Three-phase contactor with distributed U-shaped bundles of hollow-fibers for pertraction. J Membr Sci 2002; 210: 331-347.
10. Schlosser S, Rothova I. A new type of hollow-fiber pertractor. Sep Sci Technol 1994; 29: 765-780.
11. Schlosser S, Rothova I, Frianova H. Hollow-fiber pertractor with bulk liquid membrane. J Membr Sci 1993; 80: 99-106.
12. Zhivkova S, Dimitrov K, Kyuchoukov G, Boyadzhiev L. Separation of zinc and iron by pertraction in rotating film contactor with Kelex 100 as a carrier. Sep Purif Technol 2004; 37: 9-16.

13. Boyadzhiev L, Alexandrova S, Kirilova N, Saboni A. Pertraction continue de tylosine dans un contacteur a films tournants. Chem Eng J 2003; 95: 137-141.
14. Dimitrov K, Alexandrova S, Saboni A, Debray E, Boyadzhiev L. Recovery of zinc from chloride media by batch pertraction in a rotating film contactor. J Membr Sci 2002; 207: 119-127.
15. Kawasaki J, Egashira R, Kawai T, Hara H, Boyadzhiev L. Recovery of erythromycin by a liquid membrane. J Membr Sci 1996; 112: 209-217.
16. Boyadzhiev L, Dimitrov K. Recovery of silver from nitrate solution by means of rotating film pertraction. J Membr Sci 1994; 86: 137-143.
17. Lazarova Z, Boyadzhiev L. Kinetic aspects of copper(II) transport across liquid membrane containing LIX-860 as a carrier. J Membr Sci 1993; 78(3): 239-245.
18. Lazarova Z, Boyadzhiev L. Liquid film pertraction—A liquid membrane preconcentration technique. Talanta 1992; 39(8): 931-935.
19. Boyadzhiev L, Lazarova Z. Study on creeping film pertraction. Recovery of copper from diluted aqueous solutions. Chem Eng Sci 1987; 42(5): 1131-1139.
20. Boyadzhiev L. Liquid pertraction or liquid membranes—State of the art. Sep Sci Technol 1990; 25(3): 187-190.
21. Boyadzhiev L, Alexandrova S. Dephenolation of phenol-containing waters by rotating film pertraction. Sep Sci Technol 1992; 27(10): 1307-1312.
22. Boyadzhiev L. Recovery of valuable metals from diluted aqueous solutions by creeping film pertraction. In: Devis GA, ed. *Separation Processes in Hydrometallurgy*. Society of Chemical Industry, London, 1987; Pt. 3, Chapter 26: 259-287.
23. Cichy W, Schlosser S, Szymanowski J. Transport of phenol through bulk liquid membranes. In: Cox M, Hidalgo M, Valiente M, eds. *Solvent Extraction for the 21st Century. Proceedings of the ISEC'99*, Barcelona, Spain, July 1999. Society of Chemical Industry, London, 2001; 2: 1065-1070.
24. Cara G, Schlosser S, Munoz M, Valiente M. Pertraction of neodymium and holmium in a hollow fiber contactor. In: Cox M, Hidalgo M, Valiente M, eds. *Solvent Extraction for the 21st Century. Proceedings of the ISEC'99*, Barcelona, Spain, July 1999. Society of Chemical Industry, London, 2001; 2: 1023-1028.
25. Schlosser S. Pertraction through liquid and polymeric membranes. In: Belafi-Bako K, Gubicza L, Mulder M, eds. *Integration of Membrane Processes into Bioconversions. Proceedings of the 16th European Membrane Society Annual Summer School*, Veszprem, Hungary, August 1999. Kluwer Academic/Plenum Publishers, New York, NY, 2000: 73-100.
26. Schlosser S, Sabolova E, Martak J. Pertraction and membrane based extraction of carboxylic acids in hollow fibre contactors. In: Cox M, Hidalgo M, Valiente M, eds. *Solvent Extraction for the 21st Century. Proceedings of the ISEC'99*, Barcelona, Spain, July 1999. Society of Chemical Industry, London, 2001; 2: 1041-1046.
27. Wodzki R, Szczepanska G, Szczepanski P. Unsteady state pertraction and separation of cations in a liquid membrane system: Simple network and numerical model of competitive M^{2+}/H^{+} counter-transport. Sep Purif Technol 2004; 36: 1-16.
28. Wodzki R, Swiatkowski M, Lapienis G. Properties of star-shaped polymer with poly (oxyethylene) branches and monoesters of phosphoric acid end groups in pertraction of alkali, alkaline earth, and transient metal cations. React Funct Polym 2002; 52: 149-161.
29. Wodzki R, Szczepanski P. Integrated hybrid liquid membrane systems—Membrane extraction and pertraction coupled to a pervaporation process. J Membr Sci 2002; 197: 297-308.
30. Teramoto M, Takeuchi N, Maki T, Matsuyama H. Ethylene/ethane separation by facilitated transport membrane accompanied by permeation of aqueous silver nitrate solution. Sep Purif Technol 2002; 28: 117-124.

31. Teramoto M, Takeuchi N, Maki T, Matsuyama H. Gas separation by liquid membrane accompanied by permeation of membrane liquid through membrane physical transport. Sep Purif Technol 2001; 24: 101-112.
32. Teramoto M, Matsuyama H, Nakai K, Uesaka T, Ohnishi N. Facilitated uphill transport of eicosapentaenoic acid ethyl ester through bulk and supported liquid membranes containing silver nitrate as carrier: A new type of uphill transport. J Membr Sci 1994; 91: 209-213.
33. Teramoto M, Matsuyama H, Ohnishi N. Selective facilitated transport of benzene across supported and flowing liquid membranes containing silver nitrate as a carrier. J Membr Sci 1990; 50: 269-278.
34. Teramoto M, Matsuyama H, Ohnishi N. Development of a spiral-type flowing liquid membrane module with high stability and its application to the recovery of chromium and zinc. Sep Sci Technol 1989; 24: 981-999.
35. Teramoto M, *et al.* Development of spiral-type supported liquid membrane module for separation and concentration of metal ions. Sep Sci Technol 1987; 22: 2175-2201.
36. Matsuyama H, Teramoto M, *et al.* Separation and concentration of heavy metal ions by spiral type flowing liquid membrane module. Water Treatment 1990; 5: 237-252.
37. Teramoto M, *et al.* Selectivity in the extraction of metals by liquid membranes. Proc ISEC-86 1987; 1: 545-560.
38. Prasad R, Sirkar KK. Membrane-based solvent extraction. In: Ho WSW, Sirkar KK, eds. *Membrane Handbook*. Van Nostrand Reinhold, New York, NY, 1992: 727-763.
39. Kubisova L, Sabolova E, Schlosser S, Martak J, Kertesz R. Mass-transfer in membrane based solvent extraction and stripping of 5-methyl-2-pyrazinecarboxylic acid and co-transport of sulphuric acid in HF contactors. Desalination 2004; 163: 27-38.
40. Vajda M, Kosuthova A, Schlosser S. Membrane-based extraction joined with membrane-based stripping in a circulating arrangement III. Extraction of zinc. Chem Pap 2004; 58: 1-8.
41. Vajda M, Sabolova E, Schlosser S, Mikulova E. Membrane-based extraction joined with membrane-based stripping in a circulating arrangement II. Extraction of organic acids. Chem Pap 2003; 57: 3-10.
42. Kedem O, Bromberg L. Ion-exchange membranes in extraction processes. J Membr Sci 1993; 78: 255-261.
43. Bromberg L, Levin G, Libman J, Shanzer A. A novel tetradentate hydroxamate as ion carrier in liquid membranes. J Membr Sci 1992; 69: 143-153.
44. Kedem O, Eyal A, Bromberg L, Bressler E. Membrane extraction in the fermentation of carboxylic acids, 1992; Israeli Patent 101905.
45. Eyal A, Bressler EJ. Industrial separation of carboxylic and amino acids by liquid membranes: Applicability, process considerations, and potential advantages. Biotechnol Bioeng 1993; 41: 287-293.
46. Baniel A, Eyal A, Bressler E. Extraction of electrolytes from aqueous solutions, 1992; Israeli Patent 101906.
47. Wodzki R, Nowaczyk J. Propionic and acetic acid pertraction through a multimembrane hybrid system containing TOPO or TBP. Sep Purif Technol 2002; 26: 207-220.
48. Wodzki R, Szczepanski P. Integrated process of Donnan dialysis and pertraction in a multimembrane hybrid system. Sep Purif Technol 2001; 22-23: 697-706.
49. Wodzki R, Nowaczyk J, Kujawski M. Separation of propionic and acetic acid by pertraction in a multimembrane hybrid system. Sep Purif Technol 2000; 21: 39-54.
50. Wodzki R, Sionkowski G, Pozniak G. Recovery and concentration of metal ions. IV. Up-hill transport of Zn(II) in a multimembrane hybrid system. Sep Sci Technol 1999; 34: 627-649.

51. Wodzki R, Sionkowski G. Recovery and concentration of metal ions. III. Temperature and concentration effects in multimembrane hybrid system. Sep Sci Technol 1996; 31: 1541-1553.
52. Wodzki R, Sionkowski G. Recovery and concentration of metal ions. II. Multimembrane hybrid system. Sep Sci Technol 1995; 30: 2763-2778.
53. Sirkar KK. Other new membrane processes. In: Ho WSW, Sirkar KK, eds. *Membrane Handbook*. Van Nostrand Reinhold, New York, NY, 1992: 904-908.
54. Noble RD, Way JD. (Eds.). *Liquid Membranes. Theory and Applications*. ACS Symposium Series 347. ACS, Washington, DC, 1987.
55. Sirkar KK. Membrane separation technologies. Current developments. Chem Eng Commun 1997; 157: 145-184.
56. Sirkar KK. Hollow Fiber-Contained Liquid Membranes for Separations: An Overview. In: Bartsch RA, Way JD. (Eds.). *Chemical Separations with Liquid Membranes*. American Chemical Society, Symposium Series 642, 1996: 222-238.
57. Boyadzhiev L, Lazarova Z. Liquid membranes (liquid pertraction). Membr Sci Technol Ser 1995; 2: 283-352.
58. Lazarova Z, Boyadzhiev L. Modeling of copper transfer in a three-phase liquid pertraction system. Dokl Bulg Akad Nauk 1991; 44: 29-32.
59. Lazarova Z, Sapundzhiev T, Boyadzhiev L. Kinetics and mechanism of copper transfer by hydroxy-oximes in three-phase liquid systems. Sep Sci Technol 1992; 27: 493-508.
60. Alexandrova S, Saboni A, Boyadzhiev L, Mouhab N, Estel L. Recovery of substances by rotating-film pertraction. Chem Eng J (Lausanne) 2000; 79: 155-163.
61. Dindi A, Noble RD, Koval CA. An analytical solution for competitive facilitated membrane transport. J Membr Sci 1992; 65: 39-45.
62. Kislik V, Eyal A. Heavy metals removal from wastewaters of phosphoric acid production. A comparison of hybrid liquid membrane (HLM) and aqueous hybrid liquid membrane (AHLM) technologies. In: *Proceedings of the Conference on Membrane in Drinking and Industrial Water Production*, 2000; 1: 503-514.
63. Kislik V, Eyal A. Acidity dependence of Ti(IV) extraction: A critical analysis. Solv Extr Ion Exch 1993; 11: 259-272.
64. Kislik V, Eyal A. Extraction of titanium(IV) by mixtures of mono- and di-(2-ethylhexyl) phosphoric acid esters. Solv Extr Ion Exch 1993; 11: 273-281.
65. Kislik V, Eyal A. Mechanisms of titanium(IV) and some other transition metals extraction by organophosphoric acids. In: Logsdail DH, Slater MJ, eds. *Proceedings of the ISEC'93*. Elsevier Applied Science for SCI, London, 1993; 3: 1353-1361.
66. Davis S, Wydeven T, Yum A, Neukom C. A computational and experimental study of unsteady facilitated transport. J Membr Sci 1992; 69: 99-108.
67. Dindi AK, Noble RD. Experimental and modeling studies of a parasitic binding mechanism in facilitated membrane transport. J Membr Sci 1992; 66: 55-68.
68. Hu S. Selective removal of metals from wastewater using affinity dialysis. In: *Emerging Technologies in Hazardous Waste Management*. American Chemical Society, Washington, DC, 1990; 12: 187-198.
69. Danesi P, Reichley-Yinger L. Origin and significance of the deviations from a pseudo first order rate law in the coupled transport of metal species through supported liquid membranes. J Membr Sci 1986; 29: 195-211.
70. Giddings JC. *Unified Separation Science*. John Wiley & Sons, New York, NY, 1991; 17-19: 38.
71. Giddings JC. *Unified Separation Science*. John Wiley & Sons, New York, NY, 1991: 45-48.
72. Giddings JC. *Unified Separation Science*. John Wiley & Sons, New York, NY, 1991: 94.
73. Giddings JC. *Unified Separation Science*. John Wiley & Sons, New York, NY, 1991: 112-114.

74. Giddings JC. *Unified Separation Science*. John Wiley & Sons, New York, NY, 1991: 55-71.
75. Giddings JC. *Unified Separation Science*. John Wiley & Sons, New York, NY, 1991: 33.
76. Danesi P. Separation of metal species by supported liquid membranes. Sep Sci Technol 1984-1985; 19: 857-864.
77. Kislik V. Application of organic and aqueous liquid membranes in separation of metals and acids: Comparative analysis. In: *Proceedings of the Abstracts on International Conference: Advanced Membrane Technology*, Italy, October 2001: 214.
78. Wódzki R, Sionkowski G. Exchange diffusion transport of ions in liquid membranes. Part IV. Thermodynamic network analysis of nonstationary transport of divalent ions. Pol J Chem 1995; 69: 407-422.
79. Wodzki R, Sionkowski G. Exchange diffusion transport of ions through liquid membranes. Part III. Competitive transport of divalent cations. Pol J Chem 1992; 66: 351-361.
80. Wodzki R. Exchange diffusion transport of ions through liquid membranes. Part II. Stationary process mediated by a multisided carrier. Pol J Chem 1991; 65: 1715-1729.
81. Wodzki R. Exchange diffusion transport of ions through liquid membranes. Part I. Numerical studies of univalent ions transport limited by diffusion. Pol J Chem 1989; 63: 557-568.
82. Schnakenberg J. *Thermodynamic Network Analysis of Biological Systems*. Springer, New York, NY, 1977.
83. Majumdar S, Sirkar KK. Hollow-fiber contained liquid membrane. Theory. In: Ho WSW, Sirkar KK, eds. *Membrane Handbook*. Van Nostrand Reinhold, New York, NY, 1992: 777-796.
84. Schlosser S, Sabolova E, Kertesz R, Kubisova L. Factors influencing transport through liquid membranes and membrane based solvent extraction. J Sep Sci 2001; 24: 509-518.
85. Dahuron L, Cussler EL. Protein extractions with hollow fibers. AIChE J 1988; 34(1): 130-136.
86. Shinbo T, Sugiura M, Kamo N, Kobatake Y. Coupling between a redox reaction and ion transport in an artificial membrane system. J Membr Sci 1981; 9: 1-11.
87. Baker R, Blume I. Coupled transport membranes. In: Porter MC, ed. *Handbook of Industrial Membrane Technology*. Noyes Publications, Park Ridge, NJ, 1990: 511-558.
88. Lo TH, Baird MHI, Hanson C. (Eds.). *Handbook of Solvent Extraction*. John Wiley & Sons, New York, NY, 1983.
89. Rydberg J, Musikas C, Chppin GR. (Eds.). *Principles and Practices of Solvent Extraction*. Marcel Dekker, New York, 1992: 375-544.
90. Schlosser S, Sabolova E. Transport of butyric acid through layered bulk liquid membranes. Chem Pap 1999; 53: 403-411.
91. Kubisova L, Martak J, Schlosser S. Transport of 5-methyl-2-pyrazinecarboxylic acid through a layered bulk liquid membrane. Chem Pap 2002; 56: 418-425.
92. Lu H-J, Fan Y-T, Wu Y-J, Yin M-C. Tripodal lipophilic ionophores: Synthesis, cation binding and transport through liquid membranes. Polyhedron 2001; 20(28): 3281-3286.
93. Safavi A, Shams E. Selective transport of silver ions through bulk liquid membrane using Victoria blue as carrier. Talanta 1999; 48: 1167-1172.
94. Nishii Y, Kinugasa T, Nii S, Takahashi K. Transport behavior of protein in bulk liquid membrane using reversed micelles. J Membr Sci 2002; 195: 11-21.
95. Bobok D, Schlosser S, Kossaczky E. Method of solution separation by means of pertraction and equipment for application of this method, 1987; Czech Patent CS 235362 Kind B1 Date 19850515.

96. Boyadzhiev L, Dimitrov K. Silver recovery by rotating film pertraction. Process modeling. Solv Extr Ion Exch 1996; 14: 105-118.
97. Alexandrova S, Dimitrov K, Saboni A, Boyadzhiev L. Selective recovery of silver from dilute polymetal solutions by rotating film pertraction. Sep Purif Technol 2001; 22-23: 567-570.
98. Dimitrov K, Alexandrova S, Boyadzhiev L, Ruellan S, Burgard M. Recovery of copper from solutions by rotating film pertraction. Sep Purif Technol 1997; 12: 165-173.
99. Boyadzhiev L, Alexandrova S. Recovery of copper from ammoniacal solutions by rotating film pertraction. Hydrometallurgy 1994; 35: 109-121.
100. Boyadzhiev L, Atanassova I. Extraction of L-lysine from its dilute aqueous solutions by rotating film pertraction. Appl Biochem Biotechnol 1992; 37: 89-95.
101. Schlosser S. Separation method of carboxylic acids from aqueous and non-aqueous solutions. Patent SK 282775 Kind B6 Date 20021203.
102. Boyadzhiev L, Bezenshek E. Liquid film pertraction. Recovery of iodine from iodine-containing aqueous solutions. J Membr Sci 1988; 37: 277-285.
103. Boyadzhiev L, Atanasova I. Recovery of L-lysine from dilute water solutions by liquid pertraction. Biotechnol Bioeng 1991; 38: 1059-1064.
104. Boyadzhiev L, Bezenshek E, Lazarova Z. Removal of phenol from waste water by double emulsion membranes and creeping film pertraction. J Membr Sci 1984; 21: 137-144.
105. Ohki A, Takagi M, Takeda T, Ueno K. Thioether-mediated copper transport through liquid membranes with the aid of redox reaction. J Membr Sci 1983; 15: 231-244.
106. Wodzki R, Szczepanska G. Design of integrated membrane system with liquid and polymer membranes. Zeszyty Naukowe Politechniki Slaskiej, Chemia 2001; 146: 231-234 (in English).
107. Teramoto M, Matsuyama H, Yonehara T. Selective facilitated transport of benzene across supported and flowing liquid membranes containing silver nitrate as a carrier. J Membr Sci 1990; 50: 269-284.
108. Vladisavljevie G, Pavasovie V, Mitrovie M, Vulieevie I. Rev Res Work Fac Agric 1995; 40: 147-156 (in Czech).
109. Vladisavljevie G, Eokesa DM, Mitrovie M, Rajkovic MB. CHISA 96 1996.
110. Stevanovic S, Onjia A, Mitrovic M. Comparative study of arsenic removal from drinking water (precipitation, sorption and membrane extraction treatments). Zh Prikladnoi Khim (Sankt-Peterburg, Russia) 1994; 67: 1743-1744.
111. Teramoto M, Kitada S, Ohnishi N, Matsuyama H, Yonehara N. Separation and concentration of CO_2 by capillary-type facilitated transport membrane module with permeation of carrier solution. J Membr Sci 2004; 234: 83-94.
112. Teramoto M, Ohnishi N, Takeuchi N, Kitada S, Matsuyama H, Matsumiya N, Mano H. Separation and enrichment of carbon dioxide by capillary membrane module with permeation of carrier solution Sep Purif Technol 2003; 30: 215-227.
113. Teramoto M, Takeuchi N, Maki T, Matsuyama H. Facilitated transport of CO_2 through liquid membrane accompanied by permeation of carrier solution. Sep Purif Technol 2002; 27: 25-31.
114. Bartsch RA, Way JD. (Eds.). *Chemical Separations with Liquid Membranes*. ACS Sympo sium Series 642. American Chemical Society, Washington, DC, 1996.
115. Eyal AM, Kislik VS. Membrane recovery of metal pollutants from waste waters of the fertilizers industry (MERMEP). Preliminary Report of Joint Research Project, Avicenne, 1994.
116. Eyal AM, Kislik VS. Hybrid liquid membrane (HLM) process for metal recovery. In: *Membrane Recovery of Metal Pollutants from Waste Waters of the Fertilizers Industry (MERMEP)*. First Year Report of Joint Research Project, Avicenne, 1995.

117. Eyal AM, Kislik VS. Hybrid liquid membrane (HLM) and aqueous hybrid liquid membrane (AHLM) processes for metal recovery. In: *Membrane Recovery of Metal Pollutants from Waste Waters of the Fertilizers Industry (MERMEP)*. Second Year Report of Joint Research Project, Avicenne, 1996.
118. Kislik V, Eyal A. Comparison of hybrid liquid membrane (HLM) and aqueous hybrid liquid membrane (AHLM) technologies in separation of heavy metals from wastewaters. In: *Proceedings of the Abstracts EURO-MEMBRANE-2000*, Israel, 2000: 237-238.
119. Kislik V, Eyal A. Heavy metals removal from wastewaters of phosphoric acid production. A comparison of hybrid liquid membrane (HLM) and aqueous hybrid liquid membrane (AHLM) technologies. Water Sci Water Technol: Water Supply 2001; 1: 119-129.
120. Liu X, He D, Ma M. Transfer and separation of Cd(II) chloride species from Fe(III) by a hybrid liquid membrane containing tri-*n*-octylamine-secondary octylalcohol-kerosene. Chem Eng J (Netherlands) 2007; 133: 265-272.
121. Gu S, Yu Y, He D, Ma M. Comparison of transport and separation of Cd(II) between strip dispersion hybrid liquid membrane (SDHLM) and supported liquid membrane (SLM) using tri-*n*-octylamine as carrier. Sep Purif Technol 2006; 51: 277-284.
122. He D, Gu S, Ma M. Simultaneous removal and recovery of cadmium (II) and CN— From simulated electroplating rinse wastewater by a strip dispersion hybrid liquid membrane (SDHLM) containing double carrier. J Membr Sci 2007; 305(1-2).
123. Yu Y-D, Gu S-X, Luo X-J, Ma M, He D-S. Transport and separation of Cu(II) by N902-kerosene-HCl strip dispersion hybrid liquid membrane. Yingyong Huaxue 2006; 23(7): 766-769 (In Chinese).
124. Luo X-J, He D-S, Ma M, Wan Y. Study on transport and separation of copper(II) by N530-OT-kerosene-HCl strip dispersion hybrid liquid membrane system. Wuji Huaxue Xuebao 2005; 21: 588-592 (In Chinese).
125. Wodzki R, Swiatkowski M, Pretula J, Kaluzynski K. Poly[poly(oxypropylene) phosphate] macroionophores for transport and separation of cations in a hybrid: Cation-exchange polymer and liquid membrane system. J Appl Polym Sci 2004; 93: 1436-1445.
126. Wodzki R, Swiatkowski M, Lapienis G. Transport and separation properties of poly (oxypropylene) bisphosphates as macroionophores of alkali, alkaline-earth and transient metal cations in a hybrid liquid membrane system. React Funct Polym 2005; 62: 195-208.
127. Wodzki R, Szczepanski P. Simultaneous recovery and separation of Zn^{2+} and Cu^{2+} in hybrid membrane systems. Sep Purif Technol 2005; 41: 289-297.
128. He D, Liu X, Ma M. Transfer of Cd(II) chloride species by a tri-*n*-octylamine-secondary octyl alcohol-kerosene multimembrane hybrid system. Solv Extr Ion Exch 2004; 22: 491-510.
129. Wodzki R, Swiatkowski M, Lapienis G. Mobile macromolecular carriers of ionic substances, 2 transport rates and separation of some divalent cations by poly(oxyethylene) phosphates. Macromol Chem Phys 2001; 202: 145-154.
130. Wodzki R, Klosinski P. Magnesium and calcium ions transport by synthetic analogs of teichoic acids with 1,2- and 1,3-linked phosphodiester units. Makromol Chem 1990; 191: 921-931.
131. Narebska A, Wodzki R, Wyszynska A. Liquid membranes with organic polyphosphates as ionic carriers. Makromol Chem 1989; 190: 1501-1506.
132. Wodzki R, Szczepanski P, Pawlowski M. Recovery of metals from electroplating waste solutions and sludge. Comparison of Donnan dialysis and pertraction technique. Pol J Environ Stud 1999; 8: 111-124.

133. Wodzki R, Szczepanski P. Treatment of electroplating rinse solution by continuous membrane extraction and diffusion dialysis. Pol J Environ Stud 2001; 10: 101-111.
134. Wodzki R, Kaluzynski K, Klosinski P. Mobile macromolecular carriers of ionic substances. Part 1. Transport of heavy metal ions by polymer analogs of teichoic acids. Macromol Chem Phys 1997; 198: 1809-1821.
135. Shinbo T, Kurihara K, Kobatake Y, Kamo N. Active transport of picrate anion through organic liquid membrane. Nature 1977; 270: 277-278.
136. Yang Z-F, Guha AK, Sirkar KK. Simultaneous and synergistic extraction of cationic and anionic heavy metallic species by a mixed solvent extraction system and a novel contained liquid membrane device. Ind Eng Chem Res 1996; 35: 4214-4220.
137. Guha AK, Yun CH, Basu R, Sirkar KK. Heavy metal removal and recovery by contained liquid membrane permeator. AIChE J 1994; 40: 1223-1237.
138. Kertesz R, Schlosser S, Teixidor F. Pertraction of silver with octylphenylsulfide as carrier in a hollow fiber contactor. Desalination 2002; 148: 263-265.
139. Patil CB, Mohapatra PK, Singh RR, Gurba PB, Janardan P, Changrani RD, Manchanda VK. Transport of uranium from nitric acid solution by non-dispersive solvent extraction using a hollow fiber contactor. Radiochim Acta 2006; 94: 331-334.
140. Escobar A, Schimmel KA, de Gyves J, Rodriguez de san Miguel E. Hollow-fiber dispersion-free extraction and stripping of Pb(II) in the presence of Cd(II) using D2EHPA under recirculating operation mode. J Chem Technol Biotechnol 2004; 79: 961-973.
141. Marchese J, Campderrós M. Mass transfer of cadmium ions in a hollow-fiber module by pertraction. Desalination 2004; 164: 141-149.
142. Ortiz I, Bringas E, San Roman MF, Urtiaga AM. Selective separation of zinc and iron from spent pickling solutions by membrane-based solvent extraction: Process viability. Sep Sci Technol 2004; 39: 2441-2455.
143. Vajda M, Havalda I, Macek R. Membrane-based solvent extraction and stripping of zinc in a hollow-fiber contactor operating in a circulating mode. Desalination 2004; 163: 19-25.
144. Pabby AK, Haddad R, Alguacil FJ, Sastre AM. Improved kinetics-based gold cyanide extraction with mixture of LIX79-TOPO utilizing hollow fiber membrane contactors. Chem Eng J 2004; 100: 11-22.
145. Pabby AK, Melgosa A, Haddad R, Sastre AM. Hollow fiber membrane-based non-dispersive extraction of silver(I) from alkaline cyanide media using LIX 79. In: *Proceedings of the ISEC'2002*, Cape Town, South Africa, 2002: 699-705.
146. Marchese J, Campderrós M. Mass transfer of cadmium ions in a hollow-fiber module by pertraction. Desalination 2004; 164: 141-149.
147. El Bachiri A, Hagège A, Burgard M. Recovery of silver nitrate by transport across a liquid membrane containing dicyclohexano-18-crown-6 as a carrier. J Membr Sci 1996; 121: 159-168.
148. Kocherginsky NM, Zhang YK, Stucki JW. D2EHPA based strontium removal from strongly alkaline nuclear waste. Desalination 2002; 144: 267-272.
149. Yamini Y, Chaloosi M, Ebrahimzadeh H. Highly selective and efficient transport of bismuth in bulk liquid membranes containing Cyanex 301. Sep Purif Technol 2002; 28: 43-51.
150. Shamsipur M, Mashhadizadeh MH. Highly efficient and selective membrane transport of silver(I) using hexathia-18-crown-6 as a specific ion carrier. Sep Purif Technol 2000; 20: 147-153.
151. Bartsch RA, Jeon E-G, Walkowiak W, Apostoluk W. Effect of solvent in competitive alkali metal cation transport across bulk liquid membrane by a lipophilic lariat ether carboxylic acid carrier. J Membr Sci 1999; 159: 123-131.

152. Nakatsuji Y, Inoue M, Matsumoto M, Masuyama A, Kida T. Selective transport of lithium ion using noncyclic proton-ionizable ionophores through bulk and emulsion liquid membranes. J Membr Sci 1995; 104: 165-172.
153. Gong S-L, Zhong Z-L, Chen Y-Y. Calixcrown oligomers as cation carriers in a bulk liquid membrane. React Funct Polym 2002; 51: 111-116.
154. Wakita R, Matsumoto M, Nakatsuji Y, Okahara M. Selective transport of lithium ions by an acyclic carboxylic ionophore with a quinaldate moiety through a bulk liquid membrane. J Membr Sci 1991; 57: 297-306.
155. Leon G, Guzman MA. Kinetic study of the effect of carrier and stripping agent concentrations on the facilitated transport of cobalt through bulk liquid membranes. Desalination 2005; 184: 79-87.
156. León G, Guzmán MA. Kinetic study of the effect of carrier and stripping agent concentrations on the facilitated transport of cobalt through bulk liquid membranes. Desalination 2005; 184: 79-87.
157. Vajda M, Schlosser S, Kovacova K. Pertraction of silver through bulk liquid membranes. Chem Pap (Slovakia, England) 2000; 54: 423-429.
158. Berhe HG, Bourne SA, Bredenkamp MW, Esterhuysen C, Habtu MM, Koch KR, Luckay RC. High and selective Ag(I) bulk liquid membrane transport with *N,N*-diethyl-*N'*-camphanyl thiourea and structure of the complex. Inorg Chem Commun 2006; 9: 99-102.
159. Vajda M, Schlosser S, Kovacova K. Pertraction of silver through bulk liquid membranes. Chem Pap 2000; 54: 423-429.
160. Granado-Castro MD, Galindo-Riaño MD, García-Vargas M. Model experiments to test the use of a liquid membrane for separation and preconcentration of copper from natural water. Anal Chim Acta 2004; 506: 81-86.
161. Granado-Castro MD, Galindo-Riaño MD, Garcìa-Vargas M. Separation and preconcentration of cadmium ions in natural water using a liquid membrane system with 2-acetylpyridine benzoylhydrazone as carrier by flame atomic absorption spectrometry. Spectrochim Acta B 2004; 59: 577-583.
162. León G, Guzmán MA. Facilitated transport of cobalt through bulk liquid membranes containing diethylhexyl phosphoric acid. Desalination 2004; 162: 211-215.
163. Kontrec J, Kralj D, Brechevich L. Cadmium removal from calcium sulphate suspension by liquid membrane extraction during recrystallization of calcium sulphate anhydrite. Colloids Surf A 2003; 223: 239-249.
164. Al-Marzouqi MH, Hogendoorn KJA, Versteeg GF. Analytical solution for facilitated transport across a membrane. Chem Eng Sci 2002; 57(22-23): 4817-4829.
165. Gong S-L, Zhong Z-L, Chen Y-Y. Calixcrown oligomers as cation carriers in a bulk liquid membrane. React Funct Polym 2002; 51: 111-116.
166. Safavi A, Peiravian F, Shams E. A selective uphill transport of copper through bulk liquid membrane using Janus Green as an anion carrier. Sep Purif Technol 2002; 26: 221-226.
167. Gholivand MB, Khorsandipoor S. Selective and efficient uphill transport of Cu(II) through bulk liquid membrane using *N*-ethyl-2-aminocyclopentene-1-dithiocarboxylic acid as carrier. J Membr Sci 2000; 180: 115-120.
168. He D, Ma M, Zhao Z. Transport of cadmium ions through a liquid membrane containing amine extractants as carriers. J Membr Sci 2000; 169: 53-59.
169. Szpakowska M, Nagy OB. Influence of physico-chemical control and of membrane composition on the coupled transport of copper(II) ions through bulk liquid membrane. J Membr Sci 2000; 168: 183-186.
170. Brinchi L, Germani R, Mancini MV, Savelli G, Spreti N. Carrier-mediated transport of toxic heavy metal ions in bulk liquid membranes. Eur J Org Chem 2004: 1330-1335.

171. Szpakowska M, Nagy OB, Szymanowski J. On the mechanism of coupled transport of Cu(II) ions through bulk liquid membranes. Sep Sci Technol 2004; 39: 699-707.
172. Szpakowska M. Coupled transport of copper through different types of liquid membrane containing Acorga P-50 as carrier. J Membr Sci 1996; 109: 77-86.
173. Alizadeh N, Salimi S, Jabbari A. Transport study of palladium through a bulk liquid membrane using hexadecylpyridinium as carrier. Sep Purif Technol 2002; 28: 173-180.
174. Jabbari A, Mohammadi J, Shamsipur M. Selective liquid membrane transport of Pd(II) as $PdCl_4^{2-}$ ion using potassium-dicyclohexyl-18-crown-6 as carrier. Microchem J 1999; 63: 218-225.
175. Khayatian G, Shamsipur M. Highly selective and efficient membrane transport of palladium as $PdCl_4^{2-}$ ion using NH_4^+-dibenzyldiaza-18-crown-6 as carrier. Sep Purif Technol 1999; 16: 235-241.
176. Alizadeh N, Salimi S, Jabbari A. Transport study of palladium through a bulk liquid membrane using hexadecylpyridinium as carrier. Sep Purif Technol 2002; 28: 173-180.
177. Safavi A, Shams E. Selective and efficient liquid membrane transport of gold as gold cyanide using an anion carrier. J Membr Sci 1999; 157: 171-176.
178. Shamsipur M, Mashhadizadeh MH, Azimi G. Highly selective and efficient transport of mercury(II) ions across a bulk liquid membrane containing tetrathia-12-crown-4 as a specific ion carrier. Sep Purif Technol 2002; 27: 155-161.
179. Jabbari A, Esmaeili M, Shamsipur M. Selective transport of mercury as $HgCl_4^{2-}$ through a bulk liquid membrane transport using K^+-dicyclohexyl-18-crown-6 as carrier. Sep Purif Technol 2001; 24: 139-145.
180. Alpoguz HK. Bulk liquid membrane transport of Hg(II) by a crown ether derivative. J Macromol Sci A 2006; 43: 1265-1272.
181. Alpoguz HK, Kaya A, Yilmaz A, Yilmaz M. Study on the transport kinetics of Hg^{2+} through calix-oligomer bulk liquid membrane. J Macromol Sci Pure Appl Chem A 2005; 42(5): 577-586.
182. Alpoguz HK, Memon S, Ersoz M, Yilmaz M. Transport kinetics of Hg^{2+} through bulk liquid membrane using calix[4]arene ketone derivative as carrier. Sep Sci Technol 2004; 39: 799-810.
183. Alpaydin S, Yilmaz M, Ersoz M. Kinetic study of Hg(II) transport through a bulk liquid membrane containing ester derivative of bis-calix[4]arene. Sep Sci Technol 2004; 39: 2189-2206.
184. Alpoguz HK, Memon S, Ersoz M, Yilmaz M. Kinetics of mercury(II) transport through a bulk liquid membrane containing calix[4]arene derivatives as carriers. Sep Sci Technol 2003; 38: 1649-1664.
185. Ersoz M. Transport of mercury through liquid membranes containing calixarene carriers. Adv Colloid Interface Sci 2007; 134-135: 96-104.
186. Safavi A, Shams E. Selective and efficient transport of Hg(II) through bulk liquid membrane using methyl red as carrier. J Membr Sci 1998; 144: 37-43.
187. Ma M, He D, Wang Q, Xie Q. Kinetics of europium(III) transport through a liquid membrane containing HEH(EHP) in kerosene. Talanta 2001; 55: 1109-1117.
188. Gaikwad AG, Chitra KR, Surender GD, Damodaran AD. Studies on kinetics of forward and backward extraction of neodymium by using phosphonic acid monoester as an acidic extractant. Chem Eng J Biochem Eng J 1995; 60: 63-73.
189. Mishra D, Sharma U. Extraction and bulk liquid membrane transport of some main group metal ions facilitated by triethylene glycol monomethyl ether. Sep Purif Technol 2002; 27: 51-57.
190. Nanda D, Chauhan HPS, Maiti B. Transport of uranyl ion (UO_2^{2+}) across bulk liquid membrane by thenoyl trifluoro acetone (TTA). Indian J Chem Technol 2004; 11: 643-647.

191. Nanda D, Oak MS, Maiti B, Chauhan HPS, Dutta PK. Selective and uphill transport of uranyl ion in the presence of some base metals and thorium across bulk liquid membrane by di(2-ethylhexyl) phosphoric acid. Sep Sci Technol 2002; 37: 3357-3367.
192. Mohapatra PK, Lakshmi DS, Mohan D, Manchanda VK. Uranium pertraction across a PTFE flatsheet membrane containing Aliquat 336 as the carrier. Sep Purif Technol 51; 2006: 24-30.
193. Gawronski R, Religa P. Transport mechanism of chromium(III) through the unmixed bulk liquid membrane containing dinonylnaphthalenesulfonic acid as a carrier. J Membr Sci 2007; 289: 187-190.
194. Saf AO, Alpaydin S, Sirit A. Transport kinetics of chromium(VI) ions through a bulk liquid membrane containing *p*-tert-butyl calix[4]arene 3-morpholino propyl diamide derivative. J Membr Sci 2006; 283: 448-455.
195. Sahmoune A, Mitiche L. Extraction and transport of chromium(VI) through a bulk liquid membrane containing triphenylphosphine. Ann Chim (Rome, Italy) 2004; 94: 929-938 (in English).
196. Yilmaz A, Kaya A, Alpoguz HK, Ersoz M, Yilmaz M. Kinetic analysis of chromium (VI) ions transport through a bulk liquid membrane containing *p*-tert-butylcalix[4] arene dioxaoctylamide derivative. Sep Purif Technol 2008; 59: 1-8.
197. Habtu MM, Bourne SA, Koch KR, Luckay RC. Competitive bulk liquid membrane transport and solvent extraction of some transition and post-transition metal ions using acylthiourea ligands as ionophores. New J Chem (S Afr) 2006; 30: 1155-1162.
198. Svobodova E, Cibulka R, Hampl F, Smidrkal J, Liska F. Metal ion transport through bulk liquid membrane mediated by cationic ligand surfactants. Collection Czechoslovak Chem Commun 2005; 70: 441-465.
199. Szpakowksa M, Nagy OB, Matefy-Tempfli S. Chemical kinetic approach to the mechanism of coupled transport of Cu(II) ions through bulk liquid membranes. J Phys Chem A 2005; 109: 10176.
200. Fakhari AR, Khorrami AR, Shamsipur M. Selective uphill Zn^{2+} transport via a bulk liquid membrane using an azacrown ether carrier. Sep Purif Technol 50; 2006: 77-81.
201. Nanda D, Oak MS, Pravin Kumar M, Maiti B, Dutta PK. Facilitated transport of Th (IV) across bulk liquid membrane by di(2-ethylhexyl)phosphoric acid. Sep Sci Technol 2001; 36(11): 2489-2497.
202. Yaftian MR, Zamani AA, Rostamnia S. Thorium (IV) ion-selective transport through a bulk liquid membrane containing 2-thenoyltrifluoroacetone as extractant-carrier. Sep Purif Technol 2006; 49: 71-75.
203. Kazemi SY, Shamsipur M. Selective transport of lead(II) through a bulk liquid membrane using a cooperative carrier composed of benzylaza-12-crown-4 and oleic acid. Bull Korean Chem Soc 2005; 26: 930-934.
204. Zolgharnein J, Hosseini S, Sangi MR, Dadfarnia S, Shamsipur M. Dibenzopyridino-18-crown-6 as a highly selective and effective carrier for uphill transport of Pb^{2+} through a bulk liquid membrane. Chem Anal (Warsaw, Poland) 2003; 48: 65-75.
205. Farhadi K, Shamsipur M. Separation study of cadmium as CdI4(2−) through a bulk liquid membrane containing ketoconazole and oleic acid. Anal Sci 2005; 21: 501-505 (in Japanese).
206. Ramkumar J, Maiti B. Transport of molybdenum across a bulk liquid membrane using 8-hydroxy quinoline as a carrier. Sep Sci Technol 2004; 39: 449-457.
207. Wirgau JI, Crumbliss AL. Carrier-facilitated bulk liquid membrane transport of iron (III) hydroxamate complexes utilizing a labile recognition agent and amine recognition in the second coordination sphere. Dalton Trans 2003; 19: 3680-3685.
208. Wirgau JI, Crumbliss AL. Carrier-facilitated bulk liquid membrane transport of Iron (III)-siderophore complexes utilizing first coordination sphere recognition. Inorg Chem 2003; 42: 5762-5770.

209. Spreti N, Brinchi L, Germani R, Mancini MV, Savelli G. A new carrier for selective removal of heavy metal ions from aqueous solutions through bulk liquid membranes. Eur J Org Chem 2004; 18: 3865-3871.
210. Ebrahimzadeh H, Yamini Y. Highly efficient transport of indium in bulk liquid membranes containing D2EHPA as carrier. Indian J Chem A 2004; 43: 1198-1201.
211. Lee SC, Lamb JD, Cai M, Davis JT. Cs^+ and Ba^{2+} selective transport by a novel self-assembled isoguanosine ionophore through polymer inclusion and bulk liquid membranes. J Inclusion Phenom Macrocyclic Chem 2001; 40: 51-57.
212. Mashahadizadeh MH, Mohyaddini R, Shamsipur M. Selective and efficient liquid membrane transport of Au(III) by tetrathia-12-crown-4 as a specific carrier. Sep Purif Technol 39; 2004: 161-166.
213. Kislik V, Eyal A. Application of hydrophobic (water-immiscible) and hydrophilic (water-soluble) liquid membranes in the carboxylic acids separation technologies: Comparative analysis. In: *Proceedings of the Abstracts of 41st Microsymposium Polymer Membranes*, Prague, 2001: 86-87.
214. Solichien MS, O'Brien D, Hammond EG, Glatz CE. Membrane-based extractive fermentation to produce propionic and acetic acids: Toxicity and mass transfer considerations. Enzyme Microbial Technol 1995; 17: 23-31.
215. Wodzki R, Nowaczyk J. Membrane transport of organics. III. Permeation of some carboxylic acids through bipolar polymer membrane. J Appl Polym Sci 2001; 80: 2705-2717.
216. Wodzki R, Nowaczyk J. Membrane transport of organics. II. Permeation of some carboxylic acids through strongly basic polymer membrane. J Appl Polym Sci 1999; 71: 2179-2190.
217. Wodzki R, Nowaczyk J. Extraction and separation of propionic and acetic acid by permeation in a hybrid membrane system composed of liquid and ion-exchange polymer membranes. Solv Extr Ion Exch 1997; 15: 1085-1106.
218. Strzelbicki J, Schlosser S. Influence of surface-active substances on pertraction of cobalt (II) cations through bulk and emulsion liquid membranes. Hydrometallurgy 1989; 23: 67-75.
219. Basu R, Sirkar KK. Citric acid extraction with microporous hollow fibers. Solv Extr Ion Exch 1992; 10(1): 119-143.
220. Basu R, Sirkar KK. Hollow fiber contained liquid membrane separation of citric acid. AIChE J 1991; 37: 383-93.
221. Cichy W, Schlosser S, Szymanowski J. Recovery of phenol with CYANEX 923 in membrane extraction-stripping systems. Solv Extr Ion Exch 2001; 19: 905-923.
222. Zigova J, Sturdik E, Vandak D, Schlosser S. Butyric acid production by Clostridium butyricum with integrated extraction and pertraction. Process Biochem (Oxford) 1999; 34: 835-843.
223. Kubisova L, Sabolova E, Schlosser S, Martak J, Kertesz R. Mass-transfer in membrane based solvent extraction and stripping of 5-methyl-2-pyrazinecarboxylic acid and co-transport of sulphuric acid in HF contactors. Desalination 2004; 163: 27-38.
224. Kertesz R, Schlosser S. Design and simulation of two phase hollow fiber contactors for simultaneous membrane based solvent extraction and stripping of organic acids and bases. Sep Purif Technol 2005; 41: 275-287.
225. Schlosser S, Kertesz R, Martak J. Recovery and separation of organic acids by membrane-based solvent extraction and pertraction. An overview with a case study on recovery of MPCA. Sep Purif Technol 2005; 41: 237-266.
226. Teramoto M, Matsuyama H, Ohnishi N. Uphill transport of polyunsaturated fatty acid ethyl esters through liquid membranes containing silver nitrate as a carrier. In: Shallcross DC, Paimin R, Prvcic LM, eds. *Proceedings of the ISEC'96: Value Adding through Solvent Extraction*. University of Melbourne, Melbourne, Australia, 1996; 2: 917-922.

227. Teramoto M, Matsuyama H, Nakai K. Separation of ethyl esters of eicosapentaenoic acid and docosahexaenoic acid by circulating liquid membranes using silver nitrate as a carrier. Uphill transport by use of temperature and solvent dependencies of distribution ratio. Sep Sci Technol 1996; 31: 1953-1969.
228. Ma M, Chen B, Luo X, Tan H, He D, Xie Q, Yao S. Study on the transport selectivity and kinetics of amino acids through di(2-ethylhexyl) phosphoric acid-kerosene bulk liquid membrane. J Membr Sci 2004; 234: 101-109.
229. Ma M, He D, Liao S, Zeng Y, Xie Q, Yao S. Kinetic study of L-isoleucine transport through a liquid membrane containing di(2-ethylhexyl) phosphoric acid in kerosene. Anal Chim Acta 2002; 456: 157-165.
230. Uglea CV, Croitoru M. Transport of amino acids through liquid membranes. III. The alkaline ion role. J Membr Sci 1997; 133: 127-131.
231. Sahoo GC, Ghosh AC, Dutta NN, Mathur RK. Facilitated transport of 7-aminocephalosporanic acid in a bulk liquid membrane. J Membr Sci 1996; 112: 147-154.
232. Ersoz M, Vura US, Okdan A, Pehlivan E, Yildiz S. Transport studies of amino acids through a liquid membrane system containing carboxylated poly (styrene) carrier. J Membr Sci 1995; 104: 263-269.
233. Uglea CV, Zanoag CV. Transport of amino acids through organic liquid membranes. II. Aspects of the mechanism. J Membr Sci 1992; 65: 47-50.
234. Kukufuta E, Nobusawa M. Uphill transport of phosphate ion through an oxomolybdenum(v)-tetraphenylporphyrin-complex-containing bulk liquid membrane: Effect of halogen ions on phosphate ion flux. J Membr Sci 1990; 48: 141-154.
235. Scrimin P, Tonellato U, Zanta N. Cu(II) mediated selective transport of α-amino acids across a bulk liquid membrane using a chiral lipophilic ligand as a carrier. Tetrahedron Lett 1988; 29(39): 4967-4970.
236. Bucci R, Canepari S, Cardarelli E, Girelli AM, Pietrodangelo A, Valiente M. Carrier-mediated transport of amino acids through bulk liquid membranes. Sep Sci Technol 2004; 39: 3821-3838.
237. Kataev VE, Strobykina IY, Militsina OI, Korochkina MG, Fedorova OV, Ovchinnikova IG, Valova MS, Rusinov GL. Isosteviol and some of its derivatives as receptors and carriers of amino acid picrates Tetrahedron Lett 47; 2006: 2137-2139.
238. Joshi P, Joshi N, Sharma U. Extraction and carrier-facilitated transport of amino acids using synthetic non-cyclic receptors through bulk liquid membrane. Indian J Biochem Biophys 2006; 43: 323-326.
239. Ma M, Chen B, Luo X, Tan H, He D, Xie Q, Yao S. Study on the transport selectivity and kinetics of amino acids through di(2-ethylhexyl) phosphoric acid-kerosene bulk liquid membrane. J Membr Sci 2004; 234: 101-109.
240. Cichy W, Schlosser Š, Szymanowski J. Extraction and pertraction of phenol through bulk liquid membranes. J Chem Technol Biotechnol 2005; 80: 189-197.
241. Eksangsri T, Habaki H, Kawasaki J. Membrane-based extraction of cholesterol lowering drug: Effect of membrane type on extraction. Sep Purif Technol 2005; 46: 63-71.
242. Ma M, Shen A-H, Luo X-B, Zhu X-L, Chen B, Yao S-Z. Solvent extraction and liquid-membrane transport for separation and purification of Rutaecarpine from *E. rutaecarpa* var. *officinalis*. Solv Extr Ion Exch 2004; 22: 511-525.
243. Boyadzhiev L, Kirilova N. Extraction of tylosin from its aqueous solutions by rotating film-pertraction. Bioprocess Eng 2000; 22: 373-378.
244. Boyadzhiev L, Atanassova I. Extraction of phenylalanine from dilute solutions by rotating film pertraction. Process Biochem 1994; 29: 237-243.
245. Sirkar KK. Application of membrane technologies in the pharmaceutical industry. Curr Opin Drug Discover Dev 2000; 3: 714-722.
246. Basu R, Sirkar KK. Pharmaceutical product recovery using a hollow fiber contained liquid membrane: A case study. J Membr Sci 1992; 75: 131-149.

247. Dai X-P, Luo RG, Sirkar KK. A membrane-based integrated bioseparation process. Adv Filtration Sep Technol A 1999; 13: 545-549.
248. Prasad R, Sirkar KK. Hollow fiber solvent extraction: Performances and design. J Membr Sci 1990; 50: 153-175.
249. Prasad R, Sirkar KK. Hollow fiber solvent extraction of pharmaceutical products: A case study. J Membr Sci 1989; 47: 235-259.
250. Lopez JL, Matson SL. A multiphase/extractive enzyme membrane reactor for production of diltiazem chiral intermediate. J Membr Sci 1997; 125: 189-211.
251. Di Luccio M, Smith BD, Kida T, Alves TLM, Borges CP. Evaluation of flat sheet and hollow fiber supported liquid membranes for fructose pertraction from a mixture of sugars. Desalination 2002; 148: 213-220.
252. Yang ZF, Rindfeisch D, Scheper T, Schugerl K. Separation of penicillin G with hollow fiber contained liquid membrane system. In: Paterson R, ed. *Effective Membrane Processes—New Perspectives. MEP*, 1993: 49-54.
253. Rindfleisch D, Syska B, Lazarova Z, Schugerl K. Integrated membrane extraction, enzymic conversion and electrodialysis for the synthesis of ampicillin from penicillin G. Process Biochem 1997; 32: 605-616.
254. Miesiac I, Szymanowski J. Pertraction of penicillin G in hollow fiber contained liquid membranes. J Radioanal Nucl Chem 1998; 228: 77-81.
255. Sirkar KK, Luo RG, Xu Y, Dai X-P. Method and apparatus for isolation and purification of biomolecules, 2006; US Patent 2004069710 A1 20040415.
256. Kertesz R, Simo M, Schlosser S. Membrane-based solvent extraction and stripping of phenylalanine in HF contactors. J Membr Sci 2005; 257: 37-47.
257. Huang K-L, Hu W-G, Jiao F-P, Su J. Resolution of racemic clenbuterol by bulk liquid membrane and study on its kinetics. Fenxi Huaxue 2007; 35: 479-483 (in Chinese).
258. Nishii Y, Kinugasa T, Nii S, Takahashi K. Transport behavior of protein in bulk liquid membrane using reversed micelles. J Membr Sci 2002; 195: 11-21.
259. Dimitrov K, Metcheva D, Alexandrova S, Saboni A, Boyadzhiev L. Recovery of tropane alkaloids from Deadly Nightshade. I. A hybrid process of solid-liquid extraction and liquid membrane purification. Bulgarian Chem Commun 2006; 38: 231-236.
260. Boyadzhiev L, Yordanov B. Pertraction of indole alkaloids from *Vinca minor* L. Sep Sci Technol 2004; 39: 1321-1329.
261. Coucouvanis D, Rosa D, Pike J. Recognition and transport of amphiphilic molecules by a new class of inorganic ditopic receptors. The synthesis of M-$^{t}Bu_4$-salphen-3*n*-cr-*n* complexes and their use (M = Mn,Fe, *n* = 6) in the transport of tryptophan and serotonin across bulk liquid membranes. Comp Rend Chim 2003; 6: 317-327.
262. Kinugasa T, Tanahashi S, Takahashi S, Takeuchi H. Transportation of protein through reverse micellar solution layer as a liquid membrane. Proc ISEC-90 1990; B: 1839-1846.
263. Papadopoulos T, Sirkar KK. Separation of a 2-propanol/*n*-heptane mixture by liquid membrane perstraction. Ind Eng Chem Res 1993; 32: 663-673.
264. Pietraszkiewicz O, Kozbial M, Pietraszkiewicz M. Transport studies of inorganic and organic cations across liquid membranes containing Mannich-base calix[4]resorcinarenes. In: *Proceedings of the 9th International Symposium on Molecular Recognition and Inclusion*, France, 1998: 459-462.
265. Bessarabov DG, Theron JP, Sanderson RD. Novel application of membrane contactors: Solubility measurements of 1-hexene in solvents containing silver ions for liquid olefin/paraffin separations. Desalination 1998; 115(3): 279-284.
266. Krieg HM, Lotter J, Keizer K, Breytenbach JC. Enrichment of chlorthalidone enantiomers by an aqueous bulk liquid membrane containing β-cyclodextrin. J Membr Sci 2000; 167: 33-45.
267. Sahoo GC, Dutta NN, Dass NN. Liquid membrane extraction of cephalosporin-C from fermentation broth. J Membr Sci 1999; 157: 251-261.

268. Sahoo GC, Ghosh AC, Dutta NN. Recovery of cephalexin from dilute solution in a bulk liquid membrane. Process Biochem 1997; 32: 265-272.
269. Bocquet S, Gascons Viladomat F, Muvdi Nova C, Sanchez J, Athes V, Souchon I. Membrane-based solvent extraction of aroma compounds: Choice of configurations of hollow fiber modules based on experiments and simulation. J Membr Sci 2006; 281: 358-368.
270. Viladomat FG, Souchon I, Pierre F-X, Marin M. Liquid-liquid and liquid-gas extraction of aroma compounds with hollow fibers. AIChE J 2006; 52: 2079-2088.
271. Lazarova Z, Boyadzhiev L. Liquid film pertraction—A liquid membrane method for recovery and preconcentration of aromatic amines from their dilute aqueous solutions. Int J Environ Anal Chem 1992; 46: 233-244.
272. Teramoto M, Matsuyama H, Yamashiro T, Okamoto S. Separation of ethylene from ethane by a flowing liquid membrane using silver nitrate as a carrier. J Membr Sci 1989; 45: 115-136.
273. Muthuraman G, Palanivelu K. Transport of textile anionic dyes using cationic carrier by bulk liquid membrane. J Sci Ind Res (India) 2005; 64: 529-533.
274. Jeon SJ, Hong WH. Separation of ethyl (*R*) 4-chloro-3-hydroxybutyrate from racemic mixture by using bulk liquid membrane. In: *Proceedings of the 4th International Conference on Separation Science and Technology*, Nanning, China, February 18-21, 2004: 561-564.
275. Kida T, Ohe T, Higashimoto H, Harada H, Nakatsuji Y, Ikeda I, Akashi M. Amphiphilic cyclodextrins as novel monosaccharide transport carriers through a bulk liquid membrane. Chem Lett (Japan) 2004; 33: 258-259 (in English).
276. Chakraborty M, Bart H-J. Highly selective and efficient transport of toluene in bulk ionic liquid membranes containing Ag^+ as carrier. Fuel Proc Technol 88; 2007: 43-49.
277. Teramoto M, Ito N, Matsumoto H. Selective separation and concentration of carbon dioxide by flowing liquid membranes. Kagaku Kogaku Ronbunshu 1990; 16: 1234-1240.
278. Teramoto M, Ohnishi N, Kitada S, Matsuyama H, Matsumiya N, Mano H. Simultaneous recovery and enrichment of CO_2 by novel capillary membrane module with permeation of carrier solution. In: *Abstracts of Papers, 225th ACS National Meeting*, New Orleans, LA. American Chemical Society, Washington, DC, 2003: IEC-009.
279. Yeon S-H, Lee K-S, Sea B, Park Y-I, Lee K-H. Application of pilot-scale membrane contactor hybrid system for removal of carbon dioxide from flue gas. J Membr Sci 2005; 257: 156-160.
280. Lee K-H, Yeon S-H, Sea B, Park Y-I. Hollow fiber membrane contactor hybrid system for CO_2 recovery. Stud Surf Sci Catal 2004; 153(Carbon Dioxide Utilization for Global Sustainability): 423-428. Elsevier BV.
281. Basu A, Akhtar J, Rahman MH, Islam MR. A review of separation of gases using membrane systems. Petrol Sci Technol 2004; 22: 1343-1368.
282. Sirkar KK. System and method for selective separation of gaseous mixtures using hollow fibers, 2008; US Patent WO 2006049610 A1 20060511.
283. Sirkar KK. System and method for selective separation of gaseous mixtures using hollow fibers, 2008; US Patent 2006090644 A1 20060504.
284. Korikov AP, Sirkar KK, Otto H. Membrane gas permeance in gas liquid membrane contactor systems for solutions containing a highly reactive absorbent. J Membr Sci 2005; 246: 27-37.
285. Kosaraju P, Kovvali AS, Korikov A, Sirkar KK, Otto H. Hollow fiber membrane contactor based CO_2 absorption-stripping using novel solvents and membranes. Ind Eng Chem Res 2005; 44(5): 1250-1258.
286. Bao L, Goldman SL, Trachtenberg MC. CO_2 transfer across a liquid membrane facilitated by carbonic anhydrase. In: *AIChE Annual Meeting, Conference Proceedings*, 2004: 241F/1-241F/10.

287. Shalygin MG, Vorobieva EV, Teplyakov VV. Gas transport in combined membrane system with moving liquid carrier. Sep Purif Technol 2006; 57: 466-472.
288. Matsumiya N, Teramoto M, Kitada S, Matsuyam H. Evaluation of energy consumption for separation of CO_2 in flue gas by hollow fiber facilitated transport membrane module with permeation of amine solution. Sep Purif Technol 2005; 46: 26-32.
289. Poddar TK, Majumdar S, Sirkar KK. Removal of VOCs from air by membrane-based absorption and stripping. J Membr Sci 1996; 120: 221-237.
290. Yang D, Majumdar S, Kovenklioglu S, Sirkar KK. Hollow fiber contained liquid membrane pervaporation system for the removal of toxic volatile organics from wastewater. J Membr Sci 1995; 103: 195-210.
291. Yun CH, Guha AK, Prasad R, Sirkar KK. Novel microporous membrane-based separation processes for pollution control and waste minimization. In: Sawyer DT, Martell AE, eds. Industrial Environmental Chemistry. Proceedings of the Texas A&M University IUCCP 10th Annual Symposium. Plenum Publishers, New York, NY, 1992: 135-146.
292. Sirkar KK, Yang D, Majumdar S, Kovenklioglu S, Sengupta A. Hollow fiber-containing liquid membrane pervaporation for removal of volatile organic compounds from aqueous solution. US Patent 5637224 Kind A Date 19970610.
293. Majumdar S, Bhaumik D, Sirkar KK, Simes G. A pilot-scale demonstration of a membrane-based absorption-stripping process for removal and recovery of volatile organic compounds. Environ Prog 2001; 20: 27-35.
294. Sirkar KK. Gas separation using hollow fiber contained liquid membrane. US Patent 6156096 Kind A Date 20001205.
295. Kreulen H, Versteeg GF, Smolders CA, Van Swaaij WPM. Selective removal of H_2S from sour gas with microporous membranes. Part I. Application in a gas-liquid system. J Membr Sci 1992; 73: 293-304.
296. Kreulen H, Smolders CA, Versteeg GF, Swaaij WPM. Selective removal of H_2S from sour gases with microporous membranes. Part II. A liquid membrane of water-free tertiary amines. J Membr Sci 1993; 82: 185-197.
297. Meldon JH, Dutta A. Analysis of ultimate permselectivity for H_2S over CO_2 in alkaline solutions. Chem Eng Sci 1994; 49: 689-697.
298. Bhaumik D, Majumdar S, Sirkar KK. Absorption of CO_2 in a transverse flow hollow fiber membrane module having a few wraps of the fiber mat. J Membr Sci 1998; 138: 77-82.
299. Figoli A, Sager WFC, Mulder MHV. Facilitated oxygen transport in liquid membranes: Review and new concepts. J Membr Sci 2001; 181: 97-110.
300. Shanbhag PV, Guha AK, Sirkar KK. Membrane-based integrated absorption-oxidation reactor for destroying VOCs in air. Environ Sci Technol 1996; 30: 3435-3440.
301. Papadopoulos T, Sirkar KK. A modified hollow fiber contained liquid membrane technique for gas separation at high pressures. J Membr Sci 1994; 94: 163-181.
302. Majumdar S, Sengupta A, Cha JS, Sirkar KK. Simultaneous SO_2/NO separation from flue gas in a contained liquid membrane permeator. Ind Eng Chem Res 1994; 33: 667-675.
303. Guha AK, Majumdar S, Sirkar KK. Gas separation modes in a hollow fiber contained liquid membrane permeator. Ind Eng Chem Res 1992; 31: 593-604.
304. Guha AK, Majumdar S, Sirkar KK. A larger-scale study of gas separation by hollow-fiber-contained liquid membrane permeator. J Membr Sci 1991; 62: 293-307.
305. Saha S, Chakma A. Selective CO_2 separation from CO_2/C_2H_6 mixtures by immobilized diethanolamine/PEG membranes. J Membr Sci 1995; 98: 157-171.
306. Kumar PS, Hogendoorn JA, Feron PHM, Versteeg GF. New absorption liquids for the removal of CO_2 from dilute gas streams using membrane contactors. Chem Eng Sci 2002; 57: 1639-1651.

307. Lee Y, Noble RD, Yeom B-Y, Park Y-I, Lee K-H. Analysis of CO_2 removal by hollow fiber membrane contactors. J Membr Sci 2001; 194: 57-67.
308. Rangwala HA. Absorption of carbon dioxide into aqueous solutions using hollow fiber membrane contactors. J Membr Sci 1996; 112: 229-240.
309. Quinn R, Appleby JB, Pez GP. New facilitated transport membranes for the separation of carbon dioxide from hydrogen and methane. J Membr Sci 1995; 104: 139-146.
310. Barbosa VL, Tandlich R, Burgess JE. Bioremediation of trace organic compounds found in precious metals refineries' wastewaters: A review of potential options. Chemosphere 68; 2007: 1195-1203.
311. Yoon TI, Lee HS, Kim CG. Comparison of pilot scale performances between membrane bioreactor and hybrid conventional wastewater treatment systems. J Membr Sci 2004; 242: 5-12.
312. Lesjean B, Rosenberger S, Schrotter J-C, Recherche A. Membrane-aided biological wastewater treatment—An overview of applied systems. Membr Technol 2004; 8: 5-10.
313. Fernandes P, Prazeres DMF, Cabral JMS. Membrane-assisted extractive bioconversions. Adv Biochem Eng Biotechnol 2003; 80: 115-148.
314. Gastrock G, Lemke K, Metze J. Sampling and monitoring in bioprocessing using microtechniques. Rev Mol Biotechnol 82; 2001: 123-135.
315. Visvanathan C, Ben Aim R, Parameshwaran K. Membrane separation bioreactors for wastewater treatment. Crit Rev Environ Sci Technol 30; 2000: 1-48.
316. Reij MW, Keurentjes JTF, Hartmans S. Membrane bioreactors for waste gas treatment. J Biotechnol 59; 1998: 155-116.
317. Vaidya AM, Bell G, Halling PJ. Aqueous-organic membrane bioreactors. Part II. Breakthrough pressure measurement. J Membr Sci 97; 1994: 13-26.
318. Vaidya AM, Bell G, Halling PJ. Aqueous-organic membrane bioreactors. Part I. A guide to membrane selection. J Membr Sci 71; 1992: 139-149.
319. Drioli E, Iorio G, Catapano G. Enzyme membrane reactors and membrane fermentors. In: Porter MC, ed. *Handbook of Industrial Membrane Technology*. Noyes Publications, Park Ridge, NJ, 1990: 401-481.
320. Iorio G. Membrane in biotechnology: State of the art and prospects. In: Mika AM, Winnicki TZ, eds. *Advances in Membrane Phenomena and Processes*. Wroclaw Technical University Press, Wroclaw, Poland, 1989: 231-248.
321. Pagnanelli F, Beolchini F, Di Biase A, Vegliò F. Effect of equilibrium models in the simulation of heavy metal biosorption in single and two-stage UF/MF membrane reactor systems. Biochem Eng J 2003; 15: 27-35.
322. Konovalova VV, Dmytrenko GM, Nigmatullin RR, Bryk MT, Gvozdyak PI. Chromium(VI) reduction in a membrane bioreactor with immobilized *Pseudomonas* cells. Enzyme Microbial Technol 2003; 33: 899-907.
323. Tabak HH, Govind R. Advances in biotreatment of acid mine drainage and biorecovery of metals. 2. Membrane bioreactor system for sulfate reduction. Biodegradation 2003; 14: 437-452.
324. Teplyakov VV, Gassanova LG, Sostina EG, Slepova EV, Modigell M, Netrusov AI. Lab-scale bioreactor integrated with active membrane system for hydrogen production: Experience and prospects. Int J Hydrogen Energy 2002; 27: 1149-1155.
325. Hsieh Y-L, Tseng S-K, Chang Y-J. Nitrogen removal from wastewater using a double-biofilm reactor with a continuous-flow method. Bioresource Technol 2003; 88: 107-113.
326. Crespo JG, Velizarov S, Reis MA. Membrane bioreactors for the removal of anionic micropollutants from drinking water. Curr Opin Biotechnol 2004; 15: 463-468.
327. Bae T-H, Han S-S, Tak T-M. Membrane sequencing batch reactor system for the treatment of dairy industry wastewater. Process Biochem 2003; 39: 221-231.

328. Howell JA. Future of membranes and membrane reactors in green technologies and for water reuse. Desalination 2004; 162: 1-11.
329. Noronha M, Britz T, Mavrov V, Janke HD, Chmiel H. Treatment of spent process water from a fruit juice company for purposes of reuse: Hybrid process concept and on-site test operation of a pilot plant. Desalination 2002; 143: 183-196.
330. Visvanathan C, Boonthanon N, Sathasivan A, Jegatheesan V. Pretreatment of seawater for biodegradable organic content removal using membrane bioreactor. Desalination 2003; 153: 133-140.
331. Tazi-Pain A, Schrotter JC, Bord G, Payreaudeau M, Buisson H. Recent improvement of the BIOSEP process for industrial and municipal wastewater treatment. Desalination 2002; 146: 439-443.
332. Liu W, Howell JA, Arnot TC, Scott JA. A novel extractive membrane bioreactor for treating biorefractory organic pollutants in the presence of high concentrations of inorganics: Application to a synthetic acidic effluent containing high concentrations of chlorophenol and salt. J Membr Sci 2001; 181: 127-140.
333. Leon R, Prazeres DMF, Fernandes P, Molinari F, Cabral JMS. A multiphasic hollow fiber reactor for the whole-cell bioconversion of 2-methyl-1,3-propanediol to (*R*)-β-hydroxyisobutyric acid. Biotechnol Prog 2001; 17: 468-473.
334. Dai X-P, Yang Z-F, Luo RG, Sirkar KK. Lipase-facilitated separation of organic acids in a hollow fiber contained liquid membrane module. J Membr Sci 2000; 171: 183-196.
335. Craveiro AC, Lima MB. Bioreactor system with chitosan semipermeable membranes and continuous extraction process, 2000; Brazil Patent Priority: CAN 134:177457 AN 2001:164817.
336. Xudong S, Yue S, Huimin Y, Zhongyao S. Bioconversion of acrylnitrile to acrylamide using hollow-fiber membrane bioreactor system. Biochem Eng J 2004; 18: 239-243.
337. Noworyta A, Trusek-Holownia A. Modeling of enzymatic conversion in the catalytic gel layer located on a membrane surface. Desalination 2004; 162: 327-334.
338. Gebhardt W, Schröder HF. Liquid chromatography-(tandem) mass spectrometry for the follow-up of the elimination of persistent pharmaceuticals during wastewater treatment applying biological wastewater treatment and advanced oxidation. J Chromatogr A 2007; 1160: 34-43.
339. Kouakou E, Salmon T, Toye D, Marchot P, Crine M. Gas-liquid mass transfer in a circulating jet-loop nitrifying MBR. Chem Eng Sci 2005; 60: 6346-6353.
340. Jurado E, Camacho F, Luzón G, Vicaria JM. Kinetic model for lactose hidrolysis in a recirculation hollow-fibre bioreactor. Chem Eng Sci 2004; 59: 397-405.
341. Gan Q, Allen SJ, Taylor G. Design and operation of an integrated membrane reactor for enzymatic cellulose hydrolysis. Biochem Eng J 2002; 12: 223-229.
342. Slominska L, Szostek A, Grzeskowiak A. Studies on enzymatic continuous production of cyclodextrins in an ultrafiltration membrane bioreactor. Carbohydr Polym 2002; 50: 423-428.
343. Hernandez FJ, de los Rios AP, Gomez D, Rubio M, Villora G. A new recirculating enzymatic membrane reactor for ester synthesis in ionic liquid/supercritical carbon dioxide biphasic systems. Appl Catal B 2006; 67: 121-126.
344. Zhu D, Zhu Y, Cai H, Gao G, Gao C, Li B. Method for treating leather waste to produce collagen with high added value by using enzyme bioreactor and membrane separator (Shenzhen Xianke Environment Protection Co., Ltd, People's Republic of China). Patent Priority: CAN 145:194650 AN 2006:754551 (in Chinese).
345. Baumgarten S, Buer Th, Ohle P. Use of membrane separation steps in anaerobic processes. Gewaesserschutz Wasser Abwasser 2003; 190: 34/1-34/20.
346. Li H, Mustacchi R, Knowles CJ, Skibar W, Sunderland G, Dalrymple I, Jackman SA. An electrokinetic bioreactor: Using direct electric current for enhanced lactic acid fermentation and product recovery. Tetrahedron 2004; 60: 655-661.

347. Lambropoulou DA, Triantafyllos AA. Liquid-phase micro-extraction techniques in pesticide residue analysis. J Biochem Biophys Methods 2007; 70: 195-228.
348. McGowin AE. Polycyclic aromatic hydrocarbons. In: *Chromatographic Analysis of the Environment* (3rd edition). Chromatographic Science Series, 2006; 93: 555-616.
349. Joensson JA. Membrane extraction techniques in environmental analysis. Ecol Chem Eng 2005; 12: 519-537.
350. Dickert FL, Lieberzeit P, Gazda-Miarecka S, Halikias K, Mann K-J. Modifying polymers by self-organisation for the mass-sensitive detection of environmental and biogeneous analytes. Sens Actuators B 2004; 100: 112-116.
351. Chimuka L, Cukrowska E, Joensson A. Why liquid membrane extraction is an attractive alternative in sample preparation. Pure Appl Chem 2004; 76: 707-722.
352. Staniszewska M, Wolska L, Namiesnik J. Advantages and limitations of methods for determination of volatile and semivolatile organic compounds in water samples. Pol Wiadomosci Chemiczne 2003; 57: 185-222 (in Polish).
353. Jonsson JA. Liquid-membrane extraction in analytical chemistry. Dansk Kemi 2002; 83: 68-80, 72–76 (in Danish).
354. Miller KE, Synovec RE. Review of analytical measurements facilitated by drop formation technology. Talanta 2000; 51: 921-933.
355. Yatsimirskii KB, Talanova GGLV. Liquid membranes with macrocyclic ionophores and prospects for their application in analytical chemistry. Zh Anal Khim 1990; 45: 1686-703 (in Russian).
356. Lambropoulou DA, Albanis TA, Mendiguchía C, Moreno C, García-Vargas M. Separation of heavy metals in seawater by liquid membranes. Preconcentration of copper. Sep Sci Technol 2002; 46(1): 35-40.
357. Mendiguchía C, Moreno C, García-Vargas M. Separation of heavy metals in seawater by liquid membranes. Preconcentration of copper. Sep Sci Technol 2002; 46: 35-40.
358. Mendiguchía C, Moreno C, García-Vargas M. Determination of copper in seawater based on a liquid membrane preconcentration system. Anal Chim Acta 2002; 460: 35-40.
359. Yu H, Jiang G, Xu S. Liquid membrane enrichment in analytical chemistry—Liquid membrane enrichment of trace copper and its trace determination by flame atomic absorption spectrophotometry. Liaoning Shifan Daxue Xuebao, Ziran Kexueban 1994; 17: 134-7 (in Chinese).
360. Djane NK, Malcus F, Martins E, Sawula G, Johansson G. Hollow fiber cartridges for removal of particulate matter from natural waters prior to matrix isolation and trace metal enrichment using an 8-quinolinol chelating ion exchanger in a flow system. Anal Chim Acta 1995; 316(3): 305-311.
361. Jonsson JA. Membrane extraction for sample preparation—A practical guide. Chromatographia 2003; 57(Suppl.): S/317-S/324.
362. Lai B-W, Liu B-M, Malik PK, Wu H-F. Combination of liquid-phase hollow fiber membrane microextraction with gas chromatography-negative chemical ionization mass spectrometry for the determination of dichlorophenol isomers in water and urine. Anal Chim Acta 2006; 576(1): 61-66.
363. Basheer C, Lee HK, Obbard JP. Determination of organochlorine pesticides in seawater using liquid-phase hollow fibre membrane microextraction and gas chromatography-mass spectrometry. J Chromatogr A 2002; 968: 191-199.
364. Jonsson JA, Mathiasson L. Membrane-based techniques for sample enrichment. J Chromatogr A 2000; 902: 205-225.
365. Jonsson JA, Mathiasson L. Membrane extraction techniques in bioanalysis. Chromatographia 2000; 52(Suppl.): S8-S11.
366. Bazylak G, Nagels LJ. Simultaneous high-throughput determination of clenbuterol, ambroxol and bromhexine in pharmaceutical formulations by HPLC with potentiometric detection. J Pharm Biomed Anal 2003; 32: 887-903.

367. Jonsson JA, Andersson M, Melander C, Norberg J, Thordarson E, Mathiasson L. Automated liquid membrane extraction for high-performance liquid chromatography of ropivacaine metabolites in urine. J Chromatogr A 2000; 870: 151-157.
368. Megersa N, Chimuka L, Solomon T, Jonsson JA. Automated liquid membrane extraction and trace enrichment of triazine herbicides and their metabolites in environmental and biological samples. J Sep Sci 2001; 24: 567-576.
369. Norberg J, Zander A, Jonsson JA. Microporous membrane liquid-liquid extraction technique combined with gas chromatography mass spectrometry for the determination of organotin compounds. Anal Chim Acta 2000; 404: 319-328.
370. Bothun GD, Knutson BL, Strobel HJ, Nokes SE. Mass transfer in hollow fiber membrane contactor extraction using compressed solvents. J Membr Sci 2003; 227: 183-196.

CHAPTER 6

Bulk Aqueous Hybrid Liquid Membrane (BAHLM) Processes with Water-Soluble Carriers: Application in Chemical and Biochemical Separations

Vladimir S. Kislik

1. Introduction and Definitions

There are many problems associated with water-immiscible, organic hybrid bulk liquid membrane (BOHLM) systems (see Chapter 5) such as:

(1) Loss of the carrier and organic solvent through leakage and solubility in the feed and strip solutions; contamination of the products
(2) Progressive membrane instability through considerable osmotic pressure; wetting of the pores by surface-active carrier molecules resulting in contamination of organic LM by water; fouling of the membrane interfaces on the feed and strip sides by organic gel, emulsion
(3) Low selectivity and low diffusion rates due to high viscosity of concentrated organic LM, loaded by solute
(4) Pressure differential across the membrane exceeding capillary forces holding the liquid

Therefore, exploration of new LM systems that are stable and lack of the drawbacks, mentioned above, is needed.

The recently developed bulk aqueous hybrid liquid membrane (BAHLM) separation process [1-7] overcomes most of these problems. As can be seen from the scheme in Fig. 6.1, the technological concept of BAHLM transport is quite simple: an aqueous solution of a carrier, E, flows between two membrane barriers, which separate the carrier from the feed, F, and strip, R, aqueous solutions. It can be seen that the BAHLM system is similar to the BOHLM except the liquid membrane

Casali Institute of Applied Chemistry, The Hebrew University of Jerusalem, Campus Givat Ram, Jerusalem 91904, Israel

Liquid Membranes: Principles and Applications in Chemical Separations and Wastewater Treatment
DOI: 10.1016/B978-0-444-53218-3.00006-4

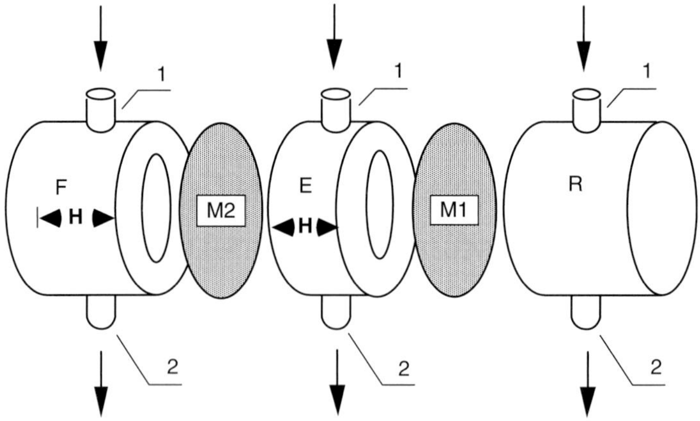

Figure 6.1 Schematic diagram of the three-aqueous phase (BAHLM) module: F, E, and R are the compartments of the feed, LM, and receiving (strip) solutions, respectively; M_1 and M_2 are the ion-exchange membranes; 1 and 2 are the inlet and outlet of the feed, LM, and strip solutions, respectively. Gaskets, made of Vytone, were inserted between compartments and membranes. H is the width of the compartment. From Ref. [5] with permission.

(LM) phase is an aqueous solution also [3-5]. Aqueous solutions of water-soluble polymers [8-10], acting as carriers, are separated from the feed and strip solutions by ion-exchange membranes (IEM). The IEM block the carriers from leakage into the feed and strip solutions. Neutral hydrophilic membranes can also be used in case of the large molecules of polymeric electrolytes, used as carriers. Suitable aqueous liquid membrane carriers are regenerable, water-soluble polyionic complexants. Polyelectrolytes typically have a very high effective concentration of charged groups and could constitute highly selective complexants [10-13].

This three-aqueous phase system is based on a combination of liquid membrane (LM) and Donnan dialysis (DD) processes in the case of ion-exchange membranes or dialysis (D) processes in the case of neutral hydrophilic membranes. On the feed-side interface, the liquid membrane is loaded by the solute; on the strip side, it is offloaded and recharged. Thus, strong acids, or bases, or salts with other cations and anions, used as respective "pump" solutes, provide a continuous process.

Feasibility tests of the BAHLM system [3-5, 7, 14-17] show promising results: relatively high transport rates of solutes at high selectivities, and a long lifetime of the membranes.

In this chapter, the basic principles of the BAHLM processes—such as theoretical considerations in the mass transfer and the transport kinetics, considerations in the process development and module design, and selected applications in chemical and biochemical products separation—will be discussed.

2. Theoretical Considerations

2.1. Background

Donnan dialysis The BAHLM systems with ion-exchange membranes, based on Donnan dialysis [18, 19], will be considered below. Donnan dialysis is a continuously operating ion-exchange process. There are many theoretical models describing transport mechanisms and kinetics of DD [18-26]. All transport kinetics models are based on Fick's or Nernst-Planck's equations for ion fluxes. In both cases, the authors introduce many assumptions and simplifications.

Polyelectrolytes Macromolecules exhibiting solubility in aqueous solutions represent a diverse class of polymers ranging from biopolymers that direct life processes to synthetic species with enormous commercial utility. Detailed description of water-soluble synthetic polymers (WSP), their synthesis, properties, and application, can be found in the literature [8–13] and in Section 3.2.

Anomalous osmosis Water flux through the IEM behaves in an anomalous manner [27, 28]. Nondiffusible ionic species, fixed in the IEM pores, affect the anomalous osmosis. The anomalous osmosis may be positive—with the flux in the direction of the activity gradient, or negative—with the flux in the opposite direction to the activity gradient. Negative osmotic flux may be greater than the flux brought about by the normal osmotic processes, resulting in an overall flow of water in the opposite direction to the activity gradient (for a more detailed description of the anomalous osmosis, see Section 3.4). Ionic species, attached to the chains of low-diffusible polyelectrolyte molecules in the boundary layers of LM solution, should behave in the same manner. The disadvantage of conventional osmosis-dilution of the product may be obviated. In some cases, anomalous osmosis can even lead to product enrichment.

2.2. Mass-transfer mechanisms and kinetics

The following theoretical analysis of the BAHLM transport mechanisms and kinetics is based on the approach:

(1) The overall mass-transfer rate can be limited by any of the diffusion resistances in the three liquid phases (diffusion-limited transport) and/or chemical reaction (complexation/decomplexation) rate resistances on the membrane-solution interfaces (reaction rate-limited transport).
(2) The aqueous film resistances may be combined with membrane pore diffusion resistance in a one-dimensional series of diffusion resistances.

The BAHLM differs from all bulk liquid membrane systems by application of polyelectrolyte aqueous solutions as carriers and charged (ion-exchange) membranes (IEM) as barriers. As can be seen in Fig. 6.2, the physicochemical aspects of the BAHLM processes are complicated. Transport of solutes or their complexes consist of the following steps (see Fig. 6.2A):

(1) Diffusion from the bulk feed through the feed-side boundary layer (h_f)
(2) Diffusion through the pores of the feed-side ion-exchange membrane (h_{mf}), denoted by Helfferich [29-31] as "interdiffusion"
(3) Interaction with polyelectrolyte on the feed-side IEM/LM solution interface, as a result of thermodynamic conditions at the F/E interface
(4) Diffusion through the feed-side (h_{fe}) and strip-side (h_{er}) LM boundary layers
(5) Interaction with the stripping agent on the strip-side IEM/LM interface, as a result of different thermodynamic conditions at the E/R interface
(6) Interdiffusion through the strip-side IEM (h_{mr})
(7) Diffusion through the strip-side boundary layer (h_r) to the bulk strip

The transport kinetics is driven mainly by Donnan equilibrium coupling in steps 1, 2, 6, and 7 and by liquid membrane facilitation (LMF) potential and Donnan equilibrium coupling in step 4.

For the BOHLM systems (see Chapter 5) with water-immiscible carriers, the concentration gradient-driven solute-solvent complexation/decomplexation interactions are the dominant driving forces. For the BAHLM systems, Donnan membrane potential [18-26, 32-36], osmotic pressure gradient [27, 37], and possibly pressure gradient [38-40], have to be added as driving forces. Therefore, the theory should take into account both diffusive and convective transport.

The model, based on linear, irreversible thermodynamics, constitutes a more general phenomenological approach, applicable to systems with either class of membranes, multiple solutes, and driving forces involved. It includes component and overall mass balances, mass-transfer rates, "local" equilibrium relations, and electroneutrality constraints.

Assumptions made when formulating the model equations are:

(1) Local steady state in the boundary layers at membrane interfaces and inside membrane pores
(2) Constant flow, density, and temperature in each stream
(3) Concentrations of species transported in the IEM pores and LM boundary layers are higher compared to those in the feed boundary layer [5-7, 24, 25]
(4) Constant concentration of fixed sites in the IEM and of the functional groups in the LM electrolyte molecules

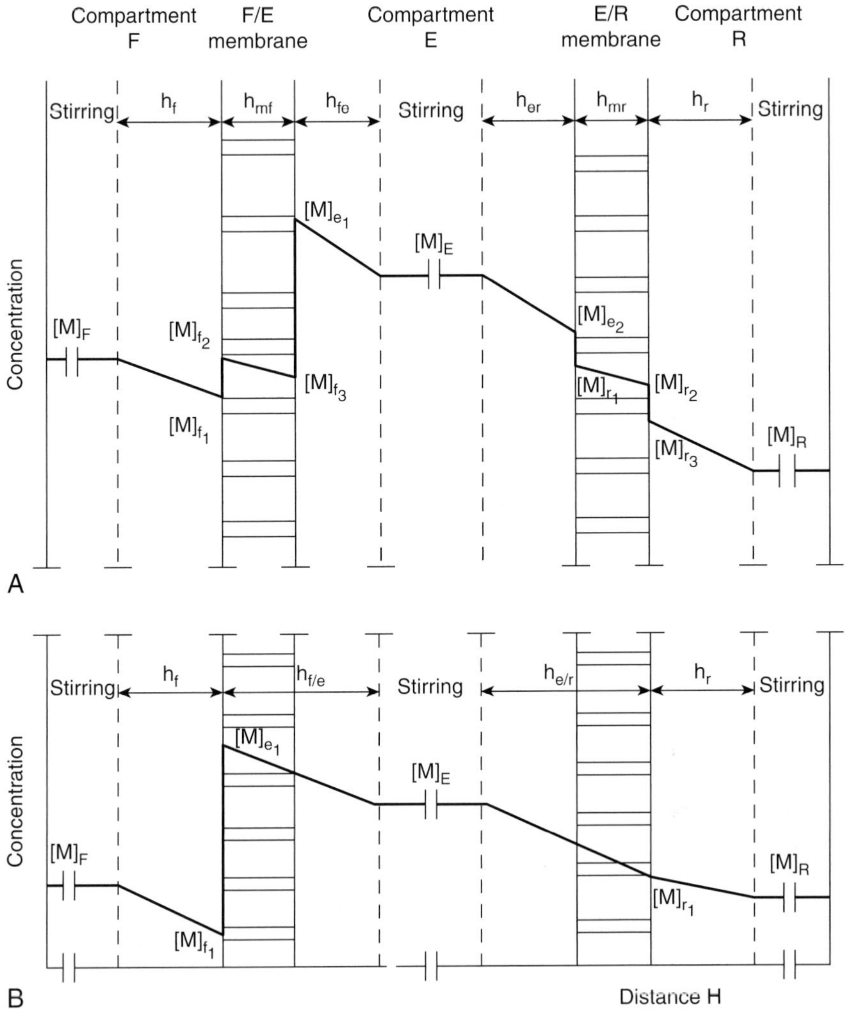

Figure 6.2 Simulation of the BAHLM transport: (A) concentration profiles in a regular scheme and (B) concentration profiles in a simplified scheme. Layers controlling the permeation rate are h_f, feed-side boundary layer; h_{mf}, feed-side IEM filled with the feed solution; h_{fe}, feed-side boundary layer of the LM solution; h_{er}, strip-side boundary layer of the LM solution; h_{mr}, strip-side IEM filled with the strip solution; h_r, strip-side boundary layer; E_F and E_R, feed-side and strip-side averaged total IEM and LMF potentials, or total feed-side and strip-side distribution coefficients, respectively; and $h_{f/e}$ and $h_{e/r}$ feed-side and strip-side averaged thickness of the IEM and the LM boundary layers, respectively. From Ref. [6] with permission.

(5) Local equilibria in the reactions at both F/E and E/R interfaces
(6) Electroneutrality in the IEM and in all liquid phases
(7) A 1:1 equivalent complexation of solute ions with IEM- and LM-charged groups
(8) Linear concentration profiles in all boundary layers and IEM pores
(9) Transfer of every counterion from the feed is considered independently: influence of each on others and competitions in mass-transfer rates are excluded
(10) The model recognizes mass-transfer resistances to diffusion in all boundary layers and in the IEM

Based on Donnan's [18, 19] and Onsager's [41, 42] fundamental works, the theories for Donnan dialysis systems were developed [20-26, 32-36]. The BAHLM system could be considered as two DD systems, operating in consecutive order, continuously in one module (see Fig. 6.2): the first is composed of feed/LM and the second is composed of LM/strip compartments, separated by ion-exchange membranes. Therefore, the Kedem-Katchalsky equations [43, 44] can be applied to our case:

$$J_W = K_P \Delta p + K_\pi(\pm\Delta\pi) \tag{1}$$

and

$$J_S = K_S \Delta c_S + c_S(1 - s)J_W, \tag{2}$$

where J_W and J_S are volumetric and diffusive solute fluxes, respectively; K_S, K_P and K_π are solute diffusive, hydraulic, and osmotic mass-transfer coefficients, respectively; Δc_S, Δp, and $\pm\Delta\pi$ are the concentration gradient, the transmembrane pressure gradient, and the osmotic pressure gradient, respectively; $\Delta\pi$ is positive when osmotic flux is in the same direction as the solute flux, and negative when the osmotic flux is in the opposite direction to the solute flux; s is the reflection coefficient [45]; and c_S is the average solute concentration within the IEM.

At the absence of pressure-driven transport, $\Delta p = 0$, $s = 0$. After some transformations, we obtain the basic equation for the BAHLM model:

$$J = K_S \Delta c_S + K_\pi c_S(\pm\Delta\pi), \tag{3}$$

where the flux J includes both diffusive and osmotic components of the transport rate.

The first component of Eq. (3), $K_S \Delta c_S$, represents the solute diffusive flux, driven by Donnan equilibrium coupling with facilitation by IEM and LMF potentials. Comparison of Eq. (3) with the equations in the model for the BOHLM systems (see Chapter 5 and [46]) shows that the diffusive mass-transfer coefficient K_S corresponds to the diffusive overall mass-transfer coefficients, $K_{F/E}$ on the feed side and $K_{E/R}$ on the strip side of the BOHLM system with hydrophilic or ion-exchange membranes:

$$K_{\mathrm{F}} = \frac{k_{\mathrm{f}} k_{\mathrm{mf}} k_{\mathrm{fe}} E_{\mathrm{F}}}{k_{\mathrm{f}} k_{\mathrm{mf}} + k_{\mathrm{f}} k_{\mathrm{fe}} E_{\mathrm{F}} + k_{\mathrm{mf}} k_{\mathrm{fe}} E_{\mathrm{F}}}, \tag{4}$$

$$K_{\mathrm{R}} = \frac{k_{\mathrm{cr}} k_{\mathrm{mr}} k_{\mathrm{r}}}{k_{\mathrm{mr}} k_{\mathrm{er}} E_{\mathrm{R}} + k_{\mathrm{r}} k_{\mathrm{er}} E_{\mathrm{R}} + k_{\mathrm{mr}} k_{\mathrm{r}}}. \tag{5}$$

The second component in Eq. (3), $K_{\pi} c_{\mathrm{S}}(\pm\Delta\pi)$, represents the solute volumetric flux, driven by osmotic pressure gradient. Osmotic mass-transfer coefficient K_{π} corresponds to the osmotic overall mass-transfer coefficients: $K_{\pi\mathrm{F}}$ on the feed side and $K_{\pi\mathrm{R}}$ on the strip side of the BAHLM system which can be determined experimentally [47-51].

Based on the principle of resistance additivity [6, 37-40], the overall mass-transfer coefficients from the feed to the carrier phases, $K_{\mathrm{F/E}}$, and from the carrier to the strip phases, $K_{\mathrm{E/R}}$, can be determined as

$$K_{\mathrm{F/E}} = \frac{K_{\mathrm{F}} K_{\pi\mathrm{F}}}{K_{\mathrm{F}} + K_{\pi\mathrm{F}}} \tag{6}$$

and

$$K_{\mathrm{E/R}} = \frac{K_{\mathrm{R}} K_{\pi\mathrm{R}}}{K_{\mathrm{R}} + K_{\pi\mathrm{R}}}. \tag{7}$$

The diffusive overall mass transfer coefficients K_{F} on the feed side and K_{R} on the strip side are related to the individual film mass-transfer coefficients, to the facilitating parameters of the membrane (IEM potential) and the aqueous LM (LMF potential). Experimental determination of some of these parameters gives major challenges.

To develop relatively simple expressions with a minimum of experimental measurements, an homogenization and averaging [6] (see Fig. 6.2B) of IEM and LMF potentials and diffusion parameters of the solute within the membrane pores and in the LM boundary layers were applied.

It is a major simplification of the real BAHLM system, but can be explained, as follows:

1. The interactions of fixed ion charges with counterions inside the pores of the IEM and in the polyelectrolyte molecules of the LM are expressed through the counterion concentration gradients (see Fig. 6.2A): $[M]_{\mathrm{f2}}$-$[M]_{\mathrm{f1}}$ and $[M]_{\mathrm{r2}}$-$[M]_{\mathrm{r3}}$ are the IEM potentials; $[M]_{\mathrm{e1}}$-$[M]_{\mathrm{f3}}$ and $[M]_{\mathrm{e2}}$-$[M]_{\mathrm{r1}}$ are the LMF potentials on the feed side and on the strip side of the BAHLM system, respectively. Known procedures, developed for determination of IEM potentials [21-25, 29-36], and LMF potentials (many authors denote the last as facilitation factors [28, 48-52]) consist of many assumptions and simplifications. They require complicated procedures for experimental measurement of some parameters. Numerical integration and a considerable amount of computation time are required for most of them.

As a rule, the rates of ion-exchange reactions, or rates of complex formation and destruction on the interfaces or within the IEM, are fast compared to diffusion rates. Thus, corresponding concentration gradients of the counterions are related through the extraction equilibrium constants [46, 53, 56]. The averaged sums of the IEM and LMF potentials can be experimentally realized through distribution coefficients at membrane-based solvent forward and backward extraction: $E_F = M_{f1}/M_{e1}$ on the feed side and $E_R = M_E/M_{r1}$ on the strip side (see Fig. 6.2B).

2. It is well known that diffusion characteristics of the polymeric electrolytes themselves in water are very low [28, 52]. On the other hand, Donnan dialysis experiments with polyanions present in the feed [5, 35, 36] show that polyelectrolytes affect the diffusion of the counterions. The diffusion mechanism inside the membrane pores is considered as "hopping steps" diffusion of the counterions through the fixed anion charges (for details, see Chapter 2). A similar diffusion mechanism may be considered in the LM boundary layers: counterions diffuse by "hopping steps" through the linear polymeric molecules with high concentration of the attached anionic functional groups. Therefore, diffusion coefficients of the same solute ions in the same solvent (water), in the IEM pore, and in the LM layer are close in magnitude. The IEM porosity has to be taken into account.

Now, all needed parameters can be obtained by relatively simple independent experimental measurements. The overall mass-transfer coefficients are calculated by equations:

$$K_{F/E} = \frac{k_f k_{f/e} E_F}{1 + k_f + k_{f/e} + k_f k_{f/e}} \tag{8}$$

and

$$K_{E/R} = \frac{k_r k_{e/r} E_R}{1 + k_r + k_{e/r} + k_r k_{e/r}}, \tag{9}$$

where k_f and k_r are individual mass-transfer coefficients (in m/s) of solute ions, diffusing through the feed-side and strip-side boundary layers, respectively; $k_{f/e}$ and $k_{e/r}$ are averaged individual mass-transfer coefficients of solute ions, diffusing through the pores of the IEM and the corresponding LM layers.

As described in the previous chapter for the BOHLM system, the BAHLM system is driven also by two main driving forces: the external driving force, derived from the coupling effect of the BAHLM system, and the internal (carrier) driving force, derived from the extraction distribution ratio of solute between the liquid membrane phase and feed, and receiving phases. K_{dF} and K_{dR} are denoted as external driving force coefficients; K_{cF} and K_{cR} are denoted as internal (carrier) driving force coefficients.

External overall mass-transfer coefficients K_{dF} on the feed side and K_{dR} on the strip side are calculated by the following equations:

$$K_{dF} = \frac{k_{d(fe)} \times k_{d(mf+ef)}}{k_{d(fe)} + k_{d(mf+ef)}} \tag{10}$$

and

$$K_{dR} = \frac{k_{d(re)} \times k_{d(mr+er)}}{k_{d(re} + k_{d(mr+er)}}. \tag{11}$$

The overall mass-transfer coefficients of the BAHLM system were calculated by equations:

$$K_F = K_{cF} \times K_{dF} \tag{12}$$

and

$$K_R = K_{cR} \times K_{dR}. \tag{13}$$

In the model for the BOHLM systems [46], only the solute concentration was stated as a time-dependent variable. In the BAHLM system, there are two time-dependent variables: solute concentration and solution volume. Transforming the two variables into one, we obtain

$$\frac{d(cV)}{dt} = \frac{d(Q)}{dt}, \tag{14}$$

where Q is the quantity of the solute in each of the feed, carrier, and strip solutions at the sampling time t_i.

Introducing the assumptions cited above, the transport rate and distribution relation equations for solute permeating species can be derived:

$$\frac{d(Q_F)}{dt} = SK_{F/E}(Q_F^o - Q_F), \tag{15}$$

$$\frac{d[Q_E]}{dt} = SK_{F/E}(Q_F^o - Q_F) - SK_{E/R}\, Q_E, \tag{16}$$

at the overall mass balance:

$$Q_F^o = Q_F + Q_E + Q_R, \tag{17}$$

where Q_F^o is the initial quantity of the solute in the feed; Q_F, Q_E, Q_R are quantities of the solute at time t in the feed, membrane, and strip phases, respectively; and S is the membrane effective working area.

The system of equations (15)-(17) results in an analytical solution:

$$Q_F = Q_F^o\left(e^{-SK_{F/E}t}\right), \tag{18}$$

$$Q_R = Q_F^o\left(1 - \frac{K_{E/R}}{K_{E/R}\text{-}K_{F/E}}e^{-SK_{F/E}t} + \frac{K_{F/E}}{K_{E/R}\text{-}K_{F/E}}e^{-SK_{E/R}t}\right), \quad (19)$$

$$Q_E = Q_F^o\frac{K_{F/E}}{K_{E/R}\text{-}K_{F/E}}(e^{-SK_{F/E}t} - e^{-SK_{E/R}t}). \quad (20)$$

This is a set of model equations for simulation and preliminary optimization of the BAHLM parameters: membrane working area; feed, carrier, strip solutions' initial concentrations; flow velocities, etc.

Using (1) Giddings' analysis [54] of the steady-state zones, formed next to the IEM-LM interfaces, and (2) theoretical considerations for one-dimensional facilitated transport of a solute, applied to the BOHLM systems [46, 55], the equations for experimental determination of individual mass-transfer coefficients (k_i) were formulated.

For the feed boundary layer,

$$k_{(f)_i} = U_{F_j}\left\{-\ln\left[\left(Q_{F_i} - \frac{J_{F_{SS}}}{U_{F_i}}\right)\left(Q_{F_{(i-1)}} - \frac{J_{F_{SS}}}{U_{F_i}}\right)\right]\right\}^{-1}. \quad (21)$$

For the IEM and LM boundary layers,

$$\frac{1}{2}k_{(f/e)i} \text{ or } \frac{1}{2}k_{(e/r)i} = U_{E_j}\left\{\pm \ln\left[\left(Q_{E_i} - \frac{J_{F_{SS}}}{U_{E_i}}\right)\left(Q_{E_{(i-1)}} - \frac{J_{F_{SS}}}{U_{E_i}}\right)\right]\right\}^{-1}, \quad (22)$$

where "plus" (positive) stands for increasing concentrations versus time and "minus" (negative) stands for decreasing ones.

For the strip boundary layer,

$$k_{(r)i} = U_{Rj}\left\{\ln\left[\left(Q_{R_i} - \frac{J_{R_{SS}}}{U_{R_i}}\right)\left(Q_{R_{(i-1)}} - \frac{J_{R_{SS}}}{U_{R_i}}\right)\right]\right\}^{-1}, \quad (23)$$

where $U_{F_j}, U_{E_j}, U_{R_j}$, are flow velocities (in m^3/s); $J_{F_{ss}} J_{E_{ss}} J_{R_{ss}}$ are fluxes (in mol/m^2 s) at steady state; Q_{F_i}, $Q_{F_{i-1}}$, Q_{E_i}, $Q_{E_{i-1}}$, Q_{R_i}, and $Q_{R_{i-1}}$ are quantities (in mol) of the solute in the feed, carrier, and strip solutions, respectively, at sampling time t_i or t_{i-1} (in s).

It must be stressed that the individual mass-transfer coefficients, determined using Eqs (21)-(23), include both the diffusive and the osmotic transport components. Q_{F_i} and $Q_{F_{i-1}}$, Q_{E_i} and $Q_{E_{i-1}}$, Q_{R_i} and $Q_{R_{i-1}}$ represent the changes in the quantities of the solute, that is, in the concentrations and volumes between sampling time t_i and t_{i-1}. Therefore, experimental determination of the osmotic pressure gradients, $\Delta\pi$, and calculation of osmotic mass-transfer coefficients, K_π, may be excluded. Overall mass-transfer coefficients of solute species through the BAHLM system are calculated using Eqs (8) and (9).

As a rule, it is more comfortable to operate with concentrations $[M]$ than with quantities (Q_M) of the solute transported. The concentrations can be obtained by equations:

$$[M]_i = Q_{M_i}/V_i, \tag{24}$$

where $[M]_i$ and Q_{M_i} are concentrations and quantities of the solute ion, and V_i is the volume of the phase ($V_{F_i}, V_{E_i}, V_{R_i}$) at the sampling time t_i;

$$V_i = V_0 + at_i^b, \tag{25}$$

where V_0 is the initial volume of the phase ($V_{F_0}, V_{E_0}, V_{R_0}$), a and b are coefficients, obtained experimentally by measuring volumes of the phases at sampling time t_i.

To develop a more adequate model, the concept of dynamic selectivity, $S_D(t)$, and the dynamic separation factor, $A_{D(M_1/M_2)}(t)$, dependent on processing time, is introduced (for details, see Section 3).

This semiempirical model may be used to minimize experimental testing at the BAHLM processes design.

3. Module Design Considerations

3.1. Module design

3.1.1. Kinetic parameters determination and preliminary optimization

Huge amounts of liquid byproduct and waste effluents in the fertilizer industry contain various heavy metals, some of which are highly toxic. Cd, Cu, and Zn are commonly encountered in these effluents and are selected for selective removal studies using liquid membrane systems [1-7, 14-17]. Below, experimental and calculated data, obtained for Cd, Cu, and Zn separation, are used for the BAHLM process design considerations.

According to the theoretical model for transport kinetics (see Section 2.2), most preliminary parameters needed for BAHLM process design and optimization may be obtained by a number of known or experimentally obtained data.

Individual mass-transfer coefficients of solute species in the feed, carrier, and strip interfacial boundary layers are determined experimentally by feed, carrier, and strip flow rate variations using Eqs (21)–(23). The correlation factor between $[Q]$ and J_{SS}/U in these equations has been examined [6] in the range of $U = 0.1$–1.5 cm^3/s. Results showed that $[Q] > J_{SS}/U$:

- In the feed solutions—more than three orders of magnitude
- In the carrier solutions—about 2.5 orders
- In the strip solutions—more than 1.5 orders

This means that the J_{SS}/U ratio may be excluded from Eqs (21)–(23). So, simplified equations are

$$k_{f_i} = \frac{U_{F_j}}{-\ln\dfrac{Q_{F_i}}{Q_{F_{(i-1)}}}} \quad \text{or} \quad k_{f_i} = \frac{U_{F_j}}{\ln\dfrac{Q_{F_{(i-1)}}}{Q_{F_{(i)}}}}, \tag{26}$$

$$\frac{1}{2}k_{f/e_i}\ (\text{or } k_{e/r_i}) = \frac{U_{E_j}}{\pm\ln\dfrac{Q_{E_i}}{Q_{E_{(i-1)}}}}, \tag{27}$$

$$k_{r_i} = \frac{U_{R_j}}{\ln\dfrac{Q_{R_i}}{Q_{R_{(i-1)}}}}. \tag{28}$$

Some examples of individual mass-transfer coefficients of metal ions at the feed, carrier, and strip flow rates' variations are shown in Table 6.1.

Internal (carrier) driving force coefficients, K_{cF} and K_{cR}, or distribution coefficients, E_F and E_R, are determined by membrane-based extraction experiments. Membrane-based forward and backward extraction is carried out in two-compartment modules using the F and R compartments, separated by the same membranes as in the BAHLM tests. The experiments lasted up to equilibrium conditions, when the concentration of solutes in every compartment does not change with time. Examples of membrane-based extraction of copper, cadmium, and zinc from the concentrated phosphoric acid solution by PVSH and backward extraction by 2 M HCl are presented in Table 6.2 [7].

Overall mass-transfer coefficients $K_{F/E}$ on the feed side and $K_{E/R}$ on the strip side were calculated by Eqs (8) and (9). Dependence of cadmium, copper, and zinc overall mass-transfer coefficients on the feed, carrier, and strip solutions' flow rates is shown in Table 6.3. Flow rates varied in the range 0.5–1.5 cm^3/s. As an example, this dependence on the feed flow rate variations is shown in Fig. 6.3.

Concentration profiles of metal species in the feed, carrier, and strip bulk solutions were simulated using model equations (18)–(20), (24) and (25). An example of the correlation between experimental and simulated data is presented in Fig. 6.4A.

Discrepancies between experimentally obtained and theoretically calculated data may be explained by:

(1) Simplifications, averaging of driving forces and diffusion parameters (see above)
(2) The lack in the theoretical consideration of mass-transfer relations of coupling ions and free carrier molecules in the opposite to the solute transport direction (external or coupling driving forces). In the model

Table 6.1 Individual mass-transfer coefficients of metal ions

Metal ion	Flow velocity (cm^3/s)	k_f^* (10^{-2} m/s)	$(k_{fe}=k_{er})^*$ (10^{-3} m/s)	k_r^* (10^{-3} m/s)	k_f^* (10^{-2} m/s)	$(k_{fe}=k_{er})^*$ (10^{-3} m/s)	k_r^* (10^{-3} m/s)	k_f^* (10^{-2} m/s)	$(k_{fe}=k_{er})^*$ (10^{-3} m/s)	k_r^* (10^{-3} m/s)
	1	2	3	4	5	6	7	8	9	10
Cd	0.5	0.29	0.34	1.2	1.8	0.3	1.0	23.7	0.44	1.5
	1.0	1.1	2.4	1.6	39.2	4.4	2.3	17.4	0.57	1.7
	1.5	2.0	4.8	2.0	57.8	2.9	2.7	14.8	0.67	1.5
Cu	0.5	28.9	0.34	0.57	8.1	0.3	0.6	12.6	0.43	0.6
	1.0	29.6	1.6	6.9	17.4	4.2	0.8	31.9	0.95	0.72
	1.5	28.9	2.8	14.5	29.6	2.1	0.98	70.4	0.36	0.87
Zn	0.5	0.15	0.05	0.10	0.13	0.04	0.06	0.19	0.05	0.08
	1.0	0.15	0.09	0.15	0.16	0.06	0.14	0.25	0.08	0.16
	1.5	0.16	0.14	0.18	0.17	0.10	0.20	0.24	0.10	0.25

Notes. (1) Initial compositions: feed: 40% H_3PO_4 WPA, containing 55.2 ppm Cd, 55.4 ppm Cu, and 290 ppm Zn; carrier: 0.5 mol/kg PVSH aqueous solution; strip: 2.0 mol/kg HCl. Membrane: Tokayama Soda cation-exchange membrane CM-2. (2) Results, presented in columns 2, 3, and 4, were obtained at various feed flow velocities (column 1) and fixed carrier and strip solutions flow velocities ($U = 0.5$ cm^3/s). Results in the columns 5, 6, and 7 were obtained at various carrier solution flow velocities and fixed feed and strip solutions' flow velocities ($U = 0.5$ cm^3/s). Results in the columns 8, 9, and 10 were obtained at various strip solution flow velocities and fixed feed and carrier solutions' flow velocities ($U = 0.5$ cm^3/s).
(3) *Source*: From Ref. [6] with permission.

Table 6.2 Distribution coefficients of copper, cadmium, and zinc ions in membrane-based solvent extraction

Metal ions	Distribution coefficients at equilibrium: Forward extraction, $E_F = [M]_E/[M]_F$	Backward extraction, $E_R = [M]_R/[M]_E$
Cadmium	4.7	0.29
Copper	41.4	0.16
Zinc	0.85	0.10

Notes. (1) Initial compositions in forward extraction: feed: 40% H_3PO_4, 52.5 ppm Cd, 50.6 ppm Cu, and 329 ppm Zn; strip: 0.5 mol/kg PVSH in water; In backward extraction: feed: 0.5 mol/kg PVSH in water, loaded by 41.3 ppm Cd, 49.4 ppm Cu, and 270 ppm Zn; strip: 2.0 mol/kg HCl. (2) Distribution coefficients are calculated in the direction of metal ions transported in the BAHLM (see Fig. 6.2B). (3) *Source*: From Ref. [6] with permission.

equations, we took into consideration only mass-transfer relations, motivated by internal driving force (forward extraction-backward extraction distribution ratio, K_c). Mass-transfer relations—motivated by external driving force (proton concentration gradients and/or anion concentration gradients) between feed and strip phases, indicated by K_d coefficient—were not considered. Interference of the different solute ions in the real BAHLM system was not considered also.

(3) Changes in mechanisms (and kinetics) of the metal ions interactions with the carrier molecules during the loading and stripping of the LM phase

These problems and correction techniques are discussed in Chapter 5, BOHLM, Section 3.4. Corrected individual mass-transfer coefficients, k_d, are evaluated using Eqs (26)–(28), where Q_{F_i}, Q_{E_i}, and Q_{R_i} were taken from model calculated data and $Q_{F_{i-1}}$, $Q_{E_{i-1}}$, and $Q_{R_{i-1}}$ were experimentally obtained data under the same conditions and sampling time, t_i (for details, see Ref. [46]).

Overall mass-transfer coefficients $K_{dF/E}$ on the feed side and $K_{dE/R}$ on the strip side were calculated by the following equations:

$$K_{dF/E} = \frac{k_{df} \times k_{d(mf+fe)}}{k_{df} + k_{d(mf+fe)}} \tag{29}$$

and

$$K_{dE/R} = \frac{k_{dr} \times k_{d(mr+r)}}{k_{dr} + k_{d(mr+r)}}. \tag{30}$$

The overall mass-transfer coefficients of the BAHLM system were calculated by equations:

Table 6.3 Overall mass-transfer coefficients of cadmium, copper, and zinc, transported through the AHLM at various flow velocities

Flow velocity (cm^3/s)		Cadmium		Copper		Zinc	
Compart		Feed side, $K_{F/E}$ (10^{-4} m/s)	Strip side, $K_{E/R}$ (10^{-6} m/s)	Feed side, $K_{F/E}$ (10^{-2} m/s)	Strip side, $K_{E/R}$ (10^{-6} m/s)	Feed side, $K_{F/E}$ (10^{-7} m/s)	Strip side, $K_{E/R}$ (10^{-9} m/s)
1	2	3	4	5	6	7	8
Feed solution,	0.5	0.046	0.118	0.289	0.019	0.635	0.800
U_F	1.0	1.22	1.10	1.50	1.08	1.14	2.16
	1.5	4.38	2.72	2.58	3.88	1.90	4.03
Carrier solution,	0.5	0.248	0.087	0.093	0.018	0.441	0.384
U_E	1.0	54.7	2.87	2.56	0.321	0.813	1.34
	1.5	46.2	2.24	1.97	0.201	1.44	3.20
Strip solution,	0.5	3.81	0.191	0.199	0.026	0.804	0.640
U_R	1.0	3.85	0.280	0.945	0.068	1.69	2.05
	1.5	3.95	0.290	0.609	0.031	2.07	4.00

Note. Overall mass-transfer coefficients of cadmium, copper, and zinc were calculated using Eqs (8) and (9) (see text). *Source*: From Ref. [6] with permission.

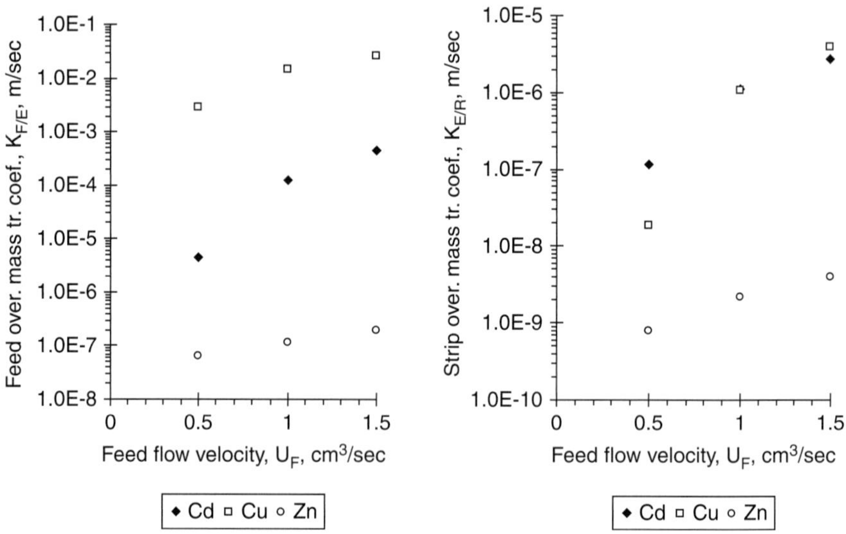

Figure 6.3 The dependence of feed-side and strip-side overall mass-transfer coefficients on feed flow velocities. Initial compositions: feed: 40% H_3PO_4 WPA, containing 55.2 ppm Cd, 55.4 ppm Cu, and 290 ppm Zn; LM: 0.5 mol/kg PVSH aqueous solution; strip: 2.0 mol/kg HCl. Membrane barriers: Tokayama Soda cation-exchange membrane CM-2. From Ref. [6] with permission.

$$K_F = K_{cF/E} \times K_{dF/E} \tag{31}$$

and

$$K_R = K_{cE/R} \times K_{dE/R}. \tag{32}$$

Corrected concentration profiles are presented in Fig. 6.4B.

Comparing the simulated results with the experimental data, it appears that at higher flow rates, where boundary layers resistance becomes less important, membrane-strip interfacial reaction kinetics dominates as a rate-controlling step for solute transport.

At low solute concentrations, there are discrepancies between the theoretical and experimental lines even after corrections. It may be explained by following. The values of distribution coefficients E_F and E_R depend on many conditions including IEM and LMF capacities. At low metal concentrations, the Donnan exclusion is pronounced and IEM potential is much higher, so its contribution to the overall BAHLM transport kinetics is more considerable than at higher concentrations. But the metal concentrations data, used for calculation of the averaged sums of IEM and LMF potentials, are much higher. So, the calculating parameters obtained do not satisfy the real transport kinetics at low concentrations.

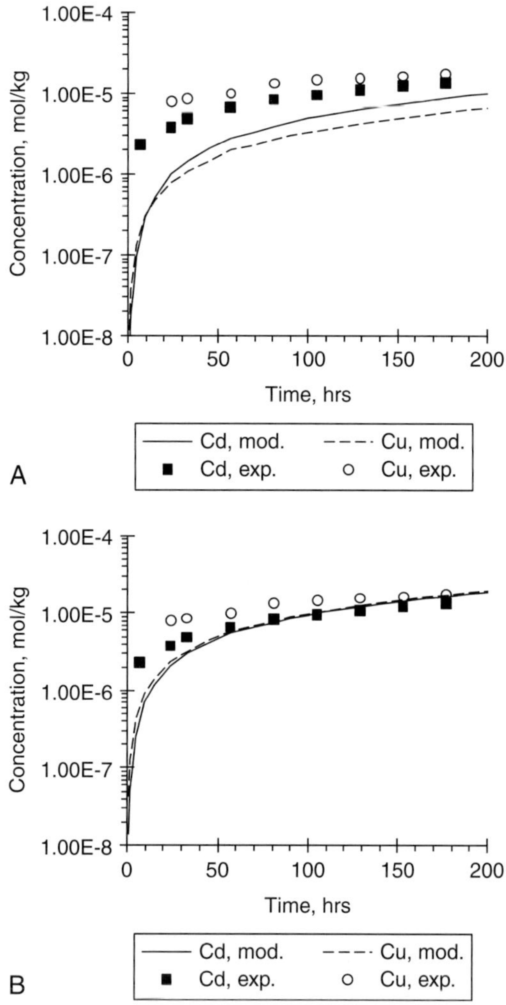

Figure 6.4 Concentration profiles of cadmium and copper in the strip phase. Initial compositions: feed: 40% H_3PO_4 WPA, containing 55.2 ppm Cd, 55.4 ppm Cu, and 290 ppm Zn; LM: 0.5 mol/kg PVSH aqueous solution; strip: 2.0 mol/kg HCl. Membrane barriers: Tokayama Soda cation-exchange membrane CM-2. (**A**) Model calculated profiles before correction of the individual and overall mass-transfer coefficients. (**B**) Model calculated profiles after correction. From Ref. [6] with permission.

The values of overall mass-transfer coefficients govern the location t_{max} of maximum of the solute quantity in the LM, Q_{Emax}. This may be found by differentiating Eq. (20) and setting $dQ_E/dt = 0$:

$$t_{max} = \frac{\ln(K_{E/R}/K_{F/E})}{(K_{E/R}-K_{F/E})}, \tag{33}$$

where t_{max} is the processing time at which maximum quantity (Q_{Emax}) of the metal species occurs in the LM solution. The Q_{Emax} may be found by combining Eqs (20) and (33):

$$Q_{Emax} = Q_F^o (K_{F/E}/K_{E/R})^{K_{E/R}(K_{E/R}-K_{F/E})}. \quad (34)$$

The greatest transport rate of the metal species to the strip solution occurs at t_{max} and Q_{Emax} (inflection point on the curve Q_R vs. time).

Let us consider two extreme cases for the Q_R curve versus time. From Eq. (19), we obtain:
(1) At $K_{E/R} \gg K_{F/E}$,

$$Q_R = Q_F^o (1 - e^{-SK_{F/E}t}). \quad (35)$$

So, the rate-controlling step of the BAHLM transport of the solute to the strip phase is determined by the $K_{F/E}$ overall mass-transfer coefficient. In this case, at designing the BAHLM module, the main attention has to be taken to the determination of optimal feed-side membrane area.
(2) At $K_{F/E} \gg K_{E/R}$,

$$Q_R = Q_F^o (1 - e^{-SK_{E/R}t}). \quad (36)$$

In this case, the rate-controlling step of the AHLM transport of the metal species to the strip phase is determined by the $K_{E/R}$ overall mass-transfer coefficient. Here, we have to optimize strip-side membrane area.

3.1.2. Evaluation of the selectivity

Selectivity of the BAHLM system is an important parameter that should be used designing module. According to the transport model equations, the selectivity of two metal species (S_{M_1/M_2}) is determined by relation:

$$S_{M_1/M_2} = \frac{K_{M_1}(Q_F^o)_{M_1}}{K_{M_2}(Q_F^o)_{M_2}} = \frac{K_{M_1}[M_1]_F^o}{K_{M_2}[M_2]_F^o}, \quad (37)$$

where subscripts M_1 and M_2 refer to the two metal species; K_{M_1} and K_{M_2} are the total overall mass-transfer coefficients; $(Q_F^0)_{M_1}$ and $(Q_F^0)_{M_2}$ are the initial quantities of two metals in the treated feed phase; and $[M_1]_F^o$ and $[M_2]_F^o$ are the initial concentrations of two metals in the same phase.

Introducing a separation factor [61], A, defined as a ratio of the total overall mass-transfer coefficients of the metal species

$$A_{M_1/M_2} = K_{M_1}/K_{M_2}, \quad (38)$$

an equation for the BAHLM system selectivity is obtained:

$$S_{M_1/M_2} = A_{M_1/M_2} \frac{[M_1]_F^o}{[M_2]_F^o}. \quad (39)$$

Based on the principle of resistance additivity, the total overall mass-transfer coefficient, K_M, of every metal passing through the BAHLM is related to the overall mass-transfer coefficients on the feed and strip sides as follows:

$$K_{\mathrm{M}} = \frac{[K_{\mathrm{F/E}}K_{\mathrm{E/R}}]_{\mathrm{M}}}{[K_{\mathrm{F/E}} + K_{\mathrm{E/R}}]_{\mathrm{M}}}. \tag{40}$$

Overall mass-transfer coefficients on the feed and strip sides are calculated using Eqs (8) and (9). E_{F} and E_{R}, or IEM + LMF potentials are determined experimentally through distribution coefficients at membrane-based equilibrium forward and backward extraction.

Thus, the separation factor of the two metal species is

$$A_{\mathrm{M_1/M_2}} = \frac{[K_{\mathrm{F/E}}K_{\mathrm{E/R}}]_{\mathrm{M_1}}[K_{\mathrm{F/E}} + K_{\mathrm{E/R}}]_{\mathrm{M_2}}}{[K_{\mathrm{F/E}}K_{\mathrm{E/R}}]_{\mathrm{M_2}}[K_{\mathrm{F/E}} + K_{\mathrm{E/R}}]_{\mathrm{M_1}}}. \tag{41}$$

For preliminary evaluation of the selectivity of two metal species separation by the BAHLM process, we can assume that in the same solution environment (water), the diffusion coefficients of these metal ions with the same charge have similar values and the diffusion coefficients of the metal-carrier complexes have similar values. Thus, substituting Eqs (8) and (9) in Eq. (41), we can represent the separation factor as dependent only on the distribution coefficients:

$$A_{\mathrm{M_1/M_2}} \approx \frac{[E_{\mathrm{F}}E_{\mathrm{R}}]_{\mathrm{M_1}}}{[E_{\mathrm{F}}E_{\mathrm{R}}]_{\mathrm{M_2}}}. \tag{42}$$

The distribution coefficient may be expressed as a function of the metal association (stability) constants in the IEM and polyelectrolyte (LM) solution, the association constants of metal ions with solvent environment in the feed and in the strip solutions, and partition coefficients of the polyelectrolyte and metal ion [60]. In this case, the separation factor can be determined by the stability constants of the metal complexes, formed with IEM and LM polyelectrolyte functional groups, if we assume that the metal ions are predominantly present (a) as free ions in the acid solution, so that complex concentrations can be disregarded, and (b) as complexes in the IEM and LM solutions, so that free ion concentrations can be disregarded. So,

$$A_{\mathrm{M_1/M_2}} \approx \frac{[\beta_{\mathrm{F}}\beta_{\mathrm{R}}]_{\mathrm{M_1}}}{[\beta_{\mathrm{F}}\beta_{\mathrm{R}}]_{\mathrm{M_2}}}, \tag{43}$$

where β_{F} and β_{R} are stability constants of the metal-polyelectrolyte complexes in equilibrium with the feed and strip solvent environment (as a rule acids for metal ions), respectively. Therefore, preliminary selectivity data of metal species separation by BAHLM may be evaluated without experimentation, if stability constants data are available in the literature.

From Eqs (42) and (43), it follows that high separation factors are favored when approaching conditions $[\mathrm{E_F E_R}]_{\mathrm{M_1}} \gg [\mathrm{E_F E_R}]_{\mathrm{M_2}}$ or $[\beta_{\mathrm{F}}\beta_{\mathrm{R}}]_{\mathrm{M}} \gg [\beta_{\mathrm{F}}\beta_{\mathrm{R}}]_{\mathrm{M_2}}$. On the other hand, the system loses its selectivity when distribution parameters of both metal species are either extremely high or extremely low [29, 58, 59, 61]. Selectivity can be increased by choosing a selective carrier with intermediate distribution data values, and adjusting its concentration, its volume, the acidity of the feed and strip solutions in such a way as to approach the above conditions.

According to model for transport kinetics, E_F and E_R are the averaged sums of IEM and LMF potentials, and $k_{f/e}$ and $k_{e/r}$ are the averaged individual mass-transfer coefficients of the metal ions through the IEM itself and the LM boundary layers on the feed and strip sides of the BAHLM system. We need to evaluate the contribution of each component in the averaged parameters. The fluxes are proportional to the total equivalent concentrations of counterions and the capacities of the fixed ionic groups in the IEM pores [29] and LM polyelectrolyte molecules. The capacity of the IEM is limited by its thickness and porosity. Therefore, the flux in the IEM is mainly dependent on the concentration of the fixed charges. The concentration profiles depend chiefly on the mobility ratio of two counterions in the IEM: the membrane accumulates the slower ion [62]. If the metal ion concentrations of the treated (feed) solution are much lower than those of the fixed charges, and the accumulation of the preferred ion does not reach the saturation level, the selectivity component of the IEM itself has to be considered. When using feed solutions with a higher concentration of metal ions, the saturation level of the metal ions in the IEM will be reached soon after the beginning of the trial; IEM potential becomes very small and Donnan exclusion is not pronounced. Thus, the contribution of the IEM in the overall BAHLM transport kinetics and selectivities can be disregarded.

This is not the case with liquid membrane capacity. The polyelectrolyte concentration is limited mainly by the viscosity of its aqueous solution, but the bulk volume of the LM, circulating through the BAHLM, can be changed in large limits. At unchangeable boundary layer thickness, a constant area exposed to the IEM interface, and constant affinity to complexation (association constant), the capacity of the LM used may be as large as necessary for efficient selectivity. Manipulation by the carrier concentration (and, therefore, by the LMF potential), by the volume of the circulating bulk LM solution, or by both, enables us to obtain selectivity that is as close to its highest level as necessary for the process developed. It is one of the biggest advantages of the BAHLM system in comparison with other liquid membrane technologies.

As was shown in the BAHLM model for transport kinetics, the values of the overall mass-transfer coefficients govern the location (t_{max}) and the maximum quantity (Q_{Emax}) of the metal species in the LM phase (see Eqs (27) and (28)). At $Q_{Emax} \gg Q_F$, the BAHLM is working mainly as a Donnan dialysis system, in which the loaded carrier solution is a treated feed. In this case, the separation factor depends only on the striping mass-transfer and distribution parameters. Therefore, here Eqs (41)–(43) may be simplified to

$$A_{M_1/M_2} = \frac{[K_{E/R}]_{M_1}}{[K_{E/R}]_{M_2}}, \tag{44}$$

$$A_{M_1/M_2} \approx \frac{[E_R]_{M_1}}{[E_R]_{M_2}}, \tag{45}$$

$$A_{M_1/M_2} \approx \frac{[\beta_R]_{M_1}}{[\beta_R]_{M_2}}. \tag{46}$$

It is known [5, 6, 29, 61, 62] that the distribution coefficients E_F and E_R and, correspondingly, the overall mass-transfer coefficients $K_{F/E}$ and $K_{E/R}$ are usually not constant but depend on experimental conditions such as metal concentrations, acidities, viscosities, IEM and LM capacities, temperature, etc. Distribution coefficients always increase with the quantity of free carrier which is not bound to the metal. Complexes formed may be changing and, therefore, stability constants may be different, as the carrier loading approaches its saturation. Thus, the separation factor is a parameter, which is not constant and may be changing during the processing time. As mentioned in Section 2.2, the idea of dynamic selectivity $S_D(t)$ and dynamic separation factor, $A_{D(M_1/M_2)}(t)$, depending on processing time, is developed. So, Eq. (40) transforms to

$$K_M(t) = \frac{[K_{F/E}(t)K_{E/R}(t)]_M}{[K_{F/E}(t) + K_{E/R}(t)]_M} \tag{47}$$

and Eq. (41) to

$$A_{D(M_1/M_2)} = \frac{[K_{F/E}(t)K_{E/R}(t)]_{M_1}[K_{F/E}(t) + K_{E/R}(t)]_{M_2}}{[K_{F/E}(t)K_{E/R}(t)]_{M_2}[K_{F/E}(t) + K_{E/R}(t)]_{M_1}}. \tag{48}$$

Individual mass-transfer coefficients are determined experimentally for every metal species in the processing time intervals between sampling (t_i-t_o), (t_j-t_i), (t_f-t_j), and so on. Then, for every metal transported, the overall mass-transfer coefficients on the feed side, $K_{F/E}(t)$, and strip side, $K_{E/R}(t)$, the total overall mass-transfer coefficients of the BAHLM system, $K_M(t)$, were calculated, using Eqs (8), (9), and (47), respectively, for every sampling time interval. Separation factors of two metal species, $A_{D(M_1/M_2)}(t)$, were also calculated separately for every sampling time interval using Eq. (48).

The dynamic selectivity (S_D) of the transfer of metal ions from the feed phase of the BAHLM system (S_{DE}) is calculated according to

$$S_{DE} = \{[(C_{M_1E_{(i)}} V_E) + (C_{M_1R_{(i)}} V_R)] - [(C_{M_1E_{(i-1)}} V_E) + C_{M_1R_{(i-1)}} V_R)]\} [(C_{M_1F_i} V_F)^{-1}]/\{[(C_{M_2E_{(i)}} V_E) + (C_{M_2R_{(i)}} V_R)] - [(C_{M_2E_{(i-1)}} V_E) + (C_{M_2R_{(i-1)}} V_R)]\}[(C_{M_2F_i} V_F)^{-1}], \quad (49)$$

where $C_{M_1F_i}$ and $C_{M_2F_i}$ are the average concentrations (in mol/kg) of the two metal ions in the feed phase during the sampling interval: $C_{MF_i} = (C_{MF_i} + C_{MF_{i-1}})/2$; $C_{M_1E_{(i)}}$, $C_{M_2E_{(i)}}$, $C_{M_1E_{(i-1)}}$, and $C_{M_2E_{(i-1)}}$ are the concentrations of metal ions in the LM solution; $C_{M_1R_{(i)}}$, $C_{M_2R_{(i)}}$, $C_{M_1R_{(i-1)}}$ and $C_{M_2R_{(i-1)}}$ are the concentrations of the metal ions in the strip solution at sampling times *(i)*, and *(i*−1*)*; V_F, V_E, and V_R are the amounts of the feed, carrier, and strip solutions (in kg), respectively.

The dynamic selectivity of the transfer of metal ions into the strip phase (S_{DR}) is calculated according to

$$S_{DR} = [(C_{M_1R_{(i)}} - C_{M_1R_{(i-1)}}) V_R][(C_{M_1F_i} V_F)^{-1}]/ [(C_{M_2R_{(i)}} - C_{M_2R_{(i-1)}}) V_R][(C_{M_2F_i} V_F)^{-1}]. \quad (50)$$

As an example, Cu/Cd dynamic selectivity at BAHLM transport is presented in Fig. 6.5.

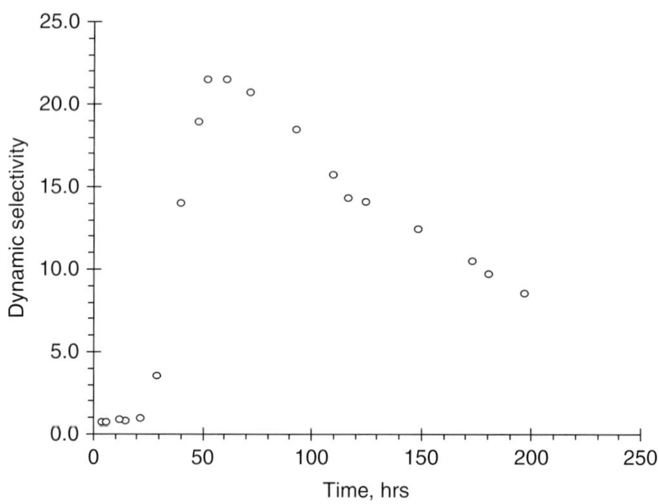

Figure 6.5 Cu/Cd dynamic selectivity at AHLM transport from the aqueous solution of their chloride salt mixtures (0.1 mol/kg Cu and 0.1 mol/kg Cd) to the sodium chloride strip solution (2.0 mol/kg sodium). Carrier: 0.5 mol/kg PVSH in water. Membrane: Tokayama Soda cation-exchange membrane CMS. From Ref. [7] with permission.

Table 6.4 Experimental determination of osmotic mass transfer of water (in g) according to Eq. (25): $V_i = V_o + a_{t_i^b}$

V_i	V_o	a	b	R^2
V_{F_i}	250	5.54	0.39	0.976
V_{E_i}	100	−3.51	0.74	0.998
V_{R_i}	250	1.56	0.81	0.998

Notes. AHLM module with initial conditions: feed: 40% H_3PO_4 WPA, containing 55.2 ppm Cd, 55.4 ppm Cu, and 290 ppm Zn; carrier: 0.5 mol/kg PVSH aqueous solution; strip: 2.0 mol/kg HCl. Membrane: Tokayama Soda cation-exchange membrane CM-2. Feed: carrier: strip ratio = 2.5:1: 2.5. Flow velocities (cm^3/min): feed = carrier = 90 and strip = 60. *Source*: From Ref. [6] with permission.

These data show that more than two times higher Cu/Cd selectivity may be obtained using maximal dynamic selectivity parameters (instead of averaged data as usually do) for designing the BAHLM technology of Cu/Cd separation from the aqueous chloride solutions.

Osmotic transfer of water through the membranes may be presented in the form of Eq. (25) coefficients. Example of data is shown in Table 6.4.

These considerations and the semiempirical model equations may be used to minimize experimental testing at the BAHLM processes design.

3.2. Polyelectrolytes as carriers in an aqueous solution

Water-soluble polymers (WSP) are classified as natural and synthetic polymers [8-10]. The natural polymer systems include starches, galactomannans, cellulose derivatives, microbial polysacharrides, gelatins, glues, etc. These are mainly nonionic WSP. Synthetic WSP are also classified as nonionic (polyoxides, polyethers, acrylic and vinyl group polymers) and ionic: polyacids (acrylic, ethylenesulfonic, phosphoric, silicic, vinylsulfuric, etc.) and their salts, and polyamines (vinylamine, ethyleneimine, 2-vinylpyridine, 4-vinylpyridine, etc.) and their derivatives. The ionic or ionizable WSP are nominated as polyelectrolytes, although many authors use this term also for neutral, nonionic polymers such as polyvinylalcohol, polyethyleneoxides, etc., without any ionizable group but partly or completely soluble in water.

Polyelectrolytes (polyanions or polycations) are an important class of functionalized polymers with charged groups attached to the chains [9-13]. The physicochemical behavior of these polymers is dominated by the attractive interactions between the fixed charges and counterions, and by the long-range repulsive interactions between the electric charges located on the macromolecular chains [63]. Electrostatic interactions between polymers and metal ions are found in many important biopolymers and synthetic polyelectrolytes. The features of coordination between polymers and metal ions may be described by the usual coordination theories, but some special aspects may be considered. The existence of a polymer domain with near constant concentration of ligands is normally recognized. This is responsible for the chelating reaction. The reactivity of the complexes is sometimes strongly affected by the polymer ligand that exists outside the coordination sphere and surrounds the metal complex. Three different modes of counterion binding may be distinguished: territorial binding, site binding, and hydrophobic binding (adsorption). Under long-range electrostatic interactions, counterions tend to be nonspecifically bound to the polyion, and they are able to move along the axis of the polymer chain. If short-range, site-specific interactions are dominant, the counterions may bind on specific sites of the polyion associated to one or more charged groups. Organic counterions may bind polyions stronger than inorganic ions due to the hydrocarbon nature of the polymer chains.

The interactions between a polyion and the counterions can affect the pK_a, conductance, colligative properties, and diffusion of the counterions [64-66]. For polyelectrolytes with weak base or weak acid functional groups, the chain conformation can be a function of the degree of ionization and, hence, the pH of the solvent system [67]. The chain conformation of polyelectrolytes may also be affected by complexation with metal ions [68]. Polyelectrolytes typically have a very high effective concentration of charged groups and could constitute highly selective complexants useful as a cast of films for ion-exchange membranes production [69-72] as aqueous solutions in ultrafiltration [27, 73] or affinity dialysis [74, 75], membrane separation of ionic species. Some polyelectrolytes used in BAHLM research and their properties are presented in Table 6.5.

Information about physicochemical and complexation properties of polyvinylsulfonic acid (PVSH) or, more correctly, polyethylenesulfonic acid (PESH) [11] is much less available than that of the polyamines. PVSH (or PESH) is a simple, strong polyacid (see Fig. 6.6), so that partial ionization and possible hydrolysis effects, encountered with weak polyacids and their salts (e.g., polyacrylic acid and its salts), do not complicate the picture [76]. Commercially available sodium salt, PVSNa, is a brittle hygroscopic solid, soluble in water, but insoluble in organic solvents (methanol, dioxane, etc.) [77]. Viscosity behavior of the alkali metal salts was studied by several groups [28, 52, 75-77], but data are too scattered to be trustworthy. The PVS polyanion shows a definite selectivity in ion binding with alkali Li, Cs metals in the presence of Na, K halides [78].

Branched polyethylenimine (BPEI) is a strong polybase even at high ionic strength [79, 80] and forms complexes with acids. Its behavior is determined by properties such as the branched structure, presence of three different types of amine groups, strong neighboring interactions between various amine groups, and the compact structure in aqueous solution.

Table 6.5 Polyelectrolytes, used in referenced articles, and their properties

Number	Name of polyelectrolyte	Monomer, M_W	Polymer, M_W	pH aqueous solution	Source
1	Branched polyethylenimine (BPEI)	44.08	50,000–60,000	10.7	Aldrich
2	Polyvinylsulfonic acid (PVSH)	123.1	30,000–60,000	3.5	Aldrich
3	Polyvinylsulfonic acid sodium salt (PVSNa)	146.1	30,000–60,000	8.2	Aldrich
4	Polyvinylpirrolidone (PVP)	111.14	40,000–70,000	7.6	Aldrich
5	Poly(trimethylene phosphate) (PTP)	123.97	2000–3000	NA	Synthesized by authors [93, 94]
6	Poly(1,2-glycerol phosphate) (PGP)	153.97	9000–10,000	NA	Synthesized by authors [93, 94]
7	Poly(1-acetoxy-2,3-glycerol phosphate) (PAGP)	195.97	14,000–16,000	NA	Synthesized by authors [93, 94]

Note. NA, data not available.

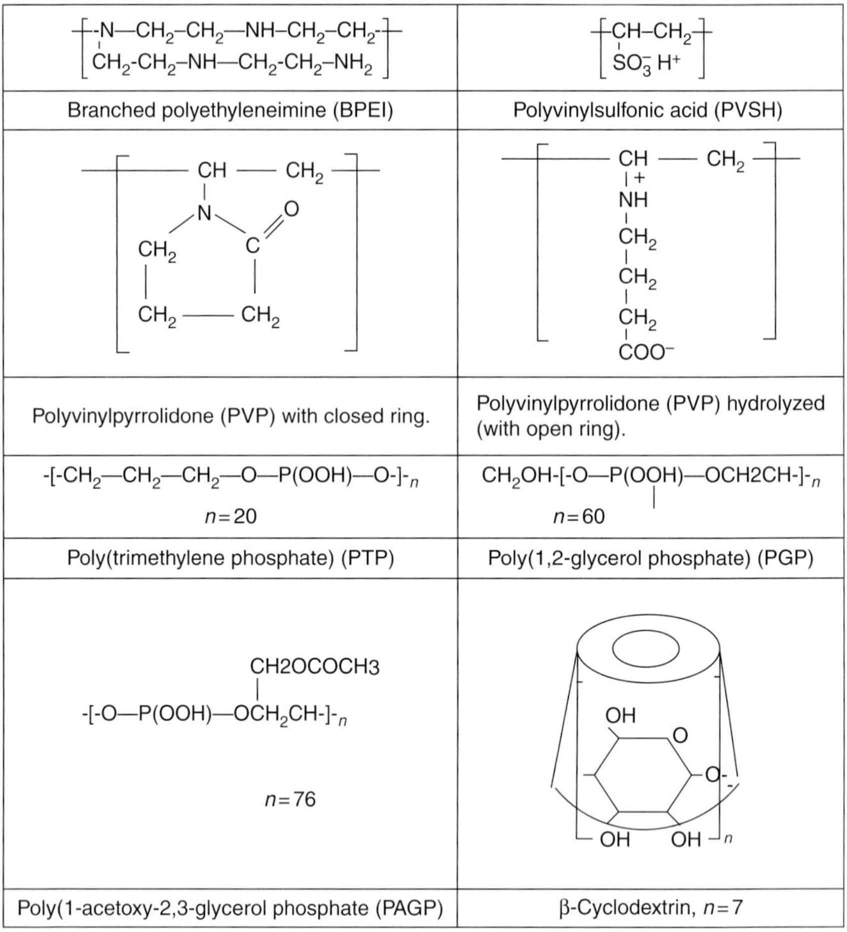

Figure 6.6 Structure schemes of polyelectrolytes used in the referenced works.

In contact with an acid, HX, BPEI protonates to form $BPEIH^{(+)}$, which forms an ion pair with the anion, $BPEIH^{(+)}X^{(-)}$.

Protonated BPEI has a great tendency to hydrogen bonding [9, 11, 80, 81]. $BPEIH(^{+})X(^{-})$ is a much weaker base ($pK_a = 5.3$) [82] than BPEI, but is capable of binding another HX molecule through hydrogen bond to form $BPEIH(^{+})X(^{-})HX$ [83]. That is the explanation for the H/BPEI molar ratios of about 2 reached in the carrier phase (see page 315 for explanation). In interaction with another acid, HY, $BPEIH(^{+})X(^{-})HX$ can be converted into any of the following complexes: $BPEIH(^{+})Y(^{-})HY$, $BPEIH(^{+})Y(^{-})HX$, and $BPEIH(^{+})X(^{-})HY$. (In some cases, a third acid molecule can be bound to form a species like $BPEI(^{+})X(^{-})(HX)_2$.)

3.3. Ion-exchange membranes as a barrier

Ion-exchange membranes consist of highly swollen polymers carrying fixed positive or negative charges. In an electrolyte solution, co-ions (ions with the same charge as the fixed on the membrane polymer) are excluded from the membrane matrix. This phenomenon is called Donnan exclusion [19]. Counterions with a higher valence will be preferentially adsorbed in the ion-exchange membranes. For counterions with equal valence, the ions with the highest polarizability or with the smallest hydration shell are preferred to be adsorbed [20]. Donnan exclusion is very effective, when the concentration of co-ions in the aqueous solution is low and that of fixed charges in the membrane is high. When the concentrations of co-ions in the membrane and in the solution have the same order of magnitude, the co-ions in the solution are not excluded anymore from the membrane. The membrane loses then its preferential selectivity for the counterions. So, for selective permeation of counterions through the membrane, the concentration of the bulk salt solution should be lower than that of the fixed ions in the membrane or in other words, the concentration of fixed ions in the membrane should be as high as possible. The charge species that are allowed to cross the membrane (counterions) will equilibrate across the membrane until the "Donnan equilibrium conditions" [19].

Membranes from Tokuyama Soda Corporation, used in the experiments, and their properties are presented in Table 6.6.

3.4. Anomalous osmosis: Ion-exchange membranes, polyelectrolytes, and osmosis

The difference between diffusion coefficients in water of low molecular weight substances (M_w = 10–200, molecular diameter 3 Å) and high molecular weight polyelectrolytes (M_w = 100,000, molecular diameter 62 Å) [52] is 1–2 orders of magnitude [28]. Inside the pores of the neutral hydrophilic or charged polymeric membranes (with the same pore size), it rises to 3–4 orders of magnitude [84]. On the other hand, dialysis experiments with polyanions present in the feed [5, 35, 36] showed that polyelectrolytes affect the diffusion of the counterions. The diffusion mechanism inside the membrane pores and in the LM boundary layers is considered as "hopping steps" diffusion of the counterions through the fixed anion charges and through the linear polymeric molecules with high concentration of the attached anionic functional groups.

The charged groups on the pore surface influence the selectivity of solute separation [85] and somewhat change the water flux in comparison with neutral hydrophilic membranes [52]. When a nonionic membrane separates two solutions, containing solute at different concentrations, water will be transferred from the side of lower to the side of higher osmotic

Table 6.6 The properties of the Tokuyama Soda Co. membranes, used in the referenced works

No.	Trademark	Type	Gel water (%)	Thickness (mm)	Permselectivity (%)	Capacity (meq/g)
1	Neosepta, CM-1	Cation-exchange, low electric resist	24	0.13–0.16	90	2.0–2.5
2	Neosepta, CM-2	Cation-exchange, low diffusion coef.	28	0.13–0.16	83	1.6–2.2
3	Neosepta, CMS	Cation-exchange, monocation permsel.	38	0.14–0.17	NA	2.0–2.5
4	Neosepta, ACH-45	Anion-exchange, low electric resist.	24	0.14–0.20	90	1.3–2.0
5	Neosepta, AFX	Anion-exchange, diffusion-dialys.	25	0.14–0.17	NA	1.5–2.0
6	Neosepta, AMX	Anion-exchange, high mechan. strength	45	0.16–0.18	NA	1.4–1.7
7	Neosepta, AM-3	Anion-exchange, low diffusion coef.	21	0.13–0.16	NA	1.3–2.0
8	Nafion-120 Du Pont, USA	Cation-exchange, low diffusion coef.	26.1	0.23	NA	0.85

Note. NA, not available.

pressure. However, with ion-exchange membranes, the water flux behaves in an anomalous manner [27, 86]. If the flux of water is in the same direction as the flux, found with a nonionic membrane of comparable porosity, that is, in the direction of the activity gradient, the increment is termed anomalous positive osmosis. On the other hand, if the flux is in the opposite direction to the activity gradient, the increment is termed anomalous negative osmosis. Negative osmotic flux may be greater than the flux brought about by the normal osmotic processes, resulting in an overall flow of water in the negative direction. However, if this anomalous negative osmotic flux is less than normal, it will only decrease the net flow.

Schlogl [86] states that the requirements for anomalous osmosis are a charged membrane and significantly different mobilities of solute ions. He postulated that the potential, arising from the difference in mobility of the diffusing ions, acted on the bound water in the membrane, with the bound water taking up the charge of the membrane counterions. This theory predicts anomalous negative osmosis for the dialysis of a strong acid using an anion-exchange membrane: the rapidly migrating protons would build up a positive potential on the dilute side of the membrane, which would exert an attractive force on the bound water containing the negatively charged counterions of the anion-exchange membrane. Thus, the bound water would be driven to the dilute side of the membrane, or in the direction reverse to normal osmotic flow, that is, negative osmosis.

Nondiffusible ionic species of the polyelectrolyte present in the carrier membrane solution of the BAHLM should affect the anomalous osmosis in both sides: in the feed-side and strip-side membranes. The disadvantage of conventional osmosis, dilution of the product, may be obviated if not turned out to the advantage of enrichment of the product.

3.5. Example of preliminary evaluation of the BAHLM system

Let us evaluate the BAHLM system for the removal of cadmium from industrial wet-process phosphoric acid (WPA) (for details, see Ref. [7]) containing 50 ppm Cd ($[\mathrm{Cd}]_{\mathrm{in}}$) and 20 ppm Cu ($[\mathrm{Cu}]_{\mathrm{in}}$), and reducing Cd to a final concentration of 1 ppm ($[\mathrm{Cd}]_{\mathrm{out}}$). We would also like to evaluate the BAHLM system for the selective separation of cadmium and copper from WPA. Using the known data presented in Table 6.7, the effective feed-side membrane area, (S_{F}), can be evaluated by equation:

$$S_{\mathrm{F}} = -\frac{U_{\mathrm{F}} \ln \dfrac{[\mathrm{Cd}]_{\mathrm{out}}}{[\mathrm{Cd}]_{\mathrm{in}}}}{K_{\mathrm{F/E}}}. \tag{51}$$

The magnitude of $S_{\mathrm{F}} = 6.9 \times 10^{-3}$ m^2 for feed flow, $U_{\mathrm{F}} = 1$ kg/h (7.25 $\times$ 10^{-4} m^3/h) or $S_{\mathrm{F}} = 9.52$ m^2 for feed flow, $U_{\mathrm{F}} = 1380$ kg/h (1 m^3/h).

Table 6.7 Data needed at preliminary evaluation of the AHLM system for removal of cadmium from industrial wet-process phosphoric acid (WPA) containing 50 ppm Cd and 50 ppm Cu

Parameters	Units	Cadmium	Copper	Reference
Distribution coefficient at forward extraction		4.7	41.4	[6]
Distribution coefficient at backward extraction		0.29	0.16	[6]
Feed-side mass-transfer coef., $K_{F/E}$	m/h	3.07	74.2	[6]
Strip-side mass-transfer coef., $K_{E/R}$	m/h	4.75×10^{-2}	8.28×10^{-2}	[6]
Maximum dynamic selectivity	Cu/Cd	23		[7]
Feed/carrier/strip volume ratios	kg	2.5/1/2.5		[6]

Notes. (1) Membrane: Tokayama Soda cation-exchange membrane CM-2. (2) Flow rates: feed = carrier = 90 cm^3/min or 5.4×10^{-3} m^3/h, strip = 60 cm^3/min or 3.6×10^{-3} m^3/h. (3) *Source*: . From Ref. [7] with permission.

Using Eq. (18), we can determine the contact time during which the cadmium concentration decreases to the magnitude of $\leq$1 ppm. At contact time = 24 h, we obtained $[Cd]_{out} = 0.3$ ppm. To decrease the contact time to 1 h, we need the effective feed-side membrane area, $S_F = 228.5$ m^2.

In the same manner, for the same contact time (1 h), we determine the effective strip-side membrane area: $S_{R\ Cd} = 305$ m^2 and $S_{R\ Cu} = 143$ m^2.

We can now design a preliminary pilot setup. We use a spiral-type, flowing liquid membrane module, developed by the Teramoto group [87, 88], in which the effective membrane area is about 40% of the total membrane area (the increase of the membrane area is mainly due to blocking of the membrane surface by spacers, and by the adhesive used to seal the sides of the module). For our system, the total feed-side membrane area is 570 m^2 and the total strip-side membrane area is 763 m^2, in which 360 m^2 is the area needed for the separation of the strip solution concentrated by copper. By designing standard, three-compartment spiral-type BAHLM modules, with 100 m^2 of the membrane on each side (feed and strip), and two-compartment modules, with 200 m^2 of the membrane, we will obtain a setup, of six standard three-compartment modules and one two-compartment module connected in consecutive order (see Fig. 6.7). After the fourth module, we will

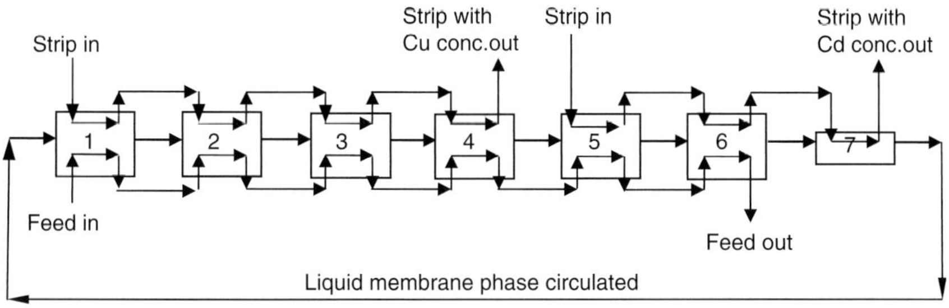

Figure 6.7 A scheme of the preliminary evaluated AHLM pilot system for treating 1 m^3/h of WPA. From Ref. [7] with permission.

obtain the strip solution, containing about 75% copper and 25% cadmium, and after the last (two-compartment) module the strip solution will contain about 25% copper and 75% cadmium.

The low mass-transfer rates of copper and cadmium on the strip side of the designed BAHLM system will lead to the accumulation of about 15% cadmium and copper in the liquid membrane phase after each cycle. Periodical regeneration of the liquid membrane solution will be needed to purify it from metal ions. Additional research for choosing polyelectrolytes with more effective transport characteristics is needed.

4. Selected Applications

4.1. Metal ions, salts separation

4.1.1. Separation with flat sheet ion-exchange membranes as barriers

Copper-cadmium separation from chloride aqueous solutions [4, 5, 14]. The copper-cadmium separation tests were carried out in a BAHLM module comprising (I) aqueous solutions of $CuCl_2 + CdCl_2$, 0.1–0.2 mol/kg each as the feed solutions, (II) an aqueous 0.5 mol/kg PVSNa solution as the liquid membrane (see Table 6.5), and (III) aqueous 0.5–2.0 mol/kg NaCl solutions as the receiving solutions [5]. Cation-exchange membranes Neosepta® CM-1 or CMS were used as barriers between the solutions (see Table 6.6).

Some experimental results are presented in Table 6.8. High dynamic selectivities were found: copper transport is up to 23 times greater than that of cadmium, combined with relatively high transport rates. The fluxes of copper are about one order of magnitude higher than those found in BOHLM systems (with water-immiscible carriers) consisting of organic carriers, such as LIX-64N [85] and Cyanex 302 [2, 14, 17, 89].

Figure 6.8A shows the fluxes of cadmium and copper ions and the Cu/Cd dynamic selectivity in various time intervals of the experiment. For details of the transport kinetics and mechanisms, see Ref. [5].

Table 6.8 Fluxes and selectivities, obtained at Cu/Cd separation by the BAHLM module

Exper. number	Compositions of the feed, membrane, and strip solutions (mol/kg)	Membrane, Tokayama Soda	Average fluxes to strip phase (mol/m^2 s)		Maximal dynamic selectivity Cu/Cd
			Cu	Cd	
1	0.2 $CuCl_2$–0.5 PVSNa–0.5 NaCl	CEM, CMS	$4.48 \times 10^{-6\ a}$		
2	0.2 $CuCl_2$–0.5 PVSNa–1.0 NaCl	CEM, CMS	$6.24 \times 10^{-6\ a}$		
3	(0.1 $CuCl_2$+0.1 $CdCl_2$)–0.5 PVSNa–1.0 NaCl	CEM, CMS	$2.61 \times 10^{-6\ b}$	$6.52 \times 10^{-7\ b}$	18
4	(0.1 $CuCl_2$+0.1 $CdCl_2$)–0.5 PVSNa–2.0 NaCl	CEM, CMS	$3.45 \times 10^{-6\ b}$	$3.40 \times 10^{-7\ b}$	23
5	(0.1 $CuCl_2$+0.1 $CdCl_2$)–0.5 BPEI–2.0 HNO_3	AEM, AM-3	$2.32 \times 10^{-8\ c}$	$3.30 \times 10^{-7\ c}$	0.04
6	(0.1 $CuCl_2$+0.1 $CdCl_2$)–(0.5 BPEI+0.5 HNO_3) –2.0 HNO_3	AEM, AM-3	$1.33 \times 10^{-8\ c}$	$3.58 \times 10^{-7\ c}$	0.02

Note. (1) Feed/carrier/strip ratio = 4/1/3. (2) Fluxes data were averaged in the experimenting time ranges (see Fig. 6.8A): [a]45–144 h; [b]30–180 h; [c]24–72 h.
(3) *Source*: From Ref. [5] with permission.

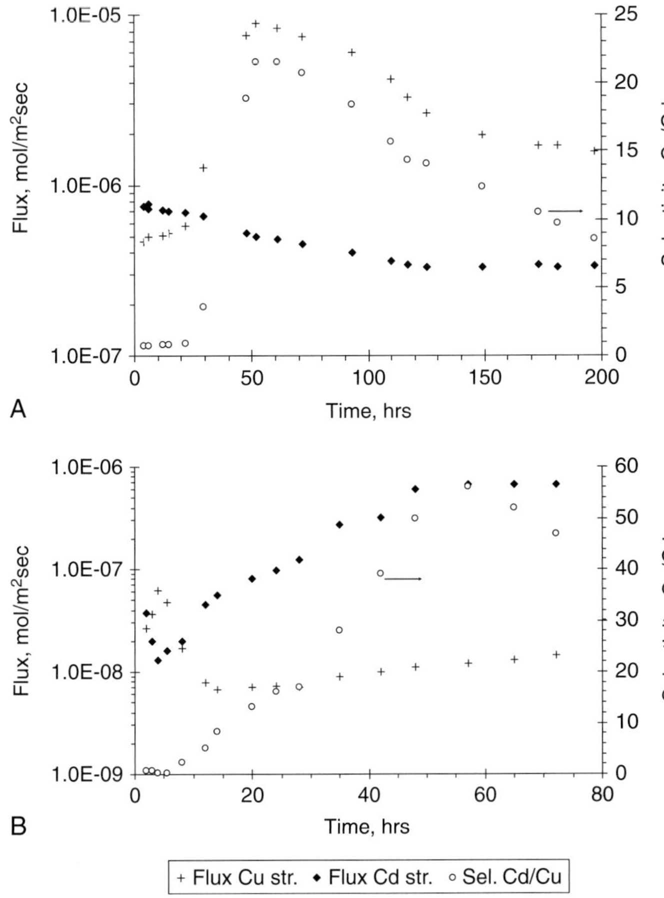

Figure 6.8 Transport rates of copper and cadmium to the strip phases and Cu-Cd selectivities. Initial feed concentration: 0.1 mol/kg $CuCl_2$ + 0.1 mol/kg $CdCl_2$. (A) Carrier: cation exchanger, 0.5 mol/kg PVSNa aqueous solution; membrane: Tokayama Soda cation-exchange CMS; strip: 2.0 mol/kg NaCl. (B) Carrier: anion-exchanger, 0.5 mol/kg BPEI aqueous solution; membrane: Tokayama Soda anion-exchange AM-3; strip: 2.0 mol/kg HNO_3. From Ref. [5] with permission.

For comparison to the above BAHLM results, Donnan dialysis tests were conducted. The fluxes were significantly higher ($J_{Cu} = 4.6 \times 10^{-5}$ mol/m^{2}s and J_{Cd}=7.7 $\times$ 10^{-6} mol/m^2 s) for the DD experiments, but the Cu/Cd dynamic selectivity is about five times lower.

Copper-cadmium separation in the anionic complex form [5]. The BAHLM copper-cadmium separation was tested using (I) an aqueous solution of $CuCl_2$ + $CdCl_2$, 0.1 mol/kg each, as the feed solution, (II) an aqueous 0.5 mol/kg BPEI solution as the LM, and (III) aqueous 1.0–2.0 mol/kg nitric acid solutions, as the receiving solution. Nitric acid was chosen for the strip solution to allow detection of the chlorine anion cotransport from the feed to the strip solution. Anion-exchange membrane, Neosepta® AM-3, was used as a barrier. Some of the experimental results are presented in Table 6.8 and Figure 6.8B.

A good cadmium-copper selectivity was obtained. The cadmium flux to the strip solution (3.3×10^{-7} mol/m^2 s) is more than an order of magnitude higher than that of copper (2.3×10^{-8} mol/m^2 s). The transport rate of copper ions in BPEI carrier-containing BAHLM is about two orders of magnitude lower than that in BAHLM with PVSH as a carrier, while that of cadmium ion is about the same.

Cadmium, copper, and zinc separation from wet-process phosphoric acid [3, 7, 15, 17]. Selective separation of copper and cadmium from the WPA containing 55.6 ppm Cd and 50.6 ppm Cu was tested in the AHLM module with 0.5 mol/kg PVSH aqueous solution as carrier, 2 mol/kg HCl as stripping solution, and Neosepta CM-2 membranes. Results are presented in Fig. 6.9.

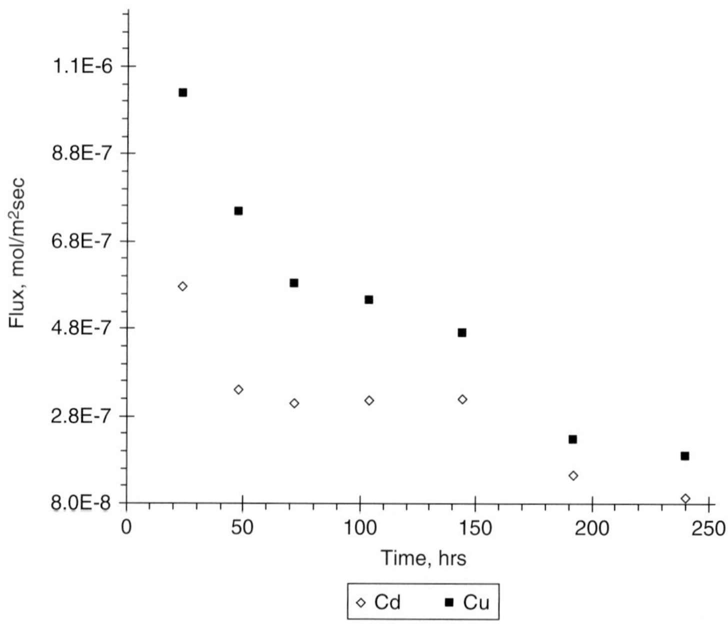

Figure 6.9 Transport rates of copper and cadmium to the strip phase at AHLM treatment of industrial WPA. Initial feed: industrial WPA composed of 55.6 ppm Cd and 50.6 ppm Cu. Carrier: 0.5 mol/kg PVSH aqueous solution. Membrane: Tokayama Soda cation-exchange CM-2; strip: 2.0 mol/kg HCl. From Ref. [7] with permission.

Cu/Cd dynamic selectivity profiles in the strip phase, obtained experimentally and calculated using model equations, are presented in Fig. 6.10. The calculated maximum selectivity of The Cu/Cd separation factor $A_{(Cu/Cd)} = 3.0$ and selectivity $S_{(Cu/Cd)} = 2.7$. The model calculated and experimentally obtained data show similar numbers. The shapes of the model calculated and the experimental curves are similar and show $S_{DR\,(Cu/Cd)} \approx 1$ after 45 h of trial (calculated) or after 140 h (experimental).

Transport of metal ions by polyphosphates [90, 91]. The scheme of experimental setup is presented in Fig. 6.11. The diaphragm cell of the tube-in-tube type, made of Teflon immersed into a glass vessel containing aqueous solution of polyphosphates, magnetically stirred at the rate 250 rpm. Solutions of polyphosphates (examples of structure schemes of polyphosphates used see in Table 6.5 and Fig. 6.6) were preliminary purified by dialysis with an ion-exchange membrane (the same as in the transport systems) to remove any low molecular weight impurities. The multimembrane hybrid system

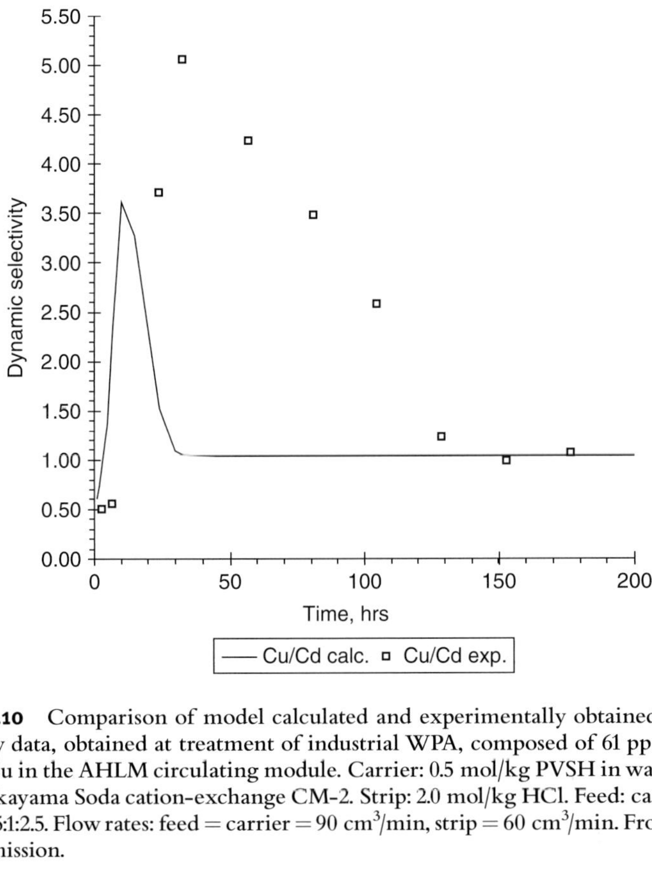

Figure 6.10 Comparison of model calculated and experimentally obtained dynamic selectivity data, obtained at treatment of industrial WPA, composed of 61 ppm Cd and 55 ppm Cu in the AHLM circulating module. Carrier: 0.5 mol/kg PVSH in water. Membrane: Tokayama Soda cation-exchange CM-2. Strip: 2.0 mol/kg HCl. Feed: carrier: strip ratio = 2.5:1:2.5. Flow rates: feed = carrier = 90 cm^3/min, strip = 60 cm^3/min. From Ref. [7] with permission.

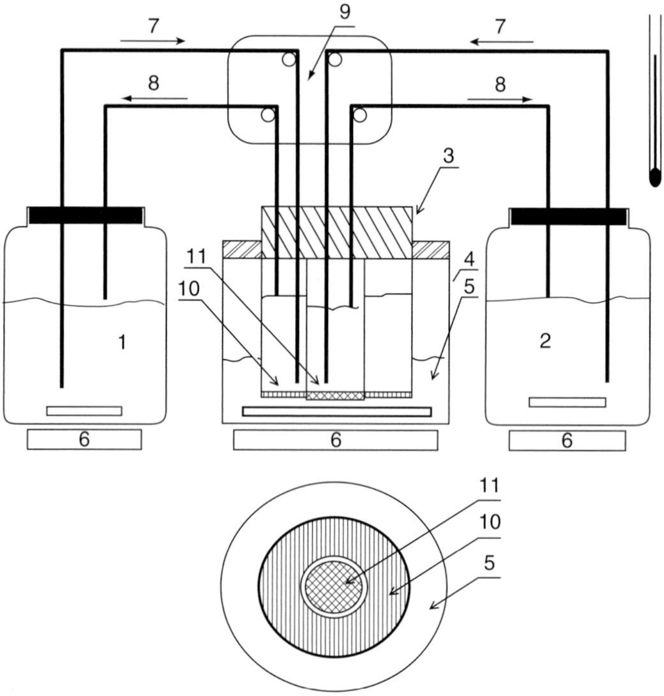

Figure 6.11 Schematic representation of experimental arrangement for a multimembrane hybrid system: feed solution (1), strip solution (2), tube-in-tube diaphragm cell (3), glass vessel (4), bulk liquid membrane (5), magnetic stirrer (6), feed and strip inlet (7), feed and strip outlet (8), peristaltic pump (9), feed (10) and strip (11) ion-exchange membranes. From Ref. [90] with permission.

(MHS) consisted of two cation-exchange polymer membranes Nafion-120. The feed chamber contained the equimolar solutions of Zn(II), Cu(II), Mn(II), Ni(II), and Co(II) sulfates, 2×10^{-3} mol/dm^3 each; the strip was 0.1 mol/dm^3 solution of sulfuric acid. The feed and strip solutions were circulated using a multichannel peristaltic pump. Experiments showed that after a transient period (up to 15 h), the output fluxes have attained their constant value and assumed to be quasistationary. Examples of the values of the fluxes are listed in Table 6.9.

The total fluxes seem typical for diffusion-limited phenomena [90]. As a rule, the systems operate under nonstationary regime or time-dependant conditions. So, the authors [90, 91] come to the conclusion of time-dependant separation factor or dynamic selectivity (see Section 3.1). Examples of nonstationary separation factors for several polyphosphates are presented in Fig. 6.12. These results demonstrate that polyphosphates preferentially transport Zn(II) and Cu(II) ions with the separation factor amounting to 2–8.

Table 6.9 Stationary fluxes in the competitive transport of divalent metal ions in the MHS by polyphosphate carriers

Carrier aqueous solution (see Table 6.5):	$10^9 \times$ Flux in mol/(cm^2 s)					
	Cu	Co	Ni	Mn	Zn	Total
PTP	1.20	1.13	1.10	1.09	0.99	5.52
PTP(Na)[a]	0.74	0.73	0.67	0.66	0.65	3.45
PGF	1.25	1.17	1.11	1.13	1.08	5.73
PAGF	0.81	0.68	0.73	0.74	0.79	3.74
Dialysis[b]	0.21	0.01	0.007	0.27	0.32	0.82

Notes. (1) [a]With Na_2SO_4 in the strip solution instead of H_2SO_4. [b]Only Nafion-120 membrane. (2) Initial composition: feed: aqueous equimolar solutions of Zn(II), Cu(II), Mn(II), Ni(II), and Co(II) sulfates, 2×10^{-3} mol/dm^3 each; carrier: 0.1 mol/dm^3 PTP, or PGF or PAGF aqueous solution; strip: 0.1 mol/dm^3 aqueous solution of sulfuric acid. Membrane barriers: cation-exchange Nafion-120. (3) *Source*: From Ref. [90] with permission.

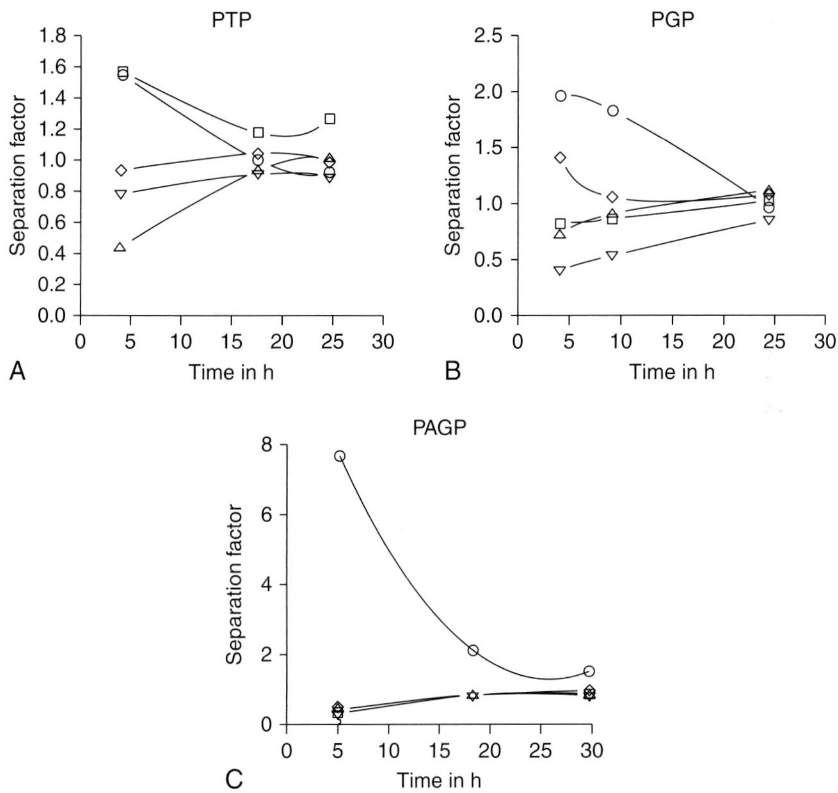

Figure 6.12 Nonstationary separation factors versus time of metal ions transport from the feed contained the equimolar solutions of Zn(II), Cu(II), Mn(II), Ni(II), and Co(II) sulfates, 2×10^{-3} mol/dm^3 each; through the LM aqueous solution of (A) PTP, (B) PGP, and (C) PAGP to the strip of 0.1 mol/dm^3 solution of sulfuric acid. Membrane barriers: cation-exchange Nafion-120. Cu(II) (□), Co(II) (Δ), Ni(II) (◇), Mn(II) (∇), and Zn(II) (○). From Ref. [90] with permission.

4.1.2. Separation with neutral hollow-fiber units [74]

Selective separation and concentration of both cations and anions using water-soluble polymer solutions LM as carriers and hollow-fiber units (artificial kidneys) as membrane barrier were tested. The authors termed the process as *affinity dialysis* [74]. Hollow fiber units of Spectrum Medical Industries, Inc. with fibers of 5000 molecular weight cutoff and 150 cm^2 surface area from Spectrapor were used in the experiments.

WSP used in this work were 2.5% w/v aqueous solution of poly (2-acrylamido-2-methyl-1-propanesulfonic acid) (poly-AMPS) for cation separations and 5% w/v solution of poly(ethylenimine) (PEI) for anion separations.

Ca/Na, and Cu/Zn mixtures were tested for the cation separation and concentration, and chromate/chloride mixture for the anion separation. For Ca/Na system, concentration factor obtained was 55 at Ca/Na selectivity equal 39. For the Cu/Zn system, the concentration factor was 50 at Cu/Zn selectivity equal 70. For potassium chromate/sodium chloride system, the concentration factor was 75 at chromate/chloride selectivity equal 30.

Figure 6.13 illustrates continuous recycle system assembled by the authors for long-term experiments. The miniplant was operated for several months separating copper from zinc or chromate from chloride.

4.2. Biotechnological separations: Carboxylic acids [5, 16]

Transport of carboxylic acids into a strong basic strip solution. BAHLM systems were tested for the transport of carboxylic acids such as lactic (HLac), citric (H_3Cit), and acetic (HAc) or their anions. In feed solutions containing a mixture of HLac and H_3Cit or HAc (or their anions), the initial molar ratio between the former and the other acid is 8/1. Aqueous 0.5 mol/kg BPEI solution was used as the carrier, anion-exchange membranes ACH-45T or AM-3 were used as the barrier between the phases, and 1 mol/kg sodium hydroxide was used as the receiving (strip) solution.

Table 6.10 presents some experimental results of separation from HLac + HAc and HLac + H_3Cit solutions. The selectivity, calculated as the flux ratios divided by the concentration ratio in the feed solution, is 1.7 in favor of HLac in transport from HLac + HAc solutions and 2.8 in favor of H_3Cit for the HLac + H_3Cit case. The transport rates are relatively high.

In BAHLM systems, with acids in the feed solution and NaOH in the strip, the driving force for transport is the acid-base neutralization. Being relatively strong, this driving force results in relatively high transport rates. The transport selectivity is affected by the interactions at the two interfaces, feed-carrier and carrier-strip.

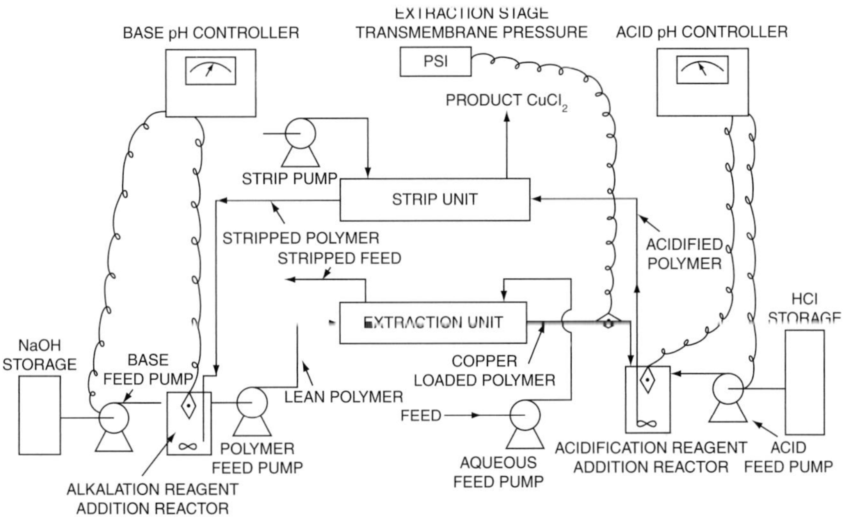

Figure 6.13 Schematic of a complete affinity dialysis miniplant with polymer solution stripping, regeneration, and recycle. From Ref. [74] with permission.

At transport from an acidic feed solution into NaOH containing strip solution, permselectivity to the stronger acid is expected, mainly due to the interaction at the feed-carrier interface [92]. Results, showing selectivity to lactic acid (pK_a of 3.86) over acetic acid (pK_a of 4.76) and to citric acid (pK_a of 3.13) over lactic acid, are in agreement with analysis, presented in Ref. [5].

Transport of carboxylic acid or its salt into a mineral acid-containing strip solution. The BAHLM system—contained 1 mol/kg NaLac feed solution, 0.5 mol/kg BPEI carrier solution, and 1.1 mol/kg HNO_3 strip solutions [5]—was tested. AM-3 membranes were used as a barrier. Flux of lactate to the strip phase versus time is presented in Fig. 6.14.

At the beginning of the trial, the main transport is of nitric acid from the strip solution into the carrier solution with formation of $BPEIH^{(+)}NO_3^{(-)}$ (I′) ion-pair complex. The fluxes of the other components are very small or negligible. It seems that there is no interaction between the Lac^- anions and the nonprotonated BPEI. At about 0.6 molar ratio of the HNO_3/BPEI, the transfer of HLac or Lac^- to the strip solution starts. The nitric acid transport into the carrier phase continues, approaching a constant level equivalent to HNO_3/BPEI molar ratio of nearly 2. At this stage, the BPEIH(+)NO_3(−) HNO_3 (II′) complex with H-bonding to the ion pair takes place at the

Table 6.10 Fluxes and selectivity, obtained for different BAHLM transport systems

Feed, carrier, and strip solution compositions (mol/kg)	Membrane type	Average fluxes (mol/m² s)					Selectivity		Selectivity counterions	
		$Na_{str.}$	Lac_{strip}	$Cit._{strip}$	Ac_{strip}	Na_{feed}	Lac/Ac	Cit/Lac	Na/Lac	Lac/Na
[0.8 HLa + 0.1 HAc] – 0.5 BPEI – 1.1 NaOH	Anion, ACH-45 T		2.60×10^{-5}		2.30×10^{-6}		1.7			
[0.8 HLa + 0.1 H_3Cit] – 0.5 BPEI – 1.1 NaOH	Anion, ACH-45 T		2.50×10^{-5}	6.70×10^{-6}				2.8		
1.0 NaLa – 0.5 PVSA – 1.1 HNO_3	Cation, CM-1	7.40×10^{-4}	4.20×10^{-6}						400	
1.0 NaLa – 0.5 BPEI – 1.1 HNO_3	Anion, AM-3	3.20×10^{-6}	2.70×10^{-5}							8.3
1.0 HLa – 0.5 BPEI – 1.1 NaOH	Anion, AM-3		4.10×10^{-5}			1.70×10^{-6}				25

Notes. (1) Feed/carrier/strip ratio = 4/1/3. (2) Fluxes data were averaged in the experimenting time ranges 24–72 h. (3) *Source*: From Ref. [5] with permission.

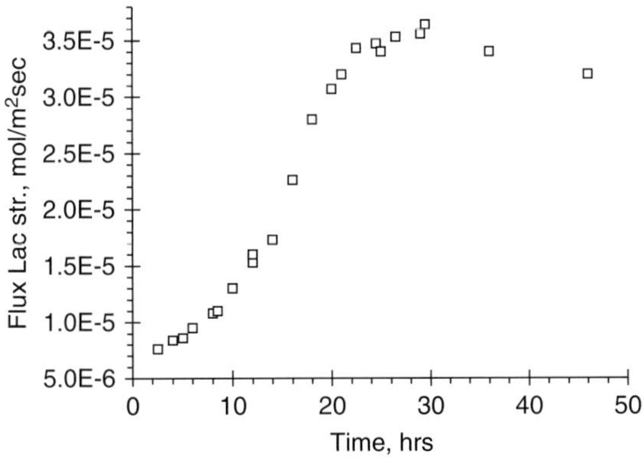

Figure 6.14 Transport rates of lactate anion into the strip solution in BAHLM system with 0.5 mol/kg BPEI aqueous solution as a carrier and Tokayama Soda AM-3 membranes. Initial compositions: feed: 1.0 mol/kg NaLac, strip: 1.1 mol/kg HNO_3. From Ref. [5] with permission.

carrier-strip interface. At the feed-carrier interface, an anion-exchange-type reaction takes place, which could result in different species, formed in the carrier solution. The flux of HLac or Lac(−) to the strip solution increases drastically and shows a maximum at maximum proton concentration in the LM phase and then levels off or slightly decreases.

Experiments in BAHLM systems comprising sodium salts of the acids in the feed solution and nitric acid in the strip solution showed 2–4 times higher selectivities than those found in systems with acid mixture feed and NaOH in the strip (see Table 6.10). The transport rates were about 40% smaller.

The selectivity results for both the lactic + acetic and lactic + citric mixtures at transport into HNO_3-containing strip solution show the greater fluxes of the stronger acid. This leads to the conclusion that the selectivity is determined by the transport from the feed to the carrier solution (for details, see Ref. [5]).

Testing the BAHLM system with weaker complexants as a carrier, for example, with polyvinylpyrrolidone (PVP), much higher selectivities were obtained (a factor of about an order of magnitude), at lower transport rates (a factor of 2–3 times) [5].

4.3. Isomer separation by LM with water-soluble polymers

4.3.1. Separation using hollow-fiber contained liquid membrane permeator (HFCLMP)[93]

Continuous separation of different isomeric mixtures of organic compounds has been studied by means of a HFCLMP. Detailed description of HFCLMP system and application data are presented in Chapter 5 and

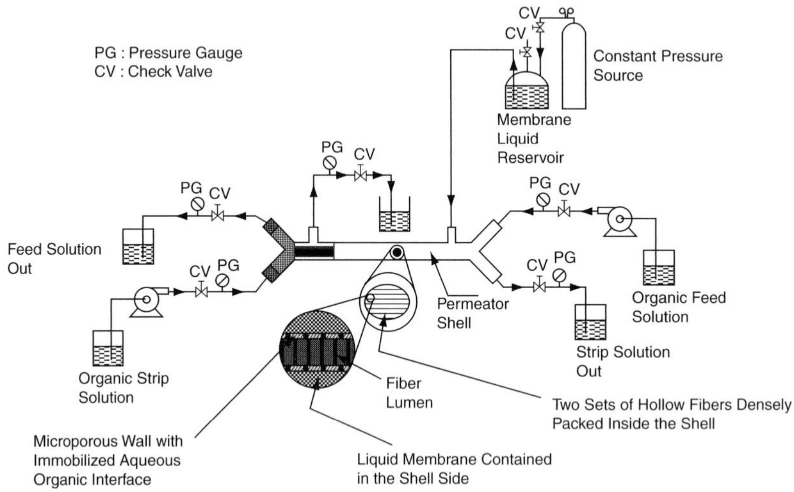

Figure 6.15 Schematic hollow-fiber contained liquid membrane permeator (HFCLMP) used for stereoisomers separation. From Ref. [93] with permission.

Ref. [57]. The difference of this application is the separation process from organic solution of isomer mixture (feed) to organic solution of strip through the aqueous solution of cyclodextrin (CD) as a carrier. Schematic diagram of the HFCLMP system is presented in Fig. 6.15.

Feed and strip organic solutions are flowing through the bores of a separate set of hollow fibers and the aqueous LM containing β-CD on the shell side. Cyclodextrins are a unique group of water-soluble host molecules. They are cyclic oligomers of glucose produced by enzymatic degradation of starch [94]. β-CD with internal diameter of about 6 Å can complex many substances, among them the species with benzene and naphthalene groups. The hydrophobic interaction between host and guest is assumed to be the major driving force for the binding ability of CD.

Separation of the stereoisomers *cis*- and *trans*-stilbene, or *p*-nitroaniline (*p*-NA) and *o*-NA was studied using 0.02–0.7 M β-CD aqueous solution as a LM. Celgard X-10 microporous hollow fibers were used for permeators design. The characteristics of the HFCLMP module see in Ref. [93]. For nitroaniline isomers separation, an equimolar 0.005 M solution of *o*-NA and *p*-NA in 80% 1-octanol and 20% heptane was used as a feed. For stilbene isomers separation, an equimolar 0.01 M solution of *cis*- and *trans*-stilbene in pure heptane was used as a feed.

For the strip solutions, the same solvent mixture or solvent was used without any isomers. Some results are presented in Table 6.11. With increasing of β-CD concentration up to 0.7 M in LM, the *p*-NA concentration in the aqueous LM increases about seven times in comparison

Table 6.11 Separation of isomers by hollow-fiber contained liquid membrane permeator (HFCLMP) using β-cyclodextrin (β-CD) as a carrier in aqueous liquid membrane

Time of trial (h)	β-CD concentration (M)	Isomer/isomer	Selectivity	Permeance of selected isomer (cm/m)
Separation of *p*-nitroaniline and *o*-nitroaniline isomers				
0.6	0.012	*p*-NA/*o*-NA	2.0	8.5×10^{-6}
1.75			2.03	1.0×10^{-5}
3.25			1.91	9.6×10^{-6}
3	0.04	*p*-NA/*o*-NA	2.83	3.22×10^{-5}
15.5			3.01	3.52×10^{-5}
3	0.7	*p*-NA/*o*-NA	4.31	2.32×10^{-5}
8			3.63	2.82×10^{-5}
24			4.63	3.52×10^{-5}
Separation of *cis*- and *trans*-stilbene isomers				
1	0.02	*Cis/trans*	1.3	3.8×10^{-5}
3			0.78	1.9×10^{-5}
7			0.9	2.6×10^{-5}
9			0.88	1.45×10^{-5}
1.8	0.2	*Cis/trans*	1.3	6.0×10^{-5}
7.5			1.44	2.2×10^{-5}
18			1.39	1.8×10^{-5}

Note. (1) Initial compositions: feed: (a) equimolar 0.005 M solution of *o*-NA and *p*-NA in 80% 1-octanol and 20% heptane, and (b) equimolar 0.01 M solution of *cis*- and *trans*-stilbene in pure heptane; strip: (a) 80% 1-octanol and 20% heptane and (b) pure heptane. Membrane: Celgard X-10 microporous hollow fibers. (2) *Source*: From Ref. [93] with permission.

with *o*-NA. The overall selectivity for *p*-NA over *o*-NA is about 4.6. The maximum selectivity of *cis*-stilbene over *trans*-stilbene was always less than 2, at β-CD concentrations increasing from 0.02 to 0.2 M. The authors suggest that the system developed has practical interest.

4.3.2. Separation with supported liquid membrane (SLM) [95]

Water-soluble cyclophanes QCP 44 or QCP 66 were impregnated as a carrier into the pores of membrane support. The SLMs showed isomer-selectivity for methylnaphthalenes (MNs), ethylnaphthalenes (ENs), dimethylnaphthalenes (DMNs), and three-ring aromatic compounds. The isomer selectivity of QCP 44 was completely opposite to that of QCP 66; the flux order of QCP 44 was 2-MN > 1-MN, 2-EN > 1-EN, 2,6-DMN > 1,5-DMN, and anthracene > phenanthrene; and that of QCP 66 was 1-MN > 2-MN, 1-EN > 2-EN, 1,5-DMN and phenanthrene > anthracene. Cyclophanes exhibited facilitated transport for aromatic but not for aliphatic compounds. The tests indicated that the rate-determining step of the transport was the diffusion process of the carrier-solute complex in the membrane.

4.4. Carrier leakage [1-7, 14-17]

Leakage of the polyelectrolyte carriers to the feed and strip solutions was tested using UV spectrophotometric techniques [26, 96]. The wavelengths used were 275 nm for PVSH, 260 nm for PVP, and 248 nm for BPEI. The reference was deionized water. The concentrations of the carrier in the feed and strip solutions were close to or under the detection level (2-3 ppm) of the technique in all measured solutions.

4.5. Membrane lifetime [1-7,14-17]

Membrane stability tests were followed through the mass-transfer rates. During 300 h of experiment, the membranes showed very stable fluxes of solutes. After washing in water, the membrane samples were used repeatedly and the fluxes were stable. The working time of some membrane samples was more than 600 h.

5. Summary Remarks

1. In comparison with other liquid membrane systems, BAHLM technology has some decisive advantages, such as:
   - A combination of relatively high transport rates and high selectivities, especially for cation/cation separation
   - No contamination of the feed and strip solutions by the water-soluble polymers used as carriers
   - Long membrane lifetime: no fouling and blinding was observed; the membranes are completely regenerable and used repeatedly
   - No special requirements for membrane resistance in organic solvents, lesser limitations on variability of the membrane pore sizes
   - The BAHLM modules may be based on commercially available ion-exchange hollow-fiber or spiral membrane modules and equipment
2. Manipulation by the carrier concentration (and therefore, by the LMF potential), by the volume of the circulating bulk LM solution, or by both, enables to obtain selectivity that is as close to its highest level as necessary for the process developed. It is one of the biggest advantages of the BAHLM system in comparison with other liquid membrane technologies.
3. The BAHLM may be particularly attractive for wastewater treatment due to its high selectivity, which allows transfer of small amounts

(50–1000 ppm) of toxic metals (instead of large volumes of solutions) and their concentration in the strip solutions. In some cases, those metal concentrates could be selectively separated.

4. The BAHLM is an environmentally friendly and economical technology due to operation at ambient temperature, low energy requirements, low chemical consumption, and no formation of parasitic byproducts.
5. Intensive research is needed in the development of:
   (a) Stable polyelectrolytes with improved properties, their regeneration from the inorganic salts
   (b) Anomalous osmosis with polyelectrolytes
   (c) Stable micro- and nanoporous ion-exchange membranes with high density of charged functional groups
   (d) Complexation chemistry and theory of the BAHLM transport, etc.
6. Many directions for the application of the BAHLM systems require development. These are mixtures after degradation of organic compounds, products of fermentation, biological mixtures (hormones, peptides, etc.), drug conversion and selective separation, catalytic reaction enhancement and selective separation, wastewater and water treatment, etc. In the authors' envision, some of these potential directions in new research and development areas are discussed in Chapter 9.

REFERENCES

1. Eyal AM, Kislik VS. Membrane Recovery of Metal Pollutants from Waste Waters of the Fertilizers Industry (MERMEP). Report of Joint Research Project, Avicenne, 1994.
2. Eyal AM, Kislik VS. Aqueous hybrid liquid membrane (AHLM) processes for metal recovery. In: Membrane Recovery of Metal Pollutants from Waste Waters of the Fertilizers Industry (MERMEP). Second year Report of Joint Research Project, Avicenne, 1995.
3. Eyal AM, Kislik VS. Aqueous hybrid liquid membrane (AHLM) process for metal recovery. In: Membrane Recovery of Metal Pollutants from Waste Waters of the Fertilizers Industry (MERMEP). Third year Report of Joint Research Project, Avicenne, 1996.
4. Eyal A, Kislik V. Separation of solutes (metal ions, acids) by hybrid liquid membrane (HLM) with water-soluble polymers as a carrier and ion-exchange membranes. In: Proceedings of International Conference on Membranes, ICOM'96, 1996; S-8-2-3: 584-585.
5. Eyal A, Kislik V. Aqueous hybrid liquid membrane: A novel system for separation of solutes using water-soluble polymers as carriers. J Membr Sci 1999; 161: 207-221.
6. Kislik V, Eyal A. Aqueous hybrid liquid membrane process for metal separation. Part I. A model for transport kinetics and its experimental verification. J Membr Sci 2000; 169: 119-132.
7. Kislik V, Eyal A. Aqueous hybrid liquid membrane process for metal separation. Part II. Selectivity of metals separation from wet-process phosphoric acid. J Membr Sci 2000; 169: 133-146.

8. Molyneux P. *Water-Soluble Synthetic Polymers: Properties and Behavior.* CRC Press, Boca Raton, FL, 1984; Vol. 1.
9. Molyneux P. *Water-Soluble Synthetic Polymers: Properties and Behavior.* CRC Press, Boca Raton, FL, 1984; Vol. 2.
10. Shalaby SW, McCormick CL, Butler GB, eds. *Water-Soluble Polymers; Synthesis, Solution Properties, and Application.* Symposium Series 467, American Chemical Society, Washington, DC, 1991.
11. Eisenberg H, Ram M. Aqueous solutions of polyvinylsulfonic acid: Phase separation and specific interaction with ions, viscosity, conductance and potentiometry. J Phys Chem 1959; 63: 671-678.
12. Von Zelewsky A, Barbossa L, Schlapfer C. Poly(ethylenimines) as Bronsted bases and as ligands for metal ions. Coordin Chem Rev 1993; 123: 229-235.
13. Chandramouli GVR, Schlapfer CW, von Zelewsky A. EPR investigation of the second sphere coordination of $[Cr(CN)_5(NO)]_3^-$ with protonated polyethyleneimine. Mag Res Chem 1991; 29: S16-S28.
14. Kislik V, Eyal A. Comparison of hybrid liquid membrane (HLM) and aqueous hybrid liquid membrane (AHLM) technologies in separation of heavy metals from wastewaters. In: Proceedings of Abstracts EURO-MEMBRANE-2000, Israel, 2000: 237-238.
15. Kislik V, Eyal A. Heavy metals removal from wastewaters of phosphoric acid production. A comparison of hybrid liquid membrane (HLM) and aqueous hybrid liquid membrane (AHLM) technologies. In: Proceedings of Conference on Membranes in Drinking and Industrial Water Production, Paris, France, 2000; Desalination Publications, Italy, 2000; Vol. I: 503-514.
16. Kislik V, Eyal A. Application of hydrophobic (water-immiscible) and hydrophilic (water-soluble) liquid membranes in the carboxylic acids separation technologies: Comparative analysis. In: Proceedings of Abstracts of 41st Microsymposium Polymer Membranes, Prague, 2001: 86-87.
17. Kislik V, Eyal A. Heavy metals removal from wastewaters of phosphoric acid production. A comparison of hybrid liquid membrane (HLM) and aqueous hybrid liquid membrane (AHLM) technologies. Water Sci Water Technol: Water Supply 2001; 1(5/6): 119-129.
18. Donnan FG. The theory of membrane equilibria. Chem Rev 1924; 1: 73-81.
19. Donnan FG. Theory of membrane equilibria and membrane potentials in the presence of non-dialising electrolytes. A contribution to physical-chemical physiology. J Membr Sci 1995; 100: 805-814.
20. Sionkowski G, Wodzki R. Recovery and concentration of metal ions. I. Donnan dialysis. Sep Sci Technol 1995; 30: 441-452.
21. Miyoshi H. Diffusion coefficients of ions through ion exchange membrane in Donnan dialysis using ions of different valence. J Membr Sci 1998; 141: 101-109.
22. Miyoshi H. Diffusion coefficients of ions through ion exchange membrane in Donnan dialysis using ions of the same valence. Chem Eng Sci 1997; 52(7): 1087-1095.
23. Miyoshi H. Donnan dialysis with ion-exchange membranes. I. Theoretical equation. Sep Sci Technol 1996; 31(15): 2117-2123.
24. Ktari T, Larchet C, Auclair B. Mass transfer characterization in Donnan dialysis. J Membr Sci 1993; 84: 53-61.
25. Ktari T, Larchet C, Auclair B. Simplified flux equation in Donnan dialysis. J Membr Sci 1987; 32: 251-259.
26. Zheleznov A, Windmoller D, Korner S, Boddeker KW. Dialytic transport of carboxylic acids through an anion exchange membrane. J Membr Sci 1998; 139: 137-146.
27. Hansen RD, Anderson ML. Anomalous osmosis in dialysis of acids with anion exchange membranes. I & E C Fundamentals 1967; 6(4): 543-551.

28. Tishchenko G. Neosepta microporous ion-exchange membranes in dialysis desalination of immunoglobulin fraction of mouse ascitic fluids. J Membr Sci 1996; 113: 237-248.
29. Helfferich F. *Ion Exchange*. McGraw-Hill, New York, 1962; Chapter 8.
30. Helfferich F. Ion-exchange kinetics. III. J Am Chem Soc 1962; 66: 39-52.
31. Helfferich F.; Ion-exchange kinetics. IV. J Am Chem Soc 1963; 67: 1157-1164.
32. Lake MA, Melsheimer SS. Mass transfer characterization of Donnan dialysis. AIChE J 1978; 24(1): 130-138.
33. Cwirko EH, Carbonell RG. A theoretical analysis of Donnan dialysis across charged porous membranes. J Membr Sci 1990; 48: 155-164.
34. Cwirko EH, Carbonell RG. Transport of electrolytes in charged pores: Analysis using the method of spatial averaging. J Coll Interf Sci 1989; 129(2): 513-519.
35. Higa M, Tanioka A, Kira A. A novel measurement method of Donnan potential at an interface between a charged membrane and mixed salt solution. J Membr Sci 1998; 140: 213-221.
36. Higa M, Tanioka A, Kira A. A study of ion permeation across a charged membrane in multicomponent ion systems as a function of membrane charge density. J Membr Sci 1990; 49: 145-151.
37. Ho WSW, Sirkar KK, eds. *Membrane Handbook*. VNR, New York, 1992; 267-273.
38. Ho WSW, Sirkar KK, eds. *Membrane Handbook*. VNR, New York, 1992; 277, 485-493.
39. Ho WSW, Sirkar KK, eds. *Membrane Handbook*. VNR, New York, 1992; 401-403.
40. Ho WSW, Sirkar KK, eds. *Membrane Handbook*. VNR, New York, 1992; 167-183.
41. Onsager L. Reciprocal relations in irreversible processes, I. J Phys Rev 1931; 37: 405-413.
42. Onsager L. Reciprocal relations in irreversible processes, II. J Phys Rev 1931; 38: 2265-2272.
43. Kedem O, Katchalsky A. Thermodynamic analysis of permeability of biological membranes to nonelectrolytes. Biochim Biophys Acta 1958; 27: 229-238.
44. Kedem O, Katchalsky A. Thermodynamics of flow processes in biological systems. Biophys J 1962; 2(Suppl.): 53-64.
45. Staverman A. Theory of measurement of osmotic pressure. Rec Trav Chim 1951; 70: 344-350.
46. Kislik V, Eyal A. Hybrid liquid membrane (HLM) system in separation technologies. J Membr Sci 1996; 111: 259-272.
47. Spriggs H, Li N. Liquid permeation through polymeric membranes. In: Mears P, ed. *Membrane Separation Processes*. Elsevier Publishers, Amsterdam, 1976: 217-225.
48. Noble RD. Analysis of facilitated transport with fixed site carrier membranes. J Membr Sci 1990; 50: 207-214.
49. Noble RD. Facilitated transport mechanism in fixed site carrier membranes. J Membr Sci 1991; 60: 297-306.
50. Noble RD, Way JD. Facilitated transport of CO_2 in ion exchange membranes. AIChE J 1987; 33(3): 480-492.
51. LeBlanc OH. Facilitated transport in ion-exchange membranes. J Membr Sci 1980; 6: 339-347.
52. Tishchenko G. Diffusion permeability of track-membranes. Dokl Akad Nauk SSSR 1989; 308: 655-664.
53. Behr JP, Kirch M, Lehn JM. Carrier-mediated transport through bulk liquid membrane: Dependence of transport rates and selectivity on carrier properties in Diffusion-limited process. J Am Chem Soc 1985; 107: 241-248.
54. Giddings JC. *Unified Separation Science*. John Wiley & Sons, 1991: 17, 38, 45-48, 94, 112-114.

55. Ho WSW, Sirkar KK, eds. *Membrane Handbook*. VNR, New York, 1992; 179.
56. Ho WSW, Sirkar KK, eds. *Membrane Handbook*. VNR, New York, 1992; 836-858.
57. Ho WSW, Sirkar KK, eds. *Membrane Handbook*. VNR, New York, 1992; 764-808.
58. Ho WSW, Sirkar KK eds. *Membrane Handbook*. VNR, New York, 1992; 907-909.
59. Danesi PR. Separation of metal species by supported liquid membranes. Sep Sci Tech 1984-1985; 19, 11, 12: 857-869.
60. IUPAC, *Equilibrium Constants of Liquid-Liquid Distribution Reactions, How to Use the Tables*. Pergamon Press, Oxford, 1978.
61. Nijenhuis WF, Buitenuis EG, et al., Calixcrowns as selective potassium cation carriers in supported liquid membranes. J Am Chem Soc, 1991; 113: 7963-7968.
62. Konishi S. Binary metal-ion sorption during permeation through chelating porous membranes. J Membr Sci 1996; 111: 1-6.
63. Molyneux P. *Water-Soluble Synthetic Polymers: Properties and Behavior*. CRC Press, Boca Raton, FL, 1984; Vol. 2: 12-49.
64. Manning GS. Limiting laws for equilibrium and transport properties of polyelectrolyte solutions. In: Seleny E, ed. *Charged and Reactive Polymers*, Reidel Dordrecht, Boston, 1974; Vol. 1: 9-18.
65. Katchalsky A. Solutions of polyelectrolytes and mechanochemical systems. J Polym Sci 1952; 7: 393-403.
66. Osada Y, Honda K, Ohta M. Control and water permeability by mechanochemical contraction of poly(methacrylic acid)-grafted membranes. J Membr Sci 1986; 27: 327-335.
67. Nishikawa H, Tsuchida E. Complexation and form of poly(vinylpyridine) derivatives with copper(II) in aqueous solutions. J Phys Chem 1975; 79: 2072-2077.
68. Wall FT, Gill SJ. Interaction of cupric ions with polyacrylic acid. J Phys Chem 1954; 58: 1128-1134.
69. Yamaguchi T, Nakao S, Kimura S. Plasma-graft filling polymerization: Preparation of a new type of pervaporation membrane for organic liquid mixtures. Macromolecules 1991; 24: 5522-5527.
70. Simons R. Pervaporation and evaporation characteristics of a new type of ion exchange membrane. J Membr Sci 1993; 78: 63-71.
71. Islam MA, Dimov A, Malinova AL. Environment-sensitive properties of poly-methacrylic acid-grafted polyethylene membranes, J Membr Sci 1993; 66: 69-75.
72. Mika AM. A new class of polyelectrolyte-filled microfiltration membranes with environmentally controlled porosity. J Membr Sci 1995; 108: 37-48.
73. Shkinev VM. Enrichment of arsenic and its separation from other elements by liquid-phase polymer-based retention. Sep Sci Technol 1987; 22, 11: 2165-2177.
74. Devis JC, Valus RJ, Lawrence EG. Affinity dialysis—A method of continuous, rapid metal ion separation using Dialysis membranes and selective water-soluble polymers as extractants. Sep Sci Technol 1988; 23, 10, 11: 1039-1066.
75. Molymeux P. *Water-Soluble Synthetic Polymers: Properties and Behavior*. CRC Press, Boca Raton, FL, 1984; Vol. 2: 18-24.
76. Breslow D, Kutner A. Polymers and copolymers of sodium ethylenesulfonate. J Polym Sci 1958; 27: 295-302.
77. Shirai M. Interaction between dyes and polyelectrolytes. Macromol Chem 1977; 178: 37-43.
78. Dialer K, Kerber R. Zur Kenntnis der Polyaethensulfonsauren. Makromol Chem 1955; 17: 56-64.
79. Molymeux P. *Water-Soluble Synthetic Polymers: Properties and Behavior*. CRC Press, Boca Raton, FL, 1984; Vol. 2: 16-18.
80. Horn D. In: Goethals E, ed *Polymeric Amines and Ammonium Salts*, Pergamon Press, Oxford, 1980: 333-339.

81. Kobayashi S, Hiroishi K, Tokunoh M, Saegusa T. Chelating properties of linear and branched poly(ethylenimines). Macromolecules 1987; 20: 1496-1503.
82. Dietrich B, Hosseini M, Lehn J, Sessions R. Anion receptor molecules. Synthesis and anion-binding properties of polyammonium macrocycles. J Am Chem Soc 1981: 103: 1282-1285.
83. Dowling S. Binding of sulfonated fluorophors by metal-poly(ethylenimine) complexes. Macromolecules 1986; 19: 344-354.
84. Cherkasov AN, Pasechnik VA. Membranes and sorbents in biotechnology. Khimiya, Leningrad 1991; 81: 36-42 (in Russian).
85. Yoou B, Lenhoff A. Computation of the electrostatic interaction energy between a protein and a charged surface. J Phys Chem 1992; 96: 3130-3139.
86. Schlogl R. Electrode Properties of Ion-exchange Membranes. Z Physik Chem (Frankfurt) 1955; 3: 73-85.
87. Teramoto M, Matsuyama H. Development of a spiral-type flowing liquid membrane module with high stability and its application to the recovery of chromium and zinc Sep Sci Technol 1989; 24(12-13): 981-997.
88. Teramoto Matsuyama H. Separation and concentration of heavy metal ions by a spiral-type flowing liquid membrane module. Water Treatment 1990; 5: 237-248.
89. Kislik V, Eyal A. Application of organic and aqueous liquid membranes in separation of metals and acids: Comparative analysis. In: Proceedings of Abstracts on International Conference "Advanced Membrane Technology", October 14-19, 2001: 203-204.
90. Wodzki R, Kaluzynski K, Klosinski P. Mobile macromolecular carriers of ionic substances. I. Transport of heavy metal ions by polymer analogues of teichoic acids. Macromol Chem Phys 1997; 198: 1809-1821.
91. Wodzki R, Swiatkowski M. Mobile macromolecular carriers of ionic substances. II. Transport rates and separation of some divalent cations by poly(oxyethylene) phosphates. Macromol Chem Phys 1998; 201: 231-241.
92. Eyal A, Canari R. pH dependence of carboxylic and mineral acid extraction by amine-based extractants: Effects of pKa, amine basicity, and diluent properties. Ind Eng Chem Res 1995; 34: 1789-1797.
93. Mandal DK, Guha AK, Sirkar KK. Isomer separation by a hollow fiber contained liquid membrane permeator. J Membr Sci 1998; 144(1-2): 13-24.
94. Wenz G, Weickenmeier M, Huff J. Association thickener by host-guest interaction of β-cyclodextrin polymers and guest polymers. In: Glass JE, ed. *Associative Polymers in Aqueous Media*. Symposium Series 765, American Chemical Society, Washington, DC, 2000; Chapter 16: 271-283.
95. Shinbo T, Shimabukuro Y, Yamaguchi T. Facilitated transport of naphthalene derivatives through a supported liquid membrane containing a water-soluble cyclophane. Separation Science and Technology 2000; 35(3): 439-445.
96. Shalygin M, Teplyakov V, Roizard D, Favre E. CO_2 transport study in combined membrane system with aqueous potassium carbonate as a liquid carrier. Desalination, 2006; V 200(1-3): 106-108.

CHAPTER 7

Liquid Membrane in Gas Separations

A. Figoli

Nomenclature

BFLM	bulk flowing liquid membrane
CLM	contained liquid membrane
FLM	flowing liquid membrane
HF	hollow-fiber membrane
HFCLM	hollow-fiber contained liquid membrane
MELM	microencapsulated liquid membrane
NF	nanofiltration membrane
PBMM	polymerized bicontinuous microemulsion membrane
SLM	supported liquid membrane
SILM	supported ionic liquid membrane
SMV	selective membrane valve

Legend

$[Car.]_o$	Carrier concentration (mol l^{-1})
D	Diffusion coefficient ($cm^2\ s^{-1}$)
f	Facilitation factor (–)
l	Effective membrane thickness (cm)
J	Permeation Flux (cm^3 (STP) $cm^{-2}\ s^{-1}$)
K	Binding constant ($cmHg^{-1}$)
p	Partial pressure ($cmHg^{-1}$)
P	Permeability (cm^3 (STP) cm $cm^{-2}\ s^{-1}\ cmHg^{-1}$)
S	Solubility coefficient (cm^3 (STP) $cm^{-3}\ cmHg^{-1}$)
α	Selectivity (–)

Subscripts

Car.*i*.	Carrier/target gas species complex / Target gas species
j	Gas species
o	Initial
tot	Total

Institute of Membrane Technology (ITM-CNR), Via P. Bucci 17/c, 87040 Rende (CS), Italy

Liquid Membranes: Principles and Applications in Chemical Separations and Wastewater Treatment
DOI: 10.1016/B978-0-444-53218-3.00007-6

1. Introduction

Gas separation membrane technologies are extensively used in industry. Typical applications include carbon dioxide separation from various gas streams, production of oxygen enriched air, hydrogen recovery from a variety of refinery and petrochemical streams, olefin separation such as ethylene-ethane or propylene-propane mixtures. However, membrane separation methods often do not allow reaching needed levels of performance and selectivity. Polymeric membrane materials with relatively high selectivities used so far show generally low permeabilities, which is referred to as trade-off or "upper bound" relationship for specific gas pairs [1].

In this context, facilitated transport of a specific gas molecule through modified polymeric membranes or liquid membranes containing mobile carriers can be employed to improve single bulk material (polymer) properties. Ceramic material is not traditionally employed as liquid membrane support due to their high cost, use of not aggressive compounds to be separated and "mild" operating conditions.

Facilitated or carrier-mediated transport is a coupled transport process that combines a (chemical) coupling reaction with a diffusion process. The solute has first to react with the carrier to form a solute-carrier complex, which then diffuses through the membrane to finally release the solute at the permeate side. The overall process can be considered as a passive transport since the solute molecule is transported from a high to a low chemical potential. In the case of polymeric membranes the carrier can be chemically or physically bound to the solid matrix (*fixed carrier* system), whereby the solute hops from one site to the other. *Mobile carrier* molecules have been incorporated in liquid membranes, which consist of a solid polymer matrix (support) and a liquid phase containing the carrier [2, 8], see Fig. 7.1. The state of the art of supported liquid membranes for gas separations will be discussed in detail in this chapter.

General advantages of facilitated transport membranes are improved selectivity, increased flux and, especially if compared with membrane contactors, the possibility to use expensive carriers. The specific prerequisites, advantages and disadvantages connected the mobile carriers, are reported in Table 7.1. So far, mainly conventional liquid membranes have been loaded with different mobile carrier systems to obtain facilitated transport properties [3]. Problems encountered are (evaporative) loss of solvent and carrier, temperature limitations, a too large membrane thickness and therefore too low permeabilities as well as a limited solubility of the carrier in the liquid medium. The low fluxes achieved have, until now, limited their application

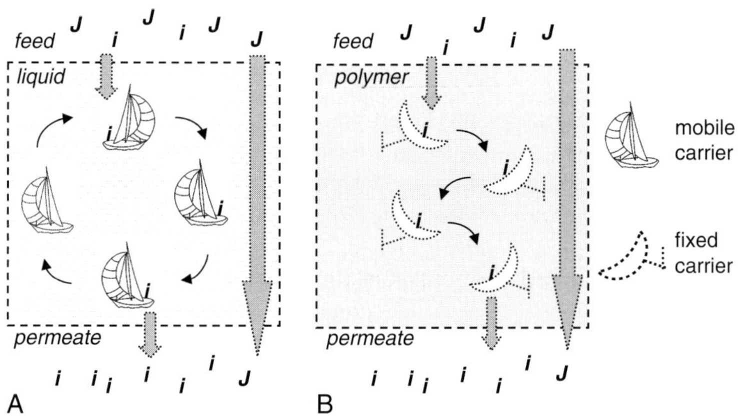

Figure 7.1 Facilitated transport of gaseous molecules by a carrier (complex) through a membrane. (A) Liquid membrane with a mobile carrier; (B) Solid membrane with a fixed carrier.

Table 7.1 The specific requirements, advantages and disadvantages connected to mobile carrier systems with respect to their selective gas transport properties

	Mobile carrier in liquid membrane
Requirements	*Membrane*: low effective thickness *Liquid medium*: low viscosity, low volatility, high compatibility with polymeric material *Carrier*: high concentration in the liquid medium, high selectivity for the gas of interest
Advantages	*High selectivity* *High diffusivity* of the permeant molecule
Disadvantages	*Loss* of membrane solvent and carrier *Low carrier concentration* *Carrier chemical degradation*

After modification from Ref. [9] Copyright 2001, reprinted with permission from Elsevier.

in industrial separation processes. In particular for oxygen carrier systems, a major problem is the instability of the carrier against irreversible oxidation. Improvements necessary for (large scale) commercial applications involve therefore the development of new membrane morphologies and configurations, as well as stable carrier systems.

2. Theory

Facilitated or carrier mediated transport is a transport process that combines a chemical reaction with a diffusion process. The solute has first to react with the carrier to form a solute-carrier complex, which then diffuses through the membrane to finally release the solute at the permeate side.

In carrier facilitated gas transport through liquid immobilized membranes, the overall process can be considered as a passive transport since the solute molecule is transported from a high to a low chemical potential. At the high-pressure feed side, the gas molecule that has to be selectively transported complexes with the carrier molecule diffuses along with the mobile carrier through the liquid membrane phase and desorbs at the low-pressure permeate side of the membrane [2, 4].

Besides transport through complexation with the carrier, it also occurs nonspecific gas transport through the solvent phase in which the carrier is dissolved. Therefore, the characteristic of facilitated transport is the occurrence of a reversible complexation process in combination with the diffusion process. Two different cases have to be taken in account: (i) diffusion is rate limiting (fast reaction), (ii) reaction is rate limiting (slow reaction and relatively fast diffusion) [2]. The second case does not occur frequently and only the first one will be considered. The selective transport of gases in facilitated transport systems can be described in analogy to a dual-mode transport mechanism [5, 6, 8].

The model describes the total flux of the gas target species to be facilitate transported, *i*, (see Table 7.2) with respect to gas species, *j*, as the sum of the ordinary solution-diffusion transport through the noncomplexing solvent phase *(1)* and facilitated transport mediated by the carrier *(2)*. Assuming a partial pressure of the target species, *i*, to equal zero on the permeate side, one can write the fluxes for each of the transport modes as:

Table 7.2 Main reactions between target species and carriers

Target gas species, *i*	Reactions
O_2	O_2 + Cobalt-base ↔ Cobalt-base (O_2)
CO_2	$CO_2 + H_2O + Na_2CO_3 \leftrightarrow 2\ NaHCO_3$
C_2H_4 or C_3H_6	$C_2H_4 + AgNO_3 \leftrightarrow AgNO_3\ (C_2H_4)$
H_2S	$H_2S + Na_2CO_3 \leftrightarrow NaHS + NaHCO_3$
CO	$CO + CuCl_2 \leftrightarrow CuCl_2(CO)$
SO_2	$SO_2 + H_2O + Na_2SO_3 \leftrightarrow 2\ NaHSO_3$

After modification from Ref. [8] Copyright 2000, reprinted with permission from McGraw-Hill.

$$J_i = S_i D_i \frac{p_i}{l} \tag{1}$$

$$J_{\mathrm{Car}.i} = \frac{[\mathrm{Car.}]_0 K}{(1 + Kp_i)} D_{\mathrm{Car}.i} \frac{p_i}{l} \tag{2}$$

where D_i is the diffusion coefficient of target species, i, in the solvent; S_i the solubility coefficient; p_i the partial pressure at the feed side of the membrane; and l the effective membrane thickness. $[\mathrm{Car.}]_0$ is the carrier concentration; K the binding constant; $D_{\mathrm{Car}.i}$ the diffusion constant of the carrier/gas target species, i, complex.

The ordinary solution-diffusion transport does not depend on the presence of the carrier and it gives rise to a background flux of solute that only depends on the experimental conditions (pressure, temperature, and solvent). It scales linearly with the driving force, the partial pressure difference across the membrane. The carrier facilitated transport strongly depends on the pressure and a Langmuir type of sorption behavior describes the complexation thermodynamics well. The permeability can be calculated from the fluxes ($J_{\mathrm{tot}} = J_i + J_{\mathrm{Car}.i}$) and equals the gas flux at standard temperature and pressure normalized for the driving force and the membrane thickness resulting in Eq. (3).

$$P_i = \frac{(J_{\mathrm{tot}})l}{p_i} = D_i S_i + \frac{[\mathrm{Car.}]_0 K}{1 + Kp_i} D_{\mathrm{Car}.i} \tag{3}$$

The flux of the other species, j, is calculated by solution-diffusion transport through the solvent phase *(4)*. Assuming a partial pressure of gas species, j, to equal zero on the permeate side, one can write the flux for the Henry's transport mode as:

$$J_j = S_j D_j \frac{p_j}{l} \tag{4}$$

The permeability is calculated from its flux (J_j) and equals the gas flux at standard temperature and pressure normalized for the driving force and the membrane thickness resulting in Eq. (5).

$$P_j = \frac{(J_j)l}{P_j} = D_j S_j \tag{5}$$

Then the selectivity (α) of the target species, i, over the species, j, $\alpha_{i/j}$, can be calculated by the following equation:

$$\alpha_{i/j} = \frac{P_i}{P_j} \tag{6}$$

and it is indicative of the membrane ability to separate species i and j.

Furthermore, the flux of the target species, i, through the liquid membrane is often described in a dimensionless fashion as facilitated transport, f, which is defined as the ratio of the permeation flux in the presence of the carrier to that in the absence of the carrier. Equation (7) gives the facilitation factor f:

$$f_i = \frac{J_{\text{tot}}}{J_i} = \frac{[\text{Car.}]_0 K}{(1 + Kp_i)S_i} \frac{D_{\text{Car}.i}}{D_i} + 1 \tag{7}$$

Thus, the facilitation efficiency of a supported liquid membrane containing the carrier depends primarily on the carrier concentration and its thermodynamic property, the binding constant K. Increasing both the values it leads to an increase in facilitation.

A second term on which facilitation depends is the ratio of the diffusion coefficients of the complex and the free target gas, i, in the solvent. Due to the larger size of the complex, its diffusion coefficient will be lower than the free target species, i, resulting in a ratio smaller than one. Hence, this ratio counteracts the facilitation. In practice, one would like to increase the concentration of carrier to its maximum solubility. But this result generally leads in an increased viscosity changing the ratio of the diffusion coefficients to even smaller values. Ultimately, one finds an optimum carrier concentration at which facilitation reaches its maximum value [7]. Equation (7) also shows that an increasing partial target species, i, pressure on the feed side results in a facilitation factor of 1 [14].

3. Modules and Design

Table 7.3 shows a classification of the liquid membranes on the basis of the configuration and module types employed in gas separation. The liquid membranes can be divided in three main classes: (i) supported liquid membrane (SLM), (ii) bulk liquid membrane (BLM), and (iii) supported ionic liquid membrane (SILM).

The most used membrane modules for carrying out liquid membrane experiments are flat-sheet (plate-and-frame and spiral-wound) and hollow-fiber (HF) type.

The SLMs usually consists of an organic or water media, containing the target gas carrier, which is immobilized in the pores of a hydrophobic/hydrophilic micropourous membrane (generally, microfiltration polymer membrane) and binds selectively the target gas at the feed side.

As an example, the SLM preparation and the experimental setup reported by Figoli *et al.* [14] are described. A flat hydrophilic polymeric membrane (polymerized bicontinuous microemulsion membranes, PBMM) was

Table 7.3 Classification of the liquid membranes in the field of gas separation according to the main configurations employed

LM-Types	Configuration	Module type
Supported liquid membrane (SLM)	• Sandwich-SLMs • Gelation • Nanoporous membranes (i.e., PBM)	Flat sheet Hollow fiber (HF)
Bulk liquid membrane (BLM)	• Flowing liquid membrane (FLM)	Flat sheet
	• Contained liquid membranes (CLM)	Plate and frame
	• Selective membrane valve (SMV)	Hollow fiber (HF)
Supported ionic liquid membrane (SILM)	• Sandwich-SiLMs • Nano-porous membranes (i.e., NF)	Flat sheet Hollow fiber (HF)

impregnated with water and porphyrin carriers (Fig. 7.2A) and the excess of the liquid was removed from the surface of the membrane by wiping it very carefully with a tissue.

The flat membrane was allocated in the permeation cell and the feed and sweep gas was recirculated on the top and bottom surface of the membrane itself as shown in Fig. 7.2B. Both streams were humidified to keep the

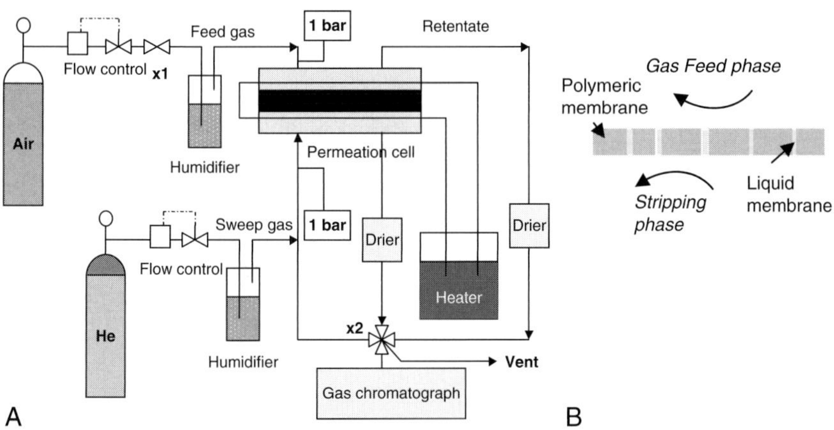

Figure 7.2 Schematic drawing of (A) a typical laboratory gas permeation setup employed for testing flat sheet supported liquid membrane (SLM), (b) the SLM located in the permeation cell.

membrane wet. The gas composition in the feed and strip phases was measured with a Gas Chromatograph.

A new type of configuration, the flowing liquid membrane (FLM) was studied by Teramoto *et al.* [20]. In this case, the membrane liquid phase is in motion as the feed and strip phase. In this type of system a plate-and-frame and spiral-wound configuration with flat membrane was used. The scheme of the FLM configuration is drawn in Fig. 7.3A. The liquid phase flows (FLM) between two hydrophobic microporous membranes. The two membranes separate the liquid membrane phase from feed and strip phases. In Fig. 7.3B, it is reported the classical plate-and-frame module employed for the separation of ethylene from ethane [20]. The liquid membrane convection increased the membrane transport coefficient in gas separation. However, the membrane surface packing density (membrane surface area/equipment volume) is much lower in spiral-wound system than in hollow fiber.

An example of hollow-fiber module design is that reported by Sirkar *et al.* [64-66]. In particular, they developed the hollow-fiber contained liquid membrane (HFCLM) configuration in which two different types of hollow fibers are displaced in the modules as shown in Fig. 7.4A. This process was achieved by packing thousands of microporous hydrophobic hollow fibers. The aqueous feed and sweep gas flow through lumen side of the fibers and, if the fibers are hydrophilic, fill the pores of the fiber or, if they are hydrophobic, the fibers are wetted by organic membrane phase (Fig. 7.4B). The fibers containing the aqueous liquid membrane on the outside were named "contained liquid membranes" (CLMs). The CLM in the shell of the HFCLM permeator is usually stationary and only the replacement of the membrane liquid cause a movement in the permeator. However, the HFCLM gives the additional possibility of recirculating the

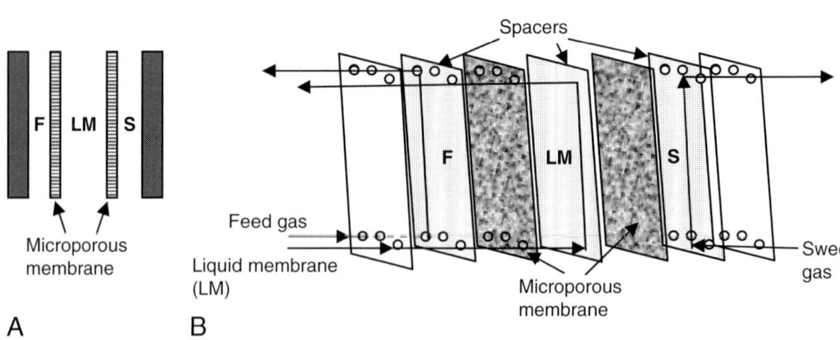

Figure 7.3 (A) Section of the membrane module and (B) plate-and-frame module of the flowing liquid membrane (FLM). F: feed gas channel, S: sweep gas channel, LM: flowing liquid membrane channel (after modification from Ref. [20] Copyright 1989, reprinted with permission from Elsevier).

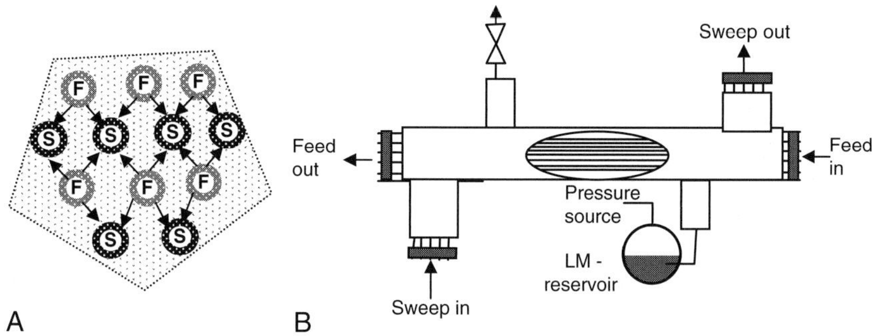

Figure 7.4 Drawing of the hollow-fiber contained liquid membrane (HFCLM): (A) configuration of HFCLM in a permeator shell and (B) side-view of the permeator (after modification from Ref. [12] Copyright 1996, reprinted with permission from Francis and Taylor).

liquid membrane which is beneficial in the case of most of the mass transfer resistance lies in the liquid membrane.

4. Stabilization of Supported Liquid Membranes and Novel Configurations

Despite their advantages, the SLMs have not been applied at industrial level, yet. The main reasons are besides low fluxes resulting from the substantial thickness (~25 μm), the short membrane stability or lifetime, which is far too low to assure reliability. In the last years, several methods have been developed to overcome the instability problems of SLMs which are depicted in Fig. 7.5.

The instability of the SLMs is due to loss of solvent and/or carrier from the membrane (Fig. 7.5A) which negatively influences the flux and selectivity of the membrane.

All the techniques used to increase the stability of the SLM, such as the gelled SLM techniques [10, 11] (Fig. 7.5B) and the addition of thin top-layer by interfacial polymerization reaction on the SLM (Fig. 7.5C) [12], are essentially applied in the removal of (metal) ions from solution. The stability of liquid membranes used for the separation of gases is more complicated. Here, the addition of a top-layer on the macroporous support can negatively influence the permeability of gases through the membrane. Therefore, a careful choice of the layer material is important because it has to be impermeable to the solvent and should posses a high permeability for the gas molecules considered. In addition, the thickness of the top-layer as well as that of the whole liquid membrane has to be minimized.

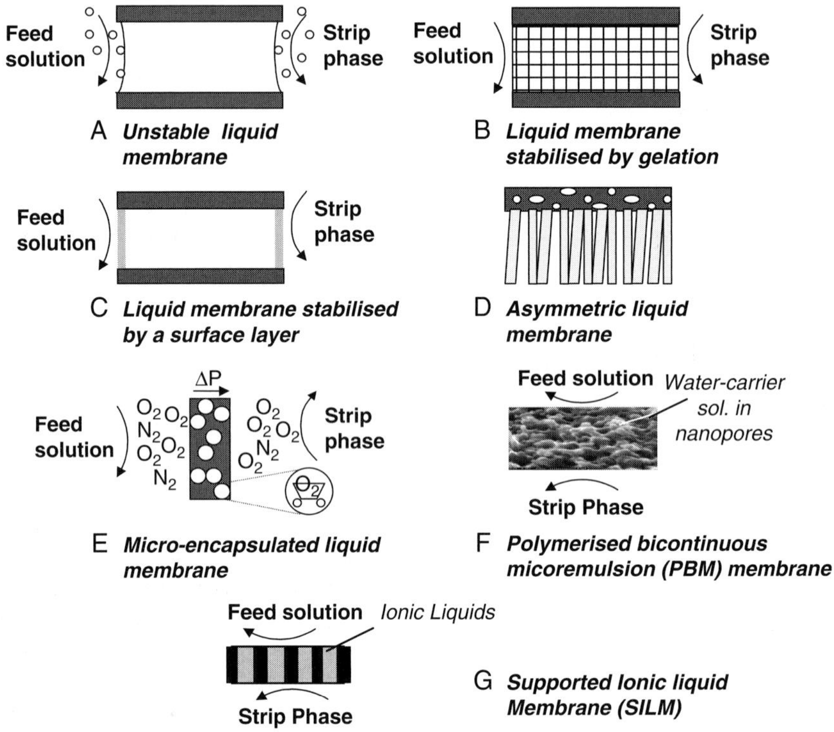

Figure 7.5 Overview on stabilization techniques for SLMs developed over the last 20 years (after modification from Ref. [9] copyright 2001, reprinted with permission from Elsevier).

In the 1990s, Bauer *et al.* [13] introduced the concept of microencapsulated liquid membrane (MELM) for gas separation application (Fig. 7.5D). They developed an asymmetric liquid membrane obtained via a modified phase inversion process, with a very thin open cell type top-layer (100–500 nm), whereby individual cells were filled with a nonvolatile solvent (oil) and the carrier molecules, allowing for high fluxes. A further advantage achieved was the prevention of carrier loss without applying supplementary coating layers, since the polymer in the top-layer completely surrounds the carrier-containing liquid phase. The important disadvantage was the low long-term stability, due to loss of the solvent as well as oxidative decomposition of the carrier complexes.

Figoli *et al.* [9] tried to overcome the limitation of the first MELM preparing well-defined capsule-containing membranes (Fig. 7.5E). These membranes were prepared adding the capsules, loaded with carrier-solvent, to the polymer membrane solution and then cast on a glass support. The gas flux through the membrane was expected to be high due to the low membrane thickness (<3 μm).

Recently, a PBMM employed as nanostructured liquid membranes for facilitated oxygen transport have been also prepared [14-15] (Fig. 7.5F). Bicontinuous microemulsions consisted of an interwoven network of water and oil domains (channels), which were stabilized by an interfacial monomolecular surfactant film. The oil (monomer) channels were polymerized to form the polymeric matrix of the liquid membranes, while the water channels remained unchanged and form, after polymerization, a network of open interconnected nanopores. By adjusting the composition of the starting microemulsion, the channel width (pore size) of the membranes could be tuned between 3 and 60 nm. The made PBM membrane had some distinct advantages: (i) The nanometer pore size of the PBM membrane is stable against significant transmembrane pressure gradients and loss of the liquid membrane phase, (ii) The PBM membranes can be manufactured into composite membranes such that the support membrane is not wetted by the liquid membrane phase and the actual resistance of the membrane is only determined by the PBM coating thickness, and (iii) Due to its potentially percolating porous network, the membrane can be reimpregnated. The PBM membranes without mobile carrier resulted stable for more than 3 weeks when in contact with a humidified gas feed stream.

In recent years, the preparation of supported liquid membranes filled with ionic liquid (SILMs) has attracted increasing interest both for gas extraction and separation [16-18]. This is mainly due to the nonvolatile and stable nature of ionic liquids which overcomes the traditional SLMs limitation such as the liquid phase evaporation. Higher SILMs stability over conventional SLMs was also attained because of their greater capillary force associated with higher viscosity of ionic liquids, which could reduce displacement of the liquid from the micron pores under pressure. Furthermore, the ionic liquid is used not only as the liquid membrane phase but also as carrier. Different types of ionic liquids can be employed depending on the gas separation. Stability up to 260 days was reported by Teramoto *et al.* [17].

Attempts to overcome the instability problems related to supported liquid membranes used novel types of liquid membranes with unique configurations such as the FLM [19-20] and HFCLM [21-23].

In FLMs, a membrane aqueous carrier solution is circulated through a thin channel between two hydrophobic microporous membranes separating feed gas and sweep gas phases from the FLM phase. The stability of this membrane was reported to be 11 days without any decrease of flux.

In the HFCLM configuration, the aqueous feed and strip solutions flow through the lumen side of the fibers. On the outside, an aqueous or organic solution was applied as the liquid membrane phase as already described. This configuration was applied for facilitated the transport of CO_2 over CH_4. Stability up to 80 days was reported by the same authors.

More recently, Teramoto *et al.* [24-25] referred to the use of a novel facilitated transport membrane for gas separation in which a carrier was supplied to the feed side (high-pressure side) and it was forced to permeate through a membrane to the permeate side (low-pressure side), and then the permeated carrier solution was recirculated to the feed side. Since the membrane was always wet with the carrier solution, the membrane became very stable with no open or unfilled pores present which usually caused membrane unselectively in traditional SLM. This new type of membrane has been named a "bulk flow liquid membrane" (BFLM). The membrane resulted to be stable over a discontinuous one-month testing period.

5. Gas Separation Applications

The concept of a molecular carrier transport involving a reversible chemical combination between permanent and mobile species was pursued and developed by Osterhout and colleagues in the early 1930s, although the principle has been demonstrated much earlier by Pfeffer in 1910 and Freudlich and Gann in 1915 [26]. The model experiments of Osterhout (1940) using quiacol, a weak organic acid, as carrier for sodium and potassium ions, clearly established the concept in the biological literature.

Apparently, the first who studied the application of liquid membranes with facilitated transport properties for gas separation were Ward and Robb in 1967 [27]. A highly porous cellulose acetate film was soaked with cesium bicarbonate solution, the liquid was kept in the membrane pores by capillary forces. A high selective permeation of CO_2 from an O_2-CO_2 feed mixture was due to its facilitated transport in the bicarbonate solution.

The number of gases for which suitable carriers are currently available is small and most effort has been devoted to the cleanup of acid gases. Many studies on facilitated transport of gases such as O_2, CO_2, H_2S, SO_2, NO, and CO through liquid membranes are reported in literature. The first studies on facilitated transport systems for different gaseous permeants are reported in Table 7.4.

In the next sections, the main gas separation applications using facilitated liquid membranes are reported. The gas permeability and selectivity in various membrane systems with facilitated transport properties are summarized in Table 7.5.

5.1. Production of oxygen-enriched air

Oxygen-enriched air is used in many industrial processes which do not require pure oxygen, for example, combustion of natural gas, coal gasification, and liquefying, as well as in the production of peroxides, in sewage treatment, in welding and in the glass production. Standard methods are cryogenic

Table 7.4 First studies on facilitated transport systems for different gaseous permeants

Year	Gas	Carrier	Applications	References
1960	O_2	Hemoglobin Fe, Co, Ru Porphyrins, Ir, Mn complexes	O_2 enrichment for medical use, combustion, sewage treatment, welding and glass production	Basset and Schultz [28]
1970	NO	Fe^{2+}		Ward *et al.* [29]
1971	CO_2	CO_3^{2-} Ethanolamines	Biogas purification, enhanced oil recovery, life support systems	Enns [30]
1974	CO	Cu^+	Synthesis gas, purification	Steighelman and Hughes [31]
1977	H_2S	CO_3^{2-}	Gasified gas, desulfurization	Matson *et al.* [32]
1981	Olefins	Ag^+, Cu^+	C_2H_4 recovery	Hughes *et al.* [33]

From Ref. [9] copyright 2001, reprinted with permission from Elsevier.

distillation and pressure swing adsorption [34]. As an alternative approach, gas separation membranes for the production of oxygen enriched air have been developed over the last 40 years based on the selective oxygen permeability of polymeric membrane materials. Existing polymeric membranes show, due to the upper bound relationship between permeability and selectivity, a selectivity which is too low to obtain the required oxygen purity in a commercially feasible single stage process. To pass the selectivity-permeability trade-off, facilitated oxygen transport has been deeply investigated.

The first experiments on facilitated oxygen transport were carried out by Scholander [35] and Wittemberg [36] in 1960 and 1966, who worked on biological systems. They showed that hemoglobin (Hb) and myoglobin could accelerate the transport of oxygen across water films and increase the interest in the synthesis of oxygen specific carriers. The first to apply synthetic oxygen carriers were Basset and Schultz in 1970 [28], who used bis(histidine)cobalt (II) as a complexing agent in an aqueous medium. Their best results gave an approximate doubling of the oxygen flux compared to water and a selectivity of 3.5.

Table 7.5 Summary of gas permeability and selectivity in various liquid membrane system with facilitated transport by mobile carrier

Gases separated	Liquid membrane: Solvent	Liquid membrane: Carrier	Module type	Permeability (P) (barrer/cm)[a]	Selectivity ($\alpha = P_{O_2}/P_{N_2}$)	References
Oxygen/Nitrogen separation						
O_2/N_2	Water	Hemoglobin (Hb)	SLM/flat.		–	[36, 63]
	Water	Co (ϕ H)$_2$	SLM/flat.		–	[28]
	Gammabutyrolactone, dymethylsulfoxide, dimethylacetamide, *N*-methylpyrrolidone	Co/Fe (dry-cave), Co(SalPr), Co(MeOsalen), Co(3-MeOsaltmen)	SLM/flat.	$P_{O_2} = 260\text{–}1000$	$\alpha = 10\text{–}30$	[4, 37]
	4-Methylanisole	(CoPIm)	SLM	$P_{O_2} = 50 \times 10^3$	$\alpha = 20\text{–}40$	[42]
	PEG 200	Co(5-NO_2-Saltmen)	MELM/flat.		$\alpha = 15.6\text{–}19.7$	[13]
	Water	Co-porphyrin	SLM PBMM/flat.		$\alpha = 2.8$	[14–15]
Carbon dioxide separations						
CO_2/O_2	Water	$CsHCO_3$ solution	SLM/flat.	$P_{CO_2} = 8.9 \times 10^4$	$\alpha = 1500$	[27]
		$CsHCO_3$ solution with sodium arsenite		$P_{CO_2} = 2.99 \times 10^5$	$\alpha = 4100$	
	Water	Carbonic anidrase (CA) + $NaHCO_3$	SLM/flat.	$P_{CO_2} = 5.3 \times 10^5$	$\alpha = 250$	[45]
	Water	Dietilammide (DEA)	CLM/HF	$P_{CO_2} = 4.18 \times 10^5$	$\alpha = 152$	[68]
		Carbonic anidrase (CA) + $NaHCO_3$		$P_{CO_2} = 1.19 \times 10^6$	$\alpha = 210$	
CO_2/N_2	Water	2-amino-methyl-1-propanol (AMP)	BFLM/HF	$P_{CO_2} = 4.42 \times 10^6$	$\alpha = 430\text{–}806$	[25]

	Water	Dietilammide (DEA)		$P_{CO_2} = 9.74 \times 10^2 - 4.8 \times 10^3$	$\alpha = 55.8\text{-}276$	[69]
	Water	DEA	BFLM/HF	$P_{CO_2} = 2.6\text{-}5 \times 10^6$	$\alpha = 596\text{-}1010$	[25]
	Water	Monoethanolamine (MEA)	BFLM/HF	$P_{CO_2} = 2.5\text{-}4 \times 10^6$	$\alpha = 570\text{-}1790$	[25]
	Glycerol	–	SLM/flat.	$P_{CO_2} = 5.6\text{-}1.1 \times 10^2$	$\alpha = 1.5\text{-}38.5$	[49]
		Na_2CO_3	SLM/HF	$P_{CO_2} = 2\text{-}6.24 \times 10^4$	$\alpha = 426\text{-}1381$	[48]
				$P_{CO_2} = 2.5\text{-}30 \times 10^4$	$\alpha = 3440\text{-}100$	
	Glycerol	Glycine-Na	SLM/flat.	$P_{CO_2} = 9.97 \times 10^2 - 1.1 \times 10^5$	$\alpha = 2000\text{-}8500$	[50]
			SLM/HF	$P_{CO_2} = 0.9 \times 10^5$	$\alpha = 1880\text{-}5830$	[48]
	Glycerol carbonate	Poly(amidoamine) dendrimer	SLM/flat.	$P_{CO_2} = 0.8\text{-} 3.80 \times 10^2$	$\alpha = 90\text{-}130$	[51]
				$P_{CO_2} = 2.5 \times 10^2 - 6.6 \times 10^3$	$\alpha = 30\text{-}1000$	
		Na-glycinate		$P_{CO_2} = 2.5 \times 10^2 - 2 \times 10^3$	$\alpha = 40\text{-}480$	
	Ionic liquids	–	SILM/flat.	$P_{CO_2} = 3.5 \times 10^2 - 1.0 \times 10^3$	$\alpha = 15$ $[Cl]^-$- 61 $[dea]^-$	[54]
CO_2/ CH_4	Water	–	BFLM/flat.	$P_{CO_2} = 1.15 \times 10^4$	$\alpha = 24\text{-}29$	[24]
	Water	Dietilammide (DEA)	BFLM/flat.	$P_{CO_2} = 1.86 \times 10^6$	$\alpha = 1970$	[46]
	Ionic liquids	–		$P_{CO_2} = 3.5 \times 10^2 - 1 \times 10^3$	$\alpha = 4$ $[Cl]^-$- 20 $[dea]^-$	[54]
	Ionic liquids		SILM/flat.	$P_{CO_2} = 7 \times 10^2$	$\alpha = 50\text{-}120$	[17]
CO_2/He	Ionic liquids		SILM/flat.	$P_{CO_2} = 7.44 \times 10^2 - 1.2 \times 10^3$	$\alpha = 3.1\text{-}8.7$	[18]
Olefins separations						
C_2H_4/ C_2H_6	Water	Silver nitrate ($AgNO_3$)	SLM/flat.	$P_{C_2H_4} = 4.6 \times 10^4$	$\alpha = 100\text{-}240$	[55]

(continued)

Table 7.5 *(continued)*

Gases separated	Liquid membrane: Solvent	Carrier	Module type	Permeability (P) (barrer/cm)[a]	Selectivity ($\alpha = P_{O_2}/P_{N_2}$)	References
	Water	$AgNO_3$	SLM/flat.	$P_{C_2H_4} = 3 \times 10^5$	$\alpha = 1000$	[56]
	Water	$AgNO_3$	FLM/HF	–	$\alpha = 460$	[20]
	Water	$AgNO_3$	BFLM/flat.	$P_{C_2H_4} = 1.19 \times 10^6$	$\alpha = 1100$	[70]
			BFLM/HF	$P_{C_2H_4} = 3.3 \times 10^6$	$\alpha = 375$	[71]
C_3H_6/C_3H_8	Water	$AgNO_3$		$P_{C_3H_6} = 3.5 \times 10^4$	$\alpha = 100$	[55]
	Triethylene glycol (TEG)	$AgBF_4$	SLM/flat.	$P_{C_3H_6} = 3 \times 10^3$	$\alpha = 110$	[57]
		$AgNO_3$			$\alpha = 40$	
C_3H_6/C_6H_{12}	Ionic liquid		SILM/flat.	$P_{C_3H_6} = 3.9 \times 10^4$	$\alpha = 1.7$-1.9	[67]
Sulfur dioxide separations						
SO_2/CO_2	Water	$NaHSO_3$	CLM/HF	$P_{SO_2} = 19.6 \times 10^4$ - 43.1×10^4	$\alpha = 138$-190	[60]
		Na_2SO_3				
	Polyethylene glycol (PEG)	DEA	SLM/flat.		$\alpha = 14$	[61]
SO_2/N_2	Water	–	SLM/flat.	$P_{SO_2} = 1.04 \times 10^3$	$\alpha > 10{,}000$	[62]
	PEG		SLM/flat.		$\alpha = 140$	[61]
Hydrogen separation						
H_2/CO	Ionic liquid	–	SILM (NF)/ flat.	$P_{H_2} = 0.15$ - 1.3×10^3	$\alpha = 4.3$	[16]

Hollow fiber = HF; flat membrane = flat.; Cobaltodihistidine = Co (ϕH)$_2$; [*N*,*N*′-bis(salicylideneimino)di-*n*-propylamine] cobalt (II) = Co(SalPr); [*N*,*N*′-bis(3-methoxysalycylidene) ethylenediamine] cobalt (II) = Co(MeOsalen); [*N*,N′-bis(3-methoxysalycylidene)-tetramethyle-thylenediamine] cobalt (II) = Co(3-Me Osaltmen); $\alpha,\alpha,\alpha,\alpha$-meso-tetrakis(*o*-pivalamidophenyl) porphynato cobalt(II) 1-methylimidazole or laurylimidazole = (CoPIm); *N*,*N*′-ethylene-bis(5-nitro-salicyliden-iminato)cobalt(II) = Co(5-NO_2-Saltmen).

[a] 1 Barrer (Ba) = 10^{-10} cm^3 (STP) cm cm^{-2} s^{-1} $cmHg^{-1}$.

The first main problem encountered in facilitated oxygen transport was the low oxygen selectivity and the instability of the carrier systems used, which tended to degrade rapidly. Due to the still significant thickness of even thin SLMs the oxygen permeability observed was too low to be of commercial interest. A substantial increase in oxygen selectivity as well as improvement in the lifetime of the carrier system was obtained by Roman and Baker [37]. In operating the membranes with a partial oxygen pressure on the product side, that was about 10 mmHg less than that of the feed stream, they obtained an O_2/N_2 selectivity of 30 and an O_2 permeability of 1×10^{-7} barrer. The energy requirement amounted to only a fraction of the costs of the cryogenic processes. Many studies on cobalt porphyrin complex carriers have been also reported by Nishide *et al.* [38-41]. The oxygen permeability was enhanced by a decrease in the upstream oxygen pressure, P_{O_2} and the oxygen transfer was analyzed by dual-model transport. In 1997, they applied for the first time a cobalt porphyrin (CoP) derivative, as an oxygen carrier, to a liquid membrane [42]. The liquid membrane of CoP was prepared by immersing a three-layer microporous flat membrane, made from porous polypropylene and polyethylene, into a 4-methylanisole solution of CoP with laurylimidazole as an axial ligand. The pores of the film were filled with the CoP solution by capillary force. A high oxygen permeability (1 barrer) and high selectivity (ranging from 20 to 40, depending on the operating conditions) was observed.

The stability and high affinity of the oxygen-carrier complexes and the control of the auto-oxidation processes are fundamental parameters for a successful application of the synthetic oxygen carriers. All different types of porphyrin-based carrier complexes used in liquid membrane are reported in a recent review [9]. These carriers have been studied for oxygen binding but no particular attention has been paid to novel developed self-assembled systems. In this case, the synthetic effort is significantly reduced widening up the possibility for practical applications. Only recently research activities have been directed to using ion-pair interactions as a valuable tool to build up molecular assemblies.

An innovative work, regarding the preparation of MELMs, as already mentioned in the previous section, was introduced by Bauer *et al.* [13]. An asymmetric membrane was prepared by a dry/wet phase inversion process whereby a carrier solution was encapsulated in closed-cell morphology within the ultrathin selective top-layer of only 0.1-0.5 μm thickness. The carrier employed was *N,N′*-ethylene-bis(5-nitro-salicyliden-iminato)cobalt (II) with dimethylpyridine (DMAP) as axial base. The preparation of this new membrane had the aim to reach high fluxes and to avoid any loss of carrier without applying additional coating layers. The selectivity (O_2/N_2) obtained with a gas permeation setup was up to 16, which is 3.5 times higher than that of the polymeric material (polyethersulone) used to entrap the carrier but the stability was very short. Furthermore, the reproducibility of the top-layer structure and thus the encapsulation was not easy to obtain.

Recently, Figoli *et al.* [15] reported the use of polymerized bicontinuous microemulsion (PBM) membranes as "nanostructured liquid membranes" for facilitated oxygen transport. The final bicontinuous microemulsion consisting of an interconnected network of water and oil channels, stabilized by the interfacial surfactant film, in which the oil (monomer) channels were polymerized to form the polymeric matrix of the liquid membranes (Fig. 7.6) and the channel width (pore size) of the membranes could be tuned between 3 and 60 nm by adjusting the composition of the co-surfactant, while the water phase remained unchanged and it was the solvent for the novel oxygen carrier.

In fact, traditional oxygen carrier molecules are dissolved in organic solvents. Only very few works focus on water-soluble carriers, such as Co-histidine which, however, suffers from very short lifetimes [28]. Therefore, a new type of water-soluble of Co-porphyrin with oxygen affinity was developed by Fiammengo *et al.* [43] for the PBM membrane. This novel carrier was based on the assembly of porphyrins with calixarenes. The self-assembly was driven by electrostatic interactions. The oxygen was assumed to complex inside the cavity of the porphyrin-calixarene assembly. Furthermore, the Co of the porphyrin was shielded from the other side by complexation with a nitrogenous base (1-methylimidazole or caffeine).

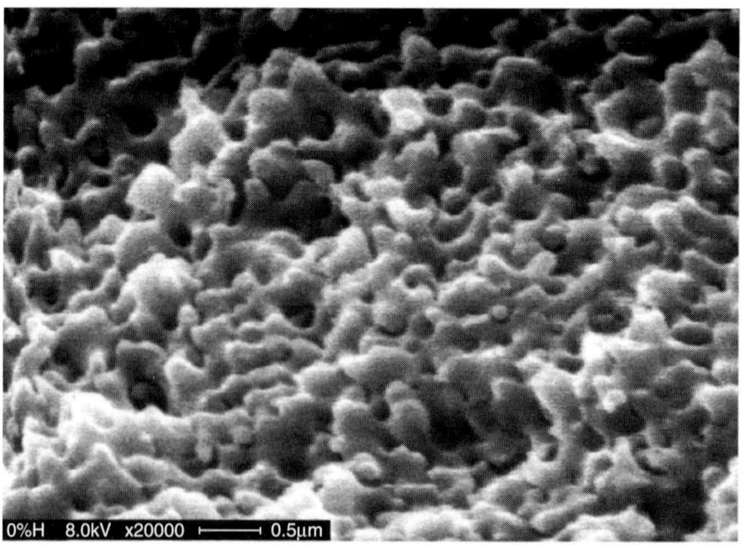

Figure 7.6 FESEM micrographs of the polymerized bicontinuous microemulsion membrane PBMM under magnification of 8 kV ×20000.

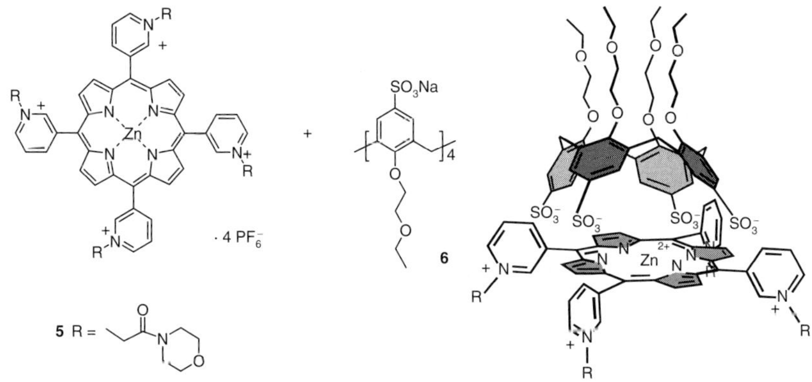

Figure 7.7 The supramolecular assembly of tetraalkyl pyridyl porphyrins and tetrasulfonato calixarene *(6)* (after modification from Ref. [43] copyright 2003, reprinted with permission from American Chemical Society).

The nitrogenous base with caffeine has the larger binding constant for the porphyrin. Together with the sulfonic acid groups on the calixarene (see Fig. 7.7, no. 6), the hydrophilic amino acid-based substituents (Fig. 7.7, no. 5) make the supramolecular assembly water soluble up to 4-5 mM. Figure 7.7 shows an example of a complex that was prepared with either Co or Zn as central metal ion [44].

The transport and separation characteristics of PBM membranes were, then, carried out hosting this new type of water-soluble Co-porphyrin with oxygen affinity. The calculated selectivity, $\alpha = 2.8$, was quite modest but it proved the potentiality of both the nanostructured PBM membrane and the novel water soluble Co-porphyrin.

5.2. Carbon dioxide separation from various gas streams

The separation of CO_2 from various gas streams, particularly flue gas or stack gas from power station, has attracted a worldwide interest to prevent global warming. It is consequent to the release of CO_2, the most important greenhouse gas, which continues to grow. Many CO_2 separation methods such as gas absorption and membrane separation are now being developed to recover and concentrate CO_2. Among membranes for CO_2 separation, facilitated liquid transport membranes have been attracting attention since very high selectivity can be obtained compared to polymeric membranes.

In this section, the major developments regarding the separation of carbon dioxide from other gases, such as H_2, O_2, CO, N_2, and CH_4, using mobile carrier systems are reported.

In 1967, Ward and Robb were the first to demonstrate the application of liquid membranes with facilitated transport properties for the separation of

CO_2 from O_2 [27]. An aqueous solution of cesium bicarbonate (6.4 M) was immobilized in a porous cellulose acetate film. For a 5% CO_2 feed gas at 1 atm and sweep side of CO_2 partial pressure at 0.005-0.026 atm the CO_2 permeance was about 8.9×10^{-4} barrer/cm and the separation factor reported was 1500. When a saturated cesium bicarbonate solution containing 0.5 M sodium arsenide was used, the permeance increased about threefold, to about 2.98×10^5 barrer/cm and selectivity rose to 4100. In another work, Suchdeo and Schultz [45] immobilized a 1.0 M sodium bicarbonate (0.5 mg/ml) in a high porous cellulose acetate membrane. The highest CO_2 performance reported was about 1.8×10^{-8} mol/m^2 s Pa (5% CO_2 in the feed) 5.3×10^5 barrer/cm and the CO_2/O_2 separation factor of about 250.

Recently, Bao *et al.* [68] compared the efficiency of facilitated transport of CO_2 across a liquid membrane by different carriers (diethanolamine (DEA) and carbonic anhydrase (CA) + bicarbonate ($NaHCO_3$) in a polypropylene HFCLM configuration. The hollow fibers used are made of polypropylene, pore size 0.04 μm. In all the experiments, the measured CO_2 permeance and selectivity (CO_2/O_2) using CA bicarbonate as carrier was higher than in the case of DEA. The separation factor (CO_2/O_2) using DEA was about 152 which are 65% lower than the selectivity calculated with CA bicarbonate.

Teramoto *et al.* [25, 47, 70, 71] reported the use of a novel liquid membrane, BFLM, for the separation of CO_2/CH_4 and CO_2/N_2 The experiments were performed using a permeation cell, where the aqueous DEA, Monoethanolamine (MEA) and 2-amino-methyl-1-propanol (AMP) (carrier) solutions and the gas mixture CO_2/CH_4 or CO_2/N_2 were supplied to the feed side (high-pressure side) and the liquid transferred to the sweep side (low-pressure side) was recycled to the feed side. An ultrafiltration polyethersulfone membrane was used as porous support. The CO_2 permeance strongly increased by the circulation of the carrier solution between the feed and sweep side, and very high CO_2 permeance was observed at high liquid circulation rate. In particular, in the case of CO_2/CH_4 the circulation rate of the carrier DEA solution was 1.9×10^{-5} m s^{-1} and CO_2 partial pressure was 4 kPa, the CO_2 permeance was 1.9×10^6 barrer/cm. This value was 10 times higher than when there was no convective flow of the carrier solution across the membrane. The selectivity (CO_2/CH_4) observed was of about 1970. Furthermore, the membrane was very stable with no decrease in gas permeance and selectivity during the performed experiments for more than 2 months. In the case of CO_2/N_2 separation the highest CO_2 permeance was obtained using AMP and DEA carriers and it was about 4×10^6 barrer/cm at 10 kPa in both cases. The separation factor of CO_2 over N_2 was ranging from 430 to 1790. Furthermore, in both cases, the membranes were very stable with no decrease in gas permeance and selectivity during the performed experiments for more than 2 months.

Several articles of Sikdar *et al.* [48-50] reported the use of nonvolatile solvent (glycerol) and nonvolatile carriers (sodium carbonate (Na_2CO_3) and sodium glycinate) for improving the stability of liquid membranes and to be used for the separation of CO_2/N_2 mixtures. Flat and hollow-fiber membranes made of hydrophilic PVDF with a pore diameter of 0.1 μm were soaked with the solvent containing carrier. Effects of experimental conditions such as carrier concentration, CO_2 partial pressure, feed stream relative humidity (RH) were investigated. The addition of carriers drastically increased the permeation rate of CO_2 through the glycerol and also the selectivity CO_2/N_2. Higher CO_2/N_2 selectivities, up to 7000, were observed using Na glycine as carrier while in the case of Na_2CO_3 the higher selectivity was 3400. This is mainly due to the fact that glycine-Na is more soluble than sodium carbonate and therefore higher carrier concentrations could be achieved. Higher selectivity and CO_2 permeance have been observed at lower feed stream humidities. The membranes were stable for 14 days in the case of Na_2CO_3-glycerol and 25 days for Na-glycine membrane.

In another work [51], glycerol carbonate was studied as a new physical solvent with and without carriers (poly(amidoamine) dendrimer and Na-glycinate) for carbon dioxide separation from CO_2/N_2 mixtures. The performance of pure glycerol carbonate appears to be independent of the CO_2 partial pressure difference and the selectivity remains constant (80-100) for any value of the feed side moisture. Addition of the carriers significantly helps CO_2 facilitation at low CO_2 partial pressures. In particular, at 0.66 kPa the presence of the dendrimer and Na-glycinate increased the selectivity (CO_2/N_2) to 1000 and 480, respectively. It was also proved that the decrease of membrane thickness did not affect the selectivity (90-100), which was similar either for 25 and 250 μm thickness membranes, but slightly increased the CO_2 permeance.

Recently, Teplyakov *et al.* [52-53] investigated the use of a selective membrane valve (SMV) which consisted of two polymeric membranes and a liquid absorbent (carrier) layer between them. Polymeric membranes were used to form and retain the liquid layer. In particular, the authors suggested the use of an asymmetric dense membrane which should solve the problem of limitation of an absolute pressure drop between gas and liquid phases (the penetration of liquid into the membrane pores and the formation of bubbles in the liquid). The membrane employed was the asymmetric polyvinyltrimethylsilane membrane; this polymer was chosen for its high permeability toward CO_2. As carrier liquid was used both distilled water and potassium carbonate aqueous solutions (K_2CO_3). The experiments were performed with three different gases, CO_2, O_2, and H_2. The experiments showed a facilitated transport of CO_2 through the membranes. Both the increase of CO_2 permeability and its experimental dependence on K_2CO_3 concentration in solution were well predicted by the model developed by the same authors.

SILMs have been also employed for the separation of CO_2/N_2 and CO/ CH_4 mixtures. Scovazzo *et al.* [54] used common ionic liquids such as imidazolium $[emim]^+$ and different water stable anions such as bis(trifluoromethanesulfonyl)amide $[Tf_2N]^-$, trifluoromethanesulfone $[CF_3SO_3]^-$, and dicianamide $[dca]^-$ in polyethersulfone as polymer substrate. It was observed an interesting increase in CO_2 permeability (up to 50%) decreasing the gas phase pressure from 19 to 1.5 kPa with an ideal selectivity (CO_2/N_2) ranging from 15 (for $[Cl^-]$) up to 61 (for $[dca]^-$) and permeabilities of 350 barrer (for $[Cl^-]$) to 1000 barrer (for $[Tf_2N]^-$). In the case of CO_2 separation from CH_4, the ideal selectivity was ranging from 4 (for $[Cl^-]$) to 20 (for $[dca]^-$) which are above the upper-bound for the CO_2/CH_4 Robson plot.

Matsuyama *et al.* [17] reported the use of SILMs for the CO_2/CH_4. Different types of ionic liquids were synthesized such as 1-butyl-3-methylimidazolium bis(trifluoromethyl-sulphonyl)imide $[C_4mim]$ $[Tf_2N]$, *N*-amino-propyl-3-methyl-imidazolium trifluoromethanesulfone $[C_3NH_2$ mim $[Tf_2N]^-$, and *N*-aminopropyl-3-methylimidazolium bis (trifluoromethylsulfonyl)imide $[C_3NH_2mim][CF_3SO_3]^-$. The porous hydrophilic polytetrafluoroethylene (PTFE) was used as support and it was soaked in the ionic liquids overnight. The highest selectivities, 120 and 100, were reached with the ionic liquids $[C_3NH_2mim][CF_3SO_3]$ and C_3NH_2mim $[Tf_2N]$, respectively. The stability of the SILMs in terms of selectivity and permeability in time was also tested. The membrane resulted stable over a period of 260 days.

In an other work, the SILMs have been employed for the selective separation of CO_2 from He [16]. The CO_2/He gas pair is useful, as more safely tested surrogate, for CO_2/H_2 which will be important in green house gas abatement from advanced coal gasification plants. The SILMs were made by soaking two different porous membranes (polysulfone and polyethersulfone, 0.2 μm pore diameter) in the ionic liquid [hmim] $[Tf_2N]$.

The prepared membranes SILMs showed high CO_2 permeability (744 barrer) and CO_2/He selectivity of 8.6. Furthermore, the stability of the membranes at higher temperature (125 °C) approached the range of interest in the capture of CO_2 from coal gasification plants. However, higher temperatures could not be reached mainly due to the support failure rather than any effect on the ionic liquid.

5.3. Olefin separation

In olefins separation the main industrial target is the separation of ethylene/ ethane and propylene/propane mixtures. Both separations are performed on high scale by distillation, but the relative volatiles of olefins and paraffins are so small that large columns are needed. Steigelmann and Hughes [55] presented the first results of a bench- and pilot-scale study of ethylene and propylene transport using microporous cellulose acetate asymmetric membranes soaked with silver nitrate solutions.

In bench-scale tests, using hollow-fiber membrane as support and a carrier concentration of 2 M the ethylene permeance was 4.6×10^4 barrer/cm with an ethylene partial pressure of 65 psia, while the selectivity C_2H_4/C_2H_6 was about 240. Same tests were carried out for separation of propylene from propane. The selectivity obtained was greater than 100 but this result was confirmed only at bench scale. In fact, in the large pilot system, the selectivity and flux declined over some weeks due to loss of solvent and carrier and to the necessity of remove hydrogen from the feed gas to prevent reduction of Ag + carrier. Despite the result, this remains the first study on the use of facilitated transport membrane for gas separations on a pilot scale.

In 1986, Teramoto *et al.* [56] reported a series of experiments on the permeation of ethylene and ethane through supported liquid membranes containing silver nitrate ($AgNO_3$) aqueous solutions at ambient temperature. It was found the highest selectivity for ethylene over ethane of approximately 1000 when silver nitrate ($AgNO_3$) concentration was 4 mol/dm^3 and an ethylene permeability of 3×10^5 barrer/cm. However, the selectivity was lost in about 200 min.

The same authors proposed a "flowing liquid membrane" [20], for the separation of ethylene over ethane, in which a liquid membrane solution flowed in a thin channel between two microporous membranes. Also in this case a silver nitrate was used as carrier of ethylene. The selectivity for ethylene over ethane was about 460 at a higher concentration of 4×10^3 mol/m^3. This type of membrane was stable for more than 11 days.

More recently, the potentiality of a novel facilitated transport membrane (BFLM) has been also applied to ethylene/ethane separation using silver nitrate as carrier [70, 71]. The experiments were performed using a permeation cell, where the carrier solution and the gas mixture were supplied to the feed side (high-pressure side) and the liquid transferred to the sweep side (low-pressure side) was recycled to the feed side. An ultrafiltration polyethersulfone membrane was used as porous support both in flat and hollow-fiber configuration. The C_2H_4 permeance increased with increasing the permeation flux. Using a flat membrane, with a permeation flux of carrier solution of 4×10^{-5} m s^{-1} and C_2H_4 partial pressure of 9 kPa, the C_2H_4 permeance was 1.19×10^6 barrer/cm and the selectivity C_2H_4/C_2H_6 was 1100. In the hollow-fiber configuration, the maximum C_2H_4 permeance was 3.3×10^6 barrer/cm and selectivity C_2H_4/C_2H_6 was about 375 at a C_2H_4 partial pressure of 164 kPa. In both cases, the membranes were very stable and the experiments could be performed for more than 2 months.

Duan *et al.* [57] reported a very high separation of propylene/propane mixture by a double layer liquid membrane using different solutions of silver salt, $AgBF_4$, or $AgNO_3$ in triethylene glycol (TEG) as facilitating agent of propylene. The selectivity (α) C_3H_6/C_3H_8 was 40 for TEG/

$AgNO_3$ and $\alpha = 110$ and permeability of 3×10^3 barrer for TEG/$AgBF_4$. Furthermore, increasing the carrier concentration, that is, $AgBF_4$ from 40 to 74% and operating with humidify gas stream, it was reported an increase of the selectivity from 100 to 400 while at dry conditions a decrease of selectivity from 100 to 40 was observed.

Recently, Krull *et al.* reported the use of SILM for the separation of propylene from propane [67]. A chemically nonreactive SILM for separation of propylene/propane mixtures has been employed by dissolving a homogeneous catalyst in the membrane phase to perform a dimerization reaction and improve the separation. The SILM used in this work consisted of IL [1-Butyl-3-methylimidazolium]-[bis(trifluoromethyl)sulfonylamide], [BMIM][BTA], immobilized in an asymmetric ceramic support. The SILMs showed stability up to 5 bar transmembrane pressure for more than 72 h of operation. The propylene permeability was of 17.7 and 26.5 m^2 h bar at 20 and 60 °C, respectively. As permeability increased with rising the temperature, propylene/propane selectivity decreased with temperature showing values from 1.89 to 1.67. The experimentally results corresponded well to the results of the simulation made by the same author.

5.4. Sulfur dioxide separation from various gas streams

Removal of SO_2 from flue gases is very important to prevent air pollution. There are many processes available for SO_2 removal from power plant flue gases. Among them, the lime/limestone process as absorbent has been widely used. All the conventional SO_2 removal processes generate sludge that requires disposal. Therefore, the use of membranes which could separate both CO_2 and SO_2 in a single step process is very attractive.

Concerning the separation of SO_2 facilitated transport membranes, Roberts and Friedlander [58-59] investigated the permeability of SO_2 through the membranes containing aqueous and alkaline salt in the range of SO_2 partial pressure from 0.01 to 0.1 kPa. High separation factors above 1000 were observed at low SO_2 partial pressure for alkaline aqueous liquid membranes containing NaOH, $NaHSO_3$, or $Na_2S_2O_5$. In another work, Segupta *et al.* [60] performed several experiments to evaluate the potential application of flue gas desulfurization using microporous polypropylene HFCLMs. The highest performance for removal SO_2 from a synthetic flue gas mixture was obtained using water and 1 N aqueous solution of $NaHSO_3$ and Na_2SO_3 as membrane liquids. Separation factors (SO_2/CO_2) in the range of 138 and 190 with corresponding SO_2 permeabilities of 19.6 $\times$ 10^4 and 43.1 $\times$ 10^4 barrer were observed.

Chakma *et al.* [61] reported the SO_2 separation from N_2 and CO_2 using polyethylene glycol (PEG 400) and ethanolamine-polyethylene glycol (DEA/PEG 400) as carrier in a porous membrane support. SO_2 was preferentially separated from N_2 and CO_2 in the PEG 400 membrane, and the best separation factors were 140 and 14, respectively, both at a low partial pressure (0.2 kPa).

More recently, Teramoto *et al.* [62] performed experiments for facilitated transport of SO_2 through a poly(vinilidene difluoride) (PVDF), used as supported membrane, containing pure water as a carrier. The permeance of SO_2 at a partial pressure of 0.003 Pa was high as 1.04×10^3 barrer/cm and the best selectivity of SO_2 over N_2 was estimated to be more than 10000.

5.5. Hydrogen separation

The separation of H_2 from CO by SLIMs is reported by Gan *et al.* [16]. Transport and separation of properties of H_2 and CO in ionic liquids are of particular interest because of the application of ionic liquids in hydrogenation, oxidation, and hydroformylation and also in fuel cell and electrochemical technologies. Different types of ionic liquids were employed. The SLIMs were prepared using a NF membrane which was soaked in a selected ionic liquid and then placed in the filtration cell. Then, a small amount of ionic liquids were add on the top of the membrane, which ensured that the gas permeation, especially H_2 which can easily penetrate most polymer, was only conducted through a thin layer of ionic liquid than through the polymer. The stability of the membrane was tested and it was proved that the NF SLIM system was very stable also at higher pressure which is more relevant for industrial application. Several factors contribute to its stability, such as the great capillary holding force associated with sub-nanometric pores and high surface tension. Furthermore, the interactions between the charged polymeric materials and the cations/anions of the ionic liquid play also an important role even if it is still not completely clear how the ionic liquid filled up the nanopores of the NF membrane structure during the wetting process. The gas permeation experiments were performed, and H_2/CO separation factor up to a maximum value of 4.3 was reached depending on the type of ionic liquid used. Both gas transport and selectivity exhibited nonlinear behavior with the gas phase pressure variation, reversing the trend of exponential increase of permeation rate.

6. Conclusion and Outlook

Since the 1960s, when the first supported liquid membranes were proposed, many efforts have been made to overcome and improve the drawbacks associated to these membrane systems. Numerous researches deal with the stability of membranes by developing new techniques such as the gelled SLM or by adding a thin top-layer by interfacial polymerization reaction on the SLM. More recently, nanoporous structures and microcapsule techniques able to entrap more efficiently the carrier solvent are only some of the examples of the studies in progress. The right choice of the carrier is also an important point to avoid the carrier degradation and therefore, loss of lifetime of the membrane. Moreover, the research of solvents with lower vapor pressure, and/or new solvents that could act also as carrier represent an other field still under investigation. Besides these efforts, the stability problems have been tried to be solved also using novel types of configurations such as the FLM and HFCLM.

The works carried out on gas separation by facilitated transport mainly refer to O_2, CO_2, H_2S, SO_2, NO, and CO. All these applications are not yet industrialized. Recently in CO_2 separation, SILMs appear to have promising properties for application at the industrial level. This potential could be further enhanced if ceramic membranes are employed, being able to withstand higher operating temperature than polymeric ones, therefore both improving the performance and enlarging the fields of application. Nevertheless, the module design for large scale application, synthesis of novel ionic liquids, also for other specific gas separations, and their limited production at bench scale are still limitations to be overcome.

The research on liquid membranes for gas separation clearly demonstrates that in recent years much progress in the field has been made. However, the scenario does not show promise for short term industrial applications. The drawbacks still present will require more developments over the long term which should also involve industries active in the field of gas separation.

REFERENCES

1. L. M. Robeson, Correlation of separation factor versus permeability for polymeric Membranes, J. Membr. Sci. 62 (1991) 165-185.
2. M. H. V. Mulder, Liquid membranes, *Basic Principles of Membrane Technology*, Kluwer, Dordrecht, Chapter VI (1996) 340-361.
3. H. Nishide, E. Tsuchida, Polymer complex membranes for gas separation, In: N. Toshima, Editor, *Polymers for Gas Separation*, VCH, New York, Chapter 6 (1992) 183-219.
4. B. M. Johnson, R. W. Baker, S. L. Matson, K. L. Smith, I. C. Roman, M. E. Tuttle, H. K. Lonsdale, Liquid membranes for the production of oxygen-enriched air: II. Facilitated transport membranes, J. Membr. Sci. 31 (1987) 31-67.
5. A. H. Chan, W. J. Koros, D. R Paul, Analysis of hydrocarbon gas sorption and transport in ethyl cellulose using the dual mode sorption/partial immobilization modes, J. Membr. Sci. 3 (1978) 117-130.
6. D. P. Paul, W. J. Koros, Effect of partially immobilising sorption on permeability and the diffusion time-lag, J. Polym. Sci., Polym. Phys. 14 (1976) 675-685..
7. X.-S. Chen, H. Nishide, K. Oyaizu, E. Tsuchida, Highly selective oxygen transport through a cobalt porphyrin liquid membrane, J. Phys. Chem. B 101 (1997) 5725-5729.
8. R. W. Baker, Carrier-facilitated transport, In: R. W. Baker, Editor, *Membrane Technology and Applications*, McGraw-Hill, New York, Chapter 11 (2000) 405-442.

9. A. Figoli, W. F. C. Sager, M. H. V. Mulder, Facilitated oxygen transport in liquid membranes: Review and new concepts, J. Membr. Sci. 181 (2001) 97-110.
10. R. Bloch, A. Finkelstein, O. Kedem, D. Vofsi, Metal-ion separation by dialysis through solvent membranes, Ind. Eng. Chem. Process Des. Dev. 6 (1967) 231-237.
11. T. Neplenbroek, SLM-stabilization by gelation, in Stability of supported liquid membranes, Ph.D. thesis, University of Twente, NL (1989) 115-144.
12. A. Kemperman, D. Bargeman, T. van Boomgaard, H. Strathmann, Stability of supported liquid membranes: State of the art, Sep. Sci. Technol. 28 (1996) 2733-2762.
13. H. Strathmann, H. Schulenberg-Schell, B. Bauer, Polymer membrane containing material specific carriers, process for its production and its utilisation for separating substance compositions, German Patent DE 42 38097 (1994).
14. A. Figoli, Synthesis of nanostructured mixed matrix membranes for facilitated gas separation, Ph.D. Thesis, Chapter 7, University of Twente, NL (2001) ISBN 90-365-1673-0.
15. A. Figoli, W. F. C. Sager, Polymerised bicontinuous microemulsion (PBM) membranes: preparation, characterisation and application, presented at Prague Meetings on Macromolecules (2001) ISBN 80-85009-41-2.
16. Q. Gan, D. Rooney, M. Xue, C. Thompson, Y. Zou, An experimental study of gas transport and separation properties of ionic liquids supported on nanofiltration membranes, J. Membr. Sci. 280 (2006) 948-956.
17. S. Hanioka, T. Maruyama, T. Sotani, M. Teramoto, H. Matsuyama, K. Nakashima, M. Hanaki, F. Kubota, M. Goto, CO_2 separation facilitated by task-specific ionic liquids using a supported liquid membrane, J. Membr. Sci. 314 (2008) 1-4.
18. J. Ilconich, C. Myers, H. Pennline, D. Luebke, Experimental investigation of the permeability and selectivity of supported ionic liquid membranes for CO_2/He separation at temperatures up to 125 °C, J. Membr. Sci. 298 (2007) 41-47.
19. M. Teramoto, N. Toho, N Oshini, H. Matsuyama, Development of a spiral-type flowing liquid membrane module with high stability and its application to the recovery of chromium and zinc, Sep. Sci. Technol. 24 (1989) 981-999.
20. M. Teramoto, H. Matsuyama,T. Yamashiro, S. Okamoto, Separation of ethylene from ethane by a flowing liquid membrane using silver nitrate as a carrier, J. Membr. Sci. 45 (1989) 115-136.
21. A. Sengupta, R. Basu, K. K. Sirkar, Separation of solutes from aqueous solutions by contained liquid membranes, AIChE J. 34 (1988) 1698-1708.
22. A. Sengupta, R. Basu, R. Prasad, K. K. Sirkar, Separation of liquid solutions by contained liquid membranes, Sep. Sci. Technol. 23 (1988) 1735-1751.
23. S. Majumdar, K. K. Sirkar, Hollow-fiber Contained Liquid Membrane in the book of W. S. W. Ho, K. K. Sirkar, Membrane Handbook, Van Nostrand Reinhold, New York, Chapter 42 (1992) 764-808.
24. M. Teramoto, N. Takeuchi, T. Maki, H. Matsuyama, Gas separation by liquid membrane accompanied by permeation of membrane liquid through membrane physical transport, Sep. Purif. Technol. 24 (2001) 101-112.
25. M. Teramoto, N. Ohnishi, N. Takeuchi, S. Kitada, H. Matsuyama, N. Matsuyama, H. Mano, Separation and enrichment of carbon dioxide by capillary membrane module with permeation of carrier solution, Sep. Purif. Technol. 30 (2003) 215-227.
26. G. M. Shean, K. Sollner, Carrier mechanisms in the movement of ions across. Porous and liquid ion exchanger membranes, Ann. N.Y. Acad. Sci. 137 (1996) 759-776.
27. W. J. Ward, L. Robb, Carbon dioxide-oxygen separation: Facilitation transport of carbon dioxide across liquid film, Science 156 (1967) 1481-1483.
28. R. J. Basset, J. S. Schultz, Nonequilibrium facilitated diffusion of oxygen through membranes of aqueous cobalthistidine, Biochim. Biophys. Acta 211 (1970) 194-215.

29. W. J. Ward, Analytical and experimental studies of facilitated transport, AIChE J. 16 (1970) 405-410.
30. T. Enns, Facilitation by carbonic anhydrase of carbon dioxide transport, Science 155 (1967) 44-47.
31. E. F. Steigelman, R. D. Hughes, Process for separation of unsaturated hydrocarbons, U.S. Patent 3, 758, 603 (1973).
32. S. L. Matson, C. S. Herrink, W. J. Ward, Progress on the selective removal of H_2S from gasified coal using an immobilised liquid membrane, Ind. Eng. Chem. Proc. Des. Dev. 16 (1977) 370.
33. R. D. Huges, E. F. Steigelman, J. A. Mahoney, Olefin separation by facilitated transport membranes, paper presented at the 1981 AIChE Spring National Meeting, Houston, Texas, April 1981, paper 1d.
34. W. F. Castle, Air separation and liquefaction: Recent developments and prospects for the beginning of the new millennium, Int. J. Refrigeration 25 (2002) 158-172.
35. P. F. Scholander, Oxygen transport through haemoglobin solutions, Science 131 (1960) 585-590.
36. J. B. Wittenberg, The molecolar mechanism of hemoglobin-facilitated oxygen diffusion, J. Biol. Chem. 241 (1966) 104-114.
37. I. C. Roman, R. W. Baker, Method and apparatus for producing oxygen and nitrogen and membrane therefore, U.S. Patent 4,542,010 (1985).
38. H. Nishide, H. Kawakami, T. Suzuki, Y. Azechi, Y. Soejima, E. Tsuchida, Effect of polymer matrix on the oxygen diffusion via a cobalt porphyrin fixed in a membrane, Macromolecules 24 (1991) 6306-6309.
39. H. Nishide, M. Ohyanagy, Y. Funada, T. Ikeda, E. Tsuchida, Oxygen transport behaviour through the membrane containing a fixed carrier and adhered to a second polymer, Macromolecules 20 (1987) 2312-2313.
40. H. Nishide, M. Ohyanagi, O. Okada, E. Tsuchida, Dual mode transport of molecular oxygen in a membrane containing a cobalt porphyrin complex, Polym. J. 19 (7) (1987) 839-844.
41. E. Tsushida, H. Nishide, M. Ohyanagi, H. Kawakami, Facilitate transport of molecule oxygen in the membranes of polymer – coordinated cobalt Schiff-base complexes, Macromolecules 20 (1987) 1907-1912.
42. X. Chen, H. Nishide, K. Oyaizu, E. Tsuchida, Highly selective oxygen transport through a cobalt porphyrin liquid membrane, J. Phys. Chem. 101 (1997) 5725-5729.
43. R. Fiammengo, K. Wojciechowski, M. Crego-Calama, P. Timmerman, A. Figoli, M. Wessling, D. N. Reinhoudt, Heme-protein active site models via self-assembly in water, Org. Lett. 5(19) (2003) 3367-3370.
44. R. Fiammengo, P. Timmerman, F. de Jong, D. N. Reinhoudt, Highly stable cage-like complexes by self-assembly of tetracationic Zn(II) porphyrinates and tetrasulfonatocalix [4]arenes in polar solvents, Chem. Commun. (2000) 2313-2314.
45. S. R. Suchdeo, J. S. Schultz, Mass transfer of CO_2 across membranes: Facilitation in the presence of bicarbonate ion and enzyme carbonic anhydrase, Biochim. Biophys. Acta 352 (1974) 412-440.
46. M. Teramoto, N. Takeuchi, T. Maki, H. Matsuyama, Facilitated transport of CO_2 through liquid membrane accompanied by permeation of carrier solution, Sep. Purif. Technol. 27 (2002) 25-31.
47. N. Matsumiya, M. Teramoto, S. Kitada, H. Matsuyama, Evaluation of energy consumption for separation of CO_2 in flue gas by hollow-fiber facilitated transport membrane module with permeation of amine solution, Sep. Purif. Technol. 46 (2005) 26-32.

48. H. Chen, G. Obuskovic, S. Majumdar, K. K. Sirkar, Immobilised glycerol-based liquid membranes in hollow-fibers for selective CO_2 separation from CO_2–N_2 mixtures, J. Membr. Sci. 183 (2001) 75-88.
49. H. Chen, A. S. Kovvali, S. Majumdar, K. K. Sirkar, Selective CO_2 separation from CO_2-N_2 mixtures by immobilised carbonate-glycerol membranes, Ind. Eng. Chem. Res. 38 (1999) 3489-3498.
50. H. Chen, A. S. Kovvali, K. K. Sirkar, Selective CO_2 separation from CO_2-N_2 mixtures by immobilised glycine-Na-glycerol membranes, Ind. Eng. Chem. Res. 39 (2000) 2447-2458.
51. A. S. Kovvali, K. K. Sirkar, Carbon dioxide separation with novel solvents as liquid membranes, Ind. Eng. Chem. Res. 41 (2002) 2287-2295.
52. M. G. Shalygin, A. Yu. Okunev, D. Roizard, E. Favre, V. V. Teplyakov, Gas permeability of combined membrane systems with mobile liquid carrier, Colloid J. 68 (2006) 518 525.
53. M. G. Shalygin, E. V. Vorobieva, V. V. Teplyakov, Gas transport in combined membrane system with moving liquid, Sep. Purif. Technol. 47 (2007) 466-472.
54. P. Scovazzo, J. Kieft, D. A. Finan, C. Koval, D. Dubois, R. Noble, Gas separations using non-hexafluorophosphate $[PF_6]^-$ anion supported ionic liquid membranes, J. Membr. Sci. 238 (2004) 57-63.
55. R. D. Hughes, J. A. Mahoney, E. F. Steigelmann, Olefin separation by facilitated transport membranes, In: N. N. Li, J. M. Calo, Editors, *Recent Developments in Separations Science*, CRC Press, Boca Raton, FL, vol. IX (1986) 173-199.
56. M. Teramoto, H. Matsuyama, T. Yamashiro, Y. Katayama, Separation of ethylene from ethane by supported liquid membranes containing silver nitrate as a carrier, J. Chem Eng. Jpn. 19 (1986) 419-424.
57. S. Duan, A. Ito, A. Ohkawa, Separation of propylene/propane mixture by a supported liquid membrane containing triethylene glycol and silver salt, J. Membr. Sci. 215 (2003) 53-60.
58. D. L. Roberts, S. K. Friendlander, Sulfur dioxide transport through aqueous solutions: Part I. Theory AIChE J. 26 (4) (1980) 593-602.
59. D. L. Roberts, S. K. Friendlander, Sulfur dioxide transport through aqueous solutions: Part II. Experimental results and comparison with theory, AIChE J. 26 (4) (1980) 602-610.
60. A. Sengupta, B. Raghuraman, K. K. Sirkar, Liquid membranes for flue gas desulfurization, J. Membr. Sci. 51 (1990) 105-126.
61. A. Chakma, Separation of CO_2 and SO_2 from flue gas streams by liquid membranes, Energy Convers. Mgmt. 36 (1995) 6-9.
62. M. Teramoto, Q. Huang, T. Maki, H. Matsuyama, Facilitated transport of SO_2 through supported liquid membrane using water as a carrier, Sep. Purif. Technol. 16 (1999) 109-118.
63. H. Nishide, H. Kawakami, T. Suzuki, Y. Azechi, E. Tsuchida, Enhanced stability and facilitation in oxygen transport through cobalt porphyrin polymer membranes, Macromolecules 23 (15) (1990) 3714-3716.
64. A. Guha, S. Majumdar, K. K. Sirkar, A larger-scale study of gas separation by hollow-fiber-contained liquid membrane permeator, J. Membr. Sci. 62 (1991) 293-307.
65. S. Majumdar, A. K. Guha, K. K. Sirkar, A new liquid membrane technique for gas separation, AICHE J. 34 (1988) 1135-1145.
66. S. Majumdar, A. K. Guha, Y. T. Lee, K. K. Sirkar, A two-dimensional analysis of membrane thickness in a hollow-fiber-contained liquid membrane permeator, J. Membr. Sci. 43 (1989) 259-276.

67. F. F. Krull, M. Medved, T. Melin, Novel supported ionic liquid membranes for simultaneous homogeneously catalysed reaction and vapor separation, Chem. Eng. Sci. 62 (2007) 5579-5585.
68. L. Bao, M. C. Trachtenberg, Facilitated transport CO_2 across a liquid membrane: Comparing enzyme, amine, and alkaline, J. Membr. Sci. 280 (2006) 330-334.
69. A. K. Guha, S. Majumdar, K. K. Sikdar, Facilitated transport of CO_2 through an immobilised liquid membrane of aqueous diethanolamine, Ind. Eng. Chem. Res. 29 (1990) 2093-2100.
70. M. Teramoto, N. Takeuchi, T. Maki, H. Matsuyama, Ethylene/ethane separation by facilitated transport membrane accompanied by permeation of aqueous silver nitrate solution, Sep. Purif. Technol. 28 (2002) 117-124.
71. M. Teramoto, S. Shimizu, H. Matsuyama, N. Matsumiya, Ethylene/ethane separation and concentration by hollow-fiber facilitated transport membrane module with permeation of silver nitrate solution, Sep. Purif. Technol. 44 (2005) 19-29.

CHAPTER 8

Application of Liquid Membranes in Wastewater Treatment

Roman Tandlich

1. Introduction

Rapid industrial development and the diversified character of the world economy today have led to large volumes of different types of wastewaters. Wastewater treatment methods include physical processes, for example, adsorption, and sand filtration, chemical processes, for example, flocculation by addition of aluminum salts, iron salts, and polymeric substances and biological processes, for example, the activated sludge process and anaerobic digestion. These processes are efficient and can provide water that meets different regulatory requirements. However, problems are encountered if toxic levels of different organic compounds and/or heavy metals are present in wastewater. This can lead to the breakdown of treatment performance, for example, in biological processes. Liquid membrane (LM) processes take advantage of the already established principles of biological treatment and solvent extraction but provide the advantage of reduction in treatment costs (see Section 2). Their significance and potential applications in wastewater treatment are discussed here.

2. Bulk Liquid Membranes (BLMs)

2.1. Two-phase partitioning bioreactors

2.1.1. General description

Many anthropogenic organic compounds are highly hydrophobic in character, that is, they possess high values of the 1-octanol/water partitioning coefficient (generally expressed as the log *P*). They can be present in different types of wastewaters. In such context, the two-phase partitioning bioreactors can be used for the elimination of these compounds from wastewater. They

Division of Pharmaceutical Chemistry, Faculty of Pharmacy, Rhodes University, P.O. Box 94, Grahamstown 6140, South Africa

Liquid Membranes: Principles and Applications in Chemical Separations and Wastewater Treatment
DOI: 10.1016/B978-0-444-53218-3.00008-8

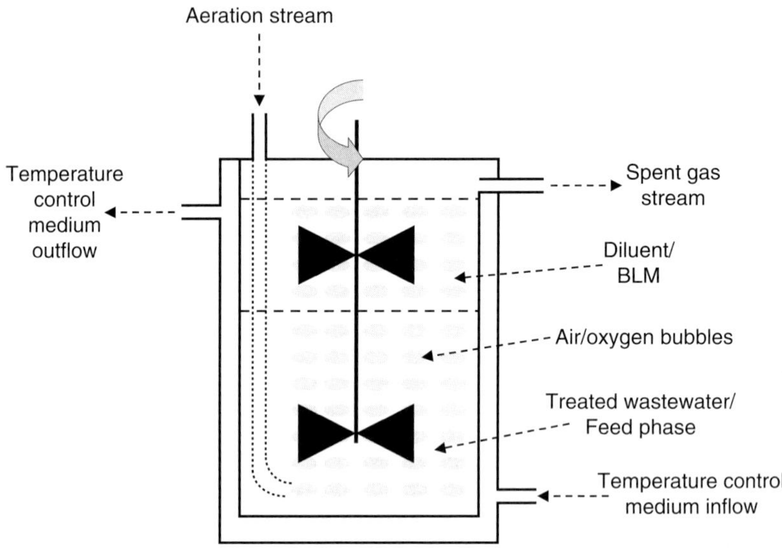

Figure 8.1 The two-phase partitioning bioreactor (for more information see [7]).

are a specific case of the BLM, where the feed and the stripping phase are one and the same, namely the treated wastewater [1]. Schematic representation of the two-phase partitioning bioreactor is presented in Fig. 8.1.

The wastewater/feed phase in question is pumped from a storage reservoir into the bottom part of the two-partitioning bioreactor where it is aerated. The aqueous phase of the system is overlaid with the diluent. Given the hydrophobic character of anthropogenic pollutants, they will tend to partition out of the wastewater in question into the diluent, for example, diethylhexyl sebacate [2], silicone oil [3], and hexadecane [4]. Organic compounds of anthropogenic origin are often toxic to the MO with the ability to biodegrade it [5]. In the two-phase partitioning bioreactor, the BLM extracts the biodegraded pollutant from the wastewater in question, leaving behind the feed phase concentrations of pollutant in the $mg.dm^{-3}$ range [6] or the $\mu g.dm^{-3}$ range [7]. Partitioning of the pollutant out of the feed phase into the diluent allows for high loadings of the organic pollutants into the system without the occurrence of inhibitory effects [8].

The low concentrations of the pollutants in the feed phase start to be biodegraded, resulting in a decrease of the pollutant concentration in the feed phase. Thus the partitioning equilibrium of the pollutant, between the diluent and the feed phase, is disturbed. The pollutant is transported back into the feed phase at finite rates determined by the rate of biodegradation there [8], as well as the respective value of $P_{\text{Diluent/Feed phase}}$ [1], as defined in Eq. (1).

$$P_{\text{Diluent/Feed phase}} = \frac{C_{\text{Diluent}}}{C_{\text{Feed phase}}} \tag{1}$$

where $C_{diluent}$ is the equilibrium BLM concentration of the pollutant ($mol.dm^{-3}$ or $g.dm^{-3}$), and $C_{feed\ phase}$ is the equilibrium concentration of the pollutant in the feed phase, that is, treated wastewater of the two-phase partitioning bioreactor ($mol.dm^{-3}$ or $g.dm^{-3}$). The transport into the feed phase leads to self-regulated dosing of the degraded pollutant to the active MO [9], thus preventing inhibition/death of the MO cells. Guidelines for the design and main literature findings are summarized below.

2.1.2. Selection of the diluent

Several criteria must be considered in the selection of the diluent solvent. The first one is biocompatibility, that is, the solvent must not be toxic to the biodegrading microorganism [10]. The second criterion is the resistance to biodegradation and/or utilization by the active microorganism used, that is, so-called (non)bioavailability [8]. Third criterion is the favorable mass-transfer characteristics for the biodegraded pollutant [1], and the fourth criterion is the mutual immiscibility of the diluent with the treated wastewater [1]. The fifth criterion is the limited volatility of the diluent [11].

The solvent should be readily available from a local supplier and be cost-effective for pilot- and/or industrial scale [7]. The solubility of the pollutant in question in the diluent should be as high as possible to allow maximum pollutant loadings in the two-phase partitioning bioreactor system, that is, make the biodegradation system perform most effectively [1]. As the last criterion, the diluent should not form emulsion with the treated wastewater. This is required to minimize the entrainment of the cells of the MO used in the organic phase of the system, potentially leading to decrease in the rate of biodegradation [12].

2.1.2.1. *Biocompatibility* Biocompatibility is generally evaluated based on the critical log P concept [10]. All solvents with a log P value higher than the critical value of the particular organism are expected to be nontoxic to the microorganism in question, that is, they will not influence its metabolic activity in comparison to a control system without the diluent [13]. The respective log P values are first obtained for given diluents. This can be done by measurement, consultation of the literature [14], or the respective value can be estimated using computational techniques, for example, ClogP [15].

Biocompatibility of the candidate diluents is then tested, for example, using the procedure of Yeom and Daugulis [13]. In this procedure, the influence of the diluent in question on metabolic activity of the chosen microorganism is examined during the utilization of an easily utilizable source of carbon and energy, for example, glucose and corn oil [13]. After the stationary phase of growth has been reached, the biomass concentration of the microorganism in question is measured, for example, as dry weight [16]. Then the value of the relative metabolic activity of the microorganism Z is calculated using Eq. (2).

$$Z = 100 \times \frac{X_{\text{solvent}} - X_{\text{negative control}}}{X_{\text{positive control}} - X_{\text{negative control}}} \qquad (2)$$

where X_{solvent} is the final/stationary phase biomass concentration of the MO used in the presence of the chosen carbon and energy source and the candidate diluent (grams dry weight.dm^{-3}; in further text referred to as gdw.dm^{-3}). At the same time, $X_{\text{positive control}}$ the final/stationary phase biomass concentration of the MO is in the positive control, that is, without the candidate solvent (gdw.dm^{-3}). Finally, $X_{\text{negative control}}$ the final/stationary phase biomass concentration of the MO is in the negative control, that is, without the candidate solvent or the carbon and energy source (gdw.dm^{-3}). The fraction in Eq. (2) is multiplied by 100 to obtain the respective Z value in percent.

Equation (2) expresses the metabolic activity of the microorganism, measured in the presence of the particular solvent, relative to the positive control, that is, when the metabolism of the microorganism is uninhibited. Both of these values are corrected for any changes in the biomass concentration that might have occurred due to stress or any other process(es), with exception of the utilization of carbon and energy source. This is done by subtracting $X_{\text{negative control}}$. After all the Z values are calculated, these are plotted as a function of the respective log P values. This is demonstrated for a hypothetical MO in Fig. 8.2.

According to the critical log P concept, an organic solvent is considered biocompatible if it does not influence the activity of the cells of the MO used [10]. As a result, the Z value for the MO used in the presence of such a biocompatible diluent will be approximately equal to 100% [10]. Based on data in Fig. 8.2, Z is equal to 100% for a diluent with the log P value of 4.5. That means that any diluent with a log P value higher than 4.5 will be biocompatible with the hypothetical MO used, and therefore log P value of 4.5 is called the critical log P value for this MO [17]. All biocompatible diluents are potential candidates for the membrane phase in the two-phase partitioning system, and the choice in a particular case must be made by taking other criteria into account (see below).

The critical values of log P reflect the composition of the cytoplasmic membrane of the particular MO. This means that the critical log P values depend on whether the particular MO is a bacterium, that is, a Gram positive or Gram negative, yeast, an archea or a fungus [18]. For example, the critical log P for *Pseudomonas putida* IFO 3738 is equal to 3.1, with the solvent being *p*-xylene, while the critical log P was shown to be equal to 4.8 for *Agrobacterium tumefaciens* IFO 3058 and the solvent was isooctane [18]. Values of critical log P ranging from 6.0 to 7.0 have been measured for

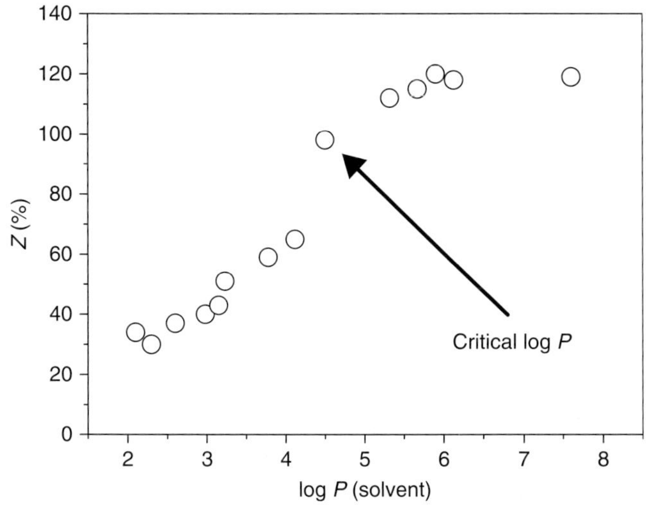

Figure 8.2 Dependence of the relative metabolic activity of a hypothetical MO on the log *P* values of a hypothetical series of diluents, tested for possible application as the BLM(s) in a two-partitioning bioreactor.

bacteria with mycolic acids in their cell wall, for example, *Corynebacterium spp.*, *Rhodococcus spp.*, and some strain of *Brevibacterium spp.* [17].

Upon contact with the microbial cells, the diluent molecules dissolve in the cytoplasmic membrane and cause its swelling [19]. This leads to disruption of the transmembrane potential, pH gradient across the cytoplasmic membrane, and it ultimately leads to disruption of the ion-active pumps [20]. Fluid-mosaic model has been widely accepted as the best description of the real-time structure of biological membranes, and it states that the cytoplasmic membrane exists in a fluid state under physiological conditions [21].

In the presence of diluent molecules in the extracellular medium, microorganisms tend to decrease the fluidity of their cytoplasmic membranes. The decreased fluidity of the cytoplasmic membrane increases its rigidity and viscosity, thus reducing the ability of the diluent molecules to penetrate into the structure of the cytoplasmic membrane [2]. Three mechanisms by which the fluidity of the cytoplasmic membrane can be decreased have been reported in the literature to date. The first one is the homeoviscous adaptation which involves changing the proportions of fatty acids in the membrane lipids [19, 22-25]. The second mechanism is the increase in the relative concentration of proteins relative to lipids in the cytoplasmic membrane [26]. The third mechanism involves the change in cell shape and has been reported for archae [18].

Shifts in the proportions of some phospholipids have also been demonstrated [27]. A *Staphylococcus haemolyticus* strain isolated from the gut of the oil fly has been shown to adapt to toxic solvents, namely toluene by increasing the proportion of the branched fatty acids [28], and anteiso fatty acids in the cytoplasmic membrane [29]. This decreases the packing order of the cytoplasmic membranes, and thus increasing the fluidity of the cytoplasmic membrane. The purpose of this change in fluidity is not clear at present.

2.1.2.2. *Bioavailability* Bioavailability is the ability of the MO used to utilize the particular diluent as the carbon and energy source [2]. Bioavailability of a particular solvent can be evaluated using the same technique as shown for biocompatibility (see section 2.1.2.1.). However, the carbon and energy source such as glucose of corn oil are omitted from the nutrient medium with diluent addition [2]. In this way, the diluent is tested as a possible carbon and energy source, and the metabolic activity/growth of the microorganism in question on the diluent is compared to the growth on glucose or corn oil. If the membrane solvent is bioavailable, then it provides an additional source of carbon and energy. In the presence of the pollutant, this could lead to diauxic growth [30], and/or co-metabolism [31].

MacLeod and Daugulis [2] studied the biodegradation of several polynuclear aromatic hydrocarbons (PAHs) by *Mycobacterium* PYR-1, and chose a bioavailable solvent, that is, bis(ethylhexyl) sebacate as the diluent. It was the cheapest solvent and had the highest solubility for the PAHs studied, that is, allowing for the highest loading of the pollutant into the two-phase partitioning bioreactor. At the same time, it had more favourable viscosity than the other three diluents tested, thus allowing for optimum mass-transfer characteristics for the PAHs studied [2].

2.1.2.3. *Other criteria* Criteria 3–6 for the selection of the diluent solvent are discussed in this section. Third criterion is the favorable mass-transfer characteristics of the diffusion of the pollutant in question through the LM. The diluent properties that are important in this context include diluent viscosity, diffusion coefficient of the pollutant in the diluent phase, and the value of $P_{\mathrm{Diluent/feed\ phase}}$ [1]. The viscosity of the LM has a strong influence on the diffusion coefficient of the pollutant in the diluent phase, thus the application of a highly viscous diluent can lead to a substantial decrease in the mass-transfer rates of the biodegraded pollutant [2]. This prolongs wastewater treatment.

The value of $P_{\mathrm{Diluent/feed\ phase}}$ determines the equilibrium concentrations of the pollutant in the LM, and the treated wastewater based on its definition in Eq. (1). The higher the $P_{\mathrm{Diluent/feed\ phase}}$ value, the higher the loading of the pollutant that can be dosed into the two-phase partitioning bioreactor system, based on the higher solubility of the pollutant in the LM. At the same time, however, the lower the rate of diffusion from the LM into the feed phase, thus the longer the time required for the elimination of the pollutant from the wastewater [1]. The values of the $P_{\text{Diluent/feed phase}}$ observed for benzene with the following diluents: 1-decanol, *n*-dodecane, *n*-hexadecane and 1-octadecene were shown to range from 118.2 to 162.5 [4]. Viscosity, the $P_{\text{Diluent/feed phase}}$ and the diffusion coefficient for pollutant in question can be determined using standard techniques [32, 33].

The fourth criterion of mutual immiscibility of the diluent and the feed phase is important for several reasons. The lower the diluent solubility in the treated wastewater, the less likely it is that diauxic growth or co-metabolism will occur [34]. The lower the solubility of the diluent in the feed phase, the higher the recovery of the diluent at the end of the treatment of one batch of the feed phase, that is, wastewater. This reduces losses of the diluent used, thus treatment costs are lower due to decreased costs of solvent procurement.

Limited mutual miscibility is also important for the extent of the emulsification of the diluent and the feed phase. The lower the tendency of the pair to form emulsions, the lower the chance of entrainment of the MO cells used in the LM [35]. The entrained biomass can be nonactive in the biodegradation of a given pollutant, and thus the emulsification can prolong the biodegradation, as well as the period of wastewater treatment, due to lower concentration of the active biomass.

The emulsification tendency for a given combination of a diluent and the particular wastewater is generally quantified using the emulsification index [36]. Several definitions of this parameter have been published in the literature [35, 36]. The one published by Cooper and Goldenberg [37] is shown in Eq. (3).

$$\mathrm{EI}_{24} = 100 \times \frac{\text{Height(emulsion layer)}}{\text{Total Height}} \tag{3}$$

where Height(emulsion layer) stands for the height of the visible emulsion layer in the test tube (cm), while Total Height is the total height of the test tube contents (cm) [37]. The lower the value of EI_{24}, the lower the chance that emulsification will occur in the particular two-phase partitioning bioreactor for a given diluent and the treated wastewater. EI_{24} has been shown to range from 0 to 70 [37]. The subscript 24 stands for the time period in hours between the mixing of the sample and the measurement of the parameters in Eq. (3).

The fifth criterion is the volatility of the diluent. It needs to be taken into account because volatile diluents can cause losses of the pollutant from the two-phase partitioning bioreactor without proper biodegradation, due to stripping of the diluent in the aeration stream [7]. It has been suggested that nonvolatile solvents have the boiling points above 150 °C [38]. The last two criteria are the price and the availability of the chosen diluent from a local supplier. These two criteria are important because of the feasibility and cost-effectiveness of the operation of any two-phase partitioning bioreactor. Given the considerations above, the selection process for the diluent to be used take all the parameters into account, and the decision about the final selection should be made by weighing all the criteria mentioned above.

2.1.3. Laboratory studies

Different types of organic pollutants have been released into the environment from anthropogenic sources. Examples include polychlorinated biphenyls (PCBs) [39], PAHs [40], and (un)substituted phenols [41]. Release is tied to specific industries where they have been used, given their specific physical and chemical properties [39]. Biodegradation of these pollutant groups possesses several challenges. The first one is the limited contact between the biodegrading MO cells and the pollutant in question [42]. This results from the low aqueous solubility of the biodegraded organic pollutant.

The second challenge originates from toxicity of the pollutant to the degrading microflora [39]. This often requires an acclimation of the MO to the biodegradation of the pollutant in question, before full operation of the two-phase partitioning bioreactor system becomes feasible. The first challenge can be overcome by the release of biosurfactants [33], or direct contact between the cells of the biodegrading organism and the crystal/liquid droplets of the toxic xenobiotic [43]. Shortening the acclimation period, and/or alleviating the toxicity, can be addressed by genetic manipulation [44], or by the application of a two-phase partitioning bioreactor.

Eibes *et al.* [45] studied the application a two-partitioning bioreactor in enzymatic degradation of sparingly soluble organic compounds. Vandermeer and Daugulis [40] used the mixed culture of *Sphingomonas aromaticivorans* B0695 and *Sphingomonas paucimobilis* EPA505, with dodecane used as the diluent, to eliminate several PAHs. Naphthalene, phenanthrene, and fluoranthene were completely eliminated within 8 h. After the same time period, the extent of pyrene biodegradation reached 64% of the initial amount. This yielded the volumetric biodegradation rate of 42.9 mg dm^{-3} day^{-1}. Under identical conditions, the extent of benzo[α]pyrene reached 11% of the initial concentration, and the volumetric biodegradation rate was equal to 7.5 mg dm^{-3} day^{-1}. The extent of biodegradation increased, but the volumetric biodegradation rates decreased when silicone oil was used as diluent instead of dodecane.

Collins and Daugulis [46] used industrial grade oleyl alcohol as the diluent in a two-phase partitioning bioreactor. This was inoculated with *Pseudomonas sp.* ATCC 55595, and the authors studied the biodegradation of benzene, toluene, and *p*-xylene, individually and in different mixtures. The authors compared the efficiency of the aromatic hydrocarbon removal in the batch mode and the fed-batch mode. In a mixture of benzene and toluene, the volumetric rates of biodegradation were 92% higher for toluene and 19% higher for benzene in the fed-batch in comparison to the batch mode operation.

For the mixture of toluene and *p*-xylene, the volumetric biodegradation rate increased by 20% for toluene and by 39% for *p*-xylene the fed-batch mode in comparison to the batch mode operation. These results indicate the extent of organic pollutant removal in the two-phase bioreactor system depends on the presence or absence of other pollutants, as well as on the feeding strategy applied. Selection of the mode operation will depend on the particular combination of diluent, microorganism, and the pollutant in question. The optimization of operating conditions for the fed-batch mode can be obtained using the approach of Cruickshank *et al.* [35], while the continuous mode of operation could be optimized using a similar model of Cruickshank *et al.* [47].

2.1.4. Biodegradation mechanisms

MOs with the required biodegradation potential can be obtained by enrichment techniques or genetic manipulation. Enrichment cultures have been used in more cases than the genetic manipulation. MOs generally require an acclimation period before optimum biodegradation performance can be achieved. The acclimation period is required for the adaptation of the MO in question to the biodegraded pollutant as the sole carbon and energy source. Once the adaptation occurred the lag phase during the growth of the MO on the pollutant is shortened, and MO multiplication takes place [48].

Adaptation can be accompanied by an increase in the rate of biodegradation [49]. The acclimation period can be caused by the time required for induction and synthesis of the necessary enzymes, a spontaneous mutation could be observed, or the genetic information can be acquired giving the microflora with the biodegradation potential [34]. Propagation of the MO can result from the multiplication of an originally small number of MO cells to produce visible growth. The length of the lag phase can also depend on the concentration of inorganic nutrients, and the presence of more easily utilizable carbon/energy substrates [48]. It is also possible that predation by protozoa, and the presence of inhibitors and/or toxins, can have an effect on the length of the lag phase [50].

Enrichment cultures have been reported to fail during biodegradation of certain substrates [3]. Application of the two-phase partitioning bioreactor can help shorten the acclimation period and can also lead to faster removal of higher loadings of the pollutant in question [3]. Enrichment can also be successful in contrast with one-phase systems (see below). Several mechanisms of pollutant uptake exist in the two-phase partitioning bioreactor. These are direct uptake of the molecules dissolved in the feed phase [43], and/or uptake of the dissolved molecule but the MO increases this by the excretion of surfactants [34]. The next possible mechanism involves the attachment of the MO cells to the aqueous side of the interface between the diluent and the feed phase [3]. The last mechanism involves the attachment of the bacteria to the organic side of the interface between the diluent and the feed phase [51].

2.1.5. Challenges to industrial applications

Aeration is required for the proper operation of aerobic wastewater treatment systems. Up to date, most of the two-phase partitioning bioreactors have been described for the biodegradation of pollutants under aerobic conditions. Oxygen limitation has been reported by many authors. Problems have been reported as a result of mechanical aeration, and these include excessive foaming and entrainment of microbial cells in the organic phase of the system [52], and the over pressure during sampling and volatilization losses of the pollutant [7]. Often pure oxygen has to be used for aeration to prevent oxygen limitation [7]. If active aeration could be omitted from the operation of the two-partitioning bioreactors, then the capital and running costs of such system would be decreased, and all of the above-mentioned problems could be eliminated.

There are two possible ways of how to achieve this. The first one is the inoculation of a mixed algal-bacterial or algal-fungal inoculum into the aqueous phase of the bioreactor, while the other is the delivery of additional oxygen through the membrane solvent. The first option was investigated by Muňoz *et al.* [8]. They studied biodegradation of phenanthrene with silicone oil or tetradecane as the diluent. The mixed culture of alga *Chorella sorokiniana* and the bacterial strain *Pseudomonas migulae*. Based on phenanthrene biodegradation, aeration in a mixed culture of *C. sorokiniana* and *P. migulae* provided lower volumetric rates of biodegradation and longer times would be required to achieve the target concentrations, in comparison to mechanical aeration [53]. At the same time, entrainment of biomass into the organic phase and other problems would be avoided.

The solubility of the oxygen in organic solvents is generally higher the respective solubility in water, and this fact has been exploited in different types of bioreactors for the enhancement of oxygen mass-transfer rates [54]. Perfluorohydrocarbons are a good example of such a group of compounds. These compounds are chemically stable, mainly due to the presence of highly inert C-F bonds [55]. The molecules of O_2 can be dissolved in the intermolecular spaces. Henry's law applies to the dissolution of oxygen because of the lack of specific interactions between the molecules of oxygen and the molecules of perfluorohydrocarbons [55]. Perfluorohydrocarbons have been applied for the oxygenation of a culture of *E. coli*, in the form of falling drops [55].

One of the disadvantages is the formation of emulsions, which can extract protein from the microbial cells, and thus leads to the rupture of cells, that is, breakdown of the bioreactor performance [56]. The extraction of protein can be avoided by the application of two-phase aqueous bioreactors, where a polymer, and/or polymer and salt, is dissolved in water and two phases are formed [56]. Such two-phase system has a lower surface tension between the two phases, that is, 0.0001–0.1 mN m^{-1} as compared to 0.5–50 mN m^{-1} in the case of an organic second phase. The lower surface tension prevents unfolding/denaturation of the protein at the interface, and protein losses are prevented.

Interfacial phenomena are important if considering the use diluents and BLM components as an aeration medium. The selection of an appropriate diluent for aeration of the feed phase can be obtained by the measurement and calculation of the spreading coefficient S_p [57]. This property is defined in Eq. (4).

$$S_p = \sigma_{WA} - (\sigma_{OA} + \sigma_{OW}) \tag{4}$$

where σ_{WA} is the interfacial (surface) tension at the interface between water and air (N m^{-1}), σ_{OA} is the interfacial (surface) tension at the interface between the diluent and air (N m^{-1}), and σ_{OW} is the interfacial (surface) tension at the interface between the diluent and water (N m^{-1}).

If the S_p value is negative for a given combination of a diluent and the wastewater in question, then the oxygen concentration in the organic phase does not influence the oxygen mass-transfer coefficient in the BLM. In this case, the presence of the diluent in the two-phase partitioning bioreactor decreases the interfacial area for the O_2 transport. Droplets of such a diluent tend to accumulate at the interface between the diluent/feed phase interface, and these block the contact between the interface and microbial cells, that is, decreasing the rate of oxygen transfer from the diluent into the feed phase [57].

If the S_p value is positive for the given combination of a diluent and wastewater, then presence of the organic phase in the system leads to an increase in the mass-transfer coefficient for oxygen, as well as the interfacial area of oxygen transport. At the BLM/feed phase interface, molecules of the organic phase form a thin film adjacent to the aqueous phase, thus leading to a decrease in the surface tension at the interface, that is, less hindered transport of oxygen molecules through the interface. The S_p value must be measured for the potential diluent and the feed phase wastewater to be treated, before any application is conducted [57].

2.1.6. Industrial applications

To date and to the best of the author's knowledge, only one commercial application has been reported for two-phase partitioning bioreactors. It is based on the work by the research group of Prof. Andrew S. Daugulis from the Queen's University in Kingston, Canada [58], which has been licensed to Xethanol Corporation [59].

2.1.7. Potential future developments

The most promising is the application of the two-phase partitioning bioreactors in the treatment of highly complex mixtures, such as mining wastewater and metal refinery wastewaters as demonstrated [7]. One potential and intriguing application could be the use of two-phase partitioning bioreactor in processes, where an inorganic anion functions as the terminal acceptor of electrons, for example, denitrification, the application of sulfate reducing bacteria, and so on. This would allow for the elimination of the operational problems related to aeration, such as foaming, and MO cell entrainment in the organic phase. At the same time, organic and inorganic pollutants, such as nitrates, would be eliminated at the same time.

2.2. Other Applications of the BLMs

The application of mercury is widespread in agriculture, for example, as insecticide in seed treatment, and different types of industry [60]. A promising method for the removal and preconcentration of mercury from wastewater has been the application of liquid membranes containing calixarenes as carriers [61]. A three-phase system for the extraction of Hg from industrial wastewater has been reviewed by Ersoz [62]. Models and implication of theoretical conclusions are presented.

3. Emulsion Liquid Membranes (ELMs)

3.1. General description

Application of emulsion liquid membranes (ELMs) in wastewater treatment is relevant when the aim of the treatment is the removal and preconcentration of ionizable organic pollutants, such as different phenols [63], and recovery/preconcentration of heavy/toxic metals [64]. In the case of the metal removal, the metal cation in question diffuses to the interface between the feed phase and the ELM. There it is complexed with an extractant molecule, and the resulting complex diffuses through the membrane toward the stripping phase droplets. Upon contact between the metal-extractant complex and the microdroplets of the stripping phase, the metal dissociates from the complex, and diffuses into the internal liquid phase, where it is sequestered at highly acidic pH value. The driving force of the metal extraction from wastewater is the chemical potential gradient of H^+ between the stripping phase and the feed solution, and so the metal ions can be extracted and transported into the internal aqueous phase against their concentration gradient [64].

In the case of the ionizable organic compounds, for example, phenolics, transport is based on the concentration gradient of the nonionized form of the organic compounds between the feed phase and the stripping phase. The feed phase pH is adjusted so that the respective molecule is in its nonionized form, and it can be dissolved in the diluent [65]. Inside the membrane phase of the ELM, the nonionized form of the pollutant diffuses through the membrane matrix, free or complexed, toward the stripping phase droplets. At the interface, that nonionized organic molecule diffuses into the stripping phase. Once inside the stripping phase microdroplet ionization occurs, and thus the activity of the nonionized form of drug is zero, and so back-transport cannot occur [66].

Important parameters that have to be considered during the preparation an ELM include the concentration of surfactant, ratio of the volumes of the diluent/surfactant mixture, and the stripping phase; the chemical nature of the solute extracted, the diluent, and the carrier is needed. The removal of metals and selected organic pollutant with ELMs from wastewater is discussed below.

3.2. Removal of metals from wastewaters using ELMs

3.2.1. Laboratory studies

ELMs have long been identified as a potential replacement of precipitation of heavy metals from wastewaters [67], and they show certain advantages in the recovery of platinum group metals from precious metal refinery and mining wastewaters [68]. Raghuraman *et al.* [67] studied the removal of Pb and Cd from their binary mixture with di-2-ethylhexyl phosphoric acid as the carrier. Further data were reported for Cu, Ni, and Zn with different carrier/extractants. A single distribution constant could be used to characterize the partitioning of the complex of Cu with LIX and Zn and Ni with di-2-ethylhexyl phosphoric acid, over the entire range of pH and ionic strength values. At higher loading values, the nonideal character of the organic phase had a significant influence on the partitioning process.

Liu *et al.* [68] described the production of nanoparticles composed of $PbCrO_4$ based on ELMs. The liquid membrane consisted of kerosene, sorbitan monooleate (SPAN 80), and N7301. Based on the results, the method could be applied to treatment of various industrial wastewaters simultaneously. Fang *et al.* [69] studied cobalt extraction by a trialkyl amine hydrochloride. Based on theory and experimental results, $CoCl_4^{2-}$ was identified as the most probable species to undergo ion pairing with protonated trialkyl ammonium cation. The ELM with trialkyl amine hydrochlorides as carriers was developed and investigated as a possible medium for the selective extraction of cobalt over nickel from a hydrometallurgical waste effluent sample [69]. The optimized version of the extraction protocol provided a 99.8% recovery of Co in the internal phase of the ELM after a one-stage extraction process. Ni was the dominant metal ion in the rafinate from the extraction.

Cr(VI) is used in many industrial fields, such as the tannery industry [70], and its removal is important to avoid detrimental impacts on environmental and human health [71]. Kulkarni *et al.* [71] studied the ELM-based preconcentration of Cr(VI) from wastewater, and then Cr(VI) was further reduced to Cr(III) by the $FeSO_4$ catalysis at pH 2. Complete reduction of Cr(VI) could be achieved with a 10% stoichiometric excess of $FeSO_4$ at 100 °C. Cr(III) is precipitated using alkaline solutions, and the process effluent is suitable for discharge. Regulatory guidelines for the residual concentrations of Cr(VI) in treated wastewaters were met after hybrid treatment [72]. Zn^{2+} can be extracted from acidic industrial effluents using ELMs [73] with the percentage removal reaching up to 99.3% after optimization. These results show the feasibility of removal and/or recovery of metal ions from different types of effluents and dilute aqueous solutions.

Breakage of the ELM globules due to swelling and shear during extraction has been reported for many decades. Extraction efficiency of the ELM procedure in question decreases as a result, and therefore breakage of the ELM globules is undesirable. The stability of an ELM can be improved by increasing the concentration of the surfactant in the ELM [74, 75] and/or manipulating the viscosity of the liquid membrane by the use of non-Newtonian liquids [75].

Several drawbacks are connected to the application of the surfactant concentration as the parameter controlling ELM stability. The first one originates from the increasing swelling of the ELM with increasing surfactant concentration, due to increasing affinity for water [76]. If the ELM is prepared from a Newtonian liquid, then the second major drawback is the decrease in the rates of mass transfer inside the ELM, due to an increase in the viscosity of the ELM [77]. If the LM is prepared from a non-Newtonian liquid, then the diffusion coefficient of the extracted solute is virtually independent of the membrane viscosity [78, 79], below the critical concentration [80]. This concentration can be calculated from Eq. (5), as derived by Skelland and Meng [81].

$$C_{\text{critical}} = 228 \times \frac{(M_{\text{sru}})^{5/3}}{V_{\text{sru}} M_{\text{p}}^{2/3}} \tag{5}$$

where M_{sru} is the molecular weight of the pure liquid monomeric unit of the polymer (single repeat unit; g mol^{-1}) and V_{sru} is the molar volume of the pure liquid monomeric unit of the polymer at its respective normal pressure and boiling point (cm^3 mol^{-1}). M_{p} stands for the average molecular weight of the polymer (g mol^{-1}).

In the non-Newtonian LM, increasing viscosity leads to reduction in the coalescence of the ELM droplets, as their movement inside the ELM is more restricted [80]. This increases the stability of the ELM at a lower concentration of the surfactant used. The reduced concentration of the surfactant leads to lower affinity of the LM for water, thus lower rates of swelling are observed. To date, most of the non-Newtonian LMs were based on the solution of a polymer in a Newtonian diluent [74, 81]. Demulsification with polymers in the LM is easier, and the LM reuse becomes more feasible, because of the lowered concentration of the surfactant, that is, lower barrier to droplet coalescence during demulsification [74]. To achieve optimum ELM performance and demulsification, it has been recommended that the membrane viscosity be kept below or maximum equal to 0.01 Pa s [82].

Skelland [83] was the first to utilize non-Newtonian liquids in the design of ELM. In this and related studies, the non-Newtonian LMs were prepared from Soltrol 220 (an isoparaffinic oil, Chevron Philips Chemical, Spring, TX, USA) as the diluent, polystyrene [81], polyisobutylene [74], and polybutadiene [80] as polymers, SPAN 80 as the surfactant and different aqueous solutions were used as stripping phases depending on the nature of the solute removed from the treated wastewaters [74]. Skelland and Meng [80] studied the removal of benzoic acid, phenol and ammonia from simulated wastewater.

In most of the ELM studies to date, the LMs and the wastewaters have been contacted in continuously stirred tanks. The problems with this design in liquid-liquid extractions originate from the fact that most of the volumetric energy dissipation occurs in a limited part of the total volume near the impeller tips [74].

Drag and large shear are exercised near and around the impeller surface, thus leading to substantial turbulence which causes the inertial breakage of the emulsion droplets [84]. Shear becomes even stronger with increasing size of the contactor, and so the scale-up of such devices (for industrial use) is difficult [85]. At the same time, substantial amounts of electrical energy are required for the complete mixing of the LM and the treated wastewater [85]. The problems encountered using the continuously stirred tanks can be avoided, if the extraction is performed in a Taylor-vortex column. This is depicted in Fig. 8.3.

Forney *et al.* [85] were among the first authors to study the potential advantages of the liquid-liquid extraction using a Taylor-vortex column. In the column, the power input is evenly distributed throughout the entire volume of the contactor, and the rotor and tank stirrers are roughly equal in diameter [74]. The maximum shear is one to two orders of magnitude lower than in a continuously stirred contactor. This leads to a between 10-fold and 100-fold increase in the area inside the continuously stirred tank that is exposed to constant maximum shear.

This is based on the friction drag on the large cylindrical surfaces near the boundaries of Taylor-Couette flow. One major advantage of

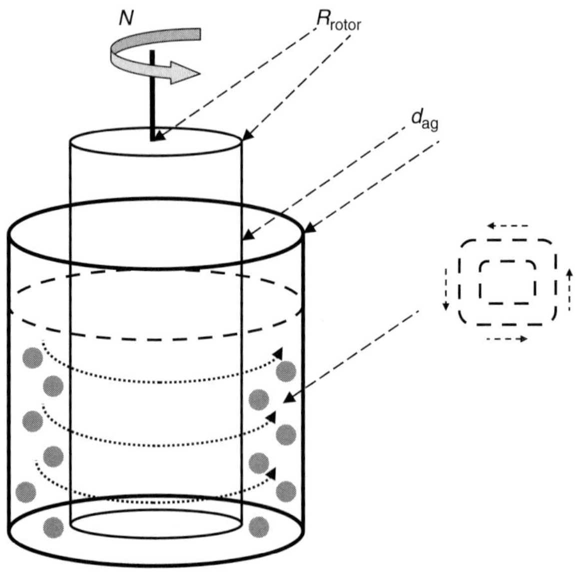

Figure 8.3 The Taylor-vortex column. All symbols have the same meaning as in Eq. (6). The membrane globules (•) are dispersed in the feed phase of the system via the rotation of the inner cylinder of the Taylor-vortex column, and the flow pattern, depicted by the dashed lines on the right-hand side of this figure, is established.

the Taylor-column type devices is the simplicity of the potential scale-up, as it only requires the constant value of the dimensionless Taylor number [85]. This criterion is defined in Eq. (6).

$$\mathrm{Ta} = \frac{2\pi N}{\phi}(R_{\mathrm{rotor}})^{1/2}(d_{\mathrm{ag}})^{3/2} \tag{6}$$

where R_{rotor} is the radius of the inner cylinder of the Taylor column, that is, rotor (m), N stands for the rotational speed of the inner cylinder (s^{-1}), d_{ag} is the diameter of the space between the inner and the outer cylinders of the Taylor-vortex column (m), φ is the kinematic viscosity of the liquid inside the Taylor-vortex column ($m^2\ s^{-1}$).

Park [74] studied the efficiency of the ELM with non-Newtonian liquids in the removal of Zn, Pb, Ni and Cd from a simulated industrial wastewater using the Taylor-vortex column. The author adapted the shrinking core mathematical model of Liu and Liu [86] for quantitative description of the mass-transfer kinetics of the process [74]. The LM was prepared by the dissolution of 5 g dm^{-3} of polyisobutylene in Soltrol 220 (see above). After complete dissolution of the polymer, the membrane phase of the ELM was prepared by addition of 50 g dm^{-3} of SPAN 80 and 0.1 M of the carrier in question [74].

The following carriers were tested: Cyanex 301, Cyanex 302, Cyanex 923, and di(2-ethylhexyl)phosphoric acid. The stripping solution was 2 M HNO_3, in a ratio of 0.33 (the volume of the stripping phase to the volume of the membrane phase). In the experiments, 200 ml of the simulated industrial wastewater and 25 ml of the ELM were mixed inside the Taylor column. The stirring speed of the central cylinder ranged from 400 to 600 rpm, and all experiments were conducted at 20 °C [74].

The leakage of the stripping solution was monitored by the measurement of the NO_3^- concentration in the feed phase of the system, using an HPLC method with conductometric detection [74]. The best removal efficiencies were obtained for all metals studied with LMs containing Cyanex 301 and all metals reacted formed the 1:2 complex with the carrier. At the stirring speed 500 rpm after 30 min 96% of Zn was removed from the feed phase with the initial Zn concentration of 0.01 M. The composition of the feed/external phase had a strong influence on the leakage of the stripping NO_3^- into the feed phase, and the minimum leakage was measured for the 0.2 M CH_3COOH feed solution. The removal of the individual metals from the feed phase was related to the values of the stability constants for the metal complexes and Cyanex 301.

The ELM pertraction has been developed as a combination of the ELM and the supported liquid membrane (SLM) in hollow-fiber configuration [73]. Hollow-fiber contactors are highly energy efficient and provide high interfacial areas for the efficient wastewater treatment [87]. The installation of these modules has been driven by tighter environmental regulations, and by the fact that hybrid processes, e.g., ELM pertraction, make meeting these regulations easier for the companies in question [87]. The ELM pertraction systems can have potential applications in the chemical, petrochemical, pharmaceutical, and galvanic industries [87]. The main advantage of combining the ELM with a hydrophobic membrane module in a hollow-fiber contactor prevents leakage of the diluent and the stripping phase into the feed phase [88]. The schematic representation of the metal removal in the ELM-SLM pertraction module can be seen in Fig. 8.4.

The treated metal-bearing wastewater solution flows along the outside of several hollow-fiber membranes. The particular ELM is pumped through the inside of the hollow fibers (Fig. 8.4A). The diluent and the dissolved extractant (H_nL) fills the pores of the membrane support, while the stripping phase is dispersed further away from the interface. The metal ion to be extracted (M^{n+}) from the wastewater diffuses to the interface between the wastewater and the hollow-fiber membrane. Here it reacts with the extractant and the metal-extractant complex, ML, is formed (Fig. 8.4B). To preserve the electroneutrality of the system, n of protons H^+ are transported into the treated wastewater. ML diffuses through the ELM matrix toward

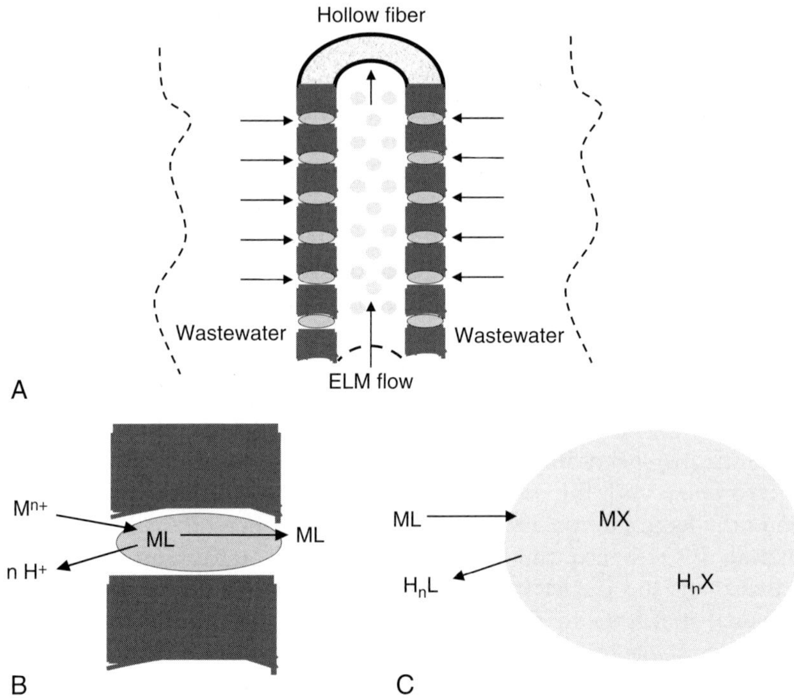

Figure 8.4 The schematic representation of the ELM pertraction process.

the stripping phase droplets (Fig. 8.4C). Here the M ion is sequestered inside the stripping phase by reaction with the molecule of a particular inorganic acid (H_nX), and the MX/salt is formed. The molecule of extractant is continuously regenerated (Fig. 8.4C). This regenaration and continuous mode of operation help maintain the optimum driving force for the metal extraction from the wastewater in question (for further details see [87]).

Ortiz *et al.* [72] studied the applicability of the ELM pertraction for the recovery of Cr(VI) from the bearing wastewater. The authors designed a countercurrent pertraction module for the treatment of a wastewater sample with the following composition (all concentrations in mg dm^{-3}): 52 of total organic carbon, 543 of calcium, 2170–2692 of Cr(VI) in the form of CrO_4^{2-}, 1735 for sulfates, 1187 for chlorides, and 14 for silicates. The pH value of the wastewater was equal to 7.30. Prior to treatment, the pH was adjusted to 1.50 with H_2SO_4. The ELM were pumped through the hollow-fiber module, and the removal of Cr(VI) occurred by the adsorption of the ELM globules to the pores of the hydrophobic membrane support [89]. The ELM was composed (all percentages are v/v) of 10% dodecanol as the modifier, 10% of Alamine 336 as the extractant, 5% of Pluronic 331 as the polymer membrane (non-Newtonian liquid), and 75% of Isopar L (an isoparaffinic hydrocarbon solvent) as the diluent.

The feed phase passed through the pertraction module once, while the ELM was recycled. The continuous recycling of the ELM leads to constant replenishment of the extractant, thus increasing the rate of Cr(VI) pertraction, due to constant supply of fresh and nonsaturated droplets of the stripping phase to the interface. NaOH solution with the concentration of 3000 molm^{-3} was used as the stripping phase [72].

Inorganic anions, mostly SO_4^{2-}, are co-pertracted during the studied ELM pertraction process. The authors proposed this to be solved by the addition of a second ELM pertraction module, leading to practically complete rejection of the $SO_4{}^{2-}$ anions from the resulting effluent [72]. The combined two-module system allowed for the removal of up to 99.5% of Cr (VI), and the concentration in the stripping phase, after phase separation of the ELM, was equal to 20 mol m^{-3}. The optimum setup and the number of modules could be optimized by the nonlinear dynamic programming as demonstrated by Fresnedo San Román *et al.* [88].

3.2.2. Industrial applications and future trends

Several successful applications of the ELM in the removal of heavy metals from industrial wastewaters have been reported in the literature. These include the Zn removal from wastewater at the viscous fiber plant located in Lenzing, AG, Austria [89]. This process can treat up to 75 m^3.h^{-1} of zinc bearing wastewater with the zinc concentration ranging from 0.3 to 200 mg dm^{-3}. Zn can be removed with up to 99.5% efficiency [89]. Other two industrial applications are also aimed at Zn removal, and the respective plants are located in Glanzstoff, AG, Austria (with the reported capacity of 700 m^3.h^{-1}), at the CFK Schwarza plant, Germany and the AKZO Ede plant in the Netherlands (both with the reported capacity of 200 m^3 h^{-1} [67]).

Commercial application of the ELM technology to the recovery of precious metals, such as the platinum group metals, has long been proposed in the literature [90]. Practical applications have, however, so far been limited by the instability of the structure(s) of the ELMs used. Several ways how to improve the stability have been suggested, for example, the use of bifunctional surfactants when one molecule acts as the emulsifier and extractant at the same time [91], and/or the application of additives that alter the elasticity of the respective LM [92].

The ELMs could be applied if the concentrations of precious metals in the wastewater range from 0.1 to 10,000 mg dm^{-3} [90]. The application of the ELMs based on non-Newtonian liquids and the Taylor-vortex column carry a lot of promise for industrial applications in the near future. Several commercial applications have been reported for ELM pertraction. Passivating is an operation commonly used in the galvanization industry to improve the resistance of metal parts to corrosion [87]. Metal parts are submerged into the passivating bath and become coated with Cr and/or Co to provide a protective layer against corrosion [87].

The process is applied, if the metal parts are made of steel, Al, Cu, Cd, Ag, Sn, Mg, and their alloys [87]. The passivating bath liquids are reused during successive runs, and so over time the concentrations of Zn and Fe increase. High levels of these metals cause coloring and decreased in the resistance of the metal parts to corrosion after coating [88]. Thus the passivating bath liquids need to be replaced once every 6 weeks, and/or when the concentration of Zn reaches 4000 mg dm^{-3} [87]. These liquids subsequently become wastewaters that require treatment.

The ELM pertractions have been established at several plants in the Netherlands by the TNO research company [87]. These include systems at Galvano Techniek Veenendaal and Loko Gramsbergen [87]. Their design capacity is equal to 15,000 dm^3 h^{-1}. During 9 months of operation, the average concentration of Zn was always below 2000 mg dm^{-3}, leading to a sixfold reduction in the volume of the wastewaters produced. The operation of the ELM pertraction system has also led to the elimination of the necessity for the operation of an incinerator, and thus savings of the cost 5 m^3 of natural gas, that is, totaling 30,000 Euros per year [87]. This can also lead to the reduction of the carbon-footprint of the company where the ELM pertraction system is installed. Details about the industrial ELM pertraction systems can be found in Klaassen *et al.* [87].

The ELM pertraction technology has good potential for more applications at industrial scale in the near future. The industries in question might include metal mining and refinery operations (precious metals and platinum group metals are good examples), tannery industry (recovery of hexavalent chromium), and processing of nuclear wastes (recovery of uranium, strontium, and other metals).

3.3. Removal of organic pollutants from wastewaters using ELMs

3.3.1. Laboratory studies

As mentioned before, the ELMs have been used for the recovery of different phenols, benzoic acids, and other compounds. Terry *et al.* [65] studied the application of LMs in the treatment of the simulated wastewaters containing phenolic compounds and organic acids, as well as the treatment of coal liquefaction plant wastewater. The best treatment results were obtained with the ELM composed of a 1:1 mixture of the 4% solution of PIBSA (a polyamine-based surfactant) in a low-odor paraffin solvent as diluent, with 10% aqueous solution of NaOH as the stripping phase.

The extraction efficiency for phenol changed as a function of time. Ninety-nine percent of phenol removal was observed after 2 min of stirring at stirring speed of 420 rpm. At this stirring rate, the ELM swelling was prevented. The optimum ratio of the wastewater to the ELM was 5:1. Treatment results for *o*-, *m*- and *p*-cresol were comparable to the observations made for phenol.

The acetic acid removal from the simulated wastewater containing 1.0–5.0 g dm^{-3} of acid was tested. The maximum removal of 99.5% was reached at the initial concentration of 3.0 g dm^{-3}. Preferential extraction of phenol over acetic acid was recorded in the mixture of the two compounds in nonionized form. This indicates that the hydrophobicity of the compound in question determines the rate of accumulation in the ELM. The pH values of the simulated wine distillery wastewater had a strong influence on the efficiency of the extraction.

For coal liquefaction wastewater, phenol and *o*-, *m*- and *p*-cresol were removed from the wastewater but no removal was observed for acetate or propionate. The results of this study could be used to design an effective extraction system for the removal of short-chain fatty acids, which have been shown to inhibit anaerobic biodegradation of different types of wastewaters [93]. This is desirable to avoid the inhibitory effect of the short-chain fatty acids on the course of the treatment process.

At treating wastewaters from petrochemical and oil refinery industries application of ELM in the extraction of 2-chlorophenol [66] has been indicated as a possible preconcentration step, and thus avoiding toxicity toward biodegrading microflora [94]. The application of ELMs for the removal of different nitrophenols has been described by Gadejkar *et al.* [95].

Kumaresan *et al.* [96] studied the applicability of ELMs in the treatment of distillery wastewater. These types of wastewaters are generally toxic to microflora, because of the high content of several phenolic compounds [93]. The optimum ELM composition was as follows (v/v): 4% of xylene, 10% of CRESLOX, and 85% of liquid paraffin, while the optimum ELM/treated wastewater ratio was 0.8. At this ratio, the chemical oxygen demand (COD) of the distillery wastewater was reduced from 6800 mg dm^{-3} in the feed solution to the final effluent concentration of 927 mg dm^{-3}. At the same time, the five-day biochemical oxygen demand (BOD_5) decreased from 3500 to 96.5 mg dm^{-3}. The recovery of acetic acid was 44.2%. These results indicate that ELMs can be quite effective in the reduction of the pollution of distillery wastewaters.

ELMs based on non-Newtonian liquids have been used for the removal of organic compounds from wastewaters. The same general guidelines apply to the preparation of the ELMs as mentioned for the removal of metals from wastewaters (see Section 3.2). Skelland and Meng [80] studied the application of the ELM in the removal of ammonia, phenol, and benzoic acid from simulated industrial wastewater. SPAN 80 was used as the surfactant, while as the stripping solution 0.5 M NaOH was used in the extraction of benzoic acid and phenol, and 0.5 M H_2SO_4 for the removal of ammonia [80]. The volume ratio between the stripping phase and the Soltrol/neutral oil mixture in the ELM was equal to 0.2. The ELM was prepared by sonication of all components in the appropriate solutions.

Treatment of 1000 ml of the simulated wastewater was conducted with 80 ml of the ELM for benzoic acid and phenol, and with 160 ml of the ELM for ammonia. The addition of the polymer into the ELM helped to prevent leakage of the stripping phase. The optimum concentration of SPAN 80 was equal to 2–4%, while the optimum concentration of polyisobutylene was equal to 0.5–2.0%. The equilibration times ranged from 4 to 36 minutes. The percentage of removal reached 99% for benzoic acid, 83% for phenol, and 97% for ammonia.

Removal of phenol from wastewaters has been studied using non-Newtonian ELMs and the Taylor-vortex column. It was found that a decrease in the surfactant concentration in the ELM leads to a lower surfactant accumulation per unit interfacial area at the feed phase/ELM interface. As a result, the higher was the efficiency of phenol removal from the wastewater [97, 98]. An increase in the volume of the stripping phase in the ELM could result in a reduction of the LM thickness [74].

Decrease in the thickness of the ELM leads to a decrease in the diffusion pathway that phenol molecules must travel inside the ELM to reach the stripping phase droplets. This leads to higher rates of the phenol mass transfer into the stripping phase, that is, across the ELM [64, 95, 99]. At the same time, more stripping phase is present in the ELM, the more feasible the demulsification process becomes [82]. Besides phenol, Park [74] also studied the application of the Taylo-vortex column for the removal of benzoic acid from synthetic industrial wastewater, with promising results for potential scale-up.

Another important group of emerging pollutants in wastewaters is antibiotics. These compounds are produced by fermentations, for example, tetracyclin [100] or by chemical synthesis, for example, chloramphenicol [101]. Separation of antibiotics from the production broth or reaction mixture generally involves two or three membrane filtration operations [102]. These are microfiltration, ultrafiltration, diafiltration, and/or nanofiltration. Effluents/wastewaters from these operations are likely to contain finite concentrations of antibiotics [103]. Because of their antimicrobial nature they might pose problems for the traditional wastewater treatment processes [104]. High capital and investment costs, along with the possibility of fouling and limited rejection rates for some inorganic ions, can prevent large scale application of the above-mentioned membrane processes in the removal of antibiotics from wastewaters. The possibility of the application of LMs instead of mentioned membrane filtration processes could lead to more cost-effective wastewater treatment operations. Some examples of the studies published on the separation of antibiotics using ELMs are discussed below.

Schügerl [105] reported several ELM-based applications for the extraction of alcohols, carboxylic acids, and antibiotics. Habaki *et al.* [106] studied the application of ELMs and SLMs in the extraction of erythromycin. The antibiotic molecules were able to cross the LMs without a carrier. Given their own data, as well as the data of other authors, Habaki *et al.* concluded that the distribution coefficient of free erythromycin between the membrane phase and the feed solution for every LM studied was constant and independent on the H^+ concentration in the feed solution. ELMs had lower extraction efficiencies for the erythromycin in comparison to SLMs [106].

With *m*-xylene as the diluent, the rate-limiting step of the overall permeation of erythromycin was the transport through the membrane and stripping phases. On the other hand, the concentration of erythromycin in the ELM with heptane as the diluent was limited by the transport through the membrane phase only.

Habaki *et al.* [107] derived an expression for the equilibrium ratio of the total concentration of erythromycin in the stripping phase and feed phase ψ, as function of concentration of hydronium cations/protons in both phase. The relationship is described in Eq. (7).

$$\psi = \frac{E_{\mathrm{Stripping;total}}}{E_{\mathrm{Feed;total}}} = \frac{K_a + [H_3O^+]_{\mathrm{Stripping}}}{K_a + [H_3O^+]_{\mathrm{Feed}}} \qquad (7)$$

where $E_{\mathrm{feed;total}}$ is the total concentration of erythromycin in the feed solution (mol dm^{-3}), while $E_{\mathrm{stripping;total}}$ is the total concentration of erythromycin in the stripping solution (mol dm^{-3}). K_a is the acidity constant for erythromycin in water (i.e., $10^{-8.8}$; dimensionless) and $[H_3O^+]_{\mathrm{Stripping}}$ is the concentration of hydronium cations/protons in the stripping solution (mol dm^{-3}). $[H_3O^+]_{\mathrm{Feed}}$ is the concentration of hydronium cations/protons in the stripping solution (mol dm^{-3}).

As it can be seen, ψ is directly proportional to the concentration of the protons in the stripping phase and indirectly proportional to the concentration of protons in the feed phase. Therefore as long as the concentration of protons is higher in the stripping phase than in the feed phase, the transport of erythromycin into the ELM will continue. At the same time, the ψ value will be independent on the actual values of the total erythromycin concentration in the feed phase or the stripping phase.

Protein-rich wastewaters have been reported in the food industry [108] and dairy industry [109]. The amino acid rich wastewaters have been reported in the amino acid producing industry [110]. Most literature references, up to date, deal with the treatment of dairy wastewater(s). Dairy wastewaters are produced at different rates and with different composition depending on the type of dairy product and the intensity of the production campaign [109]. Physical, chemical, and biological parameters of the dairy wastewaters change from one plant to another and in between individual production batches and also depend on the technology used in production [111].

Organic carbon is accounted for by lactose, fats, and proteins [112-114]. To a limited extent, protein and fat loading can be removed using precipitation [115]. Practical problems with precipitation arise from the low removal efficiencies, as well as from the need for expensive reagents [109]. Casein accounts for 80% of total protein in dairy wastewaters, and hydrolysis is the main biodegradation mechanism regardless of the concentration of oxygen, that is, whether anaerobic or aerobic treatment methods are used

[109]. This process is catalyzed by proteases, and mixtures of (poly)peptides and/or amino acids are the products of the biodegradation [109], along with NH_3/NH_4^+ [116].

Isolation of the amino acids from such wastewaters might be of commercial value. For example, L-phenylalanine has been used as a precursor in the synthesis of a variety of industrial products, for example, the artificial sweetener Aspartame [117]. Purified amino acids can, on the other hand, be used as nutritional supplements in the diet of livestock [118], as well as in human nutrition [119]. Therefore, there is potential for the extraction of amino acids from dairy and proteinaceous wastewaters for additional income of the producing industries. Application of LMs in treatment of such wastewaters could be of substantial benefit.

Reverse micelles have been applied in the separation of amino acids and proteins. The separation is based on the balance between electrostatic forces and hydrophobic interactions [120]. The pH value is a crucial parameter determining this balance. If reversed micelles are applied in LMs, then the underlying interactions are determined by interfacial partition coefficients of the amino acids/proteins separated, that is, hydrophobicity of the compounds separated, ionic strength of the feed and stripping solutions, the chemical nature of the electrolytes present, and the interfacial curvature of the amphiphilic film [121]. Changing the above-mentioned conditions, the overall charge of the reverse micelle can be altered, and so the separation conditions can be manipulated [122].

The rate-limiting step of the mass transfer for the separated compound(s) in such systems depends on the concentration of the surfactant used [121]. Hebrant and Tondre [120] studied the extraction of tryptophan and *p*-iodophenylalanine through the reverse-micelle ELM composed of isooctane and the aqueous solution of Aerosol OT, that is, the sodium salt of bis (2-ethylnexyl)sulfosuccinate, as the stripping phase. The extraction efficiencies and transport rates were shown to be amino acid specific.

The advantage of the LMs is integrating extraction and back-extraction of the desired analyte(s) into one step. Using protonation and deprotonation reactions, selected hydrophobic carriers with carboxyl groups have been shown effective in the separation of amino acids, if the carboxyl functionality was ionized [123]. Optimum values of the stability constants of the complexes between particular amino acids and carrier(s) can be found to increase extraction efficiencies. However, the kinetics of mass transfer often has a more pronounced impact on the efficiency of extraction [118].

3.3.2. Industrial applications and future trends

Results from the laboratory studies indicate that several ELM applications provide highly efficient tools for the removal of organic compounds from wastewaters. The results from the studies with phenol and substituted phenols can have wide ranging applications in wastewater treatment, mainly if reuse of the treated wastewater provides economic advantages. The highly concentrated stripping phase, containing chlorinated phenols should be treated as hazardous material. One of such hazards is the production of dioxins is possible during incineration under inadequate conditions. Therefore reuse of the ELM should be maximized.

The most promising is the application of the Taylor-vortex column with Taylor-Couette flow. Sorption of hydrophobic organic compounds can lead to under estimation of actual removal efficiency of wastewater treatment, as well as cross-contamination of the batches of the effluent due to release of the sorbed solute molecules. This is supported by the results of the laboratory studies, along with the ease of the potential scale-up. The critical part in the context of the removal of organic compounds from wastewater(s) is the material of the inner cylinder. The surface of the inner cylinder provides a potential sorption surface for hydrophobic organic molecules. Since the PTFE price could be prohibitive for the scale-up, stainless steel, or other materials should be explored as replacements for potential industrial applications.

4. Supported Liquid Membranes (SLMs)

4.1. General description

The range of applications of supported liquid membranes (SLMs) in the wastewater treatment is comparable to the ELM (Section 3), with respect to the range of chemicals that have been treated using this technology. This section, therefore, has a structure that is analogical to Section 3.

4.2. Removal of metals from wastewaters using SLMs

4.2.1. Laboratory study results

Several studies have focused on the removal of Cu from wastewaters using SLMs. Valenzuela *et al.* [124] studied the extraction of Cu from Chilean mining wastewaters using an SLM containing 5-dodecylsalicylaldoxime (LIX-860) as the extractant and Kermac 500-T as the diluent. Cataru and Badiciou [125] used a hollow-fiber SLM for the removal of copper from wastewater. The copper removal efficiency and the hollow-fiber SLM permeability were directly proportional to the flow rate and the pH value of the feed phase. Similar trend was observed for both variables as a function of the reagent concentration in the hollow-fiber SLM.

Yang *et al.* [126] used LIX54 and LIX84 as carries in another SLM for Cu removal from several wastewaters. Higher copper transmembrane fluxes were observed than with LIX84 as the carrier. LIX54 provided higher selectivities for Cu(II) over Zn(II) and Cd(II). There was also no significant carryover of ammonia into the stripping phase of the system recorded. Evaluation of membrane stabilities with LIX54 and LIX84 proved potential industrial application of the system in question feasible.

Several mass-transfer mechanisms have been reported in the literature for the transport of metals across SLMs. In one of the most common mechanisms protons from the stripping phase are transported in the opposite to metal ions direction. The principle of electroneutrality applies to the amounts of the protons and the metal ions that are exchanged and the pH of the feed phase decreases as a result [126, 127]. In such cases, transmembrane pressure or voltage do not need to be applied to facilitate the metal ion permeation. The gradient in the chemical potentials of the H^+ in the feed phase (treated wastewater) and the stripping phase is the driving force of metal permeation. As a result, copper and/or the metal ions can be extracted against the Cu ion concentration gradient [126, 127]. As the metal permeation proceeds, protons accumulate in the feed phase, thus resulting in a decrease of the feed phase pH and a decrease in the metal permeation rates [126, 127]. For the Cu (II) transport, this is the case when the conditions of Eq. (8) are fulfilled.

$$\frac{[Cu^{2+}]_{Feed}}{[Cu^{2+}]_{Stripping}} = \left\{\frac{[H^+]_{Feed}}{[H^+]_{Stripping}}\right\}^2 \tag{8}$$

where the expressions in brackets stand for the respective equilibrium concentration of either H^+ or Cu^{2+} (both in mol dm^{-3}). To allow for the metal permeation to continue, the pH of the feed solution has to be buffered, and this can be accomplished by dissolution of $NH_3/NH_4{}^+$ in the feed phase [128].

Yang and Kocherginsky [126, 127] studied the removal and recovery of copper from the spent ammoniac etching solutions from the production of the printed circuit boards. A hollow fiber SLM system was developed for the treatment of spent ammoniac etching solutions with copper concentrations around 160 g dm^{-3}. The attractiveness of the system results from the potential for high throughput and associated economic benefits. Kerosene was used as the diluent and LIX54 was used as the carrier for copper. The authors described a laboratory-scale system, with 1.4 m^2 of membrane surface area and a pilot-scale system the membrane surface area of 130 m^2 [126, 127]. The optimum stripping phase was 2 M H_2SO_4, based on the rate of permeation and the optimization of the solubility of $CuSO_4{\cdot}5H_2O$ [126, 127]. The stoichiometry of the interfacial reaction taking place at the feed phase/SLM interface can be seen in Eq. (9), and the interfacial reaction taking place at the SLM/stripping phase interface can be seen in Eq. (10).

$$Cu(NH_3)_4Cl_2 + 2\,LiX^O \rightarrow Cu(LiX)_2^O + 2\,NH_4Cl + 2\,NH_3 \tag{9}$$

$$Cu(LiX)_2 + H_2SO_4 \rightarrow CuSO_4 + 2\,LiX \tag{10}$$

The maximum value of the permeation copper flux is observed at the minimum thicknesses of the stagnant layer of adjacent to the feed phase and the stripping phase side of the respective interfaces with the hollow-fiber SLM. The residencc time of the complex of the metal in the membrane has reached optimum value when the maximum copper permeation flux is observed [126, 127]. The rate-limiting step of copper permeation through the hollow-fiber SLM was the transport of the Cu-carrier complex through the hollow-fiber SLM [126, 127].

Carryover of NH_3 into the stripping solution was observed, but it could be eliminated by an addition of HCl into the feed phase or the volatilization of NH_3 out of the feed phase by aeration. The scaled-up version of the system was used for the treatment of 200 dm^3 of the printed-circuit board etchant wastewaters. During 55 h of treatment, 96.7% of copper was possible from the initial concentration of copper was equal to 150 g dm^{-3}.

Mechanism and the transport rates were studied for the Am(III) permeation through the SLM with octyl (phenyl)-*N*,*N*-diisobutylcarbamoylmethylphosphine oxide as the carrier and diethylbenzene as the diluent. At low metal concentrations, the rate of metal diffusion of through the stagnant layer of the feed phase, adjacent to the feed phase/LM, was the rate-limiting step of metal removal from the wastewater [129]. At high metal concentration, the rate-limiting step of the removal/permeation process was the diffusion of the Am(III)-carrier complex through the LM phase [129]. The scope of the study was further widened to include all actinides and lanthanides [64]. This study also serves as an example of more complex stoichiometries of the extraction reactions.

Several considerations and types of information are necessary for the design of any online system, utilizing SLMs for the recovery of metals [130]. The first type of information needed is quantification of the effect of the metal ion concentration in the feed solution on the transmembrane flux J (metal) in mol m^{-2} s^{-1} and the permeability of the metal P(metal) [130]. This relationship can be described quantitatively using Eq. (11).

$$P(\text{metal}) = \frac{J(\text{metal})}{[\text{Metal}]_{\text{Feed}} - \dfrac{D(\text{metal})_{\text{Strip}}}{D(\text{metal})_{\text{Feed}}} \times [\text{Metal}]_{\text{Strip}}} \qquad (11)$$

where [Metal] is the (equilibrium) metal concentration in the respective phase (mol dm^{-3}). D is the respective partition coefficient of the metal between the feed phase and the LM, that is, $D(\text{metal})_{\text{Feed}}$, and the stripping solution and the SLM, that is, $D(\text{metal})_{\text{Strip}}$. Both partition coefficients are dimensionless. J(metal) can be measured as a function of the change of the metal concentration in the stripping phase, that is $[\text{Metal}]_{\text{strip solution}}$. This can be mathematically expressed by using Eq. (12).

$$J(metal) = \frac{V_{\mathrm{Strip}}}{A_{\mathrm{SLM}}} \times \frac{\mathrm{d}[\mathrm{Metal}]_{\mathrm{Strip}}}{\mathrm{d}t} \qquad (12)$$

where A_{SLM} is the area of the SLM (m^2) and V_{strip} is the volume of the stripping phase (m^3).

Five elemental processes take place during the uptake of a metal ion from the feed solution into the SLM. First, the metal must first diffuse through the feed solution to the feed phase/SLM interface [130]. The second step is the interfacial reaction of the metal ion and the carrier in question and is the formation of the complex in the SLM. The third step is the diffusion of the resulting complex through the LM. In the fourth partial step is the stripping reaction at the interface. The last partial step is the diffusion of the metal ion from this interface through the adjacent boundary layer into the bulk of the stripping solution.

Both of the interfacial reactions are generally very fast [129], and so they are rarely the rate-limiting steps of the metal permeation into the stripping phase [130]. The activity of the metal inside the stripping phase is generally kept close to zero by complexation with complexation agent that is not identical to the LM carrier [130]. This leads to fast diffusion of the metal into the bulk of the stripping solution. As a result, the rate-limiting steps of the permeation processes are either the diffusion of the metal-carrier complex through the matrix of the SLM or the diffusion of the metal through the bulk phase of the feed solution and through the boundary layer adjacent to the feed phase/SLM interface [130]. This reference can be consulted for more information upon need.

Ramadan and Danesi [131] conducted an investigation into the possible application of SLMs with synergistic crown-ether carriers for Sr(II) permeation. They demonstrated that Sr(II) can be separated from Cs(I) in a solution containing HNO_3. Vincent *et al.* [132] used Cyanex 301(R), that is, di(2,4,4-trimethylpentyl) dithiophosphinic acid as a carrier to produce a treatment system for Pd recovery from acidic solution which would be found during metal refinery [62]. Gelatin and alginate were combined with the carrier to obtain a stable SLM. This was then applied for reactive precipitation of palladium via Pd reaction with the SH group of di(2,4,4-trimethylpentyl) dithiophosphinic acid. Precipitation occurs upon ionotropic gelation of alginate.

The maximum binding capacity was found to be close to 350 mg Pd/g (Cyanex 301). Pd and Cyanex 301 were shown to form 1:1 complex, but more than one molecule of the carrier could surround the metal ion [135]. Palladium uptake was not significantly affected by the presence of competing metals namely Cu, Ni, Zn, or Pt. The association of Pd with di(2,4,4-trimethylpentyl) dithiophosphinic acid proved partially irreversible since the Pd recovery from the gel was not quantitative. This leaves as the main option for the recovery of Pd.

Chiarizia *et al.* [136] studied the application of the SLM for the removal of U(VI) from synthetic and industrial wastewaters. Bis(2,4,4-trimethylpentyl) phosphinic acid (Cyanex(TM) 272) was used as the carrier, *n*-dodecane was used as the diluent, and hollow polypropylene fibers were used as the support for the hollow-fiber SLM. This setup was chosen based on the selectivity for U(VI) over Ca^{2+} and Mg^{2+}. 1-hydroxyethane-1, 1-diphosphonic acid was added into the stripping phase as a masking agent that can increase the permeation rate by keeping the activity of U(VI) in that phase close to zero [130].

Laboratory scale hollow-fiber modules were constructed with a feed phase recycling module, and the removal efficiency of the system was tested for the treatment of synthetic and uranium bearing wastewater. The feed phase pH value was equal to 2.00. Scale-up of the system was conducted after encouraging preliminary results. During the treatment of uranium bearing wastewater, the concentration of U(VI) decreased from the initial 3.5 mg dm^{-3} to below 1.0 μg dm^{-3} in a matter of several hours, indicating the potential application of the system in the processing of metal refinery wastewaters.

Application of SLMs provides a way to circumvent the equilibrium limitations of other separation processes. Stability of SLM is, however, a major issue during operation. This results from the displacement of the membrane phase from the support pores due to osmotic pressure [138]. Nondestructive monitoring methods based on impedance spectroscopy have been reported for the monitoring of the carrier and/or diluent losses from the LM [139]. Ho and Wang [138] studied the application of several alkylated phenylphosphonic acids as carrier for the removal of ^{87}Sr and ^{90}Sr from simulated groundwater. They also reported on a system that allowed for the stabilization of the SLM against leaking of the membrane phase from the pores due to osmotic pressure, that is, the so-called SLM with strip dispersion.

In this arrangement, the solution of the surfactant and the carrier in the diluent and the stripping solution are prepared separately. Then the stripping solution is dispersed in the organic solution by vigorous mixing to obtain the particular water-in-oil liquid membrane [138]. The resulting membrane phase is then recirculated between the mixer compartment, and the shell side of a module consisting of hollow polypropylene hydrophobic support fibers. The feed solution, that is, the treated wastewater, is then pumped through the lumen side of the hollow fiber hydrophobic support module. The membrane phase readily wets and fills the pores of the hydrophobic support, and a stable SLM is formed. Overpressure of approximately 13.8 kPa is held on the lumen side of the hollow fiber module, with respect to the shell side. In this way, the leakage of the membrane phase into the feed phase is prevented [138].

Droplets of the stripping phase generally range from 80 to 800 μm in diameter, that is, they are larger than the 0.03 μm wide pore in the hydrophobic support. Therefore the leakage of the stripping phase into the feed is prevented. Both phases are continuously circulated via the action of independent peristaltic pumps. There is a constant supply of the LM components, and therefore desorption/release of the membrane phase of the SLM from the hydrophobic support pores is lowered. At the same time, concentration gradients of the metal are maintained on the feed side, and so the rates of mass transfer are maximized.

The authors tested the proposed system for the removal of ^{87}Sr and ^{90}Sr from a simulated groundwater with Zn^{2+} and Ca^{2+} [138]. The carriers studied included 2-butyloctyl phenylphosphonic acid, 2-hexyldecyl phenylphosphonic acid, 2-octyldecyl/2-hexyldodecyl phenylphosphonic acid, and 2-octyldodecyl phenylphosphonic acid. For ^{87}Sr, the removal efficiencies reached 100% with all four extractants and HCl (1 mol dm^{-3}) as the stripping solution. The overall mass-transfer coefficient was practically independent of the structure of the carrier used. This indicated that the mass-transfer rates are independent of the alkyl side chain length, and the transport through the SLM was not the rate-limiting step of the mass transfer. The rate-limiting step of the mass transfer was the rate of the interfacial complexation reaction between Sr and the respective alkylated phenylphosphonic acid. This was indicated by the independence of the metal recovery on the nature and concentration of the stripping solution, that is, the results were comparable for HCl (1 mol dm^{-3}) and H_2SO_4 (3 mol dm^{-3}).

4.2.2. Industrial applications and future trends

To the best of the author's knowledge, there are no wastewater treatment applications of SLMs in the public domain that would be aimed at the removal of metals from different wastewaters. It is possible that such applications exist, but are protected as confidential know-how. Advantages of the flat-sheet SLM arrangement in metal removal from wastewater(s) can be achieved, if low-cost extractants/diluents, such as vegetable oils, are used. However, the energy considerations will probably lead to the preferential application of hollow-fiber SLM system in the industry. The SLM with strip dispersion has been shown to improve the stability of the hollow-fiber SLM system. On the basis of the available data, the main benefits from this technology could be observed in the treatment of wastewaters from nuclear engineering facilities, thus preventing environmental contamination, decreasing the volume of the water needed for nuclear power plant operation, that is, thus providing a potential reuse and close circuit operation of the nuclear power plants with respect to water/wastewater.

4.3. Removal of organic pollutants from wastewaters using SLMs

4.3.1. Laboratory studies

Industrial wastewaters generally contain variable concentrations of a wide range of organic compounds, which are often accompanied with high concentrations of inorganic salts [140]. Several membrane-based processes have been applied in wastewater treatment and they include reverse osmosis and nanofiltration. However, both of these methods suffer from severe shortcomings during treatment of highly saline wastewaters. These problems arise from the inability of the reverse osmosis membranes to withstand high osmotic pressures encountered. On the other hand, nanofiltration membranes lack the ability to reject organic compounds with molecular weight below 200 g mol^{-1} [141]. At the same time, the installation of both above-mentioned membrane methods carries high costs and demands large area for installation.

SLM module with high molecular weight liquids could provide means to extract organic molecules at osmotic pressures of couple hundred bars, that is, provide a treatment option for the removal of organic contaminants from highly saline wastewaters. Based on costs considerations, such liquid includes polypropylene glycol and polybutylene glycol [140].

Ho *et al.* [137] tested the applicability of polypropylene glycol and polybutylene-based SLMs in the removal of *p*-nitrophenol from simulated industrial wastewater. The molecular weights ranged from 0 to 4000 g mol^{-1} for polypropylene glycol and from 500 to 5000 g mol^{-1} for polybutylene glycol. The flat-sheet SLMs were prepared with Celgard® and Gore-tex PTFE as supports [137]. The membrane/feed partition coefficient of *p*-nitrophenol was indirectly proportional to its initial concentration and to temperature and pH value of the feed phase dependence. The individual values ranged from 20.1 to 600.

The results of the FT-IR observation proved that the H-bonding with the polypropylene glycol backbone was the dominant mechanism responsible for the partitioning of *p*-nitrophenol into the SLM used. Polypropylene glycol with molecular weight of 4000 g mol^{-1} provides the best removal efficiencies, along with the best rejection rates for H^+, OH^-, SO_4^{2-}, NH_4^+, Na^+, K^+, Cl^-, and NO_3^- and permanently zwitterionic species. Up to 99% removal efficiency were observed in different experiments conducted. The optimized version of the experimental setup provided up to 90% or higher removal efficiencies benzoic acid, caproic acid, phenol, *o*- and *m*-nitrophenol.

Dastgir *et al.* [141] coated a polyvinylidenfluoride microporous support with a thin nonporous polydimethylsiloxane for SLM removal of phenol from wastewater. The choice of support material was based on favorable mass-transfer characteristics and the chemical resistance criteria.

Polypropylene glycol was used as the diluent. The final form of the flat-sheet SLM system had uniform selectivity and good operational stability during continuous operation for more than 2 months. Mass-transfer rates measured were five times the values measured usually during the operation of commercially available silicone tubing-based systems [141]. The system developed by the authors had two more advantages. These were the reduced water flux and the minimum sodium ion transfer. The authors measured the partition coefficient of phenol between polypropylene glycol and water, and determined this to be equal to 84. This value increased to 134 if the aqueous phase contained 20% KCl.

Hollow-fiber SLMs have been used in the removal of phenol from aqueous matrices. Kujawski *et al.* [142, 143] studied polypropylene membranes impregnated with methyl-terbutyl ether, cumene, and/or a mixture of hydrocarbons. With Cyanex 923 (a mixture of trialkylphosphine oxides), the recoveries of phenol reached of 98% into the stripping phase from the 0.2 mol.dm^{-3} solution of caustic soda [144, 145]. Carriers for phenol removal from wastewaters have included linear monoalkyl cyclohexane [146], *N*,*N*-di(1-methyl heptyl) acetamide [147], dibenzo-18-crown-6 [148], dodecane [149], trioctylamine [150], and *N*-octanoylpyrrolidine [151]. Many diluents and carriers are of synthetic origin, and so their application carries with issues of flammability, volatility, toxicity, and potential detrimental effects to the environment and the health of the human population [152].

Xiao *et al.* [153] reported up to 95% removal efficiencies for phenol with SLMs-based poly(dimethyl)siloxane and poly(methylvinyl)siloxane supports. Venkateswaran and Palanivelu [152] tested nine vegetable oils as carriers in flat-sheet SLM for the recovery of phenol and cresol from synthetic wastewater, as well as phenol-formaldehyde wastewater and wood-processing plant effluent from Chennai, India. PTFE was used as the hydrophobic support in the flat-sheet SLM, and the pore diameters ranged from 0.2 to 1.0 μm. The optimum phenol permeation rates were observed with the flat-sheet SLM with palm oil as the carrier, and the PTFE support with the pore diameter of 0.5 μm. The optimum pH value of the feed phase was 2.0, and no proton carryover into the stripping phase was observed. The stripping efficiency toward the phenolate anion was virtually independent of the NaOH concentration in the stripping solution between 0.1 and 1.0 mol dm^{-3}. Turbulences occurred, and this led to the stripping of the oil carrier from the pores of the PTFE support, stressing the importance of the optimization the hydrodynamic conditions during removal of phenol from wastewaters using SLM.

The maximum percentage of phenol removal was indirectly proportional to the initial phenol concentration in the feed phase. This statement is supported by the removal of 100% of phenol at the initial concentration of 100 mg dm^{-3} after 6 h and only 25% of phenol at the initial concentrations of 1250 and 1500 mg dm^{-3} after 3 h. The larger the initial concentration of phenol in the feed solution, the larger the concentration gradient of phenol between feed phase and the stripping phase, and thus the faster the saturation of the stripping phase with respect to phenol. Limited membrane area for the transport of phenol could be the explanation for the leveling off in the removal percentage from 1250 to 1500 mg dm^{-3}.

When the same experiment was repeated with cresols at the initial concentration of the individual cresols of 100 mg dm^{-3}, then complete removal of all cresols was achieved after 5–6 h. Upon the treatment of the phenol-formaldehyde wastewater, the initial concentration of phenol was equal to 1550 mg dm^{-3}. After 14 h of permeation in the experimental system, the phenol concentration in the feed solution dropped below 0.01 mg dm^{-3}, and thus 99.9% removal was achieved.

Upon the treatment of the wood-processing wastewater, the initial concentration of phenol was equal to 210 mg dm^{-3}. After 7 h of permeation in the experimental system, the phenol concentration in the feed solution dropped below 0.01 mg l^{-1}, and thus 99.9% removal was achieved. The final effluent complied with the Indian discharge limit for phenol in the treated effluent of 1.0 mg dm^{-3}.

Several examples of the applications of SLMs in the removal of antibiotics from wastewater are shown below. Kawasaki *et al.* [154] studied the application of the flat-sheet SLM in the removal of erythromycin A from aqueous matrices. The feed phase consisted of citric acid (concentration = 0.025 mol dm^{-3}), boric acid (concentration = 0.100 mol dm^{-3}), and sodium phosphate (concentration = 0.050 mol dm^{-3}). The SLM was prepared from PTFE flat sheet porous with 1-decanol as the diluent.

The 1-decanol/water distribution coefficient for the nonionized form of erythromycin A was equal to 122 and was shown to be independent of the pH of the feed and/or the stripping phase. The apparent 1-decanol/water distribution coefficient, that is, the ratio of the total concentrations of erythromycin in the aqueous phase and 1-decanol, varied strongly with the pH of the aqueous phase and decreased from 122 at pH values above 10.0 to 10 at the pH value of 7.0. When feed phase and the stripping phase compartments are both stirred independently and the stirring rates range from 250 to 1500 rpm, 60–50% of the initial erythromycin amount was recovered in the stripping phase after 10 h [154]. The low fluxes, and thus long treatment periods, were attributed by the authors to low surface to volume ratio of the SLM system used, that is, limited interface area available for the erythromycin transport into the SLM. For practical application, the interfacial area of the SLM available for the transport seems to be one of the main factors to be optimized.

Habaki *et al.* [106] found a strong dependence of the rate-limiting step of erythromycin transport, in different ELMs and SLMs, on the diluent used. In the SLM with 1-decanol as the diluent, the rate-limiting step of the overall mass-transfer rates for erythromycin was the mass-transfer resistances in the feed phase, as well as the SLM [106]. On the other hand, if heptane was used as the diluent, then the rate-limiting step of the erythromycin mass transfer was through the membrane phase of the SLM.

With *m*-xylene as the diluent, the rate-limiting step of the overall permeation of erythromycin was the transport through the membrane and stripping phases. On the other hand, the concentration of erythromycin in the ELM with heptane as the diluent was limited by the transport through the membrane phase only. Other studies on the recovery and removal of antibiotics from different matrices include Msagati and Nindi [155] and Tang *et al.* [156] and can be consulted upon need. The transport of organic acids from aqueous media was achieved using a lipase complex [157]. This mechanism carries a lot of possibilities for the future.

4.3.2. Industrial applications and future trends

But, to the best of the author's knowledge, there are no wastewater treatment applications of SLMs in the public domain that would be aimed at the removal of organic compounds. It is possible that such applications exist, but are protected as confidential know-how.

A wide variety of the flat-sheet SLM systems have been described in Section 4.2.1. The advantages of the flat-sheet SLM arrangement in the removal of organic compounds from wastewater can be achieved, if low-cost extractants/diluents, such as vegetable oils, are used. However, the energy considerations will probably lead to the preferential application of hollow-fiber SLM system in the industry. On the basis of the laboratory study results, the SLM technology provides an ideal tool for the removal of trace amount of the antibiotics, and potentially even other pharmaceutical products, from wastewaters. This will prevent the exposure of the MOs in the environment to antibiotics, thus eliminating the potential for the selection of microbial strains resistant to the respective antibiotic(s).

5. Polymer Inclusion Membranes

Applications of the polymer inclusion membranes in the wastewater treatment have so far been limited to the radioactive wastewaters. Kusumocahyo *et al.* [158] reported the application of a polymer inclusion membrane to removal of $Ce(NO_3)_3$ from wastewater with low radioactivity. The membrane was composed of cellulose triacetate (the base polymer), 2-nitrophenyl-octyl ether was added as the plasticizer, and octyl(phenyl)-*N*,*N*-diisobutylcarbamoylmethyl phosphine oxide and *N*,*N*,*N*′,*N*′-tetraoctyl-3-oxapentanediamide were used as solvent carriers.

The wastewater contained 200 mg dm^{-3} $Ce(NO_3)_3$, 0.05 M HNO_3, and 2.95 M $NaNO_3$. Distilled water was used as the receiving phase. The partitioning coefficient of Ce(III) was equal to 550 for octyl(phenyl)-*N*, *N*-diisobutylbarbamoylmethyl phosphine oxide as the carrier and to 28 *N*, *N*,*N*′,*N*′-tetraoctyl-3-oxapentanediamide as the carrier, respectively. A 2:1 complex was formed between cerium nitrate and each of the two carriers tested. Complete recovery of Ce(III) ions in the receiving phase was achieved after 60 min. The rate permeation was directly proportional to the concentrations of the plasticizer and/or the carrier within a given interval. Analogical trend was observed for the dependence of the permeation rate on the membrane thickness. The permeation rate was indirectly proportional to the temperature.

Additional applications of the polymer inclusion membrane in wastewater treatment include the studies by Kozlowski *et al.* [159], Levitskaia *et al.* [160], and Levitskaia *et al.* [161], as well as Kim *et al.* [162]. More research in this area will be required in the future.

REFERENCES

1. Daugulis, A. J. (2001). Two-phase partitioning bioreactors: A new technology platform for destroying xenobiotics *Trends in Biotechnology* 19: 457-462.
2. MacLeod, C. T., Daugulis, A. J. (2003). Biodegradation of polycyclic aromatic hydrocarbons in a two-phase partitioning bioreactor in the presence of a bioavailable solvent. *Applied Microbiology and Biotechnology* 62: 291-296.
3. Ascon-Cabrera, M., Lebeault, J.-M. (1993). Selection of xenobiotic-degrading microorganisms in a biphasic aqueous-organic system. *Applied and Environmental Microbiology* 59: 1717-1724.
4. Yeom, S. H., Daugulis, A. J. (2001). Benzene degradation in a two-phase partitioning bioreactor by *Alcaligenes xylosoxidans* Y234. *Process Biochemistry* 36: 765-772.
5. Seker, S., Beyenal, H., Salih, B., Tanyolac, A. (1997). Multi-substrate growth kinetics of *Pseudomonas putida* for phenol removal. *Applied Microbiology and Biotechnology* 47: 610-614.
6. Singh, D., Fulekar, M.H. (2009). Bioremediation of phenol by a novel partitioning bioreactor using cow dung microbial consortia. *Biotechnology Journal* 4: 423-431.
7. Dobson, R. S., Burgess, J. E. (2007). Biological treatment of precious metal refinery wastewater: A review. *Minerals Engineering* 20: 519-532.
8. Muňoz, R., Guieysee, B., Mattiasson, B. (2003). Phenanthrene biodegration by an algal-bacterial consortium in two-phase partitioning bioreactors. *Applied Microbiology and Biotechnology* 61: 261-267.
9. Guieysse, B., Autem, Y., Soares, A. (2005). Biodegradation of phenol at low temperature using two-phase partitioning bioreactors. *Water Science and Technology* 52: 97-105.
10. Laane, C., Boeren, S., Vos, K., Veeger, C. (1986). Rules for optimization of biocatalysts in organic solvents. *Biotechnology and Bioengineering* 30: 81-87.
11. Barbosa, V. L., Tandlich, R., Burgess, J. E. (2007). Bioremediation of trace organic compounds found in precious metals refineries' wastewaters: A review of potential options. *Chemosphere* 68: 1195-1203.

12. Collins, L. D., Daugulis, A. J. (1996). Use of a two phase partitioning bioreactor for the biodegradation of phenol. *Biotechnology Techniques* 10: 643-648.
13. Yeom, S. H., Daugulis, A. J. (1999). A new method for the determination of microbial activity and critical log *P* in the presence of organic solvents. *Biotechnology Techniques* 13: 549-553.
14. Brooke, D., Nielsen, I., de Bruijn, J., Hermens, J. (1990). An interlaboratory evaluation of the stir-flask method for the determination of octanol-water partition coefficients (log POW). *Chemosphere* 21: 119-133.
15. Clog*P* (2008). View at; http://www.biobyte.com/bb/prod/clogp40.html (website visited on 30th November 2008).
16. Stone, K. M., Roche, F. W., Thornhill, N. F. (1992). Dry weight measurement of microbial biomass and measurement variability analysis. *Biotechnology Techniques* 6: 207-212.
17. Inoue, A., Horikoshi, K. (1991). Estimation of solvent-tolerance of bacteria by the solvent parameter log *P*. *Journal of Fermentation and Bioengineering* 71: 194-196.
18. Usami, R., Fukushima, T., Mizuki, T., Yoshida, Y., Inoue, A., Horikoshi, K. (2005). Organic solvent tolerance of halophilic archae, *Haloarcula* strains: Effect of NaCl concentration on the tolerance and polar lipid composition. *Journal of Bioscience and Bioengineering* 99: 169-174.
19. Heipieper, H. J., Weber, F. J., Sikkema, J., Keweloh, H., de Bont, J. A. M. (1994). Mechanisms of resistance of whole cells to toxic solvents. *Trends in Biotechnology* 12: 409-415.
20. Fernandes, P., Sommer Ferreira, B., Sampaio Cabral, J. M. (2003). Solvent tolerance in bacteria: Role of efflux pumps and crossresistance with antibiotics. *International Journal of Antimicrobial Agents* 22: 211-216.
21. Voet, D., Voet, J. (2004). *Biochemistry*, 3rd International ed. Wiley., Hoboken, NJ, USA.
22. Weber, F. J., Isken, S., de Bont, J. A. M. (1994). *Cis/trans* isomerization of fatty acids as a defence mechanism of *Pseudomonas putida* strains to toxic concentrations of toluene. *Microbiology* 140: 2013-2017.
23. Heipieper, H. J., Diefenbach, R., Keweloh, H. (1992). Conversion of cis unsaturated fatty acids to trans, a possible mechanism for the protection of phenol-degrading *Pseudomonas putida* P8 from substrate toxicity. *Applied and Environmental Microbiology* 58: 1847-1852.
24. Heipieper, H. J., de Bont, J. A. M. (1994). Adaptation of *Pseudomonas putida* S12 to ethanol and toluene at the level of fatty acid composition of membranes. *Applied and Environmental Microbiology* 60: 4440-4444.
25. Straub, S. G., Yajima, H., Komatsu, M., Aizawa, T., Sharp, G. W. G. (2002). The effects of cerulenin, an inhibitor of protein acylation, on the two phases of glucose-stimulated insulin secretion. *Diabetes* 51(Suppl 1): S91-S95.
26. Dombek, K. M., Ingram, L. O. (1984). Effects of ethanol on the *Escherichia coli* plasma membrane. *Journal of Bacteriology* 157: 233-239.
27. Touchstone, J. C., Levin, S. S., Dobbins, M. F., Carter, P. J. (1981). Differentiation of saturated and unsaturated phospholipid on thin layer chromograms. *Journal of High Resolution Chromatography Chromatography Communications* 4: 423-424.
28. Nielsen, L. E., Kadavy, D. R., Rajagopal, S., Drijber, R., Nickerson, K. W. (2005). Survey of extreme solvent tolerance in Gram-positive cocci: Membrane fatty-acid changes in *Staphylococcus haemolyticus* grown in toluene. *Applied and Environmental Microbiology* 71: 5171-6176.
29. Kaneda, T. (1977). Fatty acids of the genus *Bacillus*: As an example of branched-chain preference. *Bacteriology Reviews* 41: 391-398.
30. Loomis, JR., W. F., Magasanik, B. (1967). Glucose-Lactose Diauxie in *Escherichia coli*. *Journal of Bacteriology* 93: 1397-1401.

31. Cometabolism (2008). View at: http://www.biology-online.org/dictionary/Cometabolism (website visited on 30th June 2008).
32. HSDB (Hazardous Substances Databank) (2008). View at: http://toxnet.nlm.nih.gov/cgi-bin/sis/search (website visited on 30th November 2008).
33. Tandlich, R. (2004). Microbial PCB degradation and binding to soil components. PhD thesis, North Dakota State University, Fargo, ND, USA.
34. Madigan, M. T., Martinko, J. M., Parker, J. (2000). *Brock Biology of Microorganisms*, 9th ed. Prentice Hall, Upper Saddle River.
35. Cruickshank, S. M., Daugulis, A. J., McLellan, P. J. (2000). Dynamic modeling and optimal fed-batch feeding strategies for a two-phase partitioning bioreactor. *Biotechnology and Bioengineering* 67: 224-233.
36. Abbasi, A., Amiri, A. (2008). Emulsifying behavior of an exopolysaccharide produced by *Enterobacter cloacae*. *African Journal of Biotechnology* 7: 1574-1576.
37. Cooper, D. G., Goldenberg, B. G. (1987). Surface-active agents from two Bacilllus species. *Applied and Environmental Microbiology* 53: 224-229.
38. Collins, L. D., Daugulis, A. J. (1997). Characterization and optimization of a two-phase partitioning bioreactor for the biodegradation of phenol. *Applied Microbiology and Biotechnology* 48: 18-22.
39. Vrana B., Tandlich R., Baláž Š., Dercová K. (1998). Aerobic biodegradation of polychlorinated biphenyls by bacteria. *Biológia* 53: 251-266.
40. Vandermeer, K. D., Daugulis, A. J. (2007). Enhanced degradation of a mixture of polycyclic aromatic hydrocarbons by a defined microbial consortium in a two-phase partitioning bioreactor. *Biodegradation* 18: 211-221.
41. Busca, G., Berardinelli, S., Resini, C., Arrighi, L. (2008). Technologies for the removal of phenol from fluid streams: A short review of recent developments. *Journal of Hazardous Materials* 160: 265-288.
42. Zilouei, H., Guieysse, B., Mattiasson, B. (2008). Two-phase partitioning bioreactor for the biodegradation of high concentrations of pentachlorophenol using *Sphingobium chlorophenolicum* DSM 8671. *Chemosphere* 72: 1788-1794.
43. Tandlich R., Vrana, B., Dercová, K., Baláž, Š. (2005). Mechanism of polychlorinated biphenyl (PCB) biodegradation by bacterial isolates. *In* Proceedings of the International Conference on Mine and Energy Waste Processing, Dolní Domaslavice, Czech Republic. 1st December, 2005.
44. Wittich, R.-M., Wolf, P. (2007). Growth of the genetically engineered strain *Cupriavidus necator* RW112 with chlorobenzoates and technical chlorobiphenyls. *Microbiology* 153: 186-195.
45. Eibes, G., Lopez, C., Moreira, M. T., Feijoo, G., Lema, J. M. (2007). Strategies for the design and operation of enzymatic reactors for the degradation of highly and poorly soluble recalcitrant compounds. *Biocatalysis and Biotransformation* 25: 260-268.
46. Collins, L. D., Daugulis, A. J. (1999). Benzene/toluene/p-xylene degradation. Part I. Solvent selection and toluene degradation in a two-phase partitioning bioreactor. *Applied Microbiology and Biotechnology* 52: 354-359.
47. Cruickshank, S. M., Daugulis, A. J., McLellan, P. J. (2000). Modelling of a continuous two-phase partitioning bioreactor for the degradation of xenobiotics. *Process Biochemistry* 35: 1027-1035.
48. Lewis, D. L., Kollig, H. P., Hodson, R. E. (1986). Nutrient limitation and adaptation of microbial populations of chemical transformations. *Applied and Environmental Microbiology* 51: 598-603.
49. Aelion, C. M., Swindoll, C. M., Pfaender, F. K. (1987). Adaptation to and biodegradation of xenobiotic compounds by microbial communities from a pristine aquifer. *Applied and Environmental Microbiology* 53: 2212-2217.

50. Wiggins, B. A., Jones, S. H., Alexander, M. (1987). Explanations for the acclimation period preceding the mineralization of organic chemicals in aquatic environments. *Applied and Environmental Microbiology* 53: 791-796.
51. MacLeod, C. T., Daugulis, A. (2005). Interfacial effects in a two-phase partitioning bioreactor: Degradation of polycyclic hydrocarbons (PAHs) by a hydrophobic *Mycobacterium. Process Biochemistry* 40: 1799-1805.
52. Nielsen, D. R., Daugulis, A. J., McLellan, P. J. (2007). Dynamic simulation of benzene vapor treatment by a two-phase partitioning bioscrubber Part II: Model calibration, validation, and predictions. *Biochemical Engineering Journal* 36: 250-261.
53. Muñoz R, Rolvering C, Guieysse B, Mattiasson B. (2005). Aerobic phenanthrene biodegradation in a two-phase partitioning bioreactor. *Water Science and Technology* 52: 265-271.
54. Jia, S., Li, P., Park, Y. S., Okabe, M. (1996). Enhanced oxygen transfer in tower bioreactor on addition of liquid hydrocarbons. *Journal of Fermentation and Bioengineering* 82: 191-193.
55. Damiano, D., Wang, S. S. (1985). Improvements in ethanol concentration and fermentor ethanol productivity in yeast fermentations using whole soy flour in batch, and continuous recycle systems. *Biotechnology Letters* 7: 135-140.
56. Van Sonsbeek, H. M., Beeftink, H. H., Tramper, J. (1993). Two-liquid-phase bioreactors. *Enzyme and Microbial Technology* 15: 722-729.
57. Linek, V., Benes, P. (1976). A study of the mechanism of gas absorption into oil-water emulsions. *Chemical Engineering Science* 31: 1037-1046.
58. Daugulis, A. J. *Personal website*: http://www.chemeng.queensu.ca/people/daugulis/ (visited on 15th June 2008).
59. Xethanol Corporation. (2008). View at: http://www.xethanol.com/contact/headquarters.htm (visited on 15th June 2008).
60. Katzung, B. G. (1987). *Basic and Clinical Pharmacology*, 3rd ed., p. 734-735. Appleton and Lange, New York, USA.
61. Alpoguz, H. K., Memon, S., Ersoz, M., Yýlmaz, M. (2004). Transport kinetics of Hg2+ through bulk liquid membrane using calix[4]arene ketone derivative as carrier. *Separation Science and Technology* 39: 799-810.
62. Ersoz, M. (2007). Transport of mercury through liquid membranes containing calixarene carriers. *Advances in Colloid and Interface Science* 134–135: 96-104.
63. Luan, J., Plaisier, A. (2004). Study on treatment of wastewater containing nitrophenol compounds by liquid membrane process (Short Communication). *Journal of Membrane Science* 229: 235-239.
64. Chiarizia, R., Danesi, P. R. (1987). A double liquid membrane for the removal of actinides and lanthanides from acidic nuclear wastes. *Separation Science and Technology* 22: 641-649.
65. Terry, R. E., Li, N. N., Ho, W. S. (1982). Extraction of phenolic compounds and organic acids by liquid membranes. *Journal of Membrane Science* 10: 305-323.
66. Lin, S. H., Pan, C. L., Leu, H. G. (1999). Liquid membrane extraction of 2-chlorophenol from aqueous solution. *Journal of Hazardous Materials* 65: 289-304.
67. Raghuraman, B. J., Tirmizi, N.P., Kim, B.-S., Wiencek, J. M. (1995). Emulsion Liquid Membranes for Wastewater Treatment: Equilibrium Models for Lead- and Cadmium-di-2-ethylheyl Phosphoric Acid Systems. *Environmental Science and Technology* 29: 979-984.
68. Liu, H. J., Wu, Q. S., Ding, Y. P., Liu, L. (2004). Biomimetic synthesis of metastable PbCrO4 nanoparticles by emulsion liquid membrane system with carrier and coupled treatment of Pb(II) and Cr(VI) wastewaters. *Acta Chimica Sinica* 62: 946-950.
69. Fang, J. Z., Li, M. Y., Xu, Z. H. (2003). Separation of cobalt from a nickel-hydrometallurgical effluent using an emulsion liquid membrane. *Separation Science and Technology* 38: 3553-3574.

70. Ashraf Chaudry, M., Ahmad, S., Malik, M. T. (1997). Supported liquid membrane technique applicability for removal of chromium from tannery wastes. *Waste Management* 17: 211-218
71. Kulkarni, P. S., Kalyani, V., Mahajani, V. V. (2007). Removal of hexavalent chromium by membrane-based hybrid processes. *Industrial and Engineering Chemistry Research* 46: 8176-8182.
72. Ortiz, I., Fresnedo San Román, M., Corvalán, S., Eliceche, A. M. (2003). Modeling and optimization of an Emulsion Pertraction Process for Removal and Concentration of Cr(VI). *Industrial & Engineering Chemistry Research* 42: 5891-5899.
73. Valenzuela, F., Auspont, J., Basualto, C., Tapia, C., Sapag, J. (2005). Use of a surfactant liquid membrane contactor for zinc uptake from an acid aqueous effluent. *Chemical Engineering Research and Design* 83: 247-255.
74. Park, Y. (2006). Development and optimization of novel emulsion liquid membranes stabilized by non-Newtonian conversion in Taylor-Couette flow for extraction of selected organic and metallic contaminants. Ph.D. thesis, Georgia Institute of Technology, Atlanta, GA, USA.
75. Kulkarni, P. S., Mukhopadhyay, S., Bellary, M. P., Ghosh, S. K. (2002). Studies on membrane stability and recovery of uranium (VI) from aqueous solutions using a liquid emulsion membrane process. *Hydrometallurgy* 64: 49-58.
76. Kinugasa, T., Watanabe, K., Takeuchi, H. (1989). Effect of organic solvents on stability of liquid surfactant membranes. *Journal of Chemical Engineering of Japan* 22: 593-597.
77. Lacks, D. J., Rear, D. B., Van Orman, J. A. (2007). Molecular dynamics investigation of viscosity, chemical diffusivities and partial molar volumes of liquids along the $MgO–SiO_2$ join as functions of pressure. *Geochimica et Cosmochimica Acta* 71: 1312-1323.
78. Skelland, A. H. P. (1967). Non-Newtonian Flow and Heat Transfer. Wiley, New York, NY.
79. Hopper, J. R. (1964). A study of diffusion in non-Newtonian fluids. MS thesis, University of Delaware, DL, USA.
80. Skelland, A. H. P., Meng, X. M. (1999). Non-Newtonian conversion solves problems of stability, permeability, and swelling in emulsion liquid membranes. *Journal of Membrane Science* 158: 1-15.
81. Skelland, A. H. P., Meng, X. M. (1996). A new solution to emulsion liquid membrane problems by non-Newtonian conversion. *AIChE Journal* 42: 547-561.
82. Hsu, E. C., Li, N. N. (1985). Membrane recovery in liquid membrane separation processes. *Separation Science and Technology* 20: 115-130.
83. Skelland, A. H. P. (1993). Stabilizing liquid membrane without sacrificing permeability by non-Newtonian conversion of the membrane. U.S. Patent No. 5,229,004.
84. Groeneweg, F., van Dieren, F., Agterof, W. G. M. (1994). Droplet break-up in a stirred water-in-oil emulsion in the presence of emulsifiers. *Colloids and Surfaces A: Physicochemical and Engineering Aspects* 91: 207-214.
85. Forney, L. J., Skelland, A. H. P., Morris, J. F., Holl, R. A. (2002). Taylor-vortex column: Large shear for liquid-liquid extraction. *Separation Science and Technology* 37: 2967-2986.
86. Liu, X., Liu, D. (2000). Mass transfer resistance analysis of L-tryptophan extraction in an emulsion liquid membrane system. *Separation Science and Technology* 35: 2707-2724.
87. Klaassen R., Feron, P., Jansen, A. (2008). Membrane contactor applications. *Desalination* 224:81-87.
88. Fresnedo San Román, M., Bringas, E., Ortiz, I., Grossmann, I. E. (2007). Optimal synthesis of an emulsion pertraction process for the removal of pollutant anions in industrial wastewater systems. *Computers and Chemical Engineering* 31: 456-465.

89. Marr, R. J., Draxler, J. (1992). Emulsion liquid membrane application. In: Ho, W. S., Sirkar K. K. (Eds.), *Membrane Handbook*, pp. 701-717. Chapman & Hall, New York, NY.
90. Othman, N., Goto, M., Mat, H. (2004). Liquid membrane technology for precious metals recovery from industrial waste. Presented at the Regional Symposium on Membrane Science & Technology (MST2004), Johor, Malaysia, April 21st-25th, 2004.
91. Uddin, M. S., Kathiresan, M. (2000). Extraction of metal ions by emulsion liquid membrane using bi-functional surfactant: Equilibrium and kinetic studies. *Separation and Purification Technology* 19: 3-9.
92. Stevens, G. W., Chang C., Mackay M. E. (1996). Stabilising emulsion liquid membranes. *Separation Science and Technology* 31: 1025-1033.
93. Melamane, X., Tandlich, R., Burgess, J. E. (2007). Treatment of wine distillery wastewater by high rate anaerobic digestion *Water Science and Technology* 56: 9-16.
94. Autenrieth, R. L., Bonner, J. S., Akgerman, A., Okaygun, M., McCreary, E. M. (1991). Biodegradation of phenolic wastes. *Journal of Hazardous Materials* 28: 29-53.
95. Gadekar, Pradeep, T., Mukkolath, Anil V., Tiwari, Krishna K. (1992). Recovery of nitrophenols from aqueous solutions by a liquid emulsion membrane system. *Separation Science and Technology* 4: 427-445.
96. Kumaresan, T., Meera Sheriffa Begum, K. M., Sivashanmugam, P., Anantharaman, N., Sundaram, S. (2003). Experimental studies on treatment of distillery effluent by liquid membrane extraction. *Chemical Engineering Journal* 95: 199-204.
97. Park, Y., Forney, L. J., Kim, J. H., Skelland, A. H. P. (2004). Optimum emulsion liquid membranes stabilized by non-Newtonian conversion in Taylor-Couette flow. *Chemical Engineering Science* 59: 5725-5734.
98. Mortaheb, H. R., Amini, M. H., Sadeghian, F., Mokhtarani, B., Daneshyar, H. (2008). Study on a new surfactant for removal of phenol from wastewater by emulsion liquid membrane. *Journal of Hazardous Materials* 160: 582-588.
99. Frankenfeld, J. W., Li, N. N. (1987). Recent advances in liquid membrane technology. In: Rousseau, R. W. (Ed.), *Handbook of Separation Process Technology*, pp. 840-861. Wiley, New York.
100. Yang, S.-S., Ling, M.-Y. (1989). Tetracycline production with sweet potato residue by solid state fermentation. *Biotechnology and Bioengineering* 33: 1021-1028.
101. Li, X., Cui, Y. (2008). Study on synthesis and chloramphenicol release of poly(2-hydroxyethylmethacrylate-co-acrylamide) hydrogels. *Chinese Journal of Chemical Engineering* 16: 640-645.
102. Brites Alves, A. M., Morão, A., Cardoso, J. P. (2002). Isolation of antibiotics from industrial fermentation broths using membrane technology. *Desalination* 148: 181-186.
103. Loper, C. A., Crawford, J. K., Otto, K. L., Manning, R. L., Meyer, M. T., Furlong, E. T. (2006). Concentrations of Selected Pharmaceuticals and Antibiotics in South-Central Pennsylvania Waters, March Through September 2006. *U.S. Geological Survey Data Series 300*, USGS, New Cumberland, PA, USA.
104. Gartiser, S., Urich, E., Alexy, R., Kümmerer, K. (2007). Anaerobic inhibition and biodegradation of antibiotics in ISO test schemes. *Chemosphere* 66: 1839-1848.
105. Schügerl, K. (2005). Extraction of primary and secondary metabolites. *Advances in Biochemical Engineering/Biotechnology*. Technology Transfer in Biotechnology Series, Vol. 92/2005. Springer, Berlin.
106. Habaki, H., Isobe, S., Egashira, R., Kawasaki, J. (1998). Permeation and concentration of erythromycin by supported and emulsion liquid membranes. *Journal of Chemical Engineering of Japan* 31: 47-54.
107. Habaki, H., Egashira, R., Stevens, G. W., Kawasaki, J. (2002). A novel method improving low separation performance for W/O/W ELM permeation of erythromycin. *Journal of Membrane Science* 208: 89-103.

108. Guerrero, L., Omil, F., Méndez, R., Lema, J. M. (1999). Anaerobic hydrolysis and acidogenesis of wastewaters from food industries with high content of organic solids and protein. *Water Research* 33: 3281-3290.
109. Vidal, G., Carvalho, A., Méndez, R., Lema, J. M. (2000). Influence of the content in fats and proteins on the anaerobic biodegradability of dairy wastewaters. *Bioresource Technology* 74: 231-239.
110. Rodríguez-Martínez, J., Martínez-Amador, S. Y., Garza-García, Y. (2005). Comparative anaerobic treatment of wastewater from pharmaceutical, brewery, paper and amino acid producing industries. *Journal of Industrial Microbiology and Biotechnology* 32: 691-696.
111. Rico Gutierrez, J. L., Garcia Encina, P. A., Fdz-Polanco, F. (1991). Anaerobic treatment of cheese production wastewater using a UASB reactor. *Bioresource Technology* 37: 271-276.
112. Hansen, C. L., Hwang, S. H. (1990). Use of upflow anaerobic sludge blanket (UASB) reactor to treat whey permeate. In: Proceedings of the Sixth International Symposium on Agriculture and Food Processing Wastes. Am. Soc. Ag. Eng., Chicago, Vol. I11, p. 124.
113. Ozturk, I., Eroglu, V., Ubay, G., Demir, I. (1993). Hybrid upflow anaerobic sludge blanket reactor. treatment of dairy effluent (HUASBR). *Water Science and Technology* 28: 77-81.
114. Perle, M., Kimchie, Sh., Shelef, G. (1995). Some biochemical aspects of the anaerobic degradation of dairy wastewater. *Water Research* 29: 1549-1554.
115. Ruston, B. (1993). Chemical pretreatment of dairy wastewater. *Water Science and Technology* 28: 67-72.
116. Maharaj, I., Elefsiniotis, P. (2001). The role of HRT and low temperature on the acid-phase anaerobic digestion of municipal and industrial wastewaters. *Bioresource Technology* 76: 191-197.
117. Pickering, P. J., Chaudhuri, J. B. (1997). Enantioselective extraction of (D)-phenylalanine from racemic (D/L)-phenylalanine using chiral emulsion liquid membranes. *Journal of Membrane Technology* 127: 115-130.
118. Evonik Industries (Degussa; 2008). Amino acids as feed additives website: http://www.degussa.com/degussa/en/profile/corporatestructure/feedadditives/default_2.htm, (website visited on 23rd October 2008).
119. Waliszewski, K. N., Estrada, Y., Pardio, V. (2004). Sensory properties changes of fortified nixtamalized corn flour with lysine and tryptophan during storage. *Plant Foods for Human Nutrition* 59: 51-54.
120. Hebrant, M., Tondre, C. (1998). Liquid membrane transport of amino acids mediated by reverse micelles. *Analytical Sciences* 14: 109-115.
121. Leodidis, E. B., Hatton, T. A. (1991). Amino acids in reversed micelles. 4. Amino acids as cosurfactants. *Journal of Physical Chemistry* 95: 5957-5965.
122. Wang, W., Weber, M. E., Vera, J. H. (1995). Effect of the concentration of DODMAC and 1-decanol on the behavior of reverse micelles in the extraction of amino acids. *Biotechnology and Bioengineering* 46: 343-350.
123. Behr, J. P., Lehn, J. M. (1973). Transport of amino acids through organic liquid membranes [1]. *Journal of the American Chemical Society* 95: 6108-6110.
124. Valenzuela, F., Basualto, C., Tapia, C., Sapag, J. (1999). Application of hollow-fiber supported liquid membranes technique to the selective recovery of a low content of copper from a Chilean mine water (Short communication). *Journal of Membrane Science* 155: 163-168.
125. Cataru, M. M., Badicioiu, M. (2007). Researches concerning the copper extraction with liquid membranes on solid casing support. *Revista de Chimie* 58: 1112-1116.

126. Yang, Q., Kocherginsky, N. M. (2006). Copper recovery and spent ammoniacal etchant regeneration based on hollow fiber supported liquid membrane technology: From bench-scale to pilot-scale tests. *Journal of Membrane Science* 286: 301-309.
127. Yang, Q., Kocherginsky, N. M. (2007). Copper removal from ammoniacal wastewater through a hollow fiber supported liquid membrane system: Modeling and experimental verification. *Journal of Membrane Science* 297: 121-129.
128. Noble, R. D., Way, J. D. (1987). *Liquid Membranes: Theory and Applications*, ACS Symp. Ser. American Chemical Society, Washington, D.C.
129. Rathore, N. S., Leopold, A., Pabby, A.K., Fortuny, A., Coll, M.T., Sastre, A. M. (2009). Extraction and permeation studies of Cd(II) in acidic and neutral chloride media using Cyanex 923 on supported liquid membrane. *Hydrometallurgy* 96: 81-87.
130. Uedee, E., Ramakul, P., Pancharoen, U., Lothongkum, A. W. (2009). Performance of hollow fiber supported liquid membrane on the extraction of mercury(II) ions. *Korean Journal of Chemical Engineering* 25: 1486-1494.
131. Panja, S., Mohapatra, P.K., Tripathi, S.C., Manchanda, V.K. (2009). Studies on uranium(VI) pertraction across a N,N,N′N′-tetraoctyldiglycolamide (TODGA) supported liquid membrane. *Journal of Membrane Science* 337: 274-281.
132. Ramakul, P., Prapasawad, T., Pancharoen, U., Pattaveekongka, W. (2007). Separation of radioactive metal ions by hollow fiber-supported liquid membrane and permeability analysis. *Journal of the Chinese Institute of Chemical Engineers* 38: 489-494.
133. Ura, P., Prakorn, R., Weerawat, P. (2005). Purely extraction and separation of mixture of Cerium(IV) and Lanthanum(III) via hollow fiber supported liquid membrane. *Journal of Industrial and Engineering Chemistry* 11: 926-931.
134. Danesi, P. R., Horwitz, E. P., Rickert, P. G. (1983). Rate and mechanism of facilitated Americium(III) transport through a supported liquid membrane containing a bifunctional organophosphorus mobile carrier. *Journal of Physical Chemistry* 87: 4708-4715.
135. Vincent, T., Guibal, E., Chiarizia, R. (2007). Palladium recovery by reactive precipitation using a cyanex 301-based stable emulsion. *Separation Science and Technology* 42: 3517-3536.
136. Chiarizia, R., Horwitz, E. P., Rickert, P. G., Hodgson, K. M. (1990). Application of supported liquid membranes for removal of uranium from groundwater. *Separation Science and Technology* 25: 1571-1586.
137. Teramoto, M., Fu, S. S., Takatani, K., Ohnishi, N., Maki, T., Fukui, T., Arai, K. (2000). Treatment of simulated low level radioactive wastewater by supported liquid membranes: uphill transport of Ce(III) using CMPO as carrier. *Separation and Purification Technology* 18: 57-69.
138. Ho, W. S. W., Wang, B. (2002). Strontium removal by new alkyl phenylphosphonic acids in supported liquid membranes with strip dispersion. *Industrial and Engineering Chemistry Research* 41: 381-388.
139. Zha, F. F., Coster, H. G. L., Fane, A. G. (1994). A study of stability of supported liquid membranes by impedance spectroscopy. *Journal of Membrane Science* 93: 255-271.
140. Ho, S. V., Sheridan, P. W., Krupetsky, E. (1996). Supported polymeric liquid membranes for removing organics from aqueous solutions I. Transport characteristics of polyglycol liquid membranes. *Journal of Membrane Science* 112: 13-27.
141. Dastgir, M. G., Peeva, L. G., Livingston, A. G. (2005). Performance of composite supported polymeric liquid membranes in the Membrane Aromatic Recovery System (MARS). *Chemical Engineering Science* 60: 7034-7044.
142. Kujawski, W., Warzawski, A., Ratajczak, W., Porebski, T., Capala, W., Ostrowska, I. (2004). Removal of phenol from wastewater by different separation techniques. *Desalination* 163: 287-296.

143. Kujawski, W., Warzawski, A., Ratajczak, W., Porebski, T., Capala, W., Ostrowska, I. (2004). Application of pervaporation and adsorption to the phenol removal from wastewater. *Separation and Purification Technology* 40: 123-132.
144. Cichy, W., Szymanowski, J. (2002). Recovery of phenol from aqueous streams in hollow fiber modules. *Environmental Science and Technology* 36: 2088-2093.
145. Reis, M. T. A., de Freitas, O. M. F., Ismael, M. R. C., Carvalho, J. M. R. (2007). Recovery of phenol from aqueous solutions using liquid membranes with Cyanex 923. *Journal of Membrane Science* 305: 313-324.
146. Trivunac, K., Stevanovic, S., Mitrovic, M. (2004). Pertraction of phenol in hollow fiber membrane contactors. *Desalination* 162: 93-101.
147. Wan, Y. H., De Wang, X., Zhang, X. J. (1997). Treatment of high concentration phenolic wastewater by liquid membrane with N503 as mobile carrier. *Journal of Membrane Science* 135: 263-270.
148. Park, S. -W., Kim, K. -W., Sohn, I. -J., Kaseger, C. F. (2000). Facilitated transport of sodium phenolate through supported liquid membrane *Separation and Purification Technology* 19: 43-54.
149. Le, Q. T. H., Ehler, D. S., McCleskey, T. M., Dye, R. C., Pesiri, D. R., Jarvinen, G. D., Sauer, N. N. (2002). Ultrathin gate for the transport of phenol from supported liquid membranes to permanent surface modified membranes. *Journal of Membrane Science* 205: 213-222.
150. Wang, M. L., Hu, K. H. (1994). Extraction of phenol using sulfuric acid salts of trioctylamine in a supported liquid membrane. *Industrial and Engineering Chemistry Research* 33: 914-921.
151. Li, Z., Wu, M., Jiao, Z., Bao, B., Lu, S. (2004). Extraction of phenol from wastewater by *N*-octanoylpyrrolidine. *Journal of Hazardous Materials B* 114: 111-114.
152. Venkateswaran, P., Palanivelu, K. (2006). Recovery of phenol from aqueous solution by supported liquid membrane using vegetable oils as liquid membrane. *Journal of Hazardous Materials B* 131: 146-152.
153. Xiao, M., Zhou, J., Tan, Y., Zhang, A., Xia, Y., Ji, L. (2006). Treatment of highly-concentrated phenol wastewater with an extractive membrane reactor using silicone rubber. *Desalination* 195: 281-293.
154. Kawasaki, J., Egashira, R., Kawai, T., Hara, H., Boyadzhiev, L. (1996). Recovery of erythromycin by a liquid membrane. *Journal of Membrane Science* 112: 209-217.
155. Msagati, T. A. M., Nindi, M. M. (2005). Application of Supported Liquid Membranes in the Multi-Residue Extraction of Aminoglycoside Antibiotics in Milk and Urine. *Bulletin of the Chemical Society of Japan* 78: 2135-2141.
156. Tang, K. -W., Zhou, C. -S., Jiang, X. -Y. (2003). Racemic ofloxacin separation by supported-liquid membrane extraction with two organic phases. *Science in China Series B: Chemistry* 46: 96-103.
157. Miyako, E., Maruyama, T., Kamiya, N., Goto, M. (2005). Supported liquid membrane encapsulating a surfactant-lipase complex for the selective separation of organic acids. *Chemistry - A European Journal* 11: 1163-1170.
158. Kusumocahyo, S. P., Kanamori, T., Sumaru, K., Aomatsu, S., Matsuyama, H., Teramoto, M., Shinbo, T. (2004). Development of polymer inclusion membranes based on cellulose triacetate: Carrier-mediated transport of cerium(III). *Journal of Membrane Science* 244: 251-257.
159. Kozlowski, C. A., Girek, T., Walkowiak, W., Koziol, J. J. (2005). Application of hydrophobic [beta]-cyclodextrin polymer in separation of metal ions by plasticized membranes. *Separation and Purification Technology* 46: 136-144.

160. Levitskaia, T. G., Macdonald, D. M., Lamb, J. D., Moyer, B. A. (2000). Prediction of the carrier-mediated cation flux through polymer inclusion membranes via fundamental thermodynamic quantities: Complexation study of bis(dodecyloxy)calix[4]arene-crown-6 with alkali metal cations. *Physical Chemistry and Chemical Physics* 2: 1481-1491.
161. Levitskaia, T. G., Lamb, J. D., Fox, K. L., Moyer, B. A. (2002). Selective carriermediated cesium transport through polymer inclusion membranes by calix[4]arene-crown-6 carriers from complex aqueous mixtures. *Radiochimica Acta* 90: 43-52.
162. Kim, J. S., Kim, S. K., Ko, J. W., Kim, E. T., Yu, S. H., Cho, M. H., Kwon, S. G., Lee, E. H. (2000). Selective transport of cesium ion in polymeric CTA membrane containing calixcrown ethers. *Talanta* 52: 1143-1148.

CHAPTER 9

Progress in Liquid Membrane Science and Engineering

Vladimir S. Kislik

1. Introduction

The basic properties of liquid membrane (LM) operations make them ideal for industrial production: they are simple in concept and operation; they are modular and easy to scale-up; and they are low in energy consumption with a remarkable potential for low environmental impact and energetic aspects.

While astonishing industrial progress has been made over the past two decades in membrane-based separation of gaseous and aqueous liquid mixtures and in the purification of chemical and biological products, their potential for even more extensive industrial application in such fields as food/beverage processing, wastewater reclamation, gaseous waste detoxification, large scale air-gas separation, hydrometallurgical processing, and production of gaseous and liquid fuels and petrochemicals remains unexploited. This is traceable to limitations imposed by inherent deficiencies in membrane properties and to deficiencies in membrane module design or configuration, or of fluid-management strategy.

Specifically, the limitations of greatest practical significance are the following: (1) low permeation flux, (2) inadequate permselectivity, (3) inadequate membrane durability or service lifetime, (4) membrane fouling, and (5) excessively high capital or operating costs. These deficiencies render LM processes either technically impractical or (more often) uneconomic, for such applications.

The most interesting developments for industrial LM technologies are related to the possibility of integrating in the same industrial cycle various membrane operations with overall important benefits in terms of product quality, plant compactness. Design industrial production cycles by combining various membrane separation operations and conversion units, thus realizing highly integrated membrane processes, are an attractive

Casali Institute of Applied Chemistry, The Hebrew University of Jerusalem, Campus Givat Ram, Jerusalem 91904, Israel

Liquid Membranes: Principles and Applications in Chemical Separations and Wastewater Treatment
DOI: 10.1016/B978-0-444-53218-3.00009-X

opportunity because of synergic effects that can be attained. Recent developments in LM module design, including rotational, vibrational membrane devices [1-3] and pulsed-flow fluid management [4] for polarization control, use of low-cost refractory monoliths as membrane supports [5], and use of electric potentials to minimize macrosolute polarization and fouling [6], may permit practical and economic application of membrane processes to liquid and gaseous streams which today are untreatable by such methods.

Of growing industrial importance are a family of new applications that make use of membrane structures not as intrinsic separation barriers, but as substrates for immobilization of catalysts (e.g., enzymes) [7-10] or of specific complexing agents (e.g., affinity ligands) [11-13]. The development of novel polymeric materials of unique functionality or microstructure, inorganic (ceramic) semipermeable materials [5, 14, 15], novel ultrathin-barrier laminate structures [5, 16, 17] comprised of both organic and refractory components, and interpenetrating multiphase structures with anomalous transport characteristics promises to yield membranes with superior chemical/thermal stability, fouling resistance, organic solvent resistance, and unusually high permselectivities and permeabilities. These developments should lead to important new chemical synthesis processes and to novel and efficient strategies for industrial-scale purification of complex biological products.

Exciting opportunities also exist for the imaginative marriage of membrane separation techniques with other physical or chemical/biochemical transformation/separation procedures to yield integrated processes which circumvent the limitations of the individual steps, and achieve far greater selectivity, efficiency, or economic utility than either process element alone. Examples include BOHLM processes including membrane solvent extraction [18], affinity-complexation/ultrafiltration [11-13, 19], selective LM transport/selective precipitation [20-22], and extractive, membrane-moderated immobilized-cell biotransformation [23, 24]. Lastly, the overlooked interface between electrochemistry and membrane technology [25-28] merits serious attention as a route to even more dramatic solutions to problems in industrial separations and chemical synthesis. The production of ultrapure gases, the removal of trace concentrations of toxicants or high-value substances from liquid or gaseous streams, the development of novel chemical and biochemical sensors, and the synthesis of high-value chemical intermediates via membrane-immobilized catalysts in an electrochemical cell are among the many opportunity areas for ongoing membrane process research and development.

Practically, there are a lot of opportunities for liquid membrane separation processes in all areas of the modern industry. Hopefully, the following survey may help to identify specific challenges and opportunities which particularly merit scrutiny now, and which may lead to important and profitable results within the next decades.

2. Fundamental Studies in LM Science and Engineering

Liquid membrane science remains a fruitful and exciting area for research, with promise of offering many novel solutions to critical problems of separation and chemical synthesis confronting the chemical, biological, food, metallurgical, and waste treatment industries. Continuous research work on fundamental aspects of transport phenomena in the various membrane operations is important for the future of membrane science and technology [29, 30]. There is a need for both basic and applied research to develop new membrane supports, incorporating improved transport, stability and resistivity properties with specific, selective to separating solutes affinity agents. These research efforts must take into account the studies done in other areas such as supramolecular chemistry, molecular imprint materials [31], nanotechnology [32], nonlinear optics, studies on biological membranes and biological phenomena, and so on.

Progress in better understanding of fouling phenomena [33] and their prevention, improving of membrane selectivity, and development of new enantioselective membranes [8] are some examples of sectors where basic research can contribute. More progress is still desirable on anticipating and predicting relationship between membrane chemical properties, their morphology and configuration, with the overall liquid membrane phenomena. The role of interfacial phenomena and its influence on the properties in the solution upstream and downstream the membrane faces are strongly needed.

Computational strategies will be key tools for the understanding of properties and behavior of materials, to exploit the potential use of novel highly complex composite systems, new molecules and new multifunctional materials.

The possible use of molecular dynamics tools might be an interesting area for future research activities to design new organic or inorganic materials having selectivity for different chemical species as the palladium for hydrogen [34] or the perovskite for oxygen [35].

3. Potential Advances in SLM and Selective Membrane Supports Production Technologies

Supported liquid membranes have a better chance of being applied, but the concurrent requirements for fast and selective transport as well as good membrane stability, necessitate a careful choice of carrier/solvent/

support combination. Studies have revealed that transport through the SLM systems is still often diffusion controlled, but the specific morphology of the membrane affects the properties of the supported liquid as well. To increase transport rates still further, very thin membranes are needed. This will take us into the area of kinetic rather than diffusion control.

Potential direction in the SLM gas separation and biochemical conversion/separation techniques is the immobilization by selective facilitating agents at the stage of membrane supports fabrication. There is a need for both fundamental and applied research to develop new membrane supports, incorporating improved transport, stability and resistivity properties with specific affinity to separating solutes. The main idea for the membrane production researchers and developers is to produce membrane supports not only with improved transport and stability characteristics but also permselective to different gas and liquid solutes. The lamination methods permit to obtain ultrathin films of suitably permselective material to be deposited on prefabricated porous support structures of different composition.

3.1. Facilitating membrane structures

Classical facilitated transport theory teaches that permselectivity can be achieved only if the facilitating species (and the complex formed by it) are freely diffusible across the membrane barrier (see Chapter 2). There are, however, proponents of that belief: a polymeric matrix with covalently bound to it high concentration of specific complexing agents should be capable of selectively transmitting the complexing penetrant by "stopping steps in bucket brigade" fashion, as schematically illustrated in Fig. 2.3 of Chapter 2. Such a mechanism has been invoked, incidentally, to account for selective molecular transport across cell membranes.

Increasing attention is devoted to inorganics (particularly refractory oxides). The extraordinary sorptive selectivity of the molecular sieve zeolites [14, 15] has long tantalized the membrane scientist as to their adaptability for use as membrane barriers, and techniques for fabricating thin films out of such materials (e.g., glass, silica, alumina, zirconia), as well as asymmetric microporous laminates, have become well established over the past few years. Microporous and ultramicroporous ceramic structures are now commercially available and are finding expanding use in the SLM applications where superior corrosion resistance, organic solvent resistance, high-temperature stability, and elevated mechanical strength are important. It is noteworthy that the SLM has now opened the door to processing of liquid hydrocarbons and petrochemicals.

The high-temperature-service capability of porous ceramic (or carbon) membranes opens up the opportunity of creating novel high-temperature SLM for unusual separations. For example, it should be possible to impregnate these matrices with fusible salts capable of transporting selectively reactive gaseous compounds via reversible complexation reactions [36], for example, the use of copper salts for selective transport of carbon monoxide. Such "supported-fused-salt" membranes may have unique application for the separation of alkenes and alkynes from alkanes, or the separation of aromatic from aliphatic or naphthenic hydrocarbons. Molten phases might also function as catalysts as well as selective transport media: for example, fused vanadium pentoxide to both extract sulfur dioxide from air, and convert it directly to sulfur trioxide and sulfuric acid [37].

Exciting opportunity area for development of unique interpenetrating-phase permselective SLMs is the formation of two-phase structures comprising an electronically conductive (e.g., metallic or semiconductor) phase [38] and an ionically conductive (e.g., polyelectrolyte or ceramic solid electrolyte) phase [39]. The only species directly transmittable through such a composite are either electrons (by conduction through the metallic or semiconductor phase) or ions (by diffusion through the electrolyte phase). If, however, there is present in a gas mixture contacting such structure a component which can (1) chemisorb at the interface between the two phases, (2) undergo reduction or oxidation by electron transfer to or from the electronically conductive phase, and (3) generate an ion common to (or tolerated in) the electrolyte phase, then such a gas can be selectively transported through that composite to the virtual exclusion of all others present. This is an "electrochemically permselective" SLM: entry of a gas molecule into the structure results in electron flow in the conductive phase and ion flow in the electrolyte phase, with recombination of electrons and ions to form the neutral gas molecule at the downstream membrane boundary. A membrane comprised of platinum and zirconia, for example, might function as an oxygen-permselective membrane at elevated temperature; one comprised of platinum and hydrated polystyrene sulfonic acid, as a hydrogen-selective membrane [40]. Here again, the use of a microporous ceramic matrix for deployment of both electronically and ionically conductive media might make such a SLM both technically and economically feasible. The concept is to coat the surfaces of the membrane-pores with a monolayer of metal [41] and then fill the pores with the ionically conductive phase. This becomes a triple-continuous-phase structure, the ceramic phase serving as the template for deployment of the other two phases, and also as a rugged and durable support for the functional elements.

Interesting progress has been made recently in chemically modifying the barrier-layer surface of asymmetric polymeric gas permeation membranes by reactive gaseous or liquid treatment (e.g., fluorination) to improve membrane permselectivity or stability [42]. Such surface treatments modify the ultrathin barrier layer almost exclusively and allow conversion of that layer into a compositionally different structure. The result may be a more permselective membrane without significant permeability loss, a more fouling resistant membrane.

3.2. Affinity SLM structures

Substances which form affinity complexes with specific proteins or other biopolymers, covalently coupled to the membrane surface, called "affinity membranes" [11]. Solutions which contain even trace concentrations of the affinity complement molecule, when passed through such membranes, are virtually totally stripped of that molecule, which can subsequently be quantitatively recovered in high concentration by permeation of the membrane with a displacing solution. Both flat-sheet and hollow fiber membrane structures are adaptable to affinity sorption; the former are in commercial use in diagnostic devices for determination of specific proteins (e.g., viral antigens) in body fluids [43], where very small amounts of sorptive immobilization are required. Hollow fiber devices, which contain large amounts of adsorptive surface in a quite small volume, are of particular value for both laboratory and industrial preparative separations. A representative membrane of this type is a Protein A affinity membrane [23], which selectively binds human immunoglobulins; when human blood plasma is permeated through this membrane. The immunoglobulin fraction is selectively and quantitatively adsorbed; subsequent elution of the membrane with concentrated saline solution releases the immunoglobulins in active form and in high concentration. The regenerated membrane can then be recycled and reused.

Affinity membranes can also be fabricated using a selective enzyme inhibitor as the affinity reagent; such membranes become selectively sorptive for the particular enzyme that binds to that inhibitor, and are thus useful for extracting that enzyme selectively from a complex solution mixture [44].

An obviously important requirement to an affinity membrane is that it displays very low sorptivity for solutes other than the "target" compound.

3.3. New permselective materials

The expanding interest in separation of gaseous, vapor, and liquid mixtures by membrane permeation has stimulated a search for materials of novel composition or microstructure which display enhanced sorptive or diffusive selectivity for small molecules differing in size, shape, or polarity [45]. Such structures show anomalously high permeabilities to gases and vapors—far higher than would be predicted for normal polymeric glasses. These structures are heterogeneous and contain a continuous network of submicroscopic voids which provide the pathways for small-molecule transport. Microporous inorganic glasses and porous ceramic crystals (e.g., zeolites) [14, 46] also show elevated small-molecule permeability.

Polyacetylenic glasses display also anomalous and potentially very useful property: ability to separate mixtures of high- and low-critical temperature gases with extraordinarily high permselectivities [47]. Such permselectivity-enhancement is obviously limited to certain gas mixtures, pressures, and temperatures.

The gulf between organic and inorganic solids as membrane supports is now rapidly closing as increasing attention is devoted to inorganics (particularly refractory oxides) [14, 15, 48]. The extraordinary selectivity of the molecular sieve zeolites [14] led researchers to use these materials as membrane barriers and to develop techniques for fabricating thin films [49] on these materials. Microporous and ultramicroporous ceramic membranes are now commercially available and are finding expanding use [50].

3.4. Improved thin barrier multilayer laminates

There is growing interest in the fabrication of membranes by lamination methods which permit ultrathin films of suitably permselective material to be deposited on prefabricated porous support [51, 52]. Most are formed of compositionally homogeneous polymers by phase inversion techniques. Among the more interesting substrates for lamination are the commercially available asymmetric porous ceramic (alumina) [15] and porous carbon membrane structures [53], in tubular, flat-sheet, or extruded monolith form. These have the advantage of high-temperature stability, corrosion and solvent resistance, and outstanding compressive strength. Thin barrier layers of either polymeric or ceramic materials can be deposited on such substrates by coating with polymer solutions, with liquid prepolymers which can be polymerized in situ as thin films, or by slip-casting or sol-gel coating techniques [54].

Of particular interest are laminates in which the barrier layer has been thermally transformed (by chemical reaction, decomposition, or crystallization) after deposition on the substrate: this should be an attractive route to fabricate ultrathin porous carbon membranes [55], porous glass membranes [56], or possibly polycrystalline zeolite membranes [57].

An intriguing recent development has been the discovery that porous ceramic monoliths (fabricated for use as catalyst supports) can be lamination-coated with microporous or ultramicroporous barrier layers to yield high-area membrane modules [58]. It is reducing fabrication costs of ceramic membrane devices to levels comparable to (or less than) those of polymeric membrane devices, and it opens the door to development of large-area laminate devices with utility for particulate, macromolecular, or small-molecule separations.

Finally, interesting progress has been made recently in chemically modifying the barrier-layer surface of asymmetric polymeric gas permeation membranes by reactive gaseous or liquid treatment (e.g., fluorination) to improve membrane permselectivity or stability [59]. Such surface treatments modify the ultrathin barrier layer almost exclusively, and allow conversion of that layer into a compositionally different or more refractory structure. The result may be a more permselective membrane without significant permeability loss, a more fouling resistant membrane, or a membrane far less susceptible to loss in permselectivity in the presence of polymer-solvating impurities in the feed.

3.5. Electrochemically driven techniques (fuel cells) utilizing permselective membranes

An electrochemically driven, membrane-moderated separation process in commercial use today is electrodialysis, which finds its principal applications in water desalting and electrolyte concentration. An electrochemical gas separation device of this type could be designed and used to both extract and convert to useful form some objectionable gaseous impurity such as hydrogen sulfide in stack gas [61]. In this case, the cell would consist of noble metal electrodes deposited on an anion-permeable membrane; hydrogen sulfide would be oxidized at the anode to water and sulfite ion, and the sulfite discharged into a steam or air stream at the cathode surface as sulfur dioxide (oxygen being reduced at the cathode to hydroxide ion).

Electrochemical cell laminates could be used for the selective oxidation or reduction of organic compounds in nonaqueous media. The anode (or cathode) of such a cell contacts with the solution maintaining a suitable reductant (e.g., hydrogen) or oxidant (e.g., oxygen) at the opposite electrode. By appropriate choice of electrode material, and control of impressed voltage, highly specific chemical transformations should be possible. Laminate structures of this sort, embodying high-temperature-service inorganic permionic membranes, may provide the long sought solution to the compact, high-energy density, efficient fuel cell, utilizing either gaseous or liquid fuels and air as the oxidant [62, 63]. A tubular, ceramic membrane module, with the tubes coated internally and externally with electrodes, with tube-side fuel feed and shell-side air supply, may serve as a compact power-generating fuel cell for electric vehicles. Such a prospect may be remote at present, but it does point to the remarkable range of potential applications which membrane process-integration can encompass.

4. Catalytic Membrane Reactors

Catalytic membrane reactors (CMRs) are an interesting example of integrated system in which molecular separation and chemical conversions are combined in one step [8]. The heterogenization of catalysts in membrane is particularly suitable for catalyst design at the atomic and molecular level. One of the main advantages of the membrane reactors, compared to traditional reactors, is the possibility to recycle easily the catalyst. Moreover, the selective transport properties of the membranes can be used to shift the equilibrium conversion (e.g., esterification reaction), to remove selectively products and by-products from the reaction mixture, to supply selectively the reagents (e.g., oxygen for partial oxidation reactions) [6].

The scientific literature on CMRs is significant today; however, practically only few large-scale industrial applications have been reported so far because of the relatively high price of membrane units. Current and future advancements in membrane engineering might significantly reverse this trend. CMRs are today successfully applied in small-scale operations, but for their use on large industrial scale, additional efforts mainly related to the optimization of membrane materials, modules manufacturing and reactors design, are required.

The application of CMRs appears of particular interest in several areas such as hydrogen production [59, 64], oxidation reactions [60, 65], and enantiometric productions [8, 66]. The SLM reactors in which the membrane defines the reaction volume, (providing a contacting zone for two immiscible phases, phase-transfer catalysis) make the process environmentally more attractive.

The first applications of CMRs have concerned high-temperature reactions. The employed inorganic membranes, characterized by higher chemical and thermal stability with respect to polymeric membranes, still suffer from some important drawbacks: high cost, limited lifetime, difficulties in reactor manufacturing (delamination of the membrane top-layer). On the other hand, the use of polymeric membranes in CMRs is of increasing interest [8, 9]. The cost of polymeric membranes is generally low, and the preparation protocols allow a better reproducibility; moreover, the relatively low operating temperatures are associated with a less stringent demand for materials used in the reactor construction.

4.1. Immobilized catalytic membrane reactors

For biological applications, synthetic membranes provide an ideal support for catalyst immobilization [7] due to their wide available surface area per unit volume. Enzymes that are retained in the reaction side do not pollute the products and can be continuously reused. Biocatalytic membrane reactors can also be used in production, processing, and treatment operations [6-9]. The trend toward environmentally friendly technologies makes these units particularly attractive because they do not require additives, because of their ability to operate at moderate temperature and pressure, and to reduce the formation of by-products. Enzymes, compared to inorganic catalysts, generally permit greater stereospecificity, and higher reaction rates under milder reaction conditions [7]. Relevant applications of biocatalytic membrane reactors include production of new or better foodstuffs, in which desired nutrients are not lost during thermal treatment; novel pharmaceutical products with well-defined enantiomeric compositions; wastewater treatment [7, 10, 20, 24]. The catalytic action of enzymes is extremely efficient, selective and highly stereospecific when compared with chemical catalysts; moreover, immobilization procedures have been proved to enhance the enzyme stability [10, 23]. Potential applications have been at the origin of important developments in various technology sectors, mainly concerning: induction of microorganisms to produce specific enzymes, techniques of enzymes purification, bioengineering techniques for enzyme immobilization, and design of efficient productive cycles.

One of the most interesting aspects of catalyst heterogenization in membranes is the effect of the membrane environment on the catalyst activity. Membrane composition (hydrophobic or hydrophilic characteristics of the membrane material, presence of chemical groups with acid or basic properties, etc.) and membrane structure (dense or porous, symmetric or asymmetric) can positively influence the catalyst performance not only by the selective sorption and diffusion of reagents or products but also influencing the catalyst activity by electronic (electron-donating and electron-withdrawing groups) and conformational effects (stabilization of the transition states) [6-8]. These effects are the same, occurring in biological membranes.

A membrane-induced structure-reactivity trend that may be exploited to achieve selective processes has been recently observed in polymeric catalytic membranes prepared embedding polyoxotungstares, W(VI)-oxygen anionic clusters having interesting properties as photocatalysts, in polymeric membranes [21]. These catalytic membranes have been successfully applied in the photooxidation of organic substrates in water providing stable and recyclable photocatalytic systems.

Microporous membranes can also be used as substrates for covalent attachment of enzymes or other catalysts. With substrate-containing solutions they can permit the catalyzed transformation [7-9]. The rapid exchange of substrate and product between solution and catalytic pore wall virtually eliminates diffusional resistance to reaction and limits conversion to that imposed by intrinsic reaction kinetics or thermodynamic equilibrium. Thus, high conversions are achievable with very short residence times and very high reaction capacity per unit mass of enzyme or catalyst. Flat-sheet and hollow fiber geometries are applicable. Immobilized enzyme membrane reactors of this type are now in use for such processes as the conversion of lactose into glucose and galactose [67], the isomerization of glucose into fructose [68], and the separation of chiral isomers of acylated racemic amino acids [69]. Such devices with enzyme immobilized are also under evaluation for use as components of selective coulometric or potentiometric biosensors [70].

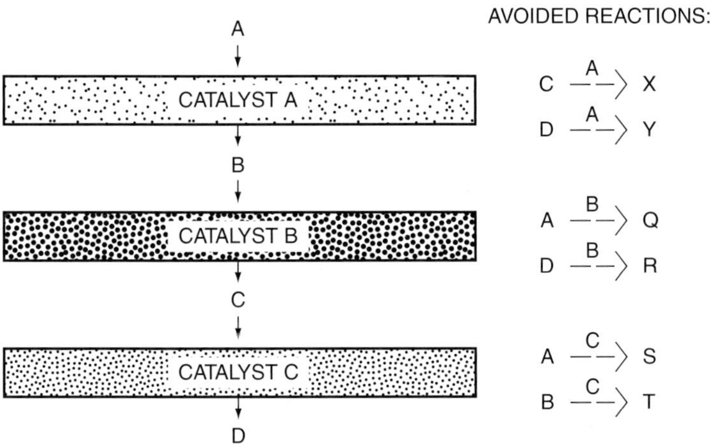

Figure 9.1 Sequential reactions using catalytic membranes.

An interesting, as yet unstudied, potential application of immobilized enzyme or catalytic membranes is their use in the conduct of sequential catalytic reactions, as illustrated in Fig. 9.1. Rapid and slow catalytic reactions with different catalysts can be conducted consequently in different converters: fluid leaving one membrane converter can be delivered to a second catalytic membrane containing a different catalyst operative on the product of the first transformation. Since there can be no back diffusion from one membrane to the other, there is no chance for products of the second reaction to be acted upon by the first catalyst or to interact with the initial substrate. Thus, cross-reaction between different intermediates and different catalysts is avoided. This may allow continuous, sequential catalytic reactions, impossible to perform concurrently, to be carried out efficiently and rapidly.

4.2. Electrochemical/catalytic membrane processes

There are some exciting prospects—as yet unexplored—for the constructive interplay between electrochemical processes and membrane transport/catalytic processes. This may lead to novel routes of organic chemical synthesis or removal of toxic substances from process streams. Many important enzymatic reactions involve oxidative or reductive transformations of the substrate. These usually require the participation of a cofactor which serves as the vehicle for transfer of electrons between the enzyme-bound substrate molecule and the participating oxidant or reductant in solution [71, 72]. For such electron transfer to take place, both substrate molecule and cofactor must be localized in close proximity at the active site of the enzyme until the transfer occurs. Cofactor-requiring enzymatic transformations are exceedingly difficult to carry out in solution because of the problem of confining and conserving the cofactor, and of maintaining a stable oxidizing or reducing environment. Such an oxidative or reductive transformation might take place at or near the anode or cathode of an electrolytic cell. The enzyme and cofactor adjacent to the electrode allow free access to substrate provide a pathway for transfer of electrons to (or from) the electrode from (or to) the cofactor. In addition, the cofactor molecule has sufficient mobility to move from the electrode surface to its binding site on the enzyme, and vice versa. An ideal matrix suitable for accomplishing this complicated series of events is illustrated in Fig. 9.2. It consists of a hydrous polymeric membrane to which both enzyme and cofactor can be covalently coupled. The fibers must be fine enough and closely enough spaced that the interfiber distance is at least of the same order as the migratory range of the cofactor molecules. The membrane is mounted on the surface of one electrode of an electrochemical cell; the assembly is immersed in a solution containing the substrate of interest, and a voltage is impressed between that electrode and an indifferent counter-electrode to complete the circuit. Electron flow from electrode to or from the microfibers, followed by shuttle transport of electrons via cofactor molecule between fibers, enzyme, and site-bound substrate, should lead to continuous transformation of the substrate. Control of the impressed potential would limit electrochemical conversion at the fiber surfaces to the cofactor alone, and current flow through the cell would be a direct measure

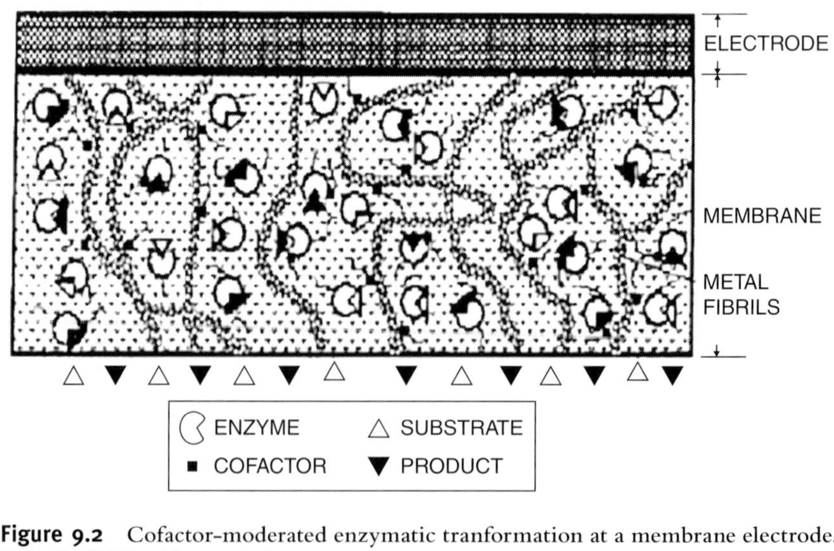

Figure 9.2 Cofactor-moderated enzymatic tranformation at a membrane electrode. From Ref. [72] with permission.

of the transformation kinetics. Such a device can be employed (1) to synthesize pharmaceutical intermediates difficult or impossible to produce by other synthetic routes and (2) to convert selectively highly toxic organic substances (such as chlorinated hydrocarbons) into harmless oxidation products. Such a system concept constitutes a new research frontier for membrane and electrochemical science and engineering. It is worthy of serious consideration.

5. Membrane-Based Gas Separation

The membrane-based gas separation (GS) processes are characterized by lower energy consume and capital investments compared to traditional gas separation systems. Both porous and nonporous membranes can be used in GS. Generally, porous membranes exhibit high level of gas flux but low selectivity. On the contrary, dense membranes generally provide high selectivity and low flux values compared to porous one. In a porous membrane, gas with different size can be separated; in a dense membrane, even gas having similar size can be separated. Membrane-based gas separation can be used as a single separation unit or as part of an integrated separation (see Fig. 9.3 [36]).

Theoretical infinite selectivity can be achieved for hydrogen separation by Pd or Pd-alloy membranes [40]. The reaction is carried out at high temperature, about above 1000 K, because of the equilibrium of the endothermic reactions involved in the process. Many studies demonstrate that Pd-based membranes with ideally infinite selectivity for H_2 can be used to increase the equilibrium conversion of methane steam to H_2 removal [36, 40]. The driving force in this process is the partial pressure difference. The high cost, limited lifetime, and low permeability are relevant limits of Pd and Pd-alloy membrane. To overcome these drawbacks, the studies have been carried out for the preparation of supported metallic membranes in which a thin metallic layer is supported on a thicker sublayer [73].

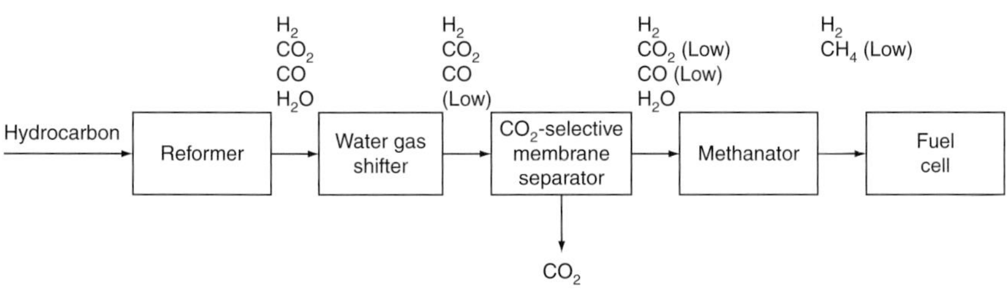

Figure 9.3 Example of the integrated system. From Ref. [36] with permission.

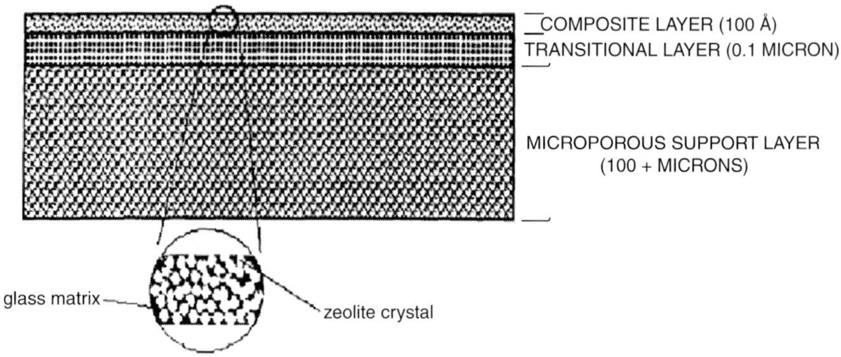

Figure 9.4 The molecular-sieve composite gas-separation membrane. From Ref. [29] with permission.

Fuel cells membrane systems are present not only as proton-exchange membrane (PEM), which allows protons to pass from the anode to the cathode, to combine with oxygen and electrons producing water, but also for the production and purification of the H_2 (see Fig. 9.4). This system provides a CO_2-selective membrane process for the purification and water-gas shift reaction of a reformed gas, generated from on-board reforming of a fuel, for example, hydrocarbon, gasoline, diesel, methanol, or natural gas, to hydrogen for fuel cell vehicles [74].

More intriguing are potential applications which employ ionically conductive membranes as the separation barrier between porous, catalytically active, electronically conductive films serving as electrodes [28, 60]. Electrical potentials can be impressed to mediate oxidative or reductive reactions at the membrane/electrode boundaries [75]. Since current flow in such a laminate is necessarily confined to the membrane between the electrodes, any fluid phase (be it gas, aqueous solution, or nonconductive organic liquid) can be processed in such a cell, delivering a reactant to (or accept a product from) the electrode. A simple example of an electrochemical cell designed to effect a highly selective separation of a gas mixture is illustrated in Fig. 9.5 [28]. Impressing a potential between the electrodes will result in selective oxidation of hydrogen to hydrogen ion, with transfer of the ion through the membrane and its reduction to hydrogen at the cathodic surface. Thus, only hydrogen will be transferred across the cell, and the pressure of the gas in the delivery compartment will be a direct function of the impressed voltage. Thus, this device is both a gas separator and a no-moving parts compressor [76]. Gas separation device of this type could be designed and used to both extract and convert to useful form, some objectionable gaseous impurity such as hydrogen sulfide in stack gas. The cell would consist of noble metal electrodes deposited on an anion-permeable membrane. Hydrogen sulfide will be oxidized at the anode to

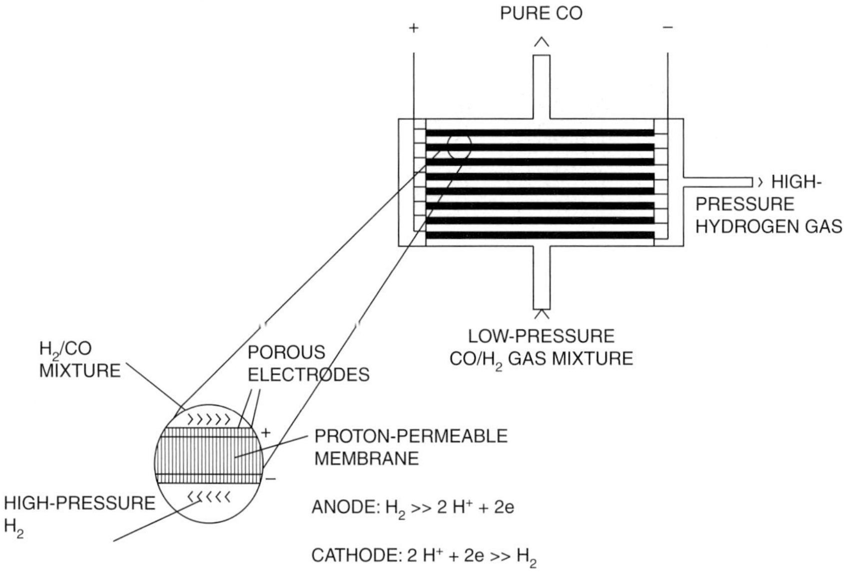

Figure 9.5 Hydrogen recovery/concentration. From Ref. [28] with permission.

water and sulfite ion, and the sulfite discharged into a steam or air stream at the cathodic surface as sulfur dioxide.

Membranes have interesting application also in oxygen separation [77]. In many gas-fired power plants for electricity generation, the oxygen for combustion is produced by very expensive cryogenic air separation techniques. This technology requires large plants to cool air to below zero to separate the component gases. Nitrogen and oxygen are then distributed to customers in liquid form by tanker trucks. About noncryogenic air separation membrane technology offers interesting opportunities. Recent advances in the development of ion-transport membranes using dense mixed-conducting ceramic membranes to separate oxygen from air [38] will contribute to the complete replacing of cryogenic air separation techniques.

6. Advances in the ELM

Membrane reactors using biological catalysts can be used in enantioselective processes. Methodologies [78] for the preparation of emulsions (sub-micron) of oil in water have been developed and such emulsions have been used for kinetic resolutions in heterogeneous systems catalyzed by enantioselective enzyme (see Fig. 9.6). A catalytic reactor containing

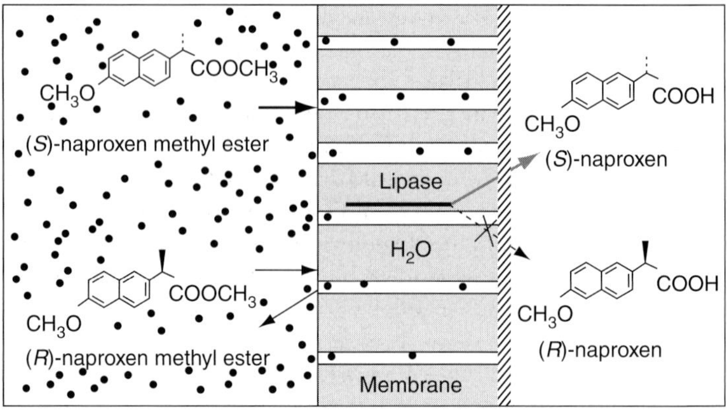

Figure 9.6 Scheme of the *S*- and *R*-naproxen isomer production in the enzyme-emulsion membrane reactor. From Ref. [8] with permission.

membrane immobilized lipase has been realized. In this reactor, the substrate has been fed as emulsion [8]. The distribution of the water-organic interface at the level of the immobilized enzyme has remarkably improved the property of transport, kinetic, and selectivity of the immobilized biocatalyst.

6.1. Reversed micellar separation

Reversed micelles, which are water-in-oil microemulsions, are capable of solubilizing a variety of biomolecules, especially proteins. In order to preserve the biological activity of the protein, it is necessary to maintain its aqueous environment by adding a surfactant that aggregates in a polar solvent and forms reversed micelles [63]. Importantly, enzymes can retain their catalytic activity when reversed micelles are used to solubilize them. In addition, because many properties of the water core of reversed micelles resemble those of water present at interfaces in biological systems, reversed micelles provide an excellent system for studying the interactions between polypeptides and interfacial water [79]. Three techniques have been applied to prepare enzyme-containing micellar solutions stirring of solid protein with a hydrocarbon solution of surfactant, injection of a few microliters of a concentrated enzyme solution into the surfactant solution, or transfer of the protein from an aqueous phase into the reversed micellar phase.

The ELM method has been found to be very successful for the extraction of proteins [80]. Solubilization of several biomolecules with surfactants into reversed micelles, only two have received sufficient attention for use: the anionic surfactant sodium bis(2-ethylhexyl)sulfosuccinate (known commercially as AOT) and trialkylmethyl ammonium salts, which are cationic amphiphilic molecules. A typical process for protein separation using reversed micelles consists of the following steps: pressurization of the liquid reversed-micellar solution with the gas, equilibration of the system, sampling of the liquid (plus any dissolved gas) phase, depressurization of the sample to ambient conditions, and sample analysis for protein, water and surfactant contents. The mechanism of protein recovery from reversed micelles is still not very well defined. Under increased pressure, nonpolar gas causes the formation of hydrates from the water pools of the reversed micelles. Reversed micelles with a molar ratio of water to surfactant of <2-3 cannot encapsulate proteins effectively. So far, separation by reversed micelles has mainly been applied in the batch mode. Attempts are needed to use this technique in the continuous mode. However, little is known about the kinetic and mass transfer parameters for continuous operations, and much work needs to be done to understand the hydrodynamics involved.

ELM has great promise for the separation of proteins and for the enhancement or inhibition of specific reactions [81-83]. It allows refolding of proteins and retention of enzyme activity after separation. So far, the process has been used only for small-scale batch operations.

6.2. Integrated liquid membrane processes

Promising results are shown by recently developed integrated SLM-ELM [84, 85] systems. These techniques are known as supported liquid membrane with strip dispersion (SLMSD), pseudo-emulsion-based hollow fiber strip dispersion (PEHFSD), emulsion pertraction technology (EPP), and strip dispersion hybrid liquid membrane (SDHLM). All techniques are the same: the organic phase (carrier, dissolved in diluent) and back extraction aqueous phase are emulsified before injection into the module and can be separated at the module outlet. The difference is only in the type of the SLM contactors: hollow fiber or flat sheet and in the liquid membrane (carrier) composition. These techniques have been successfully demonstrated for the removal and recovery of metals from wastewaters. Nevertheless, the techniques still need to be tested in specific applications to evaluate the suitability of the technology for commercial use.

Another novel technique integrates ELM with electrodialysis [86], where O/W emulsion is used as the catholyte. Results have demonstrated that the process is highly efficient in the recovery of organic acids. The author developed a model describing the processes involved. The technique has a good potential for developing selective separation processes in pharmaceutical and food industries but still needs to be tested to evaluate the suitability of the technology.

7. Advances in the BOHLM Systems

BOHLM systems represent a very important branch of new processes having wide range of applications with a high charge of innovation [3, 7-13, 16-20, 24, 87]. This field of process integration for separation/purification is one of the most fruitful areas for membrane process research and development today. BOHLM offers potential solution in a wide range of gas-liquid and liquid-liquid applications: gas adsorption and stripping, liquid-liquid extraction, dense gas extraction, fermentation and enzymatic transformation, pharmaceutical applications, protein extraction, wastewater treatment, chiral separations, semiconductor manufacturing, carbonation of beverages, metal ion extraction, and VOCs removal from waste gas (details see in Chapter 5). There are numerous opportunities for effecting molecular separations, or physical or chemical transformations, by imaginatively combining membrane-moderated processes with other biochemical or physicochemical processes to yield an integrated process which is far more efficient or economic than is possible by other processing means. What follows is an attempt to illustrate the merits of this approach by description of a few examples of successful integration now being practiced, and of potentially promising integration concepts yet to be evaluated.

7.1. Separation by liquid membrane solvent extraction

Immiscible-liquid solvent extraction is a well-established practice for recovery, concentration, and purification of organophilic solutes (e.g., antibiotics, amino acids, vitamins) present in aqueous process streams such as fermentation broths or plant or animal tissue extracts [88]. The process is, however, frequently rendered difficult or impossible by problems of emulsification, loss of entrained solvent, and contamination by particulate impurities in the feed. Integrated membrane separation with liquid/liquid extraction is illustrated in Fig. 9.7.

In Chapters 1, 2, and 5 we discussed the term membrane solvent extraction or solvent extraction using membrane contactors. It is a part of BOHLM systems, constructed using hollow fiber or capillary membranes: forward and back extraction combined in one unit or in two units separately, as shown in Fig. 9.7. In the last decade many researchers presented this process and devices as a specific. We discuss it here as one of the BOHLM systems.

The solution to be extracted is fed to the lumen side of a microporous hollow fiber membrane module, with extracting organic solvent being fed to the shell side. The membrane is selected to be preferentially wetted by one of the phases (usually the organic phase), and the pressure adjusted to prevent capillary penetration of the aqueous phase into the membrane. In this

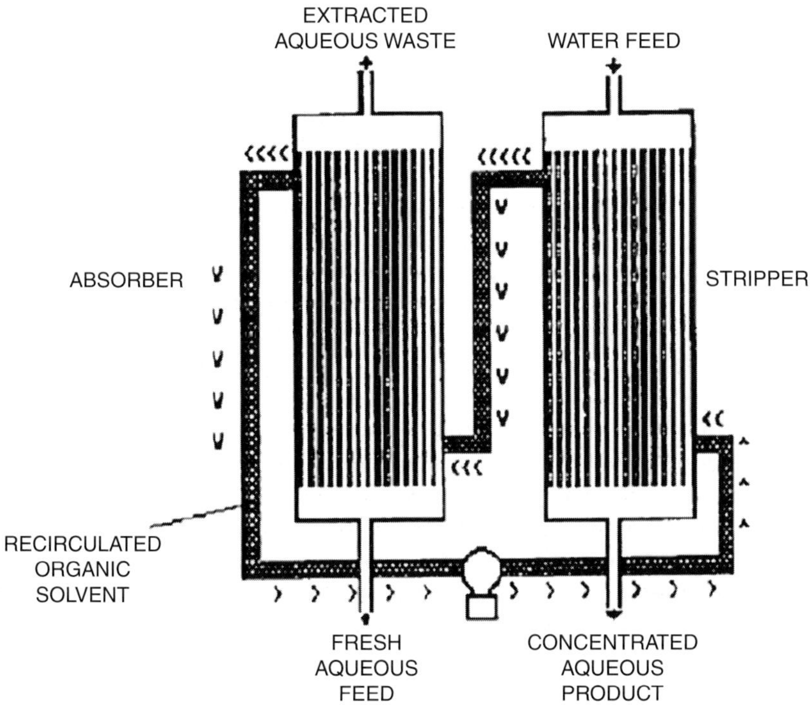

Figure 9.7 Membrane solvent extraction. From Ref. [29] with permission.

manner, the solvent/water interface is stabilized at the membrane/aqueous solution boundary, and extraction of the desired component into the organic phase takes place by transmembrane diffusion and convection. The solute-enriched organic phase is then delivered to a second membrane unit where solute is transferred into an aqueous solution in which it has elevated solubility; the solute-depleted organic is then recycled to the first unit.

This arrangement has important advantages over conventional solvent extraction: It substantially eliminates emulsification and the need for phase separation equipment such as centrifugal separators; it provides a stable, large interfacial area for interphase mass transfer; and it prevents contamination of the organic phase and extracted product by particulate impurities in the feed. Indeed, if hollow fiber membranes of suitably large lumen diameter are used, it is possible to process whole fermentation broths containing high concentrations of suspended biomass. This latter feature makes possible the process of "continuous extractive fermentation," in which the product-bearing whole broth is feed to the extraction unit, and the product depleted broth (containing viable cells) is returned to the fermenter.

A number of commercial applications of BOHLMs have been already successfully realized. A bubble-free membrane-based carbonation line, using Liqui-Cel equipment, is in operation by Pepsi in West Virginia since 1993 [89]. The systems are also used in beer production: the CO_2 removal stage is followed by nondispersive nitrogenation to obtain a dense foam head [90].

This technology clearly has potential for use in organophilic solute recovery or removal from chemical process streams and industrial or municipal wastewaters. Another important field of application of BOHLM is the production of ultrapure water for semiconductor manufacturing [91, 92].

One more interesting field in research activities may be mentioned: solution of the problem with slow kinetics of back extraction of the solutes into the strip phase, discussed in Chapter 5 in detail, may be found using mixtures of the carriers (with different complexing properties) in LM or different type of membrane supports on the strip side of the BOHLM module.

8. Potential Advances in the BAHLM System Applications

Many directions for the development of the BAHLM systems are waiting for their research. These are mixtures after degradation of organic compounds, products of fermentation, biological mixtures (hormones, peptides, etc.), drug conversion and selective separation, catalytic reaction enhancement and selective separation, wastewater and water treatment, and so on. In this section, the author's envision on some new research and application areas for the BAHLM systems is discussed briefly.

8.1. Drug separation from biochemical mixtures

In recent years, there has been an explosion of interest in water-soluble synthetic polyanions, anionic dyes as selective drug carriers and enhancers of pharmacokinetics. For example, application of pyran copolymer as a carrier of enzyme superoxide dismutase (SOD) to areas of inflammation induced by influenza virus [93], successful development of polyanions for the focused transport of proteins or peptides [94], coupling or charge complexing of antibiotics to anionic copolymers, effect of polyanions in calcium metabolism [95] with blocking calcium flux by inositol-1,4,5 triphosphate with heparin [96], and so on. Water-soluble polyethylene glycol (PEG) derivatives have a variety of biochemical and biomedical applications including aqueous two-phase partitioning, enzyme, antibody immobilization, drug modification, and so on. Detailed description of polyionic water-soluble polymers as drug carriers see in [97]. In all these works water-soluble polymers are considered as carriers for drug delivery.

Not less interesting direction is application of water-soluble polyions as carriers in the BAHLM separation and purification systems for drug production. The author did not find any works in this field; but, according to his opinion, investigation of available polyelectrolytes in selective separation of drug components from the biochemical mixtures, development of new selective stable WSP for these purposes may have enormous effect in pharmaceutical industry.

8.2. BAHLM reactors: Fermentation, catalysis, and separation with enrichment of valuable compounds

There is no commonly accepted definition of a membrane reactor but the term is applied to membrane (including liquid membrane) processes and devices whose function is to perform chemical conversion, coupling and combining chemical and transport processes, using the unique contacting features of membranes. As a rule, functional definition of this term includes fermentation, catalysis, separation of the products and their enrichment. A few published reviews at this time are available [98–104]. In most of publications the bioreactors, based on enzymes or whole cells, impregnated into the membrane pores (immobilized or supported liquid membranes) or deposited on the membrane surfaces are discussed.

Ability of WSP to interact with enzymes, drugs, selective and chelating properties of many polyelectrolytes make them very perspective in the development of liquid membrane bioreactors. It is known [103] that formation of a ternary metal ion-carrier-chelator complex at the inner vesicle wall can enhance the overall selectivity in accordance with a multiplicative, rather than additive, function of equilibrium metal-ligand binding constants. Enzyme-containing lipid vesicles (liposomes), which are lipid dispersions that contain water-soluble enzymes in the trapped aqueous space, may be named as liquid membrane micro- or nanobioreactors with intraped WSP. Preparation and properties of lipid vesicles are described in [104] review.

The use of anionic polymers coupled to catalytic enzymes or to catalytic chelating agents [105, 106], degradation-controlled drug elution, provides a complex system that behaves in catalytic conversion, selective transport enhancement and enrichment of products in similar fashion to membrane reactors. Articles, deal with WSP liquid membrane reactors, were not found, nevertheless this direction is very outlook in the application of the BAHLM systems.

8.3. Desalination of wastewater and sea water

WSP, natural and synthetic, are used in water and wastewater treatment as flocculants. It includes such processes as settling, clarification, gravity dewatering, filtration, centrifugation, and flotation [107] due to high charge and hydrogen bond densities, high molecular weights of polyelectrolytes. Examples of natural WSP application: red mud separation by starch, soda ash processing by guar, coal wash wastewater treatment by corn and potato starch, recovery of magnesium from seawater by precipitation using guar, removal of heavy metals from wastewaters by precipitation with starch xanthates, concentrating of metals in ore leachates by flotation using guars, cellulose derivatives, dextrines, and their combinations, and so on. There are hundreds of different synthetic polyelectrolytes with similar but more effective functional properties available for wastewater treatment applications.

The largest area of polyelectrolytes application is potable and waste water treatment by filtration [108]. Here the anionic, nonionic, and cationic polyelectrolytes are used. The USA, France, Germany, Japan, and other countries are widely use polyelectrolytes in direct water and wastewater filtration. Large quantities of ultra-high quality water are produced by suitably designed direct filtration equipment employing polyelectrolytes [109]. Polyelectrolytes strongly improve removal of color, oil, asbestos at treatment of water by filtration [110].

A big number of publications were appeared last years in research of WSP-metal ion interactions [111]. Interactions of metal ions with WSP are mainly due to electrostatic forces and the formation of coordinating bonds. Other weak interactions may appear such as trapping metal ions in the bulk of the polymer phase. The author stated that WSP are very perspective for selective separation and concentration (enrichment) of metal ions from the treated diluted solutions (wastewaters, desalination of sea water).

Membrane-based pretreatment, before reverse osmosis (RO), employing polyelectrolytes, is used on wastewater, brackish water, and sea water plants [109-114]. It includes ultrafiltration or nanofiltration membrane units with feed pressure from 1-1.5 to 2.5-4 MPa. RO units need much higher feed pressure, 6-8 up to 10 MPa. It makes the RO technology economically expensive. Decreasing the high feed pressure in RO plants is the main direction for designers' efforts.

Nanofiltration is one of promising technologies for the treatment of natural organic matters and inorganic pollutants. Low-pressure operation of nanofiltration is possible [115, 116]. The nanofiltration system, which has a pretreatment process of microfiltration or ultrafiltration, may be applicable to drinking water treatment.

Replacing the ultrafiltration, nanofiltration pretreatment and reverse osmosis by BAHLM processes in desalination industry is the main idea of this proposal. Rejection characteristics of natural organic matters and inorganic salts in a low pressure nanofiltration (e.g., >99% at 1.5 MPa [115]) and capacity of polyanions to complex monovalent and especially bivalent cations [92, 95, 115-117] make this idea promising.

One of the complex problems in the proposed technology is the regeneration of polyelectrolytes with separation and utilization of concentrated inorganic salts and organic matter.

Intensive research is needed in the development of

(a) Stable polyelectrolytes with improved properties, their regeneration from the inorganic salts
(b) Anomalous osmosis with polyelectrolytes
(c) Stable micro- and nanoporous ion-exchange membranes with high density of charged functional groups
(d) Complexation chemistry and theory of the BAHLM transport, and so on

The potential of the BAHLM processes in the above-mentioned and other application directions may be very interesting for young researchers.

8.4. Integrated water-soluble complexing/filtration techniques

Technique for isolation, concentration, and purification of biological products (e.g., proteins) present in complex aqueous mixtures has become the focus of intensive research by many laboratories around the world [118, 119]. An affinity complexing reagent coupled to either a water-soluble macromolecule is added to a solution containing the desired compound. Resulting dispersion of the macromolecular affinity complex is then filtered through a membrane which is retentive for the complex, but freely permeable to all other components. The retentate, comprising a concentrated, purified dispersion of the affinity complex, is then treated to dissociate the complex. Resulting solution filtered through a second membrane which is retentive for the regenerated affinity reagent, but permeable to the product. The pure-product-bearing permeate can be further concentrated if desired. The concentrated regenerated affinity reagent dispersion is then recycled for reuse. Substrates for use as the complexing reagents include water-soluble polymers such as PEG or dextran; polyacrylamide or agarose; polystyrene latices; or phospholipid vesicles (liposomes). While developed for recovery of biological macromolecules, it should be obvious that the technique is applicable to recovery of any soluble substance (including metal ions) for which a specific, complexing, macromolecular reagent can be found. The technique has promise, for example, for (1) the recovery of trace metal values, for example, silver, gold, nickel, cobalt, from hydrometallurgical process streams or water sources; (2) the removal of trace amounts of toxic metal ions (e.g., cadmium, mercury) from industrial waste streams. For these applications, water-soluble chelating polyelectrolytes are the complexing reagents of choice. These techniques may be successfully and effectively replaced by BAHLM technique with lesser capital and operational costs.

9. Potential Directions in Reducing Concentration Polarization and Fouling

Membrane lifetime and permeate fluxes are primarily affected by the phenomena of concentration polarization and fouling-scaling (e.g., microbial adhesion, gel layer formation, solute adhesion) at the membrane surface [120-122] (see Fig. 9.8). During the initial period of operation within a cycle, concentration polarization is one of the primary reasons for flux decline. The average flux under steady-state concentration decreasing from cycle to cycle suggests irreversible substrate adsorption or fouling. Accumulation of the substrates retained on a membrane surface leads to increasing permeate flow resistance at the membrane wall region. The main mechanisms of membrane fouling are adsorption of feed components, clogging of pores, chemical interaction between solutes and membrane

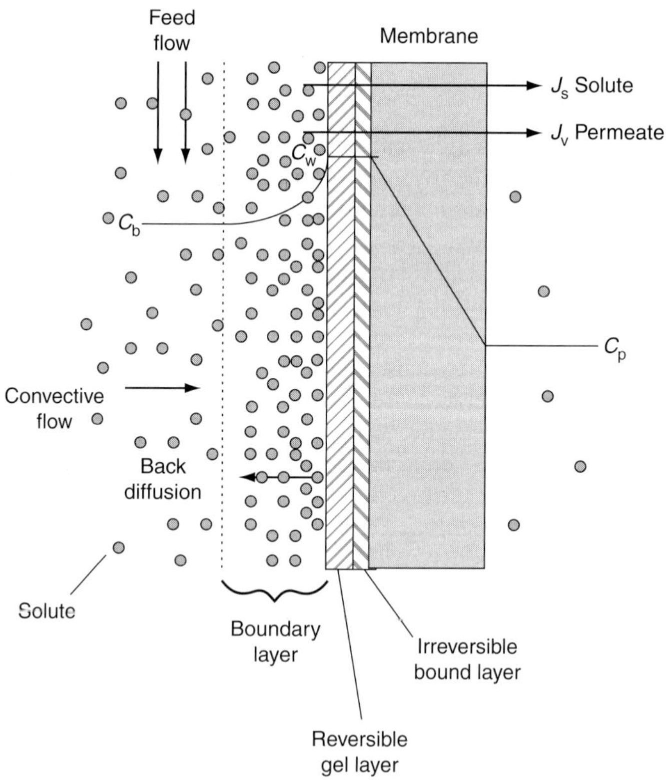

Figure 9.8 Schematic representation of concentration polarization and fouling at the membrane surface. From Ref. [122] with permission.

material, gel formation and bacterial growth [123]. Large-scale membrane systems operate in a cyclic mode, where clean-in-place operation alternates with the normal run. One of the most serious forms of membrane fouling is bacterial adhesion and growth [124]. Once they form biofilms, which are very difficult to remove, they require either disinfection or chemical cleaning. This is an area where further research is required. Fouling in gas separation processes, however, is less severe [125].

Concentration polarization is an inherent part of a membrane separation process. Equipment design and operating conditions normally take care of it. Scaling-fouling, which is triggered by concentration polarization, reduces flux that can be minimized using antiscalants. Concentration polarization is not the same as fouling except that by theory it results in flux decline due to higher osmotic pressure.

The primary aim of this section is to critically review the literature on minimization of negative influence of concentration polarization and fouling and suggest potential directions where researchers and technologists can assist in improving the performance of membrane systems through fundamental and applied research.

9.1. Manipulations with flow

Several device design/fluid management strategies are being developed which operate to reduce both concentration polarization and fouling by rejecting substrates from the membrane surface. These are manipulation with the spacer form and thickness, duration and pressure of the back pulsing, pulsatile flow, helical inserting, and so on.

The flow through the spacer-filled channel induces unsteady flows and increases local shear rates and local mixing, which reduce the boundary layer thickness and the concentration polarization. The mass transfer and pressure drop in a spacer-filled channel can be significantly affected by the geometric characteristics of the spacer [126]. Comparing the flux obtained with and without spacers, the flux was significantly higher (three to fivefold) for the filtration with spacers. The effects of spacer thickness on permeate flux showed that the observed flux decreases by up to 50% in going from a spacer thickness of 0.1168-0.0508 cm. The penalty is increased pressure loss. This leads to an optimal design and operating range that minimizes the processing costs. Chemical modification of a membrane surface can be used in combination with spacers.

Hydrophobic membranes attracted a thicker irreversible adsorption layer than hydrophilic membranes [127, 128]. Hydrophilic membranes display low sorptivity for fouling macrosolutes such as proteins. In some situations, ionically charged membranes are more fouling-resistant than electroneutral membranes. The apparent high fouling-resistance of ceramic (alumina) membranes is worthy of special note, although the explanation of this behavior is elusive. Membranes with a higher negative surface charge and greater hydrophilicity were less prone to fouling due to fewer interactions between the chemical groups in the organic solute and the polar groups on the membrane surface. Sophisticated research is needed in understanding the fouling phenomenon at the submicroscopic level.

An interesting strategy direction in reducing fouling is rapid back pulsing at varied pulsing duration and pressure [129]. The membranes are more effectively cleaned by longer back pulse durations and higher back- pulse pressures. However, trade-offs exist between longer and stronger back pulses, and permeate loss during the back pulse. Shorter, stronger back pulses resulted in higher net fluxes than longer, weaker back pulses.

Pulsatile flow can be defined as flow with a periodic pressure fluctuation wave traveling along the flow path [130, 131]. The pressure gradient along the flow path determines the instantaneous pulsatile flow rate. Significant shear is found only in the region near the wall with the liquid in the central portion of the tube virtually unsheared. Thus, the liquid behavior for pulsatile flow is rather like a solid mass sliding inside a thin layer of viscous liquid surrounding it. The higher the frequency or the larger the tube diameter the flatter is the velocity profile, implying a thinner boundary layer.

9.2. High shear devices

High liquid flow results in high pressure drop along the membrane module, leading to considerable energy consumption and nonuniform distribution over the flow path of the feed channel [132, 133]. Such a close dependence between high crossflow velocity and pressure drop limits the use of high crossflow in various commercial membrane modules. However, in high shear membrane devices, the relative movement between the liquid and the membrane surface is induced by an independent device rather than a liquid transport pump. The typical high shear membrane devices include rotating disks and vibrating dynamic systems.

9.2.1. Rotating systems

Figure 9.9 shows the typical structure of a rotating dynamic membrane system which consists of a stationary flat membrane, a rotating disk, a hollow shaft, and a cylindrical housing with feed inlet and permeate outlet. Driven by an electric motor, the rotating disk rotates at high speed (100–1000 rpm) to produce shear on the membrane surface. The centrifugal forces generated by the disk induce an outward radial flow near the rotating disk and inward radial flow toward the axis near the membrane surface. The shear on the membrane surface at radius r can be calculated based on the expressions for different flow regimes [1].

Several commercially available rotating disk dynamic membrane systems have been tested for different applications: the cross-rotational system [134].

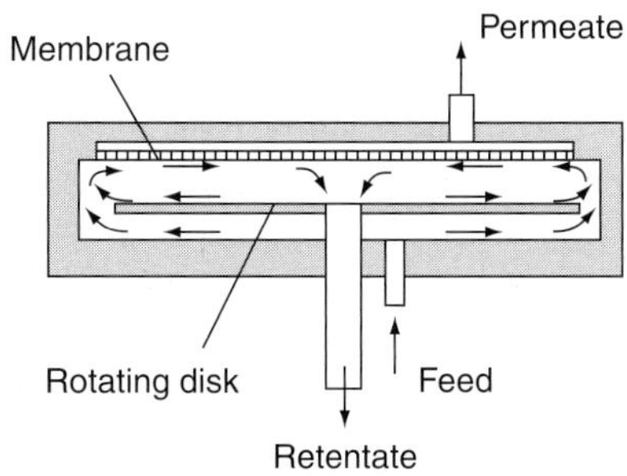

Figure 9.9 Schematic diagram of rotating disk dynamic membrane filter. From Ref. [1] with permission.

The DMF module (Pall Corp., New York) which consists of several disks mounted on the same shaft [2, 135]. The reported studies of the application of the rotating disk dynamic membrane indicate that high shear-induced filtration is much less sensitive to the solids concentration. Advantage of the rotational system is that it permits operation at both very low transmembrane pressure-drop, and low upstream mass velocity, without loss in depolarization efficiency. It is thus possible with this system to achieve cleaner separation of solute components, than is achievable with conventional systems. Equipment for this process is, unfortunately, significantly more costly, and maintenance costs much higher also, than those for conventional membrane systems.

9.2.2. Vibratory hollow fiber membranes

Vibratory mechanism was one more interesting technique for reducing fouling negative effect. Vibrations can be applied to flat sheet membrane modules, but many researchers tested the technique on performance of hollow fibers for different separation processes [3, 136]. One example of construction is shown in Fig. 9.10 [136] to assess the effect of axial membrane vibrations on mass transfer in a hollow fiber oxygenator.

The system can be operated in a frequency range of 1-18 Hz with a displacement of 0.05-4 cm. The maximal enhancement factor reached was 2.65.

9.2.3. Enhancement by gas bubbles

In the past decade, an interest in the use of gas bubbles to enhance membrane processes with tubular and submerged membrane systems was raised at the development of MBRs [137, 138]. The trends have been summarized as follows:

1. The benefit of bubbling becomes more significant at high flux, a low liquid velocity, and a high feed concentration

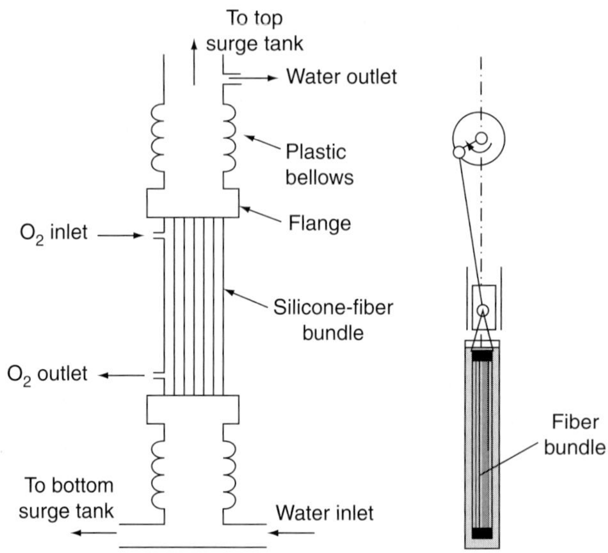

Figure 9.10 Schematic diagram of vibrating hollow fiber membrane. From Ref. [136] with permission.

2. The greater flux enhancement can be obtained with vertical tubular membranes and a moderate downward liquid flow
3. Flux enhancement by bubbling is significant in the laminar regime of liquid flow and becomes insignificant as the liquid flow Reynolds number approaches 2500–3000

9.3. Electric field enhancement

Some researchers [121, 126, 139] have demonstrated that intermittent application of AC or DC electric fields across MF or UF membranes during filtration can often promote displacement of polarization- or adherent fouling layers, with significant improvements in sustained permeation flux. The electrode is installed on either side of the membrane with the cathode on the permeate side and the anode on the feed side. Usually, the membrane support is made of stainless steel or the membrane itself is made of conductive materials to form the cathode. Titanium coated with a thin layer of a noble metal such as platinum could be one of the best anode materials [140]. The electromagnetic field reduces fouling and biofilm development.

The mechanism involved in this defouling process is not at all clear, and is the subject of continuing study.

9.4. Ultrasound enhancement

Ultrasound is typically associated with the frequency range of 20 kHz to 500 MHz. The frequency level is inversely proportional to the power output. The chemical and physical effects of ultrasound are related to the cavitation phenomenon induced by rarefaction and compression of the sound wave. When ultrasound is irradiated through a liquid medium, an alternating adiabatic compression and rarefaction cycle of the medium occurs. During the rarefaction period, negative pressures occur and microbubbles can be formed and growth at sites where there is some gaseous impurity. The formed bubbles may suddenly collapse during the compression period after a few acoustic cycles with a release of energy. This helps to generate micromixing in the liquid or form liquid microjets near a solid surface, or cause chemical changes in reactants in the cavitation bubbles. All the reported studies [141, 142] indicate that low frequency ultrasound can enhance the filtration process and improve membrane cleaning efficiency to some extent. However, reducing energy losses, increasing the efficiency of the ultrasound and system scale-up are challenges that need to be overcome before serious commercial application can be considered.

In conclusion of this section it has to be stated that most techniques presented above are developed for reducing fouling of filtration systems. In the case of liquid membranes the problems with fouling are less severe, nevertheless, they are exist and require additional studies. The majority of techniques is hydrodynamic and therefore involves additional energy input or some extra design feature. Thus in each case, there will be a trade-off between operating and capital cost.

10. Perspectives in Liquid Membrane Technology Applications

Liquid membrane separation processes are widely used in biochemical processing, in industrial wastewater treatment, in gas separations, in food and beverage production, and in pharmaceutical applications. Below the reader can find some fields where research, development and scale-up efforts are expected:

1. Desalination of sea water, removal of toxic and/or objectionable pollutants from industrial and municipal aqueous wastewaters, or from contaminated water-sources
2. Concentration and sterilization of liquid foodstuffs and beverages without flavor/fragrance deterioration, to permit packaging and long-term storage in "fresh" condition without refrigeration

3. Production of ultra-pure, submicron-particle-free water, liquid reagents, and gases employed in the processing of electronic materials and devices
4. Removal of hazardous pollutants from gaseous effluents
5. Large-scale separation of air gases
6. Large-scale separation and purification of industrial chemicals and biochemicals
7. Recovery of trace minerals from hydrometallurgical process streams and natural water sources
8. Processing, separation, and purification of gaseous and liquid fuels

REFERENCES

1. Jaffrin MY, Ding LH, Akoum O, and Brou A, A hydrodynamic comparison between rotating disk and vibratory dynamic filtration systems, J. Membr. Sci. 2004; 242: 155–167.
2. Brou A, Ding L, Boulnois P, and Jaffrin M, Dynamic microfiltration of yeast suspensions using rotating disks equipped with vanes, J. Membr. Sci. 2002; 197: 269–282.
3. Krantz WB, Bilodeau RR, Voorhees ME, and Elgas RJ, Use of axial membrane vibrations to enhance mass transfer in a hollow tube oxygenator, J. Membr. Sci. 1997; 124: 283–299.
4. Stairmand JW, and Bellhouse BJ, Mass transfer in a pulsating turbulent flow with deposition onto furrowed walls, Int. J. Heat Mass Transf. 1985; 27: 1405–1408.
5. Goldsmith RL, Cross-flow filtration device with filtrate flow conduits and method of forming, USP No 4,781,831. Nov. 1, 1988.
6. Goosen MFA, Sablani SS, Al-Hinai H, Al-Obeidani S, Al-Belushi R, and Jackson D, Fouling of reverse osmosis and ultrafiltration membranes: A critical review, Sep. Sci. Technol. 2004; 39: 2261–2298.
7. De Bartolo L, Giorno L, and Drioli E, Membrane bioreactors for biotechnology and medical applications, In: Bhattacharyya D, and Butterfield DA, eds. *New Insights into Membrane Science and Technology: Polymeric and Biofunctional Membranes*, Elsevier, Amsterdam, 2003; p. 187.
8. Li N, Giorno L, and Drioli E, Effect of immobilization site and membrane materials on multiphasic enantiocatalytic enzyme membrane reactors, Ann. N.Y. Acad. Sci. 2003; 984: 436–452.
9. Vankelecom IFJ, and Jacobs PA, Polymeric membranes in catalytic reactors, Catal. Today 2000; 56: 147.
10. Simmons DK, May SW, and Agrawal PK, Enzymes in liquid membranes, In: Hamel JFP, Hunter JB, and Sikdar SK, eds. *Downstream Processing and Bioseparation, Recovery and Purification of Biological Products. ACS Symp. Ser.* 479, ACS Press, Washington, 1990; pp. 108–129.
11. Szlag DC, Giuliano A, and Snyder SM, A low-cost aqueous two phase system for affinity extraction, In: Hamel JFP, Hunter JB, and Sikdar SK, eds. *Downstream Processing and Bioseparation, Recovery and Purification of Biological Products. ACS Symp. Ser.* 419, ACS Press, Washington, 1990; pp. 71–86.
12. Morris JE, Hoffmann AS, and Fisher RR, Affinity precipitation of proteins by poly ligands, Biotechnol. Bioeng. 1993; 41: 991–997.
13. Gupta MN, and Mattiasson B, Affinity precipitation, In: Mukherjea RN, ed. *Downstream Processing in Biotechnology*, Vendam Books Int., New Delhi, India, 1992; pp. 43–53.

14. Arrueb M, Mallada R, and Pina MP, Zeolite membranes: Synthesis, characterization, important applications, and recent advances, In: Pabby AK, Sastre AM, and Rizvi SSH, eds. *Handbook of Membrane Separations: Chemical, Pharmaceutical, and Biotechnological Applications CRC*, 2008; pp. 269–324.
15. Hsieh HP, Bhave RR, and Fleming HL, Microporous alumina membranes, J. Membr. Sci. 1988; 39: 221–241.
16. Singh PC, and Singh RK, Choosing an appropriate bioseparation technique, Trends Food Sci. Technol. 1966; 71: 49–58.
17. Drioli E, Process intensification using membrane systems, Clean Tech. Environ. Policy 2003; 5: 3.
18. Prasad R, and Sirkar KK, Membrane-based solvent extraction, In: Ho WSW, and Sirkar KK, eds. *Membrane Handbook*, van Nostrand Reinhold, New York, 1992; pp. 727–763.
19. Sirkar KK, Membrane separations: Newer concepts and applications for the food industry, In: Singh RK, and Rizvi SSH, eds. *Bioseparation Processes in Foods*, Marcel Dekker, New York, 1995; pp. 351–374.
20. Cussler EL, and Ding H, Bioseparations, especially using hollow fibers, In: Singh RK, and Rizvi SSH, eds. *Bioseparation Processes in Foods*, Marcel Dekker, New York, 1995; pp. 1–15.
21. Bonchio M, Carraro M, Scorrano G, Fontananova E, and Drioli E, Heterogeneous photooxidation of alcohols in water by photocatalytic membranes incorporating decatungstate, Adv. Synth. Catal. 2003; 345: 1119.
22. Leeper SA, Membrane separations in the recovery of biofuels and biochemicals: An Uupdate review, Sep. Purif. Technol. 1992.
23. Vijlkel W, Bosse J, Poppe W, Halwachs W, and Schiigerl K, Development and design of a liquid membrane enzyme reactor for the detoxification of blood, Chem. Eng. Comm. 1984; 30: 55–66.
24. Short J, Membrane separation in food processing, In: Singh RK, and Rizvi SSH, eds. *Bioseparation Processes in Foods*, Marcel Dekker, New York, 1995; pp. 333–350.
25. Hunter JB, In focus and on the move: Prospects for electrophoresis in the food industry, In: Singh RK, and Rizvi SSH, eds. *Bioseparation Processes in Foods*, Marcel Dekker, New York, 1995; pp. 227–295.
26. Barbic M, Magnetic wires in MEMS and bio-medical applications, J. Magn. Magn. Mater. 2002; 249: 357.
27. Huotari HM, Huisman IM, and Trägårdh G, Electrically enhanced crossflow membrane filtration of oily waster water using the membrane as a cathode, J. Membr. Sci. 1999; 156: 49–60.
28. Lee HK, Choi HY, Choi KH, Park JH, and Lee TH, Hydrogen separation using electrochemical method, J. Power Sources 2004; 132: 92.
29. Michaels AS, Membranes, membrane processes, and their applications: Needs, unsolved problems, and challenges of the 1990's, Desalination, 1990; 77: 5–34.
30. Wijmans JG, and Baker RW, The solution-diffusion model: A review, J. Membr. Sci. 1995; 107: 1.
31. Piletsky SA, Panasyuk TL, Piletskaya EV, Nicholls IA, and Ulbricht M, Receptor and transport properties of imprinted polymer membranes: A review, J. Membr. Sci. 1999; 157: 263.
32. Eijkel JCT, and van den Ber A, Nanotechnology for membranes, filters and sieves. A series of mini-reviews covering new trends in fundamental and applied research, and potential applications of miniaturised technologies, Lab. Chip, 2006; 6: 19–23.
33. Kabsch-Korbutowicz M, Majewska-Nowak K, and Winnicki T, Analysis of membrane fouling in the treatment of water solutions containing humic acids and mineral salts, Desalination 1999; 126: 179–185.

34. Tosti S, Basile A, Borgognoni F, *et al.*, Low temperature ethanol steam reforming in a Pd-Ag membrane reactor, J. Membr. Sci. 2008; 308: 258–263.
35. Caro J, Schiestel T, *et al.*, Perovskite hollow-fiber membranes in the partial oxidation of methane to synthesis gas in a membrane reactor, Desalination 2006; 129: 415–417.
36. Pez GP, Carlin RT, Laciak DV, and Sorensen JC, Gas separation using molten salt membranes, U.S. Patent US 4761164 A 08/02. 1988.
37. Jiang W, and Liang B, Fabrication of structured vanadium catalyst for SO2 conversion, Appl. Catal. A 2006; 311: 1–7.
38. Kim J, and Lin YS, Synthesis and characterization of suspension-derived, porous ion-conducting ceramic membranes, J. Am. Ceram. Soc. 1999; 82: 2641–2666.
39. Uemiya S, Kato W, Uyama A, *et al.*, Separation of hydrogen from gas mixtures using supported platinum-group metal membranes, Sep. Purif. Technol. 2001; 22–23: 309–317.
40. He RH, Li QF, Xiao G, and Bjerrum NJ, Proton conductivity of phosphoric acid doped polybenzimidazole and its composites with inorganic proton conductors J. Membr. Sci. 2003; 226: 169–184.
41. Hsu Y-J, Chang Y-H, Chai Y-L, and Chen G-J, The influence of residual stress and membrane thickness on the performance of Fe–Ni–Cr = Invar bimetal freestanding membrane for microdevices, Sens Actuators A 2002; 101: 160.
42. Qiu Z, Zhang S, and Li W, Synthesis and gas permeability of novel fluorinated poly (phenylene-co-naphthalimide)s, J. Appl. Polym. Sci. 2007; 104: 2395–2402.
43. Song X, Boga R, and Chidebelu-Eze CO, Diagnostic kit and one-step enzymatic and amine detection technique, U.S. Patent US 2006003336 A1 01/05. 2006.
44. John VT, Rao AM, Nguyen HT, and Phillips JB, A technique for protein extraction and recovery using pressurized reversed micellar solutions, In: Ataai MM, and Sikdar SK, eds. *New Developments in Bioseparation*. AfChE Symp. Ser. 290. American Institute of Chemical Engineers, New York, 1992; pp. 98–104.
45. Tong J, and Matsumura Y, Effect of catalytic activity on methane steam reforming in hydrogen-permeable membrane reactor, Appl. Catal. A 2005; 286: 226.
46. Balomenou SP, Tsiplakides D, Katsaounis A, *et al.*, Monolithic electrochemically promoted reactors: A step for the practical utilization of electrochemical promotion, Solid State Ionics 2006; 177: 2201–2204.
47. Yantovski E, Gorski J, Smyth B, and Elshof J, Zero-emission fuel-fired power plants with ion transport membrane, Energy 2004; 29: 2077.
48. Ho WSW, CO_2 selective membrane process and system for reforming a fuel to hydrogen for a fuel cell, USP No 6,579,331. B1.
49. Krystal V, Nradil J, Bernauer B, and Kocirik M, Heterogeneous membrans based on zeolites for the separation of small molecules, React. Funct. Polym. 2001; 48: 129–139.
50. Nonaka N, Nakayama K, and Yajima K, Manufacture of a zeolite membrane and its use in gas separation, US Patent, 2008047432 A1. 2008.
51. Yajima K, and Igarashi A, Nissan Motor Co., Ltd., Japan. Patent Japan JP 2007229562. A, 2007, 09/13, 14 pp.
52. de Freitas NL, Goncalves JAS, Innocentini MDM, and Coury JR, Development of a double-layered ceramic filter for aerosol filtration at high-temperatures: The filter collection efficiency, J. Hazard. Mater 2006; 136: 747–756.
53. Zhang X, Xiong G, and Yang W, Hydrogen separation from the mixtures in a thin Pd-Cu alloy membrane reactor, Stud. Surf. Sci. Catal. 2007; 167(Natural Gas Conversion VIII): 219–224.
54. Suzuki S, Wajima N, and Koga N, Asymmetric ceramic membranes, and their manufacture, Patent Japan JP 02126924. A 1990, 05/15.
55. Sano M, Sakamoto M, Hayashi Y, and Kanenori N, Polyurethane compositions with skin care properties, their membranes, laminates containing them, and their manufacture. Patent Japan JP 2005194401. A 2005, 07/21.

56. Negishi H, Oshima N, Haraya K, *et al.*, Preparation of thin and dense lanthanum cobaltite coating on porous tubular alumina supports by electrophoretic deposition, J. Ceram. Soc. Jpn. 2006; 114: 36–41.
57. Park HB, Jung CH, Kim YK, *et al.*, Pyrolytic carbon membranes containing silica derived from poly(imide siloxane): The effect of siloxane chain length on gas transport behavior and a study on the separation of mixed gases, J. Mem. Sci. 2004; 235: 87–98.
58. Kozhukharov V, Machkova M, Brashkova N, and Trapalis Ch, Sol-gel route and characterization of supported perovskites for membrane application, J. Sol Gel Sci. Technol. 2003; 26: 753–757.
59. Matus YB, De Jonghe LC, Jacobson CP, and Visco SJ, Metal-supported solid oxide fuel cell membranes for rapid thermal cycling, Solid State Ionics 2005; 176: 443–449.
60. Fujiwara N, Ohara S, and Li C-H, Manufacture of strongly adhered thin ceramic layers on porous supports. Patent Japan JP 2002-381245. 2002, 12/27.
61. Hanson EL, Organometallic films, methods for applying organometallic films to substrates and substrates coated with such films, PCT Int. Appl. 2008; 16.
62. Verweij H, Ceramic membranes: Morphology and transport, J. Mater. Sci. 2003; 38: 4677–4695.
63. Mouazer R, Elmarraki Y, Persin M, Cretin M, Sarrazin J, and Larbot A, Role of citrate and tartaric ligands for the stabilization of NASICON sols. Application to membrane preparation. Colloids Surf. A: Physicochem. Eng. Asp. 2003; 216: 261–273.
64. Barbieri G, Violante V, Di Maio FP, Criscuoli A, and Drioli E, Methane steam reforming analysis in a palladium-based catalytic membrane reactor, Ind. Eng. Chem. Res. 1997; 36: 3369.
65. Balomenou S, Tsiplakides D, Katsaounis A, *et al.*, Novel monolithic electrochemically promoted catalytic reactor for environmentally important reactions, Appl. Catal. B 2004; 52: 181–196.
66. Huang D, Huang K, Chen S, Liu S, and Yu J, Enantioseparation of racemic α-cyclohexyl-mandelic acid across hollow fiber supported liquid membrane, J. Brazilian Chem. Soc. 2008; 19: 557–562.
67. Obon JM, Manjon A, and Iborra JL, Retention and regeneration of native NAD(H) in noncharged ultrafiltration membrane reactors: Application to L-lactate and gluconate production, Appl. Catal. Biotechnol. Bioeng. 1998; 57: 510–517.
68. Howaldt M, Gottlob A, Kulbe KD, and Chmiel H, Simultaneous conversion of glucose/fructose mixtures in a membrane reactor, Ann. NY Acad. Sci. 1988; 542 (Enzyme Eng. 9): 400–405.
69. Xie R, Chu L-Y, and Deng J-G, Membranes and membrane processes for chiral resolution, Chem. Soc. Rev. 2008; 37: 1243–1263.
70. Tanaka M, and Sackmann E, Supported membranes as biofunctional interfaces and smart biosensor platforms, Physica Status Solidi A 2006; 203: 3452–3462.
71. Huang W, Yang X, and Wang E, The role of tryptophan oxygenase in regulating heme utilization when inducing cytochrome P-450 by phenobarbital in rat liver, In: *Proceedings of 7th International Conference: Biochem. Biophys. Cytochrome P-450 1992*, 1991; pp. 397–399.
72. Li G, Ji M, Gong W, *et al.*, 2008; Method for constructing human membrane cofactor protein minigene pMCP, Faming Zhuanli Shenqing Gongkai Shuomingshu 10.
73. Abate S, Centi G, Perathoner S, and Frusteri F, Enhanced stability of catalytic membranes based on a porous thin Pd film on a ceramic support by forming a Pd-Ag interlayer, Catal. Today 2006; 118: 189–197.
74. Sterte J, Hedlund J, Creaser D, *et al.*, Application of the seed-film method for the preparation of structured molecular sieve catalysts, Catal. Tod 2001; 69(1–4): 323–329.

75. Cheikhou K, and Tzedakis T, Electrochemical microreactor for chiral syntheses using the cofactor NADH, AIChE J. 2008; 54(5): 1365–1376.
76. Sullivan TB, System and method for generation of high pressure air in an integrated gasification combined cycle system. U.S. Patent Appl., US 2008115478. A1 2008; 05/22Publ2008.
77. Lin YS, De Vries KJ, Brinkman HW, and Burggraaf AJ, Oxygen semipermeable solid oxide membrane composites prepared by electrochemical vapor deposition, J. Membr. Sci. 1992; 66: 211–226.
78. Giorno L, Li N, and Drioli E, Preparation of oil-in-water emulsions using polyamide 10 kDa hollow fiber membrane, J. Membr. Sci. 2003; 217: 173.
79. Phillips JB, Nguyen H, and John VT, Protein recovery from reversed micellar solutions through contact with a pressurized gas phase, Biotechnol. Progr. 1991; 7: 43–48.
80. Drioli E, Curcio E, and Di Profio G, State of the art and recent progresses in membrane contactors, Chem. Eng. Res. Des. 2005; 83(A3): 223.
81. Li N, and Sakaki K, Performance of an emulsion enzyme membrane reactor combined with premix membrane emulsification for lipase-catalyzed resolution of enantiomers, J. Mem. Sci. 2008; 314: 183–192.
82. Yoshimoto M, and Kuboi R, Oxidative refolding of denatured/reduced lysozyme utilizing the chaperone-like function of liposomes and immobilized liposome chromatography, Biotechnol. Progr. 1999; 15: 480–487.
83. Draxler J, Furst W, and Marr R, Separation of metal species by emulsion liquid membranes, J. Membr. Sci. 1988; 38: 281–293.
84. Abellán MJ, *et al.*, Extraction and back-extraction of copper (II) with LIX622N in hollow fiber contactors, In: Mach'in AM, and Umbr'ia J, eds. *Chemical Industry and Environment IV: Proceedings of the European Meeting on Chemical Industry and Environment*, Las Palmas de Gran Canaria, Spain, February 12-14, 2003; pp. 75–84.
85. Dingsheng H, Shuxiang G, and Ming M, Simultaneous removal and recovery of cadmium(II) and CN- from simulated electroplating rinse wastewater by a strip dispersion hybrid liquid membrane(SDHLM) containing double carrier, J. Membr. Sci. 2007; 305: 36–47.
86. Lu YC, Yi SS, and Luo GS, Modeling of the mass transfer and conduction behavior in electro-electrodialysis with O/W emulsion as the catholyte, J. Membr. Sci. 2008; 322: 265–274.
87. Liu D, Yu J, Wang C, *et al.*, Process for purifying and recovering impurities from phosphoric acid by liquid membrane method, Faming Zhuanli Shenqing Gongkai Shuomingshu, 2008; 12.
88. Prasad R, and Sirkar KK, Membrane-based solvent extraction, In: Ho WSW, and Sirkar KK, eds. *Membrane Handbook*, VNR, New York, 1992; pp. 727–763.
89. Chou K-H, and Bell LN, Caffeine content of prepackaged national-brand and private-label carbonated beverages, J. Food Sci. 2007; 72: C337–C342.
90. Seike S, and Oishi K, Method of removing carbon dioxide using polymer electrolyte membrane, and apparatus and system therefore. Patent Japan JP 2003306788. A 10/31 2003.
91. Li G, Method for recycling wastewater from paper industry, Faming Zhuanli Shenqing Gongkai Shuomingshu, 2008; 9.
92. Drioli E, Criscuoli A, and Curcio E, Integrated membrane operations for seawater desalination, Desalination 2002; 147: 77.
93. Oda T, Akaike T, Hamamoto T, Suzuki F, Hirano T, and Maeda H, Relationships between water quality, morphological factors in river basins, the diversity index and the biotic index, Environ. Technol. 1991; 12: 1147–1155.

94. Maeda H, Oda T, Matsumura Y, and Kimura M, Improvement of pharmacological properties of protein-drugs by tailoring with synthetic polymers, J. Bioactive Comp. Polym. 1988; 3:27–43.
95. Mellon WS, Specificity of dye-ligand interaction with the polynucleotide binding domain of 1,25-dihydroxyvitamin D3-receptor complexes of chicken intestinal cytosol, Mol. Pharmacol. 1984; 25: 86–91.
96. Guillemette G, Lamontagne S, Boulay G, and Mouillac B, Differential effects of heparin on inositol 1,4,5-triphosphate binding, metabolism, and calcium release activity in the bokine adrenal cortex, Mol. Pharmacol. 1989; 35(3): 339–344.
97. Shalaby SW, McCormick CL, and Butler GB, eds., *Water-Soluble Polymers; Synthesis, Solution Properties, and Application, Symposium Series 467*, American Chemical Society, Washington, DC 1991; pp. 367–445. Chapters 24–28.
98. Matson SL, and Quinn JA, Membrane reactors, In: Ho WSW, and Sirkar KK, eds. *Membrane Handbook*, VNR, New York, 1992; pp 809–832.
99. Matson SL, and Quinn JA, Membrane reactors in bioprocessing, Ann. N.Y. Acad. Sci. 1986; 469: 152–165.
100. Belfort G, Membranes and bioreactors: A technical challenge in biotechnology, Biotechnol. Bioeng. 1989; 33: pp. 1047–1066.
101. Hsieh HP, Inorganic membrane reactors – A review, AIChE Symp. Ser. 1989; 85(268): 53–74.
102. Lopez JL, Matson SL, Stanley TJ, and Quinn JA, Liquid/liquid extractive membrane reactors, In: Mattiasson B, and Holst, eds. *Extractive Bioconversions Bioprocess Technologies Series*, Marcel Dekker, New York, 1990; pp. 27–66.
103. Stanish I, and Monbouquette HG, Ternary complex formation dramatically enhances metal-sorbing vesicle selectivity, J. Membr. Sci. 2001; 194: 263–272.
104. Walde P, and Ichikawa S, Enzymes inside lipid vesicles: Preparation, reactivity and applications, Biomol. Eng. 2001; 8: 143–177.
105. Glass JE, ed. *Associative Polymers in Aqueous Media, Symposium series 765*, American Chemical Society, Washington, DC, 2000; pp. 286–335. Chapter 17–19.
106. Shalaby SW, McCormick CL, and Butler GB, eds, *Water-Soluble Polymers; Synthesis, Solution Properties, and Application Symposium Series 467*, American Chemical Society, Washington, DC, 1991; pp. 468–506. Chapters 30–33.
107. Schwoyer WLK, ed. *Polyelectrolytes for Water and Wastewater Treatment*, The Permutit Company Research and Development Center: Monmouth Junction, New Jersey, 1991; pp. 48–60.
108. Schwoyer WLK, ed. *Polyelectrolytes for Water and Wastewater Treatment*, The Permutit Company Research and Development Center: Monmouth Junction, New Jersey, 1991; pp. 80–89.
109. Schwoyer WLK, ed. *Polyelectrolytes for Water and Wastewater Treatment*, The Permutit Company Research and Development Center: Monmouth Junction, New Jersey, 1991; pp. 213–242.
110. Shalaby SW, McCormick CL, and Butler GB, eds. *Water-soluble Polymers: Synthesis, Solution Properties, and Application Symposium Series 467*, American Chemical Society, Washington, DC 1991; pp. 446–465. Chapter 29.
111. Rivas BL, Pereira ED, and Moreno-Villoslada I, Water-soluble polymer–metal ion interactions, Progr. Polym. Sci. 2003; 28: 173–208.
112. Kamp PC, Kruithof JC, and Folmer HC, *UF/RO treatment plant Heemskerk: From challenge to full scale application*, Proceedings of the Conference on Membranes in Drinking and Industrial Water Production, Paris, France, October 2000, Desalination Publications, L'Aquila, Italy, 2000; Vol. 1, pp. 27–35.
113. Ventresque C, Gisclon V, Bablon G, and Chagneau G, *An outstanding feat of modern technology: The Mery-sur-Oise Nanofiltration Treatment Plant (340,000 m^3/d).* Proceedings

of the Conference on Membranes in Drinking and Industrial Water Production, Paris, France, October 2000 Desalination Publications, L'Aquila, Italy, 2000; Vol. 1, pp. 1–16.

114. Hassan AM, Farooque AM, Jamaluddin ATM, Al-Amoudi AS, Al-Sofi MAK, Al-Rubaian AF, Kither NM, Al-Tisan IAR, and Rowaili A, *A demonstration plant based on the new NF-SWRO process*. Proceedings of the Conference on Membranes in Drinking and Industrial Water Production, Paris, France, October 2000 Desalination Publications, L'Aquila, Italy, 2000; Vol. 1, pp. 313–328.
115. Levenstein R, Hasson D, and Semiat R, Utilization of the Donnan effect for improving electrolyte separation with nanofiltration membranes, J. Membr. Sci. 1996; 116: 77–92.
116. Thanuttamavong M, Oh JI, Yamamoto K, and Urase T, *Comparison between rejection characteristics of natural organic matters and inorganic salts in ultra low pressure nanofiltration for drinking water production*. Proceedings of the Conference on Membranes in Drinking and Industrial Water Production. Paris, France, October 2000 Desalination Publications, L'Aquila, Italy, 2000; Vol. 1, pp. 269–282.
117. Drioli E, Curcio E, Criscuoli A, and Di Profio G, Integrated membrane system for recovery of CaCO3 NaCl and $MgSO_4\_7H_2O$ from nanofiltration retentate, J. Membr. Sci. 2004; 239: 27.
118. Shang C, Wong HM, and Chen G, Bacteriophage MS-2 removal by submerged membrane bioreactor, Water Res. 2005; 39: 4211–4219.
119. Rivas BL, Pooley SA, Pereira ED, and Maureira A, Water – Soluble polyelectrolytes with metal ion removal ability by using the liquid phase based retention technique, In: *Proceedings of World Polymer Congress – MACRO 2006*, Macromolecular Symposia, Wiley-VCH Verlag GmbH & Co, KgaA, Weinheim, 2006; pp. 116–122.
120. Koltuniewicz A, and Noworyta A, Dynamic properties of ultrafiltration systems in light of the surface renewal theory, Ind. Eng. Chem. Res. 1994; 33: 1771–1779.
121. Sablani SS, Goosen MFA, Al-Belushi R, and Wilf M, Concentration polarization in ultrafiltration and reverse osmosis: A critical review, Desalination 2001; 141: 269–289.
122. Goosen MFA, Sablani SS, Al-Hinai H, Al-Obeidani S, Al-Belushi R, and Jackson D, Fouling of reverse osmosis and ultrafiltration membranes: A critical review, Sep. Sci. Technol. 2004; 39: 2261–2298 (Fig. 1).
123. Ridgway HF, Kelly A, Justice C, and Olson BH, Microbial fouling of reverse osmosis membranes used in Advanced Wastewater Treatment Technology: Chemical bacteriological and ultrastructural analyses, Appl. Environ. Microbiol. 1983; 46: 1066–1084.
124. Upen J, Barwada SJM, Coker SD, and Terry AR, Winning the battle against biofouling of reverse osmosis membranes, Desalination Water Reuse 2000; 10: 53–58.
125. Mulder M, *Basic Principles of Membrane Technology*, Kluwer, Dordrecht, The Netherlands, 1996.
126. Sablani SS, Goosen MFA, Al-Belushi R, and Gerardos V, Influence of spacer thickness on permeate flux in spiral-wound seawater reverse osmosis systems, Desalination 2002; 146: 225–230.
127. Kabsch-Korbutowicz M, Majewska-Nowak K, and Winnicki T, Analysis of membrane fouling in the treatment of water solutions containing humic acids and mineral salts, Desalination 1999; 126: 179–185.
128. Tu S-C, Ravindran V, Den W, and Pirbazari M, Predictive membrane transport model for nanofiltration processes in water treatment, AIChE J. 2001; 47: 1346–1362.
129. Mores WD, and Davis RH, Direct observation of membrane cleaning via rapid backpulsing, Desalination 2002; 146: 135–140.
130. Stairmand JW, and Bellhouse BJ, Mass transfer in a pulsating turbulent flow with deposition onto furrowed walls, Int. J. Heat Mass Transfer. 1985; 27: 1405–1408.

131. Millward HR, Bellhouse BJ, Sobey IJ, and Lewis RWH, Enhancement of plasma filtration using the concept of the vortex wave, J. Membr. Sci. 1995; 100: 121–129.
132. Li H, and Bertram CD, Mechanisms by which pulsatile low affects cross flow microfiltration, AIChE, J. 1998; 44: 1950–1961.
133. Gupta BB, Blanpain P, and Jaffrin MY, Permeate flux enhancement by pressure and flow pulsations in microfiltration with mineral membranes, J. Membr. Sci. 1992; 70: 257–266.
134. Dal-Cin MM, Lick CN, Kumar A, and Lealess S, Dispersed phase back transport during ultrafiltration of cutting oil emulsions with a spinning disc geometry, J. Membr. Sci. 1998; 141: 165–181.
135. Frenander U, and Jönsson AS, Cell harvesting by crossflow microfiltration using a shear-enhanced module, Biotechnol. Bioeng. 1996; 52: 397–403.
136. Genkin G, Waite TD, Fane AG, and Chang S, *The effect of axial vibrations on the filtration performance of submerged hollow fiber membranes International Congress on Membranes and Membrane Processes*, ICOM, Korea, Seoul, 2005.
137. Cui ZF, Chang S, and Fane AG, The use of gas bubble to enhance membrane processes, J. Membr. Sci. 2003; 221: 1–35.
138. Cui ZF, and Wright KIT, Flux enhancement with gas sparging in downwards crossflow ultrafiltration: Performance and mechanisms, J. Membr. Sci. 1996; 117: 109–116.
139. Akay G, and Wakeman RJ, Electric field enhanced crossflow microfiltration of hydrophobically modified water soluble polymers, J. Membr. Sci. 1997; 131: 229–236.
140. Bowen WR, Electrochemical aspects of microfiltration and ultrafiltration, In: Howell JA, Scanchez V, and Field RW, eds. *Membranes in Bioprocessing*, Chapman & Hall, Cambridge, 1993; pp. 265–292.
141. Muthukumaran S, Kentish SE, Ashokkumar M, and Stevens GW, Mechanisms for the ultrasonic enhancement of dairy whey ultrafiltration, J. Membr. Sci. 2005; 258: 106–114.
142. Juang RS, and Lin KH, Flux recovery in the ultrafiltration of suspended solution with ultrasound, J. Membr. Sci. 2004; 243: 115–124.

INDEX

Note: Page numbers followed by "*f*" and "*t*" indicate figures and tables, respectively.